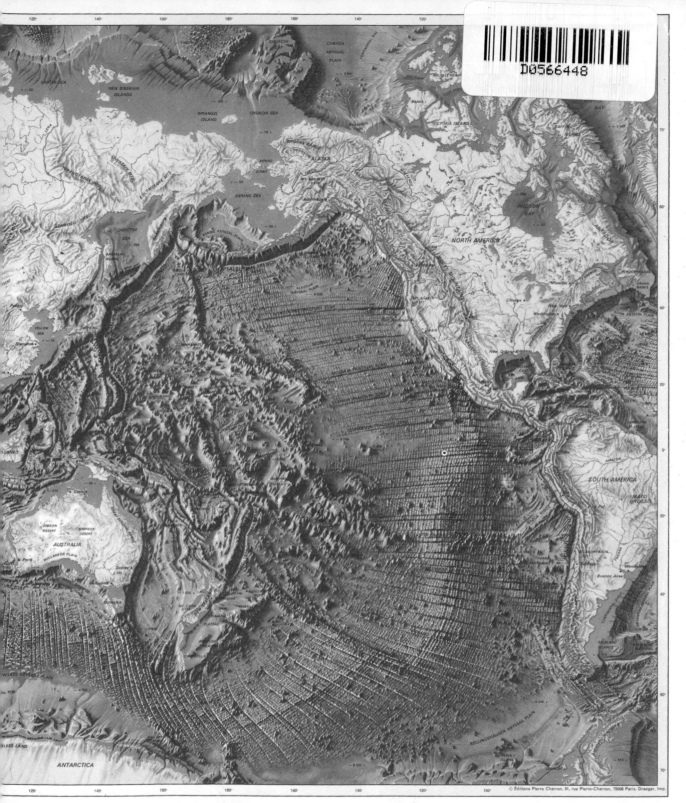

Based on Bathymétric studies by
Bruce C. Heezen and Marie Tharp
of the Lamont Doberty Geological Observatory
Columbia University Palisades, New York, 10964
SUPPORTED BY THE UNITED STATES NAVY
OFFICE OF NAVAL RESEARCH

Copyright © by Marie Tharp.

Mercator Projection 1:48,000,000 at the Equator.
Depth and Elevations in Meters.

OCEAN
SCIENCE

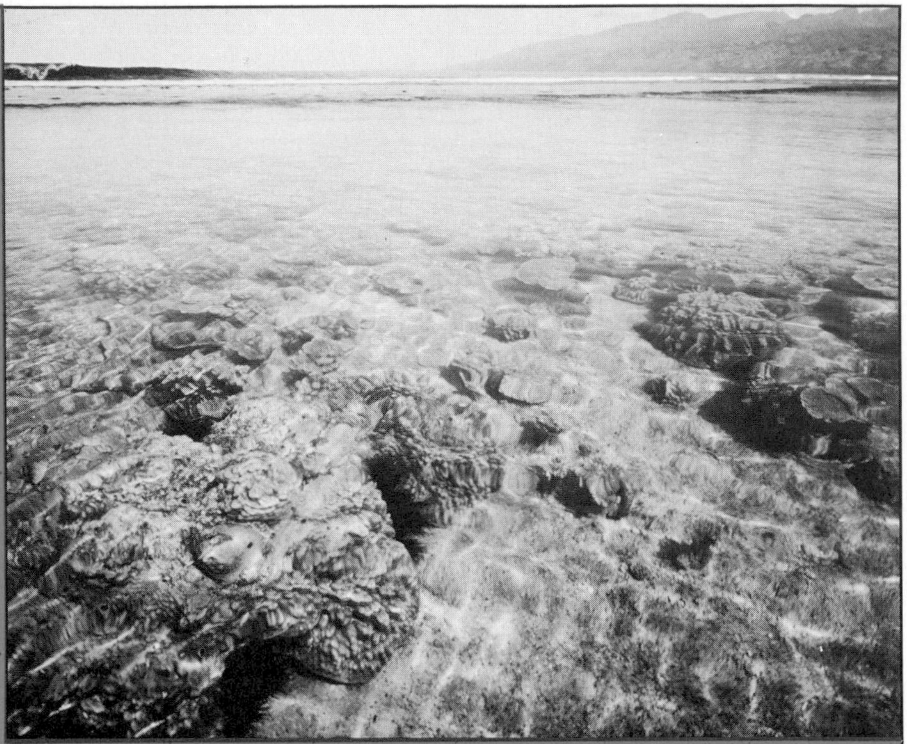

OCEAN SCIENCE

Second Edition

Keith Stowe

California Polytechnic State University

John Wiley & Sons, Inc.

New York   Chichester   Brisbane   Toronto   Singapore

COVER PHOTO AND FRONTISPIECE
Coral Heads in Tahiti by Jack Fields/Photo Researchers.

Library of Congress Cataloging in Publication Data:

Stowe, Keith S., 1943–
Ocean Science.

Includes bibliographies and indexes.
1. Oceanography.   I. Title
GC16.S76   1983   551.46   82–16120
ISBN 0-471-86719-5

Printed in the United States of America

10 9 8 7 6 5 4 3 2 1

This book was developed as an introductory text for college students having little background in the sciences. Most of these students will not be pursuing technical careers, and this course may be one of our last opportunities to share with them the intrigue and excitement of science in general, or that of oceanography in particular. I have written this book with the idea that we might be able to use this opportunity to impart to students something that will continue to enrich their lives long after details have faded from their memories.

It is the overriding purpose of this book to convince the students that the world around them is neither chaotic nor irrational. There are good reasons for things being the way they are, and most of these reasons are easily understood through everyday examples and common sense. It is hoped that the student will develop an appreciation for the curiosity of the scientists, who are constantly probing and scrutinizing the world about them, and an appreciation of our optimism that we will find and understand the reasons behind what we observe.

My motivation for writing this book was that I felt that the several books that were accurately written in terms of describing the properties and behaviors of the oceans tended to be somewhat lacking in their explanation of the *causes* of these phenomena. Without the organizing framework provided by an understanding of the underlying causes, the student tends to become lost in a forest of details. It is doubtful that many of these particulars would otherwise be retained very long past the appropriate exam date. Even worse, the student may develop a misconception of what oceanographic research is all about. The student may see it as some irrational search for data and details rather than seeing that these data are merely a tool used toward the more profound goal of a basic understanding of nature.

For this reason, I have tried in this book to answer not only "what?" but also "how?" and "why?" The causes of the ocean's various properties and behaviors are usually understandable in terms of our everyday experiences—in a coffee cup, the kitchen, a bathtub, the sky, and so on. These explanations constantly would be reinforced by the students' experiences throughout their lifetimes. In addition to retaining their understanding of the oceans, it is hoped this might also help the students master other problems they might encounter, which are not covered in this book. Hopefully, it would leave the student with the impression that most natural phenomena have simple explanations that can be understood easily, without scientific sophistication. With this attitude the students would be more eager to question and probe the world around them, and in this way develop a first-hand appreciation for what science is all about.

In this second edition I have rewritten many of the chapters extensively in an effort to improve clarity and accuracy. The material has also been updated, especially in the more rapidly changing areas such as plate tectonics and marine resources. The treatments of seas and estuaries, and of marine biology, have been expanded, and two new chapters on marine ecology have been added.

Following suggestions from students and a colleague, several of the longer chapters have been split, so that the material may be studied in smaller, more manageable packets. Also, I have tried to make the chapters as independent as possible, so that the course instructors may have maximum flexibility to choose and order topics as they see fit. The order of topics in this book simply reflects my own preference for the courses I teach.

At the end of each chapter are many questions—nearly one for each paragraph in the text. For most of these, the answers can be found in the text in the same order as the questions are numbered. This abundance of questions reflects my own feeling that most students can learn more from a paragraph if they are actively searching for answers to questions, than if they are just passively letting their eyes follow the words.

At the end of each question set are several more thought-provoking questions, indicated by asterisks. These are extensions of the material learned in the text, and require the students to use their own ingenuity and imagination to come up with answers. For these questions, the emphasis should be on the students' thought processes, and not so much on whether they actually come up with the "correct" answers. (Some of the answers are not yet known with certainty.)

I would like to express my appreciation to the many students, friends, and colleagues who have helped me with this book. I am particularly indebted to Professor C. S. Clay and Professor David Roach for their work and suggestions for improving this second edition. My colleagues Lawrence Balthaser, Raymond Bauer, David Chipping, and Kenneth Hoffman were also instrumental in helping me get this project off the ground, as were Donald Deneck and the staff at John Wiley.

Finally, I would like to apologize to my wife and two sons for the time this book has taken from them.

Keith Stowe

CONTENTS

ix

1
INTRODUCTION

H.M.S. Challenger

The most unique feature of the earth is its oceans. Among other things, they are responsible for life. They cover most of the earth, and so quite naturally they have a lot to tell us.

People who study the oceans are called oceanographers, and the various fields within oceanography are frequently placed into the four broad categories of geology, chemistry, physics, and biology. Of course, the oceans are much too complex to allow their study to be neatly categorized. Some important aspects of oceanographic studies that cannot be placed easily into one of these categories include geophysics, biophysics, nutrition, petrology, anthropology, meteorology, and pharmacology, just to name a few. In fact, some of oceanography involves studies outside the realm of the pure sciences, such as history, law, or sociology. In short, describing a person as an "oceanographer" may say very little about what he or she actually does for a living.

One result of the complexity and diversity of the field is that it is impossible in one course, or even in a finite number of courses, to teach all there is to know about the oceans. An author or teacher who would claim to do so should be as suspect as one who would purport to explain all there is to know about life. No one is qualified to make such a claim. This book concentrates on the scientific aspects of oceanography, and gives an overview of the fields of study receiving the greatest attention.

Although it would be a most welcome by-product, the primary objective of this book is not to relay information. It is rather to foster an attitude that will hopefully remain long after the details in this book are forgotten. As a course in history may allow us a deeper appreciation of various social institutions, or a course in art or music may allow us new enjoyment of these expressive forms, a course in the sciences should offer us a new way of looking at our world for greater appreciation of it.

The world is a masterpiece. Every scene is more beautiful than the finest painting, and every piece displays an intricacy and harmony finer than Beethoven's best. The more we ask how something is made, how it works, or how it evolved, the more we are in awe of the magnificence of nature. This is the beauty that science teachers would like their students to appreciate. It is everywhere, and its enjoyment requires no scientific sophistication. The inquisitiveness of a child is all it takes to turn our everday world into a magical kingdom.

In this particular book, the vehicle is the study of oceanography. Hopefully, the student will be able to catch glimpses of how personally rewarding this inquisitive attitude can be.

A. OVERVIEW

We begin our studies in this chapter with a brief history of the earth. Our own lifetimes are very short in comparison, and so we must change **3** our perspective if we're going to study events in the evolution of the earth, such as the formation of continents or the birth and death of oceans. That is followed by a brief history of oceanographic exploration as a prelude to our own study of the oceans.

We begin in Chapters 2 and 3 with a study of the earth in general. We will discuss how oceans evolve, and how the continents arrived at their present configuration. We will get a feeling for the excitement of a revolution of ideas about our earth, which has happened in the past three decades.

Then we'll look at the ocean bottom in Chapter 4. We'll see what is beneath those features appearing at the surface, and what the ocean would look like with all the water removed. We'll find it has a complex and exciting geography, much of which reflects geological processes studied in the preceding chapter.

In Chapters 5 and 6 we'll examine the sediments covering the ocean bottom. We'll study the weathering of rocks and the formation of sedimentary deposits from them. We'll also see that skeletons of organisms are a large component of the sediment in many areas, and we'll learn of sediments from other, less likely, sources as well.

Our study of waves ranges from the tiny capillary waves we make when blowing on hot soup to the gigantic tides that span entire oceans. Much of Chapters 7 and 8 concentrate on the familiar wind-driven waves that toss us about in the surf. Tides and tsunamis are also investigated.

Waves end their short lives on beaches, which are familiar recreation areas. In Chapter 9 we ask questions such as, "What can beaches be made of?" "Why isn't the sand quickly washed to sea?" and "What causes the various beach features we see?"

We then turn our attention toward the water itself. We'll ask where it came from, and what magical properties it has that make the earth such a pleasant place to live. We'll also study the salts, what they are, where they came from, and what role they play in important chemical and biological processes.

The behaviors of the ocean and atmosphere are stongly interrelated. This is because they interact across a surface of contact that is enormous in comparison to their volumes. (See Figure 1.1.) In fact, the ratio of surface to volume (or area to thickness) for each is roughly that of a sheet of paper. For this reason, the behavior of neither ocean nor atmosphere can be understood without reference to the other.

In Chapter 12 we study the weather and how the ocean controls our climate. In the following three chapters we study the circulation of the ocean and atmosphere. The first of these deals mostly with overall effects, such as how the earth's rotation causes peculiar behavior in large-scale motions of wind and water, and how this, combined with solar heating, drives the atmospheric circulation. In the next chapter, we zero in on ocean surface currents, including those driven by wind and those driven by gravity. In Chapter 15 we study deeper ocean currents and the properties of the various large deep water masses.

FIGURE 1.1 *View to the south across Florida, taken by the crew of Apollo 9. On the horizon it can be seen that the atomosphere is very thin in comparison to its lateral extent. The same is true of the oceans. Because the two are in contact over such a large area, the behaviors of the atmosphere and the ocean are intimately interrelated.*

The following chapter is a study of seas and estuaries. The emphasis is on the dominant circulation patterns, their importance to life, and problems raised by pollution.

Chapters 17 through 20 study life in the oceans. The first examines plants and plant productivities, on which the entire biological community depends. General questions are asked about the overall livelihood and biological welfare of the area, irrespective of what particular organisms live there. The following chapter presents an overview of the different forms of life in the ocean. Chapters 19 and 20 deal with marine ecology, or how the organisms function and relate to their marine environment.

Finally, we look at ocean resources. Those who expect the oceans to solve our world hunger problem will be disappointed, but some things can be done to help alleviate the suffering. Also, we will see that the ocean does have a wealth of mineral and energy resources, although we must yet learn how to extract them economically, in most cases.

FIGURE 1.2 *M16 and the Nebula of Serpens. The Solar System was born of a region such as this.*

B. PERSPECTIVE IN TIME

The earth condensed from a cloud of interstellar gas and dust (Figure 1.2) about 4.6 billion years ago, along with the sun and the rest of our Solar System. Throughout vitually all of this time, the earth and her oceans have gone unmolested by humans. We appeared only very recently. With curiosity appropriate to our "newborn" status here, we are vigorously probing this fascinating world.

We are wise enough to know that what we see today is a result of what has happened in the past. To understand our present world, we must also probe its history.

For very short-term changes, this is an easy task. For example, we are familiar with the disappearance of beach sand in the winter and its reappearance in the summer. Similarly, the extension of a river delta is observed within a lifetime. However, with regard to long-term changes, we are essentially taking a snapshot in time, and in that snapshot we're looking for clues regarding the evolution and behavior of the earth and its oceans.

We find our ability to comprehend our earth's history falls far short of our desire. The average person travels a significant fraction of the earth's circumference during his or her lifetime, yet we have difficulty in comprehending even the distance across one continent. Much greater is our difficulty in comprehending the earth's evolutionary time scales, since our lifetimes, and even the entire history of humankind, is such a minute fraction of the history of the earth.

It is relatively easy to demonstrate that the earth is quite alive geologically. Dating a typical rock on the ocean bottom yields an age of a few hundred million years at most. A glance at Figure 1.3 will convince you

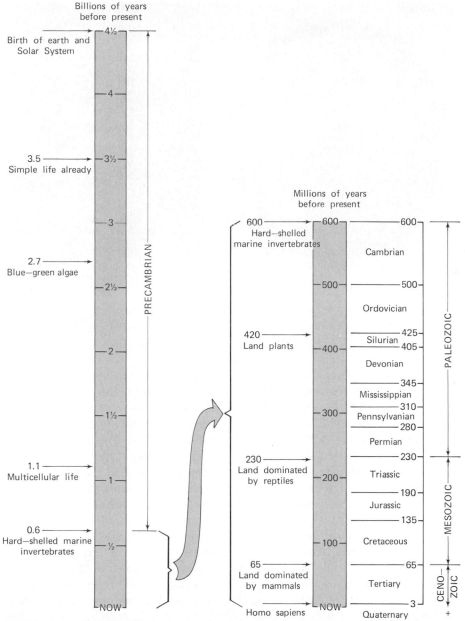

FIGURE 1.3 *The development of life on earth.*

that the age of such a rock is only a few percent of that of the earth itself, giving us a feeling for the rate at which oceans are being created and destroyed. The Atlantic Ocean, for example, has been around much longer than most species of mammals, but not as long as reptiles.

For the first 90% of the earth's history, any life that existed was oceanic. We have found the remains of single-celled marine organisms

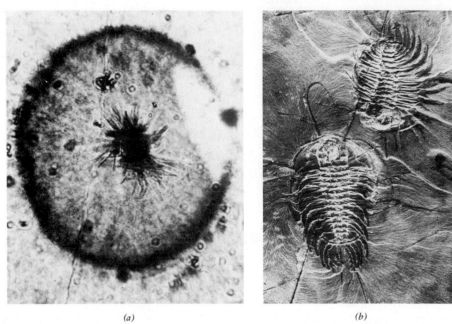

(a) *(b)*

FIGURE 1.4   (a) *Precambrian fossil: a blue-green algae of about 2 billion years ago.*
(b) *Cambrian fossil: two trilobites.*

from 3.5 billion years ago, so life here must be at least as old as that. Blue-green algae (Figure 1.4*a*) existed as early as 2 billion years ago.

Rather suddenly (on geological time scales), many forms of hard-shelled invertebrates began to appear in the oceans about 600 million years ago. This marked the beginning of the Cambrian Period. The entire 4 billion years of earth history previous to this is called the Precambrian, but since the beginning of the Cambrian, many relatively short time divisions have been made according to changes in life forms that we find in the fossil record. In order from oldest to youngest, these divisions are called the Cambrian, Ordovician, Silurian, Devonian, Mississippian, Pennsylvanian, Permian, Triassic, Jurassic, Cretaceous, Tertiary, and Quaternary periods. These 12 periods are also grouped into three broader divisions, called "eras." The first seven periods belonged to the Paleozoic Era, which lasted from 600 million to 220 million years before the present. The Mesozoic Era encompassed the Triassic, Jurassic, and Cretaceous periods, and extended from 220 million to 65 million years before the present. The last 65 million years belong to the Cenozoic Era, which includes the Tertiary and Quaternary periods.

There are two reasons for the proliferation in the names of the periods since the beginning of the Cambrian Period. They both have to do with our ability to detect changes rather than with nature's inclination to produce them. The first reason is that the more recent history is more easily discovered, since less of its remains have been destroyed or irrecoverably buried by subsequent geological processes. Consequently, we

FIGURE 1.5   *Restoration of a scene in western United States during Late Jurassic times (about 140 million years ago).*

know recent history in much greater detail, facilitating the much more elaborate classification scheme. Second, organisms with hard shells make better fossils (Figure 1.4b). They preserve better and are easier to recognize. The absence of well-preserved fossils in Precambrian rocks means that more inaccurate and unreliable techniques must be employed to date these geological events. For this reason, our understanding of Precambrian events is not very precise, and dividing the Precambrian into detailed time bins would be inappropriate.

In addition to the appearance of hard skeletons, there have been some other developments since the beginning of the Cambrian that should be mentioned here. One of these is the appearance of simple land plants during the Paleozoic Era, about 420 million years ago. (Imagine the rates of erosion before that!) Although reptiles had appeared first during the Paleozoic, they were the dominant form of land animal during the Mesozoic Era (Figure 1.5). Finally, the Cenozoic is marked by the dominance of the mammals.

Erect, two-legged primates seem to have appeared about five million years ago. This is quite recent on geological time scales, being about one tenth of 1% of the history of the earth. Consequently, hominoids would have always found the world geographically much like it is today. The biggest change they could have witnessed would be the elevation of the Isthmus of Panama above sea level.

Although our relatives have been around for five million years, stable, organized, agrarian societies seem to have begun only about 6000 years ago, probably corresponding to the stabilization of sea level at its present position at the end of the last ice age. Records of exploratory oceanographic voyages exist from about 3000 years ago.

C.   A BRIEF HISTORY OF OCEANOGRAPHY

One of the things learned in the previous section is that humankind is a very recent development. The systematic exploration of the seas is even more recent. If the entire history of the earth were condensed into 24 hours, hominoids would have first appeared less than two minutes ago, and the first recorded explorations of the oceans began in the last few hundredths of a second.

Most of history has gone unrecorded. When the Europeans began long voyages of exploration in the late 15th and early 16th centuries, they found the "newly discovered" lands had already been discovered and populated ages earlier by people of different cultural backgrounds. Frequently, these people had been there so long that even legends of their arrival had faded from memory, and we have no direct way of establishing when these lands really were first discovered.

It is also unfortuante that much recorded history has been lost. The West is particularly guilty of negligence in its cultural heritage. After the Dark Ages we had to relearn much of what had been available to us centuries earlier.

This section presents a brief overview of some of the highlights of oceanographic exploration from the point of view of western societies. We do so with apologies to those enlightened societies of past times whose records have not survived.

C.1 Ancient Civilizations

Before railroads, societies relied on waterways for the shipment of heavy cargoes. This included large shipments of food, dry goods, building materials, troops, and military hardware among other things. Clearly water was a lifeline and not a barrier to great, ancient civilizations such as those of the Egyptians, Greeks and Romans.

The Phoenicians of the eastern Mediterranean had a reputation for being very skilled merchant sailors. As early as 1500 B.C., Phoenicians were trading with people in the Persian Gulf area to the east. We also know they explored to the west beyond the Straits of Gibraltar, and as far north as Great Britain. It appears that around 600 B.C., Phoenician sailors employed by King Necho of Egypt sailed completely around the African continent, leaving by way of the Red Sea and returning through the Straits of Gibraltar.

A century later, a Carthaginian admiral named Hanno led an expedition through the Straits of Gibraltar and south along the African coast, nearly as far as the Congo River. This was the last great expedition to the south for nearly two millenia, possibly because of disappointing prospects for prosperous trade with people in that direction. Subsequently, the people in the Mediterranean area turned their attention toward the East, where trading became a thriving business. Such things as spices, silks, and gems from the East were especially prized during Roman times.

The sailors of those times usually stayed within sight of land, although they sometimes sailed by the stars at night. Their progress in the north–south direction could be determined from the stars, but their east–west progress had to be estimated from travel times. The resulting distortion of distances in the east–west direction can be seen on ancient maps.

In the 4th century B.C., Pythaes was able to relate tides to the motion

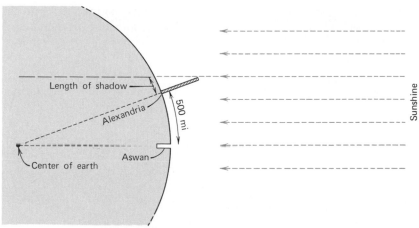

FIGURE 1.6 *Eratosthenes (276 to 195 B.C.) determined the circumference of the earth by measuring the length of the shadow cast by a vertical wall in Alexandria at midday during the summer solstice. On this same day, the sun was known to be directly overhead in a town (near modern Aswan) some 500 miles to the south, since the sunshine illuminated the bottom of vertical wells.*

of the moon. He also charted latitude and longitude on a voyage he made to the British Isles.

Erathosthenes (276 to 195 B.C.), a librarian in Alexandria, made a remarkably simple, elegant, and accurate measurement of the circumference of the earth. (See Figure 1.6.) This, combined with the writings of earlier Greeks such as Aristotle and Aristarchus of Samos, demonstrated that the Greeks had a remarkable knowledge of their world, especially when compared with our ignorance a millenium later.

The Romans were aware that the earth was active geologically. They saw how volcanoes added material to the land, and how erosion removed it. Seneca (54 B.C. to 30 A.D.) tackled the problem of why the continuous flow of large rivers into the oceans did not cause them to overflow. He explained what we now call the "hydrologic cycle," in which fresh water is removed from the oceans through evaporation, is deposited on the continents through various forms of precipitation, and returns to the sea in rivers and streams. The net change in the ocean is zero.

A Greek geographer named Strabo (63 B.C. to 24 A.D.) also made some oceanographic observations at this time. He recorded data on the tides, and made soundings with a weighted line down to a depth of 2 km.

A map by Ptolemy (Figure 1.7) summarizes the geographical knowledge in Roman times. Although later Arabic maps display similar precision, knowledge in the West deteriorated greatly (Figure 1.8).

C.2 The Dark Ages and Activity in the North

During the Dark Ages only the northern Europeans seemed to display some inquisitiveness about their greater environment, although they were considered "barbarians" by their less inquisitive neighbors to the south.

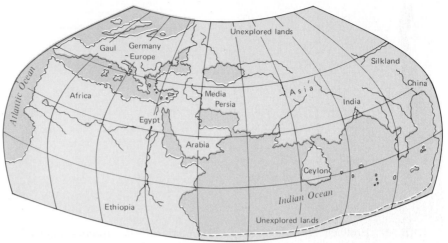

FIGURE 1.7 *The world, according to Ptolemy (approx. 140 A.D.).*

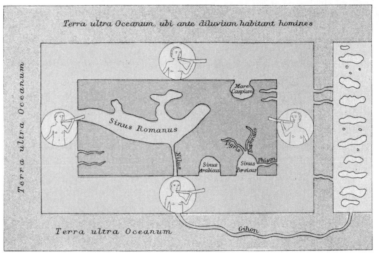

FIGURE 1.8 *A 6th century map of the world.*

Benign climatic conditions encouraged the exploration of northern waters in these times.

Iceland was visited by Picts and Celts in the 7th century, and a Celtic colony was established there in the following century. Irish priests visited Iceland in the period after about 750 A.D.

In the 10th century, the Vikings began moving westward (Figure 1.9). During a temporary forced expatriation, an outlaw named Eric the Red sailed as far as Baffin Island in Canada, and 3 years later he established a colony in Greenland. His son Leif Ericson, did more extensive investigation of the Canadian coast, spending the winter of 995 A.D. in Newfoundland. The next decade brought many additional Viking explorers into northern and eastern Canadian waters. Due to the deterioration of climatic conditions around 1200 A.D., the Vikings lost touch with their

FIGURE 1.9   A Viking ship viewed from the stern.

FIGURE 1.10   A late Viking map of the Atlantic (dated 1570).

western colonies. A Viking map of a much later date is shown in Figure 1.10.

One interesting navigational aide was introduced in Europe during the Dark Ages. This was the magnetic compass. Arabic traders obtained lodestone (a magnetic rock that can be used to magnetize iron needles) from China around 1200 A.D. Subsequently, crusaders brought it to

Europe, where its usefulness in navigation overcame superstitious fears associated with these magical stones and needles.

C.3 The Age of Discovery

Toward the end of the 15th century A.D., the Mediterranean countries began again to explore the world's oceans. Two things had great influences on this turn of events. The first was the establishment of a school of sailors by Prince Henry the Navigator of Portugal in Sagres in 1420. He hired skilled Italian navigators and map makers to help increase the skills of Portuguese sailors, who previously were reluctant to sail beyond sight of land.

The second event was the capture of Constantinople by Sultan Mohammed II in 1453. With this, the Turks had cut off the last of the overland trade routes between West and East, giving economic incentive for searching for alternate routes. In addition, Greeks expelled from Constantinople at this time, brought with them a knowledge of the globe that had long been forgotten in the West.

With their new skills, the Portuguese sailed south. Diego Cao reached the mouth of the Congo River. Bartholomew Diaz rounded the Cape of Good Hope in a voyage of 1487 to 1488. Then in 1498, Vasco da Gama sailed all the way to India, going around the Cape of Good Hope and up along the eastern coast of Africa to about the equator before crossing over to India.

Meanwhile, some famous voyages were undertaken under the Spanish flag. In 1474, a Florentine astronomer named Toscanelli wrote to the King of Portugal, suggesting a route to the East Indies by sailing west around the globe. He attached a map that greatly underestimated the distance, placing the East Indies just slightly west of where the then unknown America lay. On request, Christopher Columbus obtained a copy of that letter, and persuaded Spain to finance the suggested voyage. As everyone knows, he reached first a small island in the Bahamas, and initially erroneously assumed he'd arrived in the East Indies. He was an exceptionally skillful sailor, but unfortunately his skills didn't extend to administration, and some controversies involving settlements he established in the New World were to cause him troubles later in life.

In 1497 John Cabot sailed a northern route to the northeastern coast of what is now the United States. From 1500 to 1520, there were numerous Spanish and Portuguese expeditions to the New World, exploring most of the Atlantic Coast of South America, Central America, and up through Florida in North America.

This exploratory period culminated in the historic circumnavigation of the globe by the Magellan expedition in the years 1519 to 1522 (Figure 1.11). It started out with five ships and 230 sailors. The sailors on one ship mutinied and escaped in the night in the Straits of Magellan. Other ships had to be cannibalized to keep the remaining ships going, and when the expedition arrived at Seville on September 8, 1522, only one ship and 18 bedraggled sailors remained. Magellan was not among them. He had been

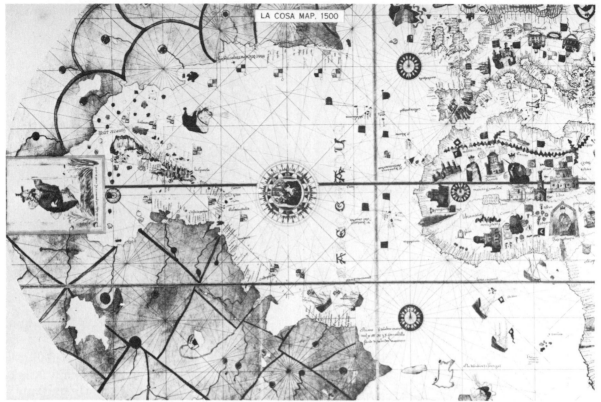

FIGURE 1.11 *This is about the best map that would have been available for the Magellan expedition. No chart of the Pacific Ocean yet existed, of course, and most knowledge of the Indian Ocean was being guarded by the Portuguese.*

killed in an avoidable fight in the Philippines. The journey was completed under the command of Sabastian del Caño.

In those days the life expectancy of a sailor was very short, due to what we now recognize as a dietary problem. The most grueling and costly part of the Magellan expedition turned out to be the slow and tedious trip north and west across the Pacific. The trip from the East Indies home should have been fairly routine, due to the pioneering work of Portuguese sailors in the previous decades. However, the remaining ship was hardly seaworthy, the remaining crew was sickly, and they had to sail a route that would keep them out of sight of any possible Portuguese ships. Portugal felt it owned the monopoly on the southern route to the East Indies since it had pioneered that course.

C.4 The Seventeenth and Eighteenth Centuries

The British continued the exploration of the eastern and northeastern parts of North America, and European nations colonized and set up trade with their new colonies. There was beginning to be felt a need for some

FIGURE 1.12 *Benjamin Franklin's chart of the Gulf Stream, published in London in 1786.*

systematic studies in order to improve the speed and safety along trade routes.

In 1700, W. Dampier published *A Discourse of the Winds,* which was a compilation of information intended to aid sea travel. Benjamin Franklin, who had been Postmaster General for the colonies prior to the American Revolution, published a chart of the Gulf Stream in 1786, of which he knew from experience with mail ships going to and from the colonies (Figure 1.12).

The first recorded voyages with primarily scientific objectives were those under Captain James Cook of the British Royal Navy (Figure 1.13), in the period of 1768 to 1779. He had previously been commissioned to map the area of the St. Lawrence River and the Newfoundland Coast, and from this work he gained a reputation for being skillful in mapping, mathematics, and astronomy. At the request of the Royal Society, The Navy sent a scientific expedition to Tahiti in 1768 in order to observe the transit of Venus. Cook was chosen to captain that voyage, and he was sent back to the Pacific twice more before being killed by Hawaiians in 1779 in a skirmish over a stolen boat.

One of his assignments was to search for the undiscovered "great southern continent." Although ice prevented him from actually ever reaching Antarctica, he was able to deduce that such a continent did exist on the basis of two pieces of evidence. First, the ice was salt free. Therefore, it formed on land, not on the ocean. Second, he observed two species of birds that required land for nesting.

Another of Captain Cook's major accomplishments was conquering the dreaded sailor's disease, scurvy, by experimenting with diets. The adoption of his practice of having crews drink lime juice led to the name "limey" for British sailors.

FIGURE 1.13   *Captain James Cook, Royal Navy (1728–1779).*

Captain Cook was one of the first to have very accurate timepieces aboard ship. This enabled him to determine longitude very accurately by timing the overhead passage of the sun or stars. By then, the technique of determining latitude by the height above the horizon of celestial bodies was already well known. By the end of the voyages of Captain Cook, the general geography of the world was fairly well known, except for Antarctica and other regions beyond 70° North or South. The arrangement of the continents and major islands on globes after that was pretty much as we find them today.

C.5   The Nineteenth Century

In the next century there was considerable incentive for more scientific ventures into the oceans. Then, as today, the motivation was largely economic.

One incentive was preparation for laying transatlantic cables. The first was laid in 1858. Naturally, much information was needed concerning the temperatures, depths, and general environments of the ocean bottom, including whether there existed organisms that might damage the cable.

Until this time, people had always had trouble making good soundings. Because of the strength needed for weighted lines long enough to reach the bottom, they were so heavy that it was frequently not possible from aboard ship to "feel" when the weight hit bottom. Sir James Clark Ross was the first to start making reliable, repeatable soundings of the deep ocean around 1848. He also found organisms on the deep ocean bottom in the Southern Hemisphere that were identical to those found by his uncle, Sir John Ross, in the northern Atlantic some years earlier. He reasoned that since these particular organisms were very sensitive to changes

FIGURE 1.14 *Photo of New York Harbor in 1887.*

in environment, the entire deep ocean bottom must have water of similar characteristics. Otherwise the migration of those species wouldn't be possible.

Another incentive for scientific investigations was the flourishing sailing trade (Figure 1.14). Then, as now, high premiums were placed on safe, efficient trade routes. This need was answered by an American naval officer named Matthew Fontaine Maury (Figure 1.15). He was permanently crippled early in his career in a stagecoach accident, and was no longer fit for sea duty. He was subsequently placed in charge of the Depot of Charts and Instruments in Washington, D.C. His misfortune was the good fortune of other sailors, as he was able to systematically analyze the log books and other records of thousands of voyages, which were stored at the depot. From this he published charts and sailing instructions, detailing patterns of wind, currents, and states of the sea throughout the welltravelled oceans. He organized an international marine conference in Brussels in 1853, where conventions were agreed upon for making routine records of the state of the air and sea during all oceanic voyages. With some modifications, these procedures are still used today. He also published in 1855 what amounts to the first real textbook in oceanography entitled, *The Physical Geography of the Sea.*

Another motivation for scientific analysis of the sea was the erratic catches of European fisheries. Some international cooperation was initiated to study the problem, although international cooperation in finding and enforcing solutions to these problems has always been somewhat lacking.

Christian Gottfried Ehrenberg (1795 to 1876) had showed that some sedimentary deposits were made up of myriads of microscopic skeletons of tiny organisms, establishing that these microscopic forms could be very

FIGURE 1.15   *Commodore Matthew Fontaine Maury, U. S. Navy (1806–1873).*

important in the ecology of the oceans. Later, it was found that the health of the harvested crop of some species could be traced back to their survival as youth, which fed on microscopic plankton. Therefore, the planktonic community in one year influences the harvest of fish a few years later.

A fourth motivation for oceanic investigations was scientific curiosity and international scientific rivarly. One question involved whether there was life in deep ocean waters. Edward Forbes (1815 to 1854) was a very distinguished scientist who held the Chair of Natural History at the University of Edinburgh. From samples he collected, he noted that the abundance of life diminished with depth. He conjectured that since plants required the sunlight of surface waters, and since the entire biological community was ultimately dependent on them for food, then probably no life existed below a depth of about 600 m. Others cited the great pressures at depth as a possible deterrent to life. Because of his good reputation, he had a large following, even though Sir John Ross had pulled up worms and a brittle star from a depth of 1800 m many years earlier.

Another interesting ingredient to debates of the time was Thomas Henry Huxley's observation of a gelatinous ooze containing calcium carbonate particles in some samples retrieved from deep water. He suggested it might be the remains of some unknown deep sea organism, which he called "bathybus." Later, it was found to be the result of a chemical reaction initiated by adding too much preservative to some samples. Later in the century, when cables were pulled up for repair, they were found covered with a variety of deep sea animals.

Charles Darwin was a naturalist aboard the *H.M.S. Beagle* during its voyage along the South American Coast. The ship rounded Cape Horn

FIGURE 1.16 *Photo of the Fram as it left Oslo on June 14, 1893.*

and sailed up as far as the Galapagos Islands (near Ecuador) during the years 1831 to 1836. His acute observations formed the basis of his ideas on evolution and natural selection, which he published in his famous *Origin of the Species.* He also postulated what has proven to be the correct explanation for the formation of many coral reefs.

A Danish chemist named Johann Forchammer analyzed seawater samples from a wide range of ocean areas. He found that the major salts are always present in the same ratios. Only the relative amount of fresh water in the mixture changes. Denmark today is still a center for research into the chemistry of sea water.

You can see that there was incentive for numerous trips to sea to make various types of observations. But in the 1870s some truly major scientific expeditions were undertaken. In 1874 to 1875 the Americans were making systematic measurements in the Pacific aboard the *Tuscarora* in preparation for laying a transpacific cable to Japan. Among other things, the greatest depth yet known was measured in the Kurile Trench just east of Japan. From 1874 to 1876, the German ship *Gazelle* studied part of the Atlantic, Indian, and Pacific oceans.

One of the most ambitious voyages of all times was made by the British ship, *H.M.S. Challenger,* in the years 1872 to 1876. It was sponsored by the British Admiralty, with encouragement from the Royal Society, and international rivalry was undoubtedly an important motivational factor. A team of scientists was commissioned and put under the direction of C. Wyville Thompson to probe all possible physical, chemical, and biological aspects of the oceans. The voyage lasted 3½ years, and covered the Atlantic, Pacific, and "southern" oceans. The data gathered took an additional 19 years for scientists to organize and edit, and eventually made up 50 large volumes, 29,500 pages in all.

FIGURE 1.17 *FLIP, a 355-ft Floating Instrument Platform, designed and developed by Marine Physical Laboratory at Scripps Institution of Oceanography, University of California, San Diego. In the vertical position (inset) FLIP gives scientists an extermely stable platform from which to carry out underwater acoustic and other types of oceanographic research. FLIP has no motive power of its own and must be towed (above) to a research site in the horizontal position. Once on station, the ballast tanks are flooded and the platform "flips" vertically, leaving 55 ft above water. The work completed, the water is forced from the tanks, and FLIP resumes the horizontal position.*

Near the end of the 19th century, the Norwegian ship *Fram* (Figure 1.16), under Fridtjof Nansen, explored the Arctic Ocean. The motivation for this voyage was inspired partly by the discovery near Greenland of driftwood from an American ship that had been crushed by ice north of Siberia some years earlier. Nansen advocated building a special ship that would not be crushed, and letting it flow with the ice across the Arctic Ocean. Eventually, he got the ship built, which was a three-masted schooner with a 1.3-m-thick reinforced hull. It sailed to the Siberian Sea, where it froze in the ice pack.

Its slow journey in the frozen ice pack took 3 years, and it did not pass directly over the North Pole as Nansen had hoped. Consequently, Nansen and a companion set out from the ship by dog sled, headed for that destination. They were not able to make it and suffered some hardships on the way back, including having to winter on one of the Franz Joseph Islands. They eventually encountered a British expedition, which they joined, and which returned them home one week before their ship *Fram* arrived.

Soundings from the *Fram* established that the Arctic Ocean was deep like the other oceans. Water samples were retrieved from depth with a device still in use today, called a "Nansen bottle," after its inventor. Careful records revealed that the ice had drifted in a direction to the right of the direction of the wind, which was later explained by V. W. Eckman as being caused by the earth's rotation.

Since the beginning of this century, there have been an increasing number of oceanographic voyages, using more modern techniques, which facilitate the collection of more data, and more accurate data, with less effort. We will not continue to chronicle these more numerous and more modern expeditions. Instead, we will get a feeling for some of the investigations carried out as we read the rest of the book.

Today, there are many oceanographic research institutions around the world, and even more research ships. Still, the enormity of the oceans ensures that we have just begun to explore them.

D. SUMMARY

Oceanography is a very diverse field, employing a wide variety of specialists. This book will focus on the scientific aspects of oceanography, with the primary objective of fostering an inquisitive attitude toward the world.

The earth is extremely old by human standards. Life here began in the ocean over 3 billion years ago. Life on land began less than ½ billion years ago. Hominoids appeared about 5 million years ago, and the first recorded voyages of ocean exploration took place about 3 thousand years ago.

Waterways were a lifeline to great ancient civilizations in the Mediterranean area. The Phoenicians were an early civilization with accomplished sailors. The Greeks had a remarkably accurate picture of the world, which was lost in the West during the Dark Ages.

Although the Vikings had done some exploration during the Dark Ages, the western Mediterranean countries didn't begin exploratory voyages until the late 15th century, after conventional trade routes with the East had been severed. The Portuguese sailed around the Cape of Good Hope, and the Spaniards "discovered" the New World. This period was culminated by the circumnavigation of the world by the Magellan expedition in 1519 to 1522.

Colonization and trade was a large motivation for increasing understanding of the oceans in the 17th and 18th centuries. Captain James Cook made several scientific expeditions during which he confirmed the existence of Antarctica and conquered scurvy, among other things.

In the 19th century oceanic science received attention for several reasons, including the laying of transoceanic cables, flourishing trade, an interest in improving fisheries, scientific curiosity, rivalry, and misconceptions concerning life in the deep ocean. Some famous names from the era are Ross, Maury, Ehrenberg, Huxley, Darwin, Forchhammer, Thompson, and Nansen. The voyage of the *H.M.S. Challenger* was extremely ambitious, and that of the *Fram* was unique.

1. What are some specific things that you think oceanographers might want to investigate, and why? What is the name of the particular field of science that each of these belongs to?

2. How old is the earth? How does this compare with the time that *Homo sapiens* have been around? How does it compare with your life expectancy?

3. What are some changes in the environment that we can easily notice happening within a lifetime? What are some we cannot?

4. How does our experience and knowledge of the earth's spatial extent compare to that of its temporal extent?

5. Suppose the Atlantic Ocean is 200 million years old. What percent of the history of earth is that?

6. What forms of life do we know existed earlier than 1 billion years ago? Give dates. What advancement do you suppose is represented by blue-green algae over previous life forms?

7. What development indicates the beginning of the Cambrian Period? Why do you suppose Precambrain fossils are extremely rare compared to post-Cambrian fossils? (There are several reasons.)

8. What are the times spanned by the Paleozoic, Mesozoic, and Cenozoic eras?

9. What are two justifications for having so many smaller divisions of time for events since the beginning of the Cambrian Period, and not having these small divisions before the Cambrian Period?

10. What were the major developments in terrestrial (land) organisms in the Paleozoic, Mesozoic, and Cenozoic eras?

11. About how many years ago did relatively stable, organized, agrarian human societies develop? How might this be related to the oceans? Do you suppose similar things happened earlier at the end of previous ice ages?

12. If the entire history of the earth were condensed into one day, how long ago did hominoids appear? How long ago did recorded oceanographic exploration begin?

13. Did the Europeans really discover America? The Pacific islands? Who did?

14. Was water a lifeline or a barrier to ancient civilizations? Explain.

15. Who were the first to circumnavigate Africa (we think), and when did they do it? What other areas did they explore beyond the Mediterranean?

16. Did ancient sailors have more difficulty in establishing latitude or longitude? Why?

17. What did Pythaes do?

18. Can you describe how Eratosthenes was able to determine the circumference of the earth without leaving Egypt?

19. What was the problem that Seneca solved? How did he explain it?

20. Who do we think made the first deep water soundings? What is a "sounding"?

21. How does knowledge of the earth at the time of Ptolemy (around 140 A.D.) compare with the knowledge in the West during the Dark Ages?

22. Briefly discuss the exploratory activity in northern Europe during the Dark Ages.

23. How did the compass get to Europe?

24. What two events in the 15th century were to help motivate the famous voyages of discovery at the end of that century? Explain why.

25. What trade route was pioneered by the Portuguese? What are the names of the explorers credited with this?

26. How was an Italian astronomer involved in the discovery of America?

27. What were some of the hardships endured by the Magellan expedition? When it arrived back home, how did its condition compare with that when it left Spain three years earlier?

28. In 1786 Benjamin Franklin published a chart of what? How did he get his information?

29. How did Captain Cook deduce the existence of Antarctica?

30. Why are British sailors called "limeys"?

31. What device enabled Captain Cook to make accurate determinations of longitude? Expalin the relationship.

32. Explain why making deep-sea soundings with a weighted line was difficult.

33. Why were we interested in conditions on the ocean bottom in the 19th century?

34. Why did Sir James Clark Ross think that the deep-sea environment must be fairly uniform the world over?

35. Explain how Matthew Fontaine Maury's misfortune became the good fortune of other sailors.

36. Who discovered the abundance of microscopic marine organisms? How are these microscopic organisms related to the size of the fish catch?

37. What led Edward Forbes to believe there was no life at great depths?

38. What was bathybus?

39. What was the *Beagle*?

40. What did Johann Forchhammer do?

41. During what years did the *H.M.S. Challanger* sail? Who directed this scientific project? How many volumes do the *Challenger Reports* encompass?

42. Describe the *Fram* and its voyage. Who directed this voyage? Why did he not return with it?

43. What were some important contributions to our knowledge made by the voyage of the Fram?

*44. Can you think of any justification for studying our history and cultural heritage, even though we know large and important segments are missing from our records?

*45. What possible explanations can you invent for the apparent loss of interest in sailing southward subsequent to Hanno's voyage?

*46. Estimate how far you walk in a day. In a lifetime. How does that distance compare to the distance around the earth?

*47. An average human lifetime is about what fraction of the age of the earth?

*48. Why do you suppose life developed on the earth rather than the moon or other planets? Explain why these factors should be important.

*49. How could you determine the age of a fossil if you found one?

*50. If you were an ancient sailor, how would you prevent your ship from running aground into islands or shorelines at night or in fog?

*51. If you were building a ship that would not be crushed as ice froze and expanded around it, how would you design the hull?

*52. If you were aboard the *Challenger,* how would you get water samples from great depth? How about life samples from the ocean bottom?

*53. How do you suppose large commercial sailing ships were maneuvered in and out of tight spots, such as busy harbors and docking areas?

SUGGESTIONS FOR FURTHER READING

1. Herbert S. Bailey, Jr., "The Voyages of the Challenger," *Scientific American* (May 1953).

2. L. Anthony Leicester, "Columbus's First Landfall," *Sea Frontiers,* **26,** No 5 (1980), p. 271.

3. M. Sears and D. Merriman, *Oceanography: The Past,* Springer-Verlag, New York, 1980.

4. Gilbert L. Voss, "On the Track of the Challenger," *Sea Frontiers,* **22,** No. 1 (1976), p. 35.

2
THE EARTH

Visitors approaching the Solar System from outer space would be particularly interested in the earth (Figure 2.1). Due to the abundance of surface water, it is unique among the planets and moons of the Solar System. The earth is dominated by the rich, deep blue color of the oceans and the patches of bright white clouds, whereas the other terrestrial planets and moons are mostly shades of reds and browns. This unique and intriguing coloration could be spotted from distances of hundreds of millions of kilometers, a point not overlooked by many writers of science fiction. If there were interstellar travelers with any curiosity at all, they would undoubtedly want to investigate the planet Earth.

Peering between occasional patches of clouds, the visitors would probably first notice the large, dark blue regions covered by oceans and the smaller, more reflective continents, just as our first glance at the moon reveals its darker maria and brighter, more reflective highlands, which make the features of "the man in the moon." The interstellar visitors would also notice the bright, white polar caps, as we notice those on Mars.

On further inspection, the visitors would notice some features that might further tantalize their curiosities. The Pacific Ocean dominates one entire hemisphere; virtually all the land mass is found in the other hemisphere (Figure 2.2). Why this asymmetry in the distribution of the continents? They would also probably notice the symmetry of the Atlantic Ocean, where both margins have the same profile, as if Europe and Africa had been somehow torn off of the Americas and Greenland.

We do not have to rely entirely on conjecture regarding the way a hypothetical interstellar traveler would begin the study of our planet, because we ourselves are becoming explorers of nearby space, and we can trace the development of our studies. We are particularly attracted to unique features on planets and satellites, such as radio signals from Jupiter or the polar caps of Mars. The heavy Venusian atmosphere, the Caloris Basin on Mercury, the rings of Saturn, and the volcanic activity on one of Jupiter's moons are also among the distinctive features that draw our attention.

We also are interested first in explaining large features, with the expectation that the correct explanation of the gross features will also explain many details. For example, we are interested in the cause of the mare on the moon (Figure 2.3). Were they made by slow lava seepage from a molten interior through fissures in the crust? If so, what caused the fissures? Why was the interior hot? Could the maria somehow have been created through impacts from very large meteors? Have the highlands a volcanic origin? Perhaps they were the wrinkles in the surface of a cooling, shrinking moon. Similarly, the huge "Caloris Basin" on Mercury and the gigantic canyon "Valle Marineras" on Mars

28

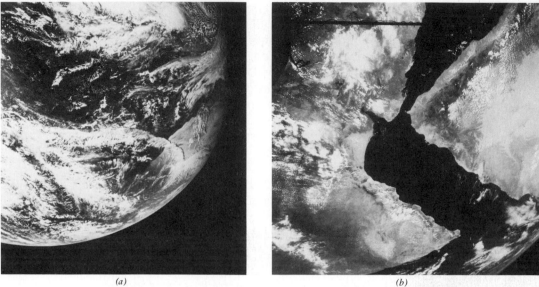

(a) (b)

FIGURE 2.1 *The earth as viewed from outer space. (a) Its distinctive markings would be noticeable from great distances. (b) Closer inspection would reveal additional curious features.*

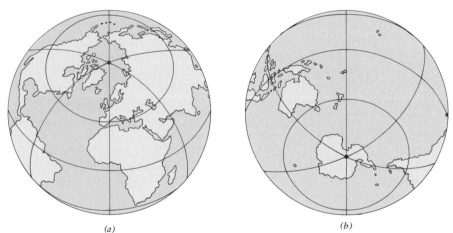

(a) (b)

FIGURE 2.2 *The distribution of continents and oceans is quite asymmetrical. One hemisphere (a) contains most of the land mass. The other hemisphere (b) is dominated by the ocean.*

are some further examples of large features whose explanations undoubtedly will come from an understanding of the geological histories of these planets. We also feel that the same geological scenario that explains one of these things may also explain many other features.

In contrast, the program through which we have undertaken the study of our own planet is quite different from that taken by a hypothetical interstellar investigator. It is also quite different from our own program for studying other nearby astronomical bodies. Rather than initiating the

FIGURE 2.3   *Full moon, as seen from Apollo 11 on its journey back home. Notice the darker maria and the more reflective highlands.*

study through a telescope from great distances, we have understandably begun our studies more through a microscope and at small distances.

One result of our great attention to detail is that we've recorded copious amounts of data and many local theories to explain local events. This abundance of detailed data has made survival difficult for global theories, which would attempt to explain gross features of the earth, such as its oceans. The earth is so complex that whatever global theory you wish to propose, it cannot explain everything. Some data will contradict it.

To support any global theory takes some courage. It will come under criticism, there will be lots of data refuting it, and it will probably fail. This is actually a strength of the sciences and not a weakness. Proponents of revolutionary ideas know they will have to do much more work and have much more thorough data and analysis than do proponents of conventional ideas. Revolutionary ideas will be subjected to very heavy scrutiny, and will be very heavily tested before being accepted by the general scientific community. If we didn't do this, we'd open ourselves up to irresponsible or poorly thought out ideas, and we'd waste a lot of time finding that they lead nowhere. In short, we expect the proponents of a new theory to do enough work to be able to back it up well.

In this chapter and the next, we present a theory that survived this uphill struggle and emerged even stronger than it began. Many geologists who had intended to write the familiar type of article that essentially states, "I doubt that such-and-such a theory is correct because my data contradict it," found that this particular theory was actually supported by

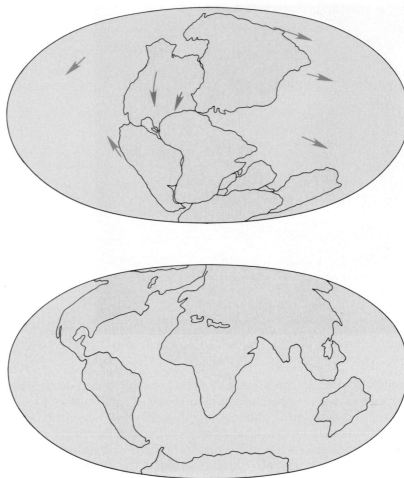

FIGURE 2.4 *(top) The supercontinent, Pangea, as it appeared 200 million years ago. The subsequent motion of the various continental fragments is indicated by the arrows. (bottom) The continents as they appear today. The continental shelves are included as part of the continents, for reasons outlined in the next two chapters.*

their data. It has revolutionized the way we think about our earth, and has made the last three decades in geology among the most exciting eras any of the sciences has enjoyed. The theory is commonly called "plate tectonics," although it must ultimately involve the entire earth, not just the behavior of the plates of the earth's surface. Sometimes it is referred to as "the new global geology."

It should be said once again that not all data support the theory of plate tectonics, because the earth is so complex that no single theory could possibly explain everything. Nonetheless, data supporting the theory have been suprisingly strong from a large number of independent sources and from a wide variety of scientific fields. The case of circumstantial evidence has become so strong that most earth scientists think the responsi-

FIGURE 2.5 *Alaskan coastline. Rugged mountains such as these seem so immutable to us. Yet evidence of change is clearly visible.*

bility for the "burden of proof" has now shifted from the proponents of the theory to its critics.

An extra ingredient in our initial reluctance to accept this new theory is that it seems to violate our "common sense." It also seems too "sensational" for the taste of most scientists, and even seems wilder than what would be ventured by respectable writers of science fiction. Scientists are not as methodical and unfeeling as cartoonists would sometimes make them appear. We realize that sometimes sensational things really do happen, and that sometimes "common sense" may be wrong. But through hard experience, we've learned to be skeptical of sensationalistic theories and to respect our common sense.

The theory of plate tectonics incorporates the idea of continental drift. The continents are moving relative to each other. They haven't always been where they are today, and will be elsewhere in the future (Figure 2.4).

"Common sense" would tell us something quite different. From our trips to the mountains we know they do not change noticeably during our lifetimes. They seem immutable to us (Figure 2.5). From our silent hours at the seaside, we feel the ocean is "eternal." "Solid as the Rock of Gibraltar," is an often-heard expression. No wonder we are at first skeptical of a theory that would send continents spinning and sliding across the face of the earth, oceans opening and closing, surface plates slipping into the bowels of the earth in some regions, and being reborn from those bowels in other regions.

Is this idea sensationalistic? Does it seem to contradict our common sense? The answer to both is "Yes." But back in our memories somewhere, we may remember seeing stratified sedimentary rocks that were lifted high in the mountains, and we may even remember seeing sedimentary

FIGURE 2.6 Originally laid down in horizontal layers beneath the ocean, these strongly folded mountaintop sediment strata are a testimony to the earth's geological activity. (Rocky Mountains, British Columbia)

strata folded into contortions quite distinct from the nice, smooth, horizontal beds in which they were first laid down (Figure 2.6). We may have felt an earthquake tremor or heard grumblings from a volcano, telling us the earth is not completely still. Undoubtedly, each of us can recount from our own experiences some pieces of evidence suggesting earth activity. If the continents are in motion, there must be many different kinds of evidence of that motion that can be analyzed and interpreted.

A. HOW RIGID IS THE EARTH?

One of our difficulties in accepting what must initially seem to be a preposterous theory is our knowledge of rock. It is extremely rigid, and during our lifetime we don't see it bend at all. But this doesn't mean that rocks don't bend. It simply means that our lives are extremely short on geological time scales. Whether something is rigid or plastic depends on the time scales involved. We all are familiar with glass. If we strike glass with a rock, it will shatter as the glass is quite rigid over the time period of the blow. Yet over the years, glass flows. We all have looked through "wavy" glass in the windows of old houses, which has flowed under the influence of gravity.

Geological time scales are much longer than human lifetimes, and things that seem quite rigid to us my have been quite plastic over the time scales of the earth's evolution. The average age of an ocean, for example, which is only a small fraction of the total age of the earth, is millions of times longer than human life expectancy. Typical rates of motion of con-

FIGURE 2.7   *Foreground: The wrinkled edges and overthrusting of the thin plates
of ice result from the ice being more rigid than the water on which
it floats. Background: "Splinter bergs" such as this one are "calved"
from glaciers that flow into the oceans.*

tinents are in the range of 1 to 4 cm/year, although rates up to 16 cm/yr
have been reached on occasion. This is extremely slow on human scales,
10,000 times slower than the hour hand on your wristwatch. Usually, a
continent moves as a block, but even where there is relative motion within
one continent, that motion is still almost imperceptible. Los Angeles is
moving toward San Francisco, being on opposite sides of the San Andreas
Fault. But it will take several hundred thousand generations of people
before they are side by side.

Although the continents are moving very slowly, they have traveled
large distances altogether. If continents have always been moving at
roughly their present speed, then the total distance traveled by a continent
since the earth's beginning would be equivalent to several trips around
the earth. The average continent may have traveled farther in its lifetime
than do average persons in theirs.

Heat and pressure have a large influence on the plasticity of materials. Most tend to become more plastic when heated, and more rigid
when compressed.[1] With reference to the earth, the surface plates are
cooler and more rigid than the material immediately beneath. Therefore,
they cannot duplicate the more fluid motion of the material they ride on
without cracking and grinding along faults or wrinkling into occasional
mountain ranges. An analogy would be the rigid plates of ice floating on
a river when the ice goes out in the spring. They are not as plastic as the
water beneath, so they grind against each other, cracking into smaller
pieces, wrinkling the edges in collision, and sometimes sliding over or
under the edge of another (Figure 2.7).

[1] *One important exception is ice. It may melt under pressure, therefore losing rigidity in the
process.*

B. PROBING THE EARTH'S
INTERIOR

For a greater appreciation of the evolution of the earth and oceans as explained with plate tectonics, we must first understand the earth's structure. Studying the structure is fairly difficult because we must rely mainly on indirect techniques. Our mine shafts and boreholes only go down a few kilometers at most, so we only scratch the surface with our direct observations. Even if we could drill down to arbitrary depth, the material extracted would be in a completely different environment and, therefore, have very different properties than it had before we removed it. We sometimes try to study the properties of materials under high temperatures and pressures by putting a sample in a pressure vessel along with an explosive such as TNT. Even if we could sustain the high temperature and pressure for months, this time period would be "instantaneous" in comparison to the geological time scales of these environments within the earth. Slow transformations into more stable crystalline forms would not have sufficient time to take place, and so the materials would still have considerably different properties than their counterparts within the earth.

Most of our information concerning the earth's interior is gleaned indirectly, using such things as the earth's gravitational field, or analyzing the travels of seismic waves within the earth.

B.1 Gravity Studies

The force of gravity between two bodies depends on their masses, m_1 and m_2, and on the distance between them, r, according to

$$\text{Force} = G\,\frac{m_1 m_2}{r^2},$$

where "G" is just a constant of proportionality, called the "gravitational constant." This relationship says that the attractive gravitational force between two masses increases if either mass (m_1 or m_2) increases. For example, your weight would increase if either: (1) you increased your mass, or (2) you went to a more massive planet. The relationship also indicates that the strength of the gravitational force between two objects *decreases* if their separation (r) is increased (Table 2.1).

"Gravitometers" are instruments made to measure small changes in the earth's gravity from its average value. Since the force of gravity is dependent on both masses and distances, an array of gravitometers can detect differences both in subsurface masses and in their distributions (Figure 2.8). These instruments are remarkably sensitive. For example, a change in elevation of just a few centimeters from the earth's center can easily be detected. If you've ever wondered how the elevation of a mountain above sea level could be so accurately measured, even if the mountain is a thousand kilometers inland, this is one way it is done.

Earth-orbiting satellites are also used to study the earth's internal structure. The orbits of the satellites are carefully observed. Any distri-

TABLE 2.1   Variation of Gravitational
Acceleration with Altitude
above Sea Level

Altitude (km)	g m/s²
0	9.806
1	9.803
10	9.775
100	9.60
1000	7.41
10,000	1.48
100,000	0.036

FIGURE 2.8   *Subsurface mass distribution can be studied by its influence on the gravitational field at the surface. In the above sketch, the actual gravitational force (black arrow) differs from the expected gravitational force (colored arrow) by a small amount due to the extra massive subsurface areas.*

bution of mass within the earth that differs from a purely concentric, spherically symmetric arrangement, will cause detectable changes in the satellite's orbit and speed. Satellites may also carry very sensitive altimeters that give us extremely accurate mappings of the topography of the earth's surface.

Of course, just one measurement with one gravitometer or with the overhead passage of one satellite will not uniquely determine subsurface mass distributions. But people trained in appropriate mathematical techniques can use a large number of measurements to give us a fairly accurate picture of mass distributions within the earth.

B.2   Seismic Studies

Seismic waves are also an important tool used to probe the internal structure of the earth. Seismic waves originate in localized disturbances, usu-

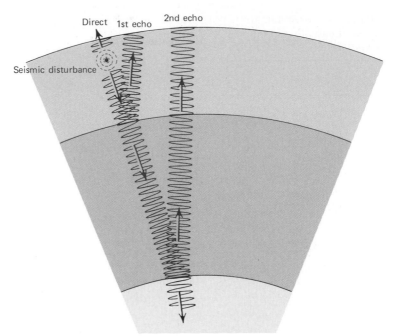

FIGURE 2.9   Seismometers record echos of the original seismic disturbance because seismic waves reflect from subsurface interfaces. Later echos come from greater depths.

ally within the earth's outer crust, such as from slippage along an earthquake fault or from an underground nuclear explosion. These vibrations travel through the earth and are received at seismographic stations on the surface in various parts of the world.

Seismic waves have two properties that make them extremely fruitful in revealing internal earth structure. One is that some of them are reflected when crossing an interface between two different materials. Light waves do the same thing, and we can see reflections on glass because some light is reflected at the interface between air and glass. This means, for example, that if you were manning a seismographic station above a local seismic disturbance, you would record a series of echoes returning from various interfaces deep within the earth (Figure 2.9). Successive echoes would correspond to successively deeper interfaces between layers within the earth.

If you wished to determine the depth of any particular interface, you would need to know both how long you had to wait for that echo and the speed with which the seismic wave traveled. The first item could be read from your seismograph, but there is no direct way to know the speed of seismic waves deep within the earth. Fortunately, this information may be inferred from data, due to the other important property of waves.

This second property is that they tend to bend, or "refract," toward regions where they move more slowly.[2] In analogy, when the wheels on

[2]*This is true of all waves. In the chapter on ocean waves, we'll see this again.*

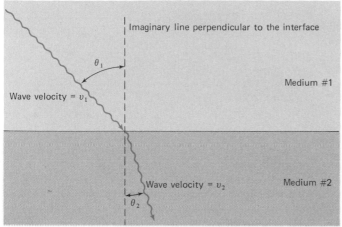

Imaginary line perpendicular to the interface

θ_1

Medium #1

Wave velocity = v_1

Wave velocity = v_2 Medium #2

θ_2

FIGURE 2.10 *According to Snell's Law, if a wave travels from one medium to another, and if its velocity in the first medium is* v_1 *and its velocity in the second medium is* v_2, *then the amount it is bent is determined by the following equation.*

$$\frac{\sin\theta_1}{\sin\theta_2} = \frac{v_1}{v_2}$$

one side of a car slip off the highway and onto the soft shoulder, the additional drag on these wheels makes them tend to slow down, and the car tends to steer toward this lower velocity region, unless the driver is alert and forces it back.

When a wave passes between two materials, there is an exact, known relationship between how much the wave gets bent and the speeds of the wave in the two materials. This is known as "Snell's law" (Figure 2.10). Through calculus, it can be generalized to find the wave trajectory through a material where the speed change is gradual and continuous. The result is that when a seismic wave travels through the earth, its trajectory is not a straight line, but rather is curved. The trajectory bends sharply at the interface between two materials, and slowly and continually within one material at various depths (Figure 2.11).

When a seismic wave arrives at a station a long way from the seismic disturbance, the time for arrival of that wave depends both on the trajectory of the wave through the earth and on the speed of the wave along that trajectory. But the two are not independent. The trajectory depends on the distribution of wave speeds within the earth. So the time for arrival of a seismic signal at a station gives a great deal of information on possible distributions of wave speed with depth in the earth. Knowing times of arrival at many different stations scattered throughout the world determines uniquely how the speed of seismic waves varies with the depth. From this information we learn something about the materials and environments at various depths. We can also use it to infer the depth of various interfaces from the delay times for echoes, as described previously.

Seismic waves traveling through the earth can be put into two categories (Figure 2.12). Those whose vibrations are parallel to the direction

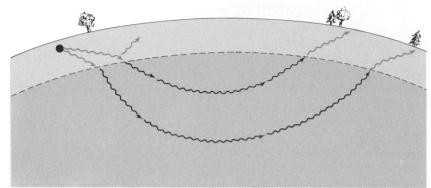

FIGURE 2.11   *Seismic waves travel curved trajectories through the earth because the wave speed changes with depth. Changes in wave speeds are caused by changes in materials, or by changes in temperature and pressure on the materials.*

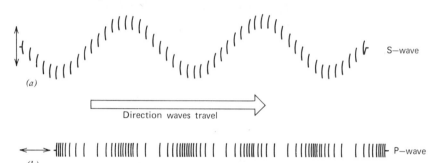

FIGURE 2.12   *Waves in a spring or "slinky." (a) When the motion of the individual links is perpendicular to the direction that the waves travel, they are S-waves. (b) When the links move back and forth in the same dimension the waves travel, they are P-waves.*

of propagation are "P-waves." Those whose vibrations are perpendicular to the direction of propagation are called "S-waves." The "P" is for "primary," as the P-waves are fastest and reach the sensors before the slower, secondary "S"-waves. On a long spring or "slinky," a P-wave can be demonstrated by quickly moving one end in and out. An S-wave occurs when the end is wiggled sideways.

The P-waves can cross an interface between a solid and a liquid. For example, if you push the side of a container of water in and out, it will push the water in and out too. However, S-waves do not cross the interface. If you jiggle the side of a container of water back and forth, the water won't move; there's too little friction between the two. You can test this by spinning a glass of water back and forth and noting that although the solid container spins, the water doesn't. Liquids will not support shear stress.

It is found that stations on the opposite side of the earth from a disturbance receive only P-waves. This indicates the earth has a liquid core

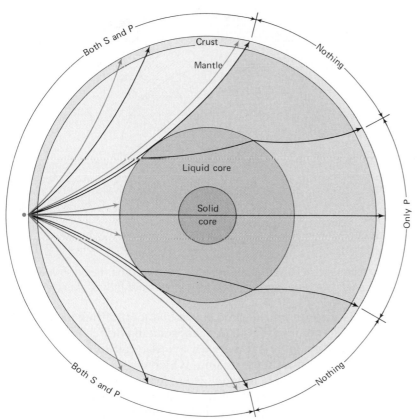

FIGURE 2.13 *Schematic diagram of the paths followed by P-waves (black lines) and S-waves (colored lines) through the earth's interior, after being generated at the left-hand side of the figure by some seismic disturbance. Notice that because the S-waves cannot cross the interface between the solid mantle and the liquid core, they will not be detected on the opposite side of the earth, arriving directly from the disturbance. In addition, due to refraction of the waves, there will be shadow zones where neither type of wave will be received directly.*

that prevents the direct arrival of S-waves through the earth's center. Of course, some will eventually arrive by traveling through the crust all the way around, or by some other circuitous route avoiding the core, but these will arrive much later, having traveled farther. By determining where S-waves are and are not received on the opposite side from a disturbance, we can map the outer border of the liquid core (Figure 2.13).

C. THE SHAPE OF THE EARTH

We know the shape of the earth from the use of conventional surveying techniques, and various types of altimeters. Recent advances in gravito-

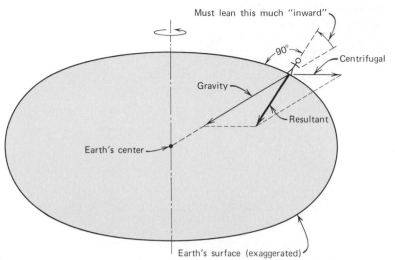

FIGURE 2.14 *The earth's center is not directly beneath you. Because it is spinning, you must lean slightly inward to avoid falling over.*

meters, satellite altimeters, and seismic ocean bottom profiling techniques, have added remarkable precision to our mapping of surface topography.

The earth is nearly spherical, with an average radius of 6370 km. It has a slight equatorial bulge due to its spin, which means that the distance from the earth's center to a pole is about 24 km less than the distance from the center to the equator. This is only about a ⅓% change, and so it couldn't be noticed just by looking at a photo of the earth taken from outer space.[3]

The fact that the earth is spinning means that a hanging plumb bob will not point exactly toward the center of the earth. You would not expect a plumb bob dangling from a spinning merry-go-round to point straight downward. It should swing outward to some degree, depending on how fast the merry-go-round is spinning. Since the earth is spinning, the same thing happens to a plumb bob dangling on the earth. (See Figure 2.14.)

Fluids flow downhill, and so do the rocks of the "solid" earth over geological time scales. This means that the surface of the earth has evolved to the point where it is everywhere perpendicular (ignoring "minor" bumps like mountains) to this "resultant" gravitational force, which is a combination of gravity and this "apparent" centrifugal force.[4] Notice that this "resultant" or "effective" gravity will be strongest at the poles and weakest at the equator. That means your weight will depend on latitude, and gravitometers have to be adjusted for latitude (Table 2.2). The change from equator to pole amounts to about ½% altogether. The

---

[3]*In contrast, the equatorial bulges of Saturn and Jupiter are easily seen on photographs.–*

[4]*This is called an "apparent" force because it is an aberration caused by our rotating frame of reference. Nothing is really pushing the plumb bob out. Rather, it is simply trying to go in a straight line. It is really the earth and the observer that are turning and being accelerated.*

TABLE 2.2   Variation of the Earth's Gravity and Radius with Latitude

Latitude	Radius(km)	Acceleration of Gravity(cm/sec²)
0°	6378	978.04
45°	6366	980.53
90°	6357	983.22
Average	6371	981.00

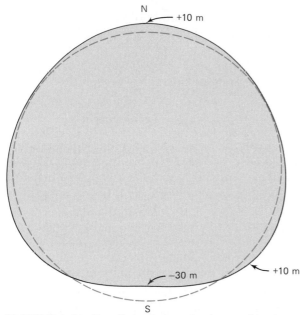

FIGURE 2.15   *Small deviations in the earth's shape from the theoretical oblate spheroid, as measured by satellite.*

slightly "squashed" spherical shape of the earth is more properly called a slightly "oblate spheroid."

In addition to the small equatorial bulge, the earth is found to be very slightly pear-shaped as well (Figure 2.15). However, these deviations are on a scale 1000 times smaller than that of the equatorial bulge, and are measured in meters rather than kilometers.

D.   THE STRUCTURE OF THE EARTH

The interior of the earth may be placed into three or four distinct categories on the basis of composition. The thin, outer layer is called the "crust." Beneath it lies the "mantle," and beneath the mantle is the "core." The core is further divided into the outer core, which is liquid, and the

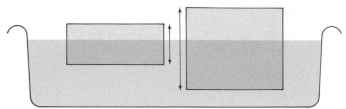

FIGURE 2.16 *Two blocks of wood floating in a bathtub. The one sticking up into the air the farthest also sticks down into the water farthest.*

solid inner core. It may be that the main difference between these two regions of the core is only their physical state and not their chemical composition. (Notice that we don't know for sure!) Throughout the long history, the densest materials have tended to sink toward the earth's interior, and the lighter materials have worked their way to the surface, although this process of gravitational separation is by no means yet complete.

D.1 The Crust

The lightest materials are in the crust, and have densities in the range of 2.6 to 3.1 g/cm³.[5] The crust beneath the ocean basins is the thinnest, being typically 5 to 6 km thick. Beneath the continents it is considerably thicker, averaging roughly 40 km in thickness, and reaching 70 km deep beneath some mountain ranges. This is what we should expect if the crust really is somehow "floating" on the material below. (See Figure 2.16.)

Consider, for example, ice cubes floating on water, or icebergs floating on the ocean. Those that stick up farthest above the water must also extend farthest down below the water's surface in order to obtain the required buoyancy. A floating object displaces an amount of the fluid equal to its own weight (Archimede's principle), which means that more massive objects must displace more fluid in order to float. Since ice is 90% as dense as water, for example, 90% of an iceberg must be below water level if it is to displace an amount of water equal to its own weight. For every meter that a tabular (i.e., flat) iceberg extends up from water level, it extends nine meters down. Similarly, if the crust is "floating" on the mantle, then wherever its upper surface extends farther upward, its lower surface must also extend farther downward into the mantle (Figure 2.17). Thicker, more massive portions must extend deeper into the mantle, displacing more of the mantle material, in order to obtain the required buoyancy. This concept is referred to as "isostasy." An "isostatic adjustment" is a rising or sinking of a block of crust in order to obtain the required buoyancy.

The crustal materials seem to be of two distinct types: those associated with the ocean bottoms and those of the continents. (See Figure 2.18.) The continental crust tends to be slightly lighter than oceanic crust in both density and in the colors of the minerals. As can be seen on maps, the

[5]*For comparison, the density of water is 1 g/cm³, or about ⅓ as dense as crustal materials.*

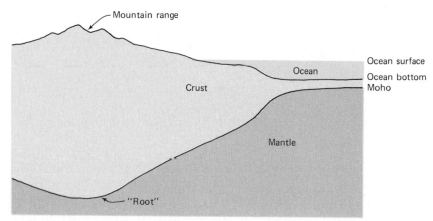

FIGURE 2.17 *Where the crust sticks up the farthest, it also sticks down the far-thest. The crust floats on the mantle like sawdust on water.*

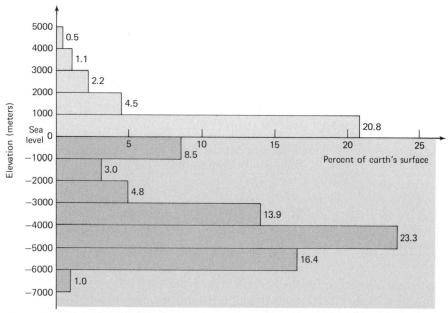

FIGURE 2.18 *Plot of the percentage of crustal surface located at various eleva-tions, in 1-km intervals. (From Sverdrup, Johnson, and Flemming, The Oceans: Their Physics, Chemistry, and General Biology, Pren-tice-Hall, © 1942, renewed 1970.)*

thicker continental materials tend to come in localized "globs," which have higher surface elevations.

Although these distinctive characteristics seem curious and in need of explanation, an even more startling result of recent decades was the discovery that oceanic crust was extremely young. Although much continental material has an age that is a significant fraction of the age of the

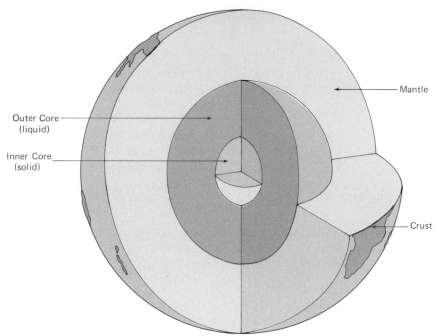

Mantle

Outer Core
(liquid)

Inner Core
(solid)

Crust

FIGURE 2.19   *The earth's interior.*

TABLE 2.3   Interior Regions of the Earth

Region	Depth Below the Earth's Surface (km)	Density (g/cm³) Range	Average
Crust	0 to 30	2.5 to 3.3	2.8
Mantle	30 to 2900	3.3 to 5.7	4.5
Outer core	2900 to 5100	9.4 to 14.2	11.8
Inner core	5100 to 6370	16.8 to 17.2	17.0
Entire earth	0 to 6370	2.5 to 17.2	5.5

earth, this is not true for the oceans, whose ages are only a few percent of that of the earth. Clearly, our present oceans are very recent features.

It should be emphasized that the crust is very thin, amounting to only about the outer 0.3% of the earth on the average (Figure 2.19 and Table 2.3). This means that if the earth were scaled down to the size of an average globe, the crust would be only as thick as about five pages of this book. Consequently, whenever we illustrate features of the earth's crust in this book such as ocean bottom features or continental margins, we'll have to use a great deal of vertical exaggeration in order to make these features visible.

D.2   The Mantle
The interface between the crust and mantle shows up clearly on seismic profiles of the earth's interior. It is called the "Moho," which is short for "Mohorovicic discontinuity," and probably marks a sudden change in

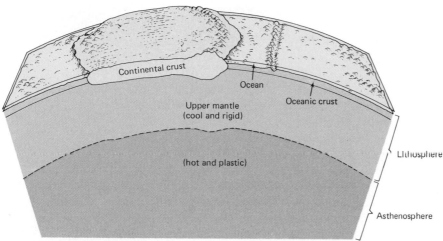

FIGURE 2.20 *The cool, rigid crust and upper mantle move together, floating on the more plastic portions of the mantle beneath.*

material composition, although its exact nature is still not thoroughly understood. The mantle extends nearly halfway to the earth's center and encompasses 84% of the earth's volume.[6] Consequently, when we make comparisons of the "earth" with other astronomical bodies, such as the stony meteorites we encounter, it is fairest to compare them with our mantle material. Certainly, our surface materials, including crust, ocean, and atmosphere, are not typical.

The mantle is thought to be fairly uniform in terms of its chemical composition, although there are large variations in its physical properties. This is understandable, since the temperatures and pressures must increase greatly with depth. Different crystalline structures would be stable in different environments.

The upper 80 to 100 km are still cool enough to be quite rigid, as can be inferred by the high velocity of seismic waves in this region. It seems to "stick" to the bottom of the crust, and it participates in the plate motions along with the crust. This combined rigid region makes up the "plates" of plate tectonics, and is called the "lithosphere."

Beneath the lithosphere, at a depth extending approximately between 100 and 200 km, is the "low velocity zone," a name derived from the low velocity of seismic waves in this region. From this we infer that the region must be much more plastic than the lithosphere. The increased temperature here undoubtedly causes some "partial melting" of the materials, which means that some minerals melt under these high temperatures and lubricate the motion of the grains of more resilient minerals. This more plastic region on which the lithosphere "floats" is also called the "asthenosphere." (See Figure 2.20.)

---

[6]*The volume of an object is proportional to width × height × thickness. If you double the dimensions of an object, its volume increases by a factor of 2 × 2 × 2 = 8. Therefore, the inner half of a sphere contains only ⅛ of the total volume, the outer half containing the remaining ⅞.*

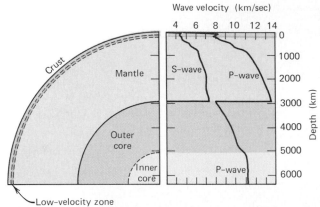

FIGURE 2.21 *Plots of wave velocities vs. depth in the earth. Notice the abrupt change in P-wave velocity between mantle and core. Also notice that S-waves don't penetrate the core at all. (From Peter J. Wyllie, The Way the Earth Works, John Wiley & Sons, 1976.)*

Beneath about 200 km in depth, the speed of seismic waves again increases, and it continues to increase with depth all the way down to the core (Figure 2.21). You might think that increased temperature with depth would make the mantle more plastic than rigid. But the pressure is also increasing, and it is the combination of the two that is important. A few small but discrete changes in the velocity of seismic waves with depth suggest that in different environments the material adopts different crystalline structures.

D.3   The Core

Slightly less than halfway to the center of the earth, the core begins. The outer core is molten. S-waves do not penetrate it, and the velocity of P-waves is greatly reduced from that in the mantle just above. P-waves regain a higher velocity in the central part of the core, which indicates it is a more rigid solid.

The composition of the core may be inferred from two approaches. In one approach we can infer the mass and density of the material of the core from the earth's gravitational field. We know what the total mass of the earth is, and we know roughly how much of it is in the mantle, from studies in mantle materials in our laboratories. The missing mass must be in the core, and only a few materials would provide the correct mass and density in the core environment.

The other approach involves studying the materials in other astronomical bodies, such as meteorites (Figure 2.22) or the sun, and comparing them to the materials of our mantle. What's missing from the mantle must be in the core.

From both these approaches we arrive at the same conclusion. The core is mostly an iron-nickel alloy, although undoubtedly many other materials are present in smaller amounts.

FIGURE 2.22 *The relative abundances of the elements in stony meteorites is similar to that in the earth's mantle.*

E. SUMMARY

The earth is clearly distinctive among members of the Solar System, and would undoubtedly arouse the interest of any interstellar travelers. Our approach to studying this planet has been quite different from our approach to other planets, as here we've historically given a great deal of attention to local details, and overall, global theories have had difficulty surviving.

In understanding the earth's geological evolution we must realize that things seeming rigid to us may not be rigid over geological time scales. Heat and pressure influence the plasticity of a material. The outer skin of the earth tends to be cooler and more brittle than the material beneath it.

To study the earth's interior we must rely on indirect techniques, such as gravity studies and seismic studies. The gravity studies give us information on the distribution of subsurface masses, and the seismic studies give us information on their rigidities as well. There are two types of seismic waves; one cannot propagate through the liquid outer core. They reflect off boundaries, and they refract in a changing medium.

The earth is rather spherical, with a slight equatorial bulge due to its spin. It has a thin outer shell called the "crust," beneath which lies the voluminous mantle. For plate movement it is more appropriate to talk of the "lithosphere," which includes both the crust and the upper, more rigid portion of the mantle as well. This rides atop the more plastic asthenosphere. The core has an outer, molten region, and an inner, solid part. It contains large amounts of iron and nickel.

1. How is the general appearance of the earth different from that of most other terrestrial bodies in the Solar System?

2. Do the continents or oceans reflect more sunlight?

3. What are some of the overall features of the continents and oceans that an extraterrestrial visitor might first notice as being strange?

4. How does our approach to the study of other planets differ from our approach to the study of our own?

5. Why do large amounts of detailed data inhibit the formulation of global theories? Is that necessarily bad? Whatever your answer to this last question, defend it.

6. Why do we subject revolutionary ideas to much heavier scrutiny than conventional ones?

7. Is the theory of plate tectonics universally accepted?

8. Were there both subjective and objective reasons for our reluctance to accept this new theory? Explain.

9. Describe how our common sense deceives us with regard to the permanence and immutability of some of the earth's features.

10. What are some of the things that you can think of that should suggest to us that the earth's surface is not entirely still?

11. Explain how the adjectives "rigid" and "plastic" depend on time scales. Give some everyday examples.

12. How does the age of an average ocean compare with the age of the earth? With human life expectance?

13. How does the relative speed of continents compare with the speed of the hour hand on a watch?

14. Can you estimate how far you've traveled in your life? Is this more or less than the distance your continent has traveled during its lifetime?

15. How does the plasticity of most materials vary with temperature? With pressure?

16. Why are the earth's surface plates more rigid than the material below?

17. Why do you suppose that nearly all earthquakes originate in and among the surface plates, and very few in the material beneath?

18. Explain why we must rely on indirect techniques to study the earth's interior.

19. How do we try to simulate conditions deep within the earth? Why is this not very satisfactory?

20. How far must one be from the center of the earth if the earth's gravitational pull is only $\frac{1}{25}$ of that on the surface?

21. The moon has ⅛₀ as much mass as the earth and ¼ the radius. How much would you weigh on the moon?

22. Briefly describe how gravitometers can be used to measure subsurface mass distributions.

23. How can satellites be used to get information regarding subsurface mass distributions?

24. Briefly explain how "echoes" can be used to measure the depths of discontinuities in the earth's internal structure.

25. Using echoes as above, what two things must be known before the depth to any discontinuity can be determined?

26. When a wave crosses the boundary between two materials, how are the respective wave velocities related to the direction the wave will bend?

27. As a seismic wave travels deep into the earth's mantle, there is a tendency for it to bend back toward the surface. What does this tell us about how the wave velocity changes with depth in this region?

28. Other things being equal, do you think that seismic waves should travel faster or slower in more rigid materials? (*Hint:* More rigidity implies neighboring atoms are bound more tightly together.)

29. How are the particle vibrations related to the direction of propagation for S-waves and P-waves. Which do you suppose is an example of a "longitudinal" wave and which a "transverse" wave? Which travels faster in the earth?

30. Why can's S-waves travel straight through the center of the earth?

31. What is the radius of the earth? By about how much does this vary from the equator to pole? Why?

32. What is a plumb bob? (Look it up in a dictionary, if you don't know.)

33. Explain why a hanging plumb bob won't point directly toward the earth's center in general. Where on earth would it point directly at the earth's center?

34. Explain why your weight depends on latitude? Would your distance from the earth's center also have some effect? Compare these effects at the equator to their effects at a pole.

35. If the equator is 24 km "uphill" from the poles, why is there water at the equator? That is, why doesn't it all run "downhill" to the poles?

36. Compare the size of the equatorial bulge with the "pear-shaped" deviations of the earth's surface.

37. What are the three main divisions of the earth's interior? Where do you expect the lightest minerals to be found? Why?

38. What is a typical thickness of the oceanic crust? Of the crust beneath the continents?

39. Explain isostasy and how it is related to the thickness variations of the crust.

40. What are the two distinct crustal types, and how do they compare?

41. Is oceanic crust old or young? How long is 1% of the age of the earth?

42. How thick is the crust compared to the rest of the earth?

43. How can the mantle encompass 84% of the earth's volume if its thickness is less than half of the earth's radius?

44. Which region of the earth is most "typical" of the earth? Explain.

45. If the mantle has a uniform composition, why do its properties, such as density and rigidity, vary so much with depth?

46. What is the "lithosphere"?

47. What is the "low velocity zone"? The asthenosphere"? Why do we think it is quite plastic?

48. If the mantle becomes hotter with depth, why doesn't it also become more plastic with depth?

49. What happens to the velocity of P-waves as they enter the outer core? What do we conclude?

*51. Notice that the indentations in the pear shape are where the continents are found (Northern Hemisphere and around Antarctica). Can you think of any possible explanation for this?

*52. How would you go about proving that the earth's center isn't hollow?

*53. If denser things tend to sink, why are lead and gold to be found in the earth's crust?

*54. Why is the moon covered with craters and not the earth?

*55. If the plates of the lithosphere each move as a unit, what do you suppose is the difference between the crust, and the upper 80 km or so of the mantle? (We know there must be a difference because of the Moho discontinuity showing up on seismic records.)

*56. Balsa wood is 40% as dense as water. How much water does 1 m^3 of balsa wood displace when it floats?

SUGGESTIONS FOR FURTHER READING

Peter J. Wyllie, *The Way the Earth Works*, Wiley, New York, 1976.

3
PLATE TECTONICS

San Andreas Fault.

With this background in the earth's internal structure, we are now prepared to understand the revolutionary new theory of the dynamics of the earth. Among other things, the theory explains the origin and evolution of our oceans, and much of the most convincing evidence supporting this theory has been extracted from the oceans by geophysicists and other oceanographers.

The manner in which this fledgling theory slew Goliaths on its road to general acceptance warms the hearts of those romantics among us who like to root for underdogs, and who like to believe that truth (or at least good data) will eventually win over all odds. The story illustrates good science at its best, both on the part of proponents and critics. We have fresh in our memories the details of a recent scientific revolution, with its psychological and sociological implications as well. I take advantage of this opportunity to tell the story.

A. ALFRED WEGENER AND CONTINENTAL DRIFT

One of the earliest and most influential proponents of the idea of continental drift was a German astonomer, meteorologist, and explorer named Alfred Wegener. In his book entitled, *The Origin of Continents and Oceans,* published in 1915, he advanced good arguments to support the idea that at one time there was no Atlantic Ocean, and the Americas were welded with Europe and Africa into one large continent, which he named "Pangea."

His strongest argument seemed to be the remarkable similarity of the continental margins on opposite sides of the Atlantic. But he also looked at similarities in ancient geological structures, similarities in the fossil records, and similarities in ancient climates to support his hypothesis. He pointed out that some unique geological structures in the Americas terminate abruptly in the Atlantic Ocean and then continue again on the other side of the Atlantic, as if the structures had formed when the continents were together and then were cleaved as the continents tore apart. He showed that the fossil records on both sides of the Atlantic were similar up to roughly a hundred million years ago, and then showed divergence, indicating that the continents had become isolated from each other then. Finally, using fossils and sediments, he was able to demonstrate that ancient climates on both sides of the Atlantic were similar, although they had been quite different from those found presently in the respective regions.

There is no denying that it was an interesting hypothesis, and good work, but we have seen that the scientific community is understandably very cautious at accepting revolutionary ideas. For the next half century, the topic was considered interesting material for conversation at cocktail parties or in informal chats among old friends at scientific meetings, but it was not the sort of material that one could expect to be published in reputable scientific journals. The subject was seriously debated at a meeting of the American Association of Petroleum Geologists in New York in 1926, where the critics won the day. This seemed to quell serious consideration of the ideas in North America for the next three decades. Alfred Wegener died on the Greenland ice sheet in 1930, far short of the time when new evidence would force reconsideration of his ideas and eventual vindication.

B.   PALEOMAGNETISM

The first evidence for reconsideration of ideas on continental drift came during the 1950s, from the unlikely field of study called "paleomagnetism." Paleomagnetism is the study of magnetization in rocks, and most people thought it to be a rather esoteric study primarily of interest to a few specialists. No one anticipated this field of study could be capable of providing the impetus for a complete revolution in our ideas on the evolution of the earth.

B.1   How a Rock Gets Magnetized

Within each atom are circulating electric charges, which tend to make each atom behave as a tiny electromagnet. This is due mostly to the circulation of negatively charged electrons, although motions in the positively charged nuclei also have a small, but noticeable, effect. When various atoms combine to form minerals, the circulation of electrons in the mineral give it magnetic properties. Certain iron-containing minerals are especially magnetic. The magnetization of the minerals in rock is not strong enough to pick up thumbtacks from the table, but it is often strong enough to be measured with an instrument called a "magnetometer."

For each mineral, there is a temperature above which these atomic magnets are free to rotate and below which they are "frozen" in their orientation. This is called the "Curie point," and is usually around 600°C for common minerals. Usually, the Curie point is below the freezing point of the mineral, so there will be a range of temperatures between these two points where the atoms are frozen into place, but these atomic magnets are still free to rotate.

All crustal materials have a volcanic or plutonic origin. This means they were formed from material that was originally very hot. As they cooled they were sitting in the earth's magnetic field, and so the atomic magnets tended to line up with that field. Once they cool past the Curie point, they are frozen in orientation, and so a record of the earth's magnetic field is frozen in that rock. (See Figure 3.1.) When we measure the magnetization of that rock today, we may find it may have a different orientation than the present magnetic field of the earth. This would tell us that either the rock has moved since its birth, or the earth's magnetic field has changed, or both.

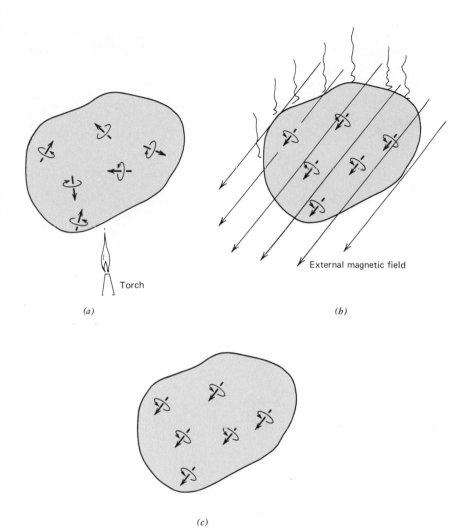

(a)

(b)

External magnetic field

Torch

(c)

FIGURE 3.1 *How to magnetize a rock. (a) When a mineral is heated to a temperature above its Curie point, the little atomic magnets are free to
change their orientation. (b) If the mineral is placed in an external
magnetic field, the atomic magnets will tend to line up with this field.
As it cools below the Curie point, the atomic magnets become frozen
in this new orientation. (c) A record of that magnetic field is then
"frozen" into that mineral, and will remain even after the external
field has been removed or changed.*

B.2 Tracing the Motions of the Continents

We can date the formation of rocks using radioactive dating techniques.
By taking many rock samples of various ages, we can find exactly how a
continent has moved relative to the earth's magnetic field over the years.
But how do we know which moved? Was it the continent or was it the
magnetic field?

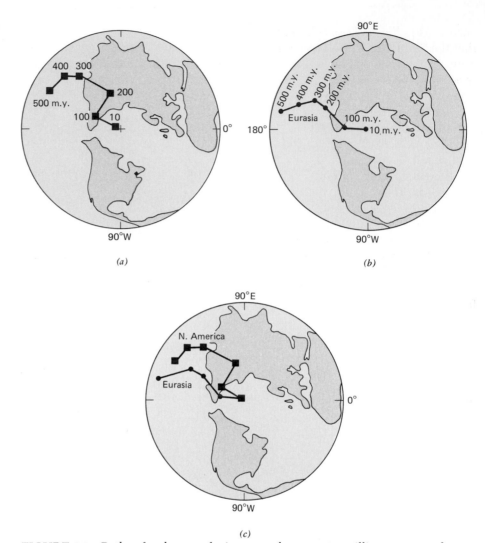

(c)

FIGURE 3.2 *Paths of polar wandering over the past 500 million years, as deter-*
mined from crustal rocks in (a) North America, and (b) Eurasia.
When combined, as in (c), it is clear that they disagree where the
magnetic North Pole has been, which indicates at least one of the
continents must have moved. (From Peter J. Wyllie, The Way the
Earth Works, John Wiley & Sons, 1976.)

To answer this question, paleomagnetists in the 1950s made charts
depicting "polar wandering." For the sake of argument, they assumed,
that the North American continent, for example, didn't move. Then, by
studying the magnetization of ancient rocks over the years, they traced
how the position of the magnetic North Pole must have shifted. (See Figure 3.2.)

Then they repeated the measurements for the rocks of a different continent. Again they traced the path of wandering for the magnetic pole

according to the rocks of this second continent. But the amazing result was that the two paths of polar wandering were not the same. Since the magnetic North Pole could not be in two different places at once, this was proof that the original assumption was false. The continents were not sitting still.

Although this type of magnetic data can establish that the continents had moved, it cannot completely determine the motion. It can be used to measure rotation of the continents and changes in latitude, but not changes in longitude.

The amount of rotation is easy to determine. If you find a crustal rock whose trapped magnetic field is pointing east, then you know that the continent has rotated 90° to the east since that rock has formed.[1]

The latitude at which a particular rock was first formed can be verified by comparing the angle that its magnetization makes with the horizontal. Notice in Figure 3.3 that the earth's magnetic field only parallels the surface at the equator. In the Southern Hemisphere it points up out of the surface, and in the Northern Hemisphere it points down into the surface. The angle it makes with the earth's surface is called the "dip." At higher latitudes the dip is greater. By studying the dip of the fossil magnetization of the rocks of various ages in India, for example, we can determine that about 180 million years ago it was near the Antarctic Circle. (The dip recorded in the rocks of that time was about +65°.) Then it very gradually started moving northward, speeding up for a period to a speed of about 16 cm per year. It crossed the equator about 50 million years ago (the dip in the fossil magnetization of the rocks born in this period was 0°), and now it has slowed down to a speed of about 2 cm per year as it collides with Asia, forming the Himalayas. (See Figure 3.4.)

C. OTHER SURPRISES OF THE FIFTIES

Also in the 1950s, data started coming in from other sources that seemed to revive voices from the past. Soundings of the ocean bottom revealed a gigantic undersea mountain range, called the "oceanic ridge," which wandered through all the earth's oceans and had a total length of about 65,000 km. (See Figure 3.5.) It covered almost as much area as the continents, and had relief comparable to the largest terrestrial mountain ranges. It had a deep rift running down the center, as if it were being torn apart in this region.

In the Atlantic Ocean, this ridge ran exactly down the center. In addition to the geometrical similarity of the eastern and western margins of the ocean, there was now also this ridge that exactly bisected it. This was

---

[1]*We have to be careful, of course. We have to study the geology of the region to make sure that local processes haven't caused significant local changes since the rock was born. We also have to check that the polarity of the earth's magnetic field was the same then as now. Both these things can be done.*

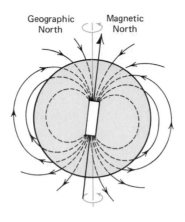

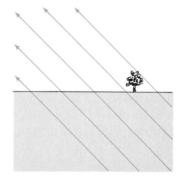

Northern Hemisphere   Equator   Southern Hemisphere

FIGURE 3.3   *Illustration of how magnetic dip varies with latitude. In the Northern Hemisphere, the earth's magnetic field points downward, into the surface. In the Southern Hemisphere, it points upward, out of the surface. By the direction and amount of dip, you can determine your magnetic latitude.*

just too much symmetry to be attributed to coincidence, and most people began thinking that it deserved some explanation.

It was also discovered that heat flow through the ridge was surprisingly high, being much greater than that through continental mountain ranges. From volcanic activity we know that the earth's interior must be hot, but why should this heat flow preferentially out through the ridge system?

Finally in the 1950s it was discovered that the ocean bottom was surprisingly young, with ages of crustal rocks being measured in terms of tens of millions of years, which is only ¼% of the age of the earth.

You can see that these developments would cause some uneasiness

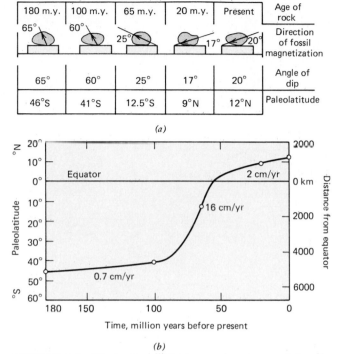

180 m.y.	100 m.y.	65 m.y.	20 m.y.	Present	Age of rock
65°	60°	25°	17°	20°	Direction of fossil magnetization
65°	60°	25°	17°	20°	Angle of dip
46°S	41°S	12.5°S	9°N	12°N	Paleolatitude

(a)

(b)

FIGURE 3.4 *Illustration of how measurements of the dip angle of the fossil magnetism in ancient volcanic rocks in India has revealed the northward movement of that continent. (a) Illustration of how the magnetic dip in the rocks of various ages has determined the magnetic
latitude of India in those ages. (b) Plot of the magnetic latitude of
India vs. time for the last 180 million years. (From Peter J. Wyllie,
The Way the Earth Works, John Wiley & Sons, 1976.)*

among earth scientists, and would put the idea in the back of our minds
that perhaps all was not well with the *status quo*. Perhaps there was a need
for a more comprehensive idea on how the earth works.

D. THE GEOPOETRY OF H. H. HESS

In 1960, H. H. Hess of Princeton presented a paper orally, which was published 2 years later, entitled "History of the Ocean Basins." He realized
that it was a little short on data to be defended as aptly as a revolutionary
theory should be, and he referred to it as "geopoetry." But it had many of
the elements that we now believe to be correct, as has been demonstrated
with subsequent data analysis.

He suggested that crustal plates may be moving, and this motion may
be driven by convection in the earth's interior. This would be like chips
of wood floating on a simmering pot of water. When water simmers, the

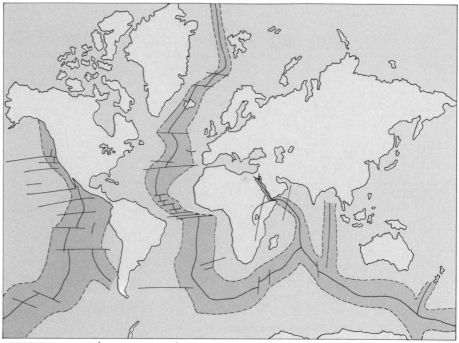

FIGURE 3.5 *The oceanic ridge system. Its boundaries are indicated by dashed lines. The ridge crest and major fracture zones are indicated by the solid lines.*

water on the bottom of the pot is heated by the fire. Being hotter the water is also lighter, and so it rises to the surface. At the surface it cools, becomes denser, and sinks again, in the familiar turbulent pattern. Chips of wood would be carried with the surface water away from regions where the water was rising and spreading out, and towards regions where it was sinking.

H. H. Hess suggested that the oceanic ridge may be the region where hotter materials surface from depths. This would account for its being pushed upward in the form of a mountain range, and for the extra heat flow through the crust in this region. (See Figure 3.6.)

According to Professor Hess, after the convective material rose beneath the ridge, it would then spread out to the sides beneath the floating lithosphere, dragging the lithosphere with it on either side of the ridge. This would tear the lithosphere down the middle and account for the large rift valley down the center of the ridge. The cooling and solidification of molten material seeping into this rift from beneath would account for the constant generation of new oceanic crust and explain why the oceanic crust near the ridge is so extremely young.

Of course, if the plates of the lithosphere are spreading away from some areas, they must be colliding in others. Professor Hess pointed to the deep ocean trenches as regions where one plate was sliding beneath another and into the earth's interior. (See Figure 3.7.)

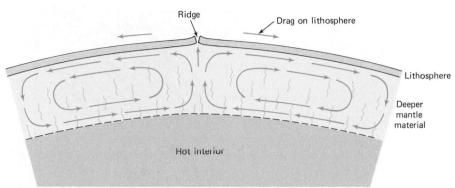

FIGURE 3.6   *If convection cells exist in the mantle, then we might expect them to push up a ridge where they rise, and drag the lithosphere away from the ridge as they flow to the side.*

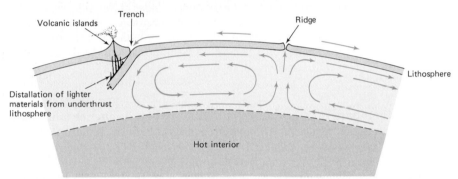

FIGURE 3.7   *Subduction zone (left). One plate is thrust under another. The increased temperatures and pressures cause the more volatile materials to bubble back to the surface, creating volcanic islands just outside the trenches and manufacturing new continental crust.*

You can see that this is the kind of sensationalistic "earth crunching" theory of which we are understandably suspicious. But you can also see that we were aware that we needed some explanations at the time. So Hess's ideas were received more sympathetically than Wegener's had been, and in just a few years, the needed confirming evidence began pouring in.

E.   THE CURIOUS BEHAVIOR OF THE EARTH'S MAGNETIC FIELD

E.1   Magnetic Anomalies

The next well-timed observations also came from the field of paleomagnetism, and also came from the ocean bottom. By dragging sensitive magnetometers either along the ocean bottom or near the ocean bottom, we can measure very accurately the magnetic field in that region. In the same

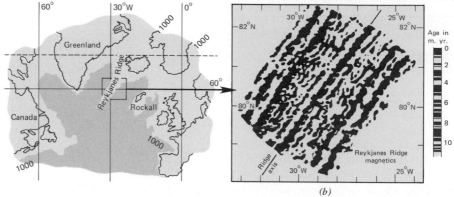

FIGURE 3.8 (a) *The location of the region enlarged in* (b) *along the Reykjanes Ridge in the northern Atlantic. The contour is at 1000 fathoms.* (b) *The magnetic anomaly patterns on both sides of the ridge axis. Black areas are regions of positive magnetic anomaly (in the direction of the earth's present field), and negative anomalies are the regions in between. (From F. J. Vine, Journ. Geol. Education, **17**, No. 1 (1969).)*

way that a compass needle changes direction when a magnet is brought nearby, so do measurements of the earth's magnetic field deviate from the expected values when magnetized rocks are nearby. These deviations in the magnetic field are called "magnetic anomalies," and are typically less than 1% of the total magnetic field. Nevertheless, these small deviations are easily detected.

The first surprising results came from the northeast Pacific Ocean, where magnetic anomaly measurements revealed a curious pattern in the magnetization of the crustal rocks of the ocean bottom. The magnetization of rocks in any area is directed either toward the north or toward the south, but not in any other directions. Furthermore, the magnetization pattern comes in long, narrow strips, each strip typically only 30 km wide. In one strip the magnetization is toward the north, and in the next strip it is toward the south. The pattern repeats itself in these adjacent strips of alternate magnetic polarity. (See Figure 3.8.)

In 1963, two British geophysicists, F. J. Vine and D. H. Matthews, made a daring suggestion for the cause of this pattern. A similar suggestion was made independently by a Canadian, L. W. Morley. They suggested that the earth's magnetic field may have switched its polarity intermittently throughout history, and these strip anomalies on the ocean bottom simply reflect the polarity of the earth's magnetic field at the time when that particular strip of material was being born in the rift valley of the appropriate oceanic ridge. This was quite bold, because there had been no previous suggestion or hint that the earth's magnetic field had ever switched polarity.

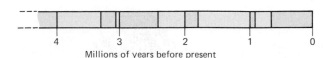

Millions of years before present

FIGURE 3.9 *Periods of "normal" (grey) and "reversed" magnetic polarity over the last 4 million years, as determined from various dated volcanic rocks in many different parts of the world.*

E.2   Records in the Rocks

But soon evidence began accumulating to support this hypothesis. By dating and measuring the magnetization of rock from ancient continental volcanic eruptions, it was soon established that the earth's field had indeed switched polarity throughout history (Figure 3.9), and in fact it seems to have had "reversed polarity"[2] about as long altogether as "normal polarity."

Notice that if we can date the reversals, then we can determine the rate at which the crust is spreading. For example, if a strip anomaly is 30 km wide, and the period of that particular polarity lasted 1 million years, then the crust was being born and sliding away from the ridge at a rate of

$$\frac{30 \text{ km}}{1,000,000 \text{ yrs}} = 3 \text{ cm/yr}$$

during that era.

Using lava flows to date the reversals of the earth's magnetic field is only useful for the very recent past—roughly the last 5 million years or so. One reason for this is that lava flows are usually instantaneous and not continuous events. Therefore, they tell us what the polarity was at that particular time, but do not give us a continuous record of the change. Another fault with this technique is the errors associated with rock dating techniques. For times more ancient than the last few million years, the errors in dating are greater than the expected length of a period of given polarity.

E.3   Records in the Sediments

The dating of reversals in the more distant past must be accomplished by looking at cores of ancient sediments. Most of the crystals in grains of sediment had their ultimate origin in some volcanic or plutonic continental process. When they cooled, they trapped a record of the earth's magnetic field, as we have discussed. This may have happened eons before they were weathered, broken up, eroded, and deposited in the sedimentary deposit where we find them today.

Many of these tiny crystals are still weakly magnetized, and as they settle out of the water they tend to align themselves with the earth's mag-

[2]*Today's polarity is defined as "normal."*

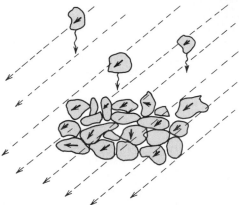

FIGURE 3.10   *As magnetized sediment particles settle out of the water and join the bottom sediments, they tend to align themselves with the earth's magnetic field. Consequently, the sediments retain a record of the earth's magnetic field at the time they were deposited.*

netic field (Figure 3.10). Of course, as they settle they push against each other, and may get pushed around by bottom dwelling organisms. But these are random processes, pushing as many grains one way as the other, and so the net magnetization in a sample of sediment will still tend to point toward the direction of the earth's magnetic field at the time when that sediment was laid down. Typically, the magnetization of a sediment sample is 100 times weaker than the magnetization of a rock sample, but it is still measureable.

The advantage of using sediment cores (Figure 3.11) from carefully selected environments for dating ancient polarity reversals is that they provide a continuous record, unlike the volcanoes which only erupt occasionally. Sediments deeper in the column indicate earlier dates, so the chronology of events is readily established.

Unfortunately, the absolute dating of a certain sediment strata is quite difficult. Radioactive techniques would give the dates when the crystals first formed, not the date that they became sediment. In some very select environments we might be able to justify assuming the rate of sedimentation has been the same as it is today. Then the absolute age of a sample would be directly proportional to its depth in the sediment column. Sometimes fossils can be used to date the sediment, and sometimes layers of volcanic ash or other peculiar sediment from a well-dated local event may be used to establish the age of certain depths in the sediment column.

This study of the earth's magnetic field has yielded many interesting results, including frustrating us in our attempts to explain this curious behavior. But for this book it is only necessary to point out that the periods of polarity reversals have been recorded and dated, and that we can use these known reversals both to confirm that the lithosphere does spread from the ridge and to calculate the rate at which spreading is taking place now and in more ancient times. Typically, the length of any one period of dominant polarity is in the range of 500,000 to 1,000,000 years at present,

FIGURE 3.11   *Sediment cores that have been split in half. One half is studied, and the other archived for possible future study.*

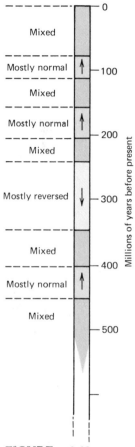

FIGURE   3.12
Illustration of the longer-term pattern of magnetic reversals. We are now in a "mixed" period, where the magnetic field switches polarity every few hundred thousand years or so. But there have been substantial periods of tens of millions of years where the field had predominantly one orientation only.

although even this pattern has changed on time scales of 50 to 100 million years (Figure 3.12).

F. PLATE MOTION

Once we have established the polarity of the earth's magnetic field during the various periods of earth history, we can use the fossil magnetism contained in rocks to trace the motion of various portions of the lithosphere. Their present on-going motion is most conveniently studied from an analysis of earthquakes. Although they don't leave permanent records of ancient activity as do volcanoes, for instance, seismic signals generated by earthquakes can be analyzed to give us fairly accurate descriptions of the direction and extent of motion along any plate boundary during such an event. Coupling these studies with some of the other techniques described earlier for studying the earth's interior, we are arriving at a fairly good general description of the dynamics of the earth.

F.1   Plate Boundaries

With the various plates in motion and grinding against each other, it is quite natural that the vast majority of earthquakes will occur along plate boundaries. In fact, we can use a map that has charted on it the locations of all recently recorded earthquakes to help us identify the present bound-

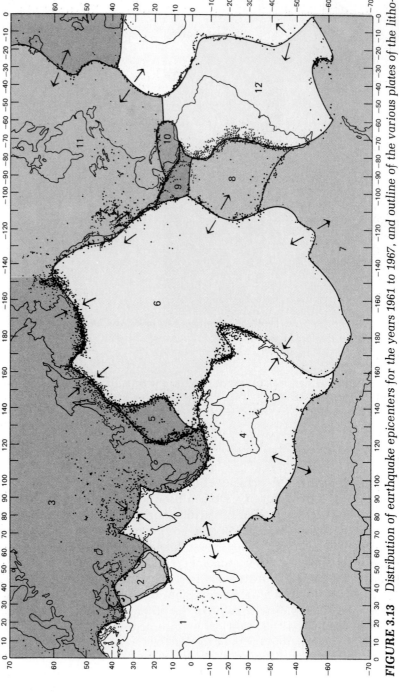

1 = African Plate   5 = Philippine Plate   9 = Cocos Plate
2 = Arabian Plate   6 = Pacific Plate   10 = Caribbean Plate
3 = Eurasian Plate   7 = Antarctic Plate   11 = North American Plate
4 = Australian Plate   8 = Nazca Plate   12 = South American Plate

FIGURE 3.13 *Distribution of earthquake epicenters for the years 1961 to 1967, and outline of the various plates of the lithosphere. Directions of relative movement along plate boundaries are indicated. (After a plot by M. Barazangi and J. Dorman, Columbia University.)*

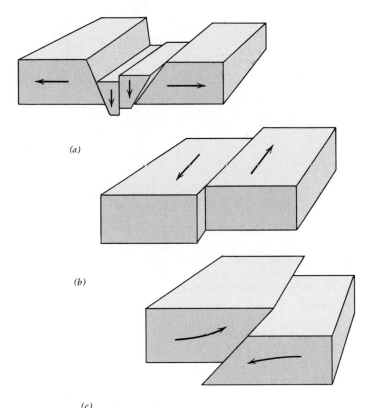

(a)

(b)

(c)

FIGURE 3.14 *Types of plate boundaries:* (a) *divergent,* (b) *lateral, and* (c) *convergent.*

aries of all plates. (See Figure 3.13.) There seem to be seven major plates and several smaller fragments.

Not all earthquakes occur at plate boundaries. It seems that occasionally the buildup of stress is more effectively relieved by some minor adjustment in the plate's interior. Occasional earthquakes in the Rocky Mountains or along the St. Lawrence Seaway are evidence of this in the North American Plate.

The types of earthquakes recorded tend to be different along the three different types of plate boundaries. These three types are where plates move away from each other as if under tension, where plates slip sideways or laterally with respect to each other, and where plates are in collision with each other. They are referred to as divergent, lateral, and convergent plate boundaries, respectively (Figure 3.14).

F.2 Divergent Boundaries
Divergent boundaries are identified by the oceanic ridge. Earthquake activity is heavy along the rift valley where the plates are being torn apart.

The rift valley tends to be oriented perpendicular to the direction of relative plate motion. This is frequently not parallel to the overall direc-

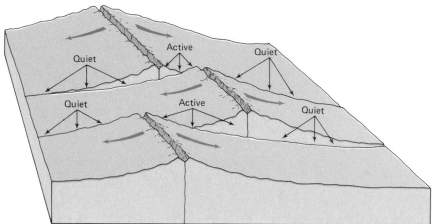

FIGURE 3.15   *Regions between displaced ridge crests are generally seismically active, due to opposite motions of the crust on opposite sides of the fracture zone. Outside these regions the crust is relatively quiet seismically, as the different crustal segments are all moving in the same direction.*

tion of the oceanic ridge, and so the ridge crest frequently is displaced sideways along transform faults called "fracture Zones." (See Figure 3.15.) Along these transform faults and between the displaced ridge crests, there is considerable earthquake activity as the plates on opposite sides of such a fault are moving in opposite directions. Outside this region, however, where the two sections of plate are moving in the same direction, there is lesser earthquake activity.

The elevated ridge exists both because it is getting pushed by the upward-flowing currents in the asthenosphere below, and because it is hotter than neighboring regions of the ocean floor. Hotter rock is less dense, and would therefore float higher on the asthenosphere below even in the absence of upward-forcing currents.

Along active ridges, magma chambers are found beneath the rift valley. They are typically 2 to 13 km below the surface, a few kilometers deep, and 10 to 20 km wide. They are fed magma from below the central portion of the chamber, and the magma solidifies on the chamber's cooler outer wings (Figure 3.16). The spreading of the crust causes cracks in the cap rock above the chambers and resulting zones of magma intrusion. Occasionally, these cracks may extend entirely through the cap rock, and the magma intrustion may result in volcanism in the rift valley on the sea floor.

Heat is removed from the ridge by its contact with seawater. This cooling of the hot rock by seawater also undoubtedly causes cracks and fractures through which more seawater or magma may flow. As seawater trickles down through these fissures it gets heated, sometimes shooting back to the surface in hot water jets called "hydrothermal vents" (Figure 3.17). The temperature of the water leaving these vents depends on the depth within the rock that it has reached, and the amount of dilution by

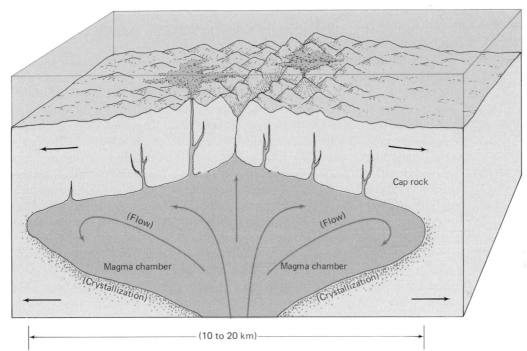

FIGURE 3.16 *Schematic cross section of region beneath the rift valley of an active oceanic ridge, showing cap rock, zones of intrusion and volcanism, magma chamber, and flow of magma into chamber from below, with subsequent cooling and crystallization on outer portions of the chamber.*

FIGURE 3.17 *A "black smoker" hydrothermal vent on the East Pacific Rise at a depth of 2.8 km.*

water that has not gone as deep. Water up to 350°C has been observed in some of these jets.[3]

The study of these hydrothermal vents by submersible research vehicles has turned up some interesting surprises. One is that they have a significant impact on the chemistry of seawater. The chemical content of the water exiting these vents has been significantly altered by its high temperature and interaction with hot crustal rock. The dissolved minerals tend to be in different proportions and in a more reduced (i.e., less oxidized) state than is normally found. Although the total flow rate from these hydrothermal vents is only about ½% as large as the flow rate of continental rivers,[4] the hydrothermal vents have about equal impact on the ocean's chemistry. This has been good news to chemical oceanographers who had previously been unable to square some details of the ocean's chemical makeup with previously known natural processes.

Bordering these vents are found rich deposits of metal sulfides (e.g., iron, copper, zinc) and other reduced minerals, and colonies of living organisms. The amazing thing about this is that right here on our very own earth, in the dark depths of our oceans, we find colonies of living creatures who are *not* dependent on sunlight for their existence.

It appears that the primary producers in this dark environment are forms of bacteria, which probably use oxidation of hydrogen sulfide (H_2S) or unoxidized metals in the vent water as their source of energy for carbon fixation and food production. This warm vent water contains typically 10^5 or 10^6 bacteria per milliliter (cubic centimeter).

The local animal community is dependent on these primary producers for their food supply. They are mainly filterfeeders, with variations in species at various distances from the vent. The warm waters near the mouth of the vent are favored by limpets and large tube worms, called "vestimentiferan worms," which can be up to 3 m long and have bright red plumes protruding from the ends of their tubes. In various zones farther from the vents are found anemones and dandelion-like relatives of theirs, long thin white worms that look like spaghetti, crabs and some shrimp-like crustaceans, and some very large clams and mussels. At the outer limit of a vent community are found corals, anemones, and sea cucumbers.

The larvae of these animals seem to be quite long-lived. This is probably a necessary adaptation in response to the rather temporal and widely dispersed nature of the vents. Larvae may need prolonged periods of time to find a new home.

From heat flow studies, the extent of the sulfide deposits, and from the ages of the biological communities, we find these hydrothermal vents

---

[3]*This is still less than the boiling point for water under the tremendous pressures at these depths.*

[4]*From observed flow rates and heat studies, it is estimated that a water volume equal to that of the entire ocean flows through these vents in 8 million years. This is about ½% the total flow rate for continental rivers.*

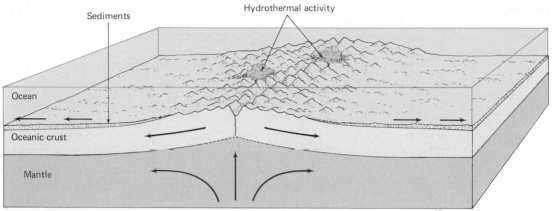

FIGURE 3.18 *Hydrothermal activity is most common near the ridge crest where the rock is hottest and sediment cover is thinnest. The oceanic crust is of rather uniform thickness across the entire ocean bottom, showing little evidence of either tension or compression.*

are rather short-lived on geological time scales. Although any one vent is short-lived, there are large numbers of vents altogether, and hydrothermal activity can be found over large portions of the ridge. Going away from the ridge crest, increasing sediment cover (older rocks accumulate more sediment cover) and cooler rocks cause a reduction of hydrothermal activity.

Going away from the ridge crest the thickness of the oceanic crust remains fairly uniform (6 to 8 km) across the entire ocean bottom (Figure 3.18). This implies that the crust is neither under a great deal of compression nor tension, probably flowing along with the asthenosphere below.

F.3 Lateral Boundaries

Along lateral plate boundaries, the two adjacent plates of the lithosphere are slipping sideways relative to each other. One famous plate boundary of this type is marked by the San Andreas Fault in California. (See the title page for this chapter.) A small portion of California is on the Pacific plate, moving northwest toward the Aleutian Trench from the East Pacific Rise. The North American plate is moving westward from the Mid-Atlantic Ridge. This means that there is a net lateral motion along the boundary between these two plates, with the Pacific side moving northward relative to the North American side. Los Angeles, on the west side of the fault, is slowly moving northward towards San Francisco, on the east side of the fault.

Along some sections of this fault, the motion is quite smooth. However, there are some regions where the two plates get hung up. Pressure builds up, and when they eventually do break loose, the built up pressure makes the resulting motion catastrophic. Unfortunately, the two most densely populated areas of California are near such regions.

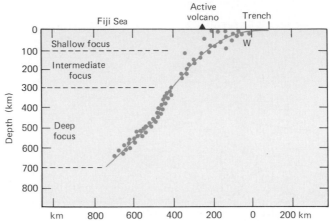

FIGURE 3.19 *Plot of the location of earthquake foci as a function of depth in a subduction zone, beginning at the Tonga Trench, and extending beneath the Fiji Sea. (From Peter J. Wyllie, The Way the Earth Works, John Wiley & Sons, 1976.)*

F.4 Convergent Boundaries

The third type of plate boundary is most significant in explaining the origin and evolution of our continents, oceans, and atmosphere. When a plate carrying a continent collides with an oceanic plate, the oceanic plate subsides. This is probably a reflection of the greater density and smaller buoyancy of the oceanic plate materials.

These areas of subsidence are marked on the surface by ocean trenches, deep creases in the ocean floor whose explanation eluded oceanographers before the theory of plate tectonics was introduced. We can use earthquakes and seismic waves to trace the path of the cool rigid oceanic plate as it plunges through the trench and on into the asthenosphere below. Careful measurements of the exact locations of earthquake foci indicate oceanic plates descend typically at inclines in the range of 35° to 45° (Figure 3.19).

The material of the descending plate is crustal material, which is quite volatile compared to the materials of the earth's interior. As a crustal plate descends into a subduction zone, it appears that many of the minerals cannot tolerate the great heat and pressures encountered. Some partial melting occurs, and many of the more volatile materials shoot back up to the surface through volcanoes. These materials include nitrogen, chlorine, carbon dioxide, and water, which leave as gases, as well as some lighter minerals, which solidify in lava flows.

This explains many things. It explains the origin of our atmosphere, the water and salt of our oceans, and it even explains the origin of much of our continental materials. These are primarily products of the "distillation" of oceanic crust in subduction zones. The reason continental materials tend to be lighter than the materials of the oceanic crust is that con-

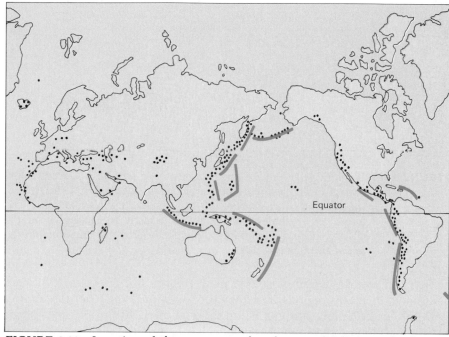

FIGURE 3.20 *Location of the ocean trenches (heavy solid lines) and volcanoes (dots). Notice that the trenches ring the Pacific, and that the majority of the earth's volcanoes are concentrated just outside these trenches.*

tinental materials are born of those minerals in the oceanic crust that are less tolerant of the harsh environment of the earth's interior.

We really don't need sophisticated seismographic equipment to convince ourselves of these processes in subduction zones. On a world map (Figure 3.20), we notice that vitually all the deep ocean trenches are found around the perimeter of the Pacific Ocean, where the Pacific Ocean bottom disappears beneath the North and South American plates on the north and east, and beneath the Asian and Indian plates on the west. The majority of all the earth's recent volcanic activity occurs just outside this perimeter of trenches. Among others this includes the Aleutian Islands, Kamchatka, the Japanese Islands, the Philippines, and the Andes Mountains of South America. In fact, the perimeter of the Pacific Ocean is sometimes called the "ring of fire" because of the volcanic activity.

It was originally thought that the oceanic plate was being forced downward through collision with the more buoyant continental plate. However, recent studies indicate that in many presently active subduction zones, the oceanic plate is descending "voluntarily." Evidence for this includes the notable lack of sediment accumulation along the margins of continental plates in subduction zones. If the two plates were really engaging in forced collisions, then we'd expect the sediments to be

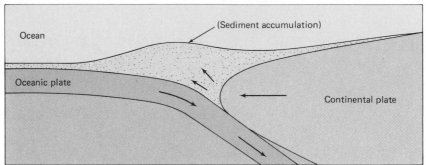

Ocean

(Sediment accumulation)

Oceanic plate

Continental plate

FIGURE 3.21 *If the two plates were undergoing forced collision, then we might expect to find sediment accumulation in subduction zones, where the sediments are scraped off the descending oceanic plate. The lack of such sediment accumulations in many areas suggests that the oceanic plate is descending "voluntarily" rather than being forced under through collision.*

scraped off the descending oceanic plate (Figure 3.21). The lack of such sediments in many regions indicates the sediments may be descending with the oceanic plate, and not being scraped off. (Offscraping does occur in some regions, but it is not as extensive as might be expected.) Also, studies of the plate materials underlying the "backarc basins," such as between the Asian continent and the subduction zones to the east, indicate these plate materials to be thin and quite young. The conclusion is that the continental plate is being stretched (rather than compressed) and augmented with new materials, in order to go out and meet a receding subduction zone (Figure 3.22).

Although these studies are demonstrating that oceanic plates are frequently descending "voluntarily," they do not tell us why. Ideas that are being explored include the possibility that the oceanic plates are cooler and denser than the material beneath, thereby being pulled down by gravity. It is also possible that they are being dragged down by the motion of the material in the asthenosphere. Perhaps both or neither of these explanations are correct.

Continental materials are sufficiently buoyant that when two continents collide, neither subsides. Mountain ranges are lifted up as the crust is compressed, wrinkled, and thickened in the region of the collision. The Himalayas are still rising today as India presses on into Asia. Older mountain ranges, mellowed by time and erosion, still mark the margins of former pairs of continents that underwent collisions.[5]

There are many local elevated portions of the ocean floors, called "plateaus," (Figure 3.23) that are found to be made of thicker, more buoyant crustal materials, like those of the continents. Some may be submerged continental fragments and others are products of volcanism on the ocean

[5]*The uplifting of the Rocky Mountains of western North America is a bit of a puzzle. We have not yet been able to associate their formation with any continental collisions.*

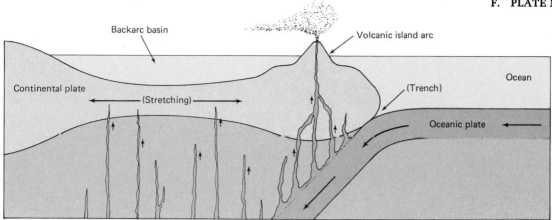

FIGURE 3.22 *In many backarc basins, the underlying crustal materials are being stretched rather than compressed as would be expected if the two plates were undergoing forced collision.*

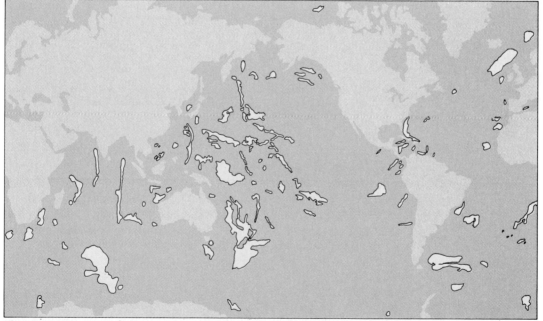

FIGURE 3.23 *Distribution of oceanic plateaus.*

bottom. Like the continents themselves, plateaus do not descend into subduction zones, so they are effectively "scraped off" of the descending oceanic plate as they approach a subduction zone (Figure 3.24). The plateau collides into the continent, and the trench reappears behind the plateau, where the oceanic plate resumes its descent.

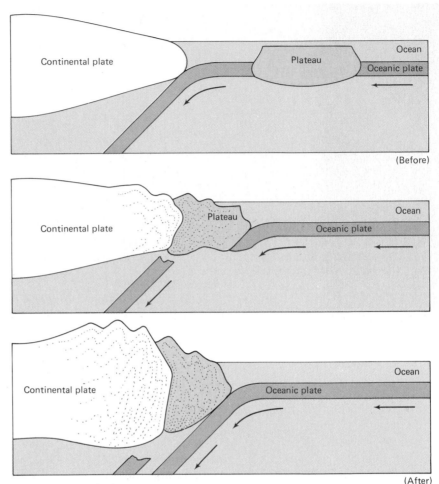

FIGURE 3.24 *Oceanic plateaus do not descend with the oceanic plate in subduction zones. Rather, they get "scraped off," causing some uplifting and mountain building as they collide with the continent. The trench reappears behind them after they join the continent.*

G. THE MORE DISTANT PAST

Although the ocean bottoms have been instrumental in revealing the earth's recent tectonic history, they cannot help us understand earlier times. The oceanic plates are conveyer belts, carrying a record of their history from the ridge where they are born, across the ocean bottom and into a subduction zone, where they are destroyed and their information is lost. This journey across the ocean bottom takes less than 200 million years, typically, which is 4% of the age of the earth. All oceanic crust is young, and therefore unable to give us clues into the earth's earlier history.

In contrast, continental materials do not descend into subduction

FIGURE 3.25   *During the period of the "late bombardment," which lasted from about 4.2 to 3.9 billion years ago, the Solar System was bombarded by a great deal of meteoritic debris. Evidence of this is still clearly visible on the heavily cratered surface of the planet Mercury.*

zones. Once created, they remain "afloat" on the asthenosphere, preserving records for us of the earth's earlier history. Continental rocks as old as 3.8 billion years have been found. This coincides with the end of the "late bombardment" of the solar system by debris from outer space (Figure 3.25). The evidence of this late bombardment is still clearly visible on the surfaces of several objects, including our moon and the planet Mercury, where lack of atmosphere and erosion have preserved much of the myriad of craters created during this period. This late bombardment probably obliterated the surface features on our own planet like it did on these others, so records of the first 800 million years of earth history (between 4.6 and 3.8 billion years ago) are unlikely ever to be found.

With the exception of this period preceding the late bombardment, we should be able to find clues of the bulk of the earth's early history still existing in the continents. Unfortunately, because of their age and the earth's geological history, the continental materials have undergone a great deal of transformation and alteration since their formation, making their history difficult to decipher.

Old continents combine through collision, and new ones form as ridges appear and divide them, with the resulting opening of new oceans. Continental margins are continually changing shape and nature with the accretion of ocean plateaus and the volcanism associated with subduction zones on their margins.

Most mountain ranges were formed during collision between continents, or the collision of plateaus with continents, and to a lesser extent through the volcanism in subduction zones. This means that although most mountain ranges mark the boundaries between former continents or continental fragments, the exact history of these boundaries is difficult to decipher. There has been a great deal of compression and distortion of their features. The ocean may have reopened and closed several times. Sediments and plateaus trapped between the colliding continents join the newly formed mountain ranges in folded and distorted shapes. Even portions of the ancient ocean bottoms sometimes are trapped and lifted up into the mountain ranges by these colliding continents. Consequently, when we now go into these mountains to investigate what went on back then, we find a variety of geological structures that confuse the overall picture.

Our present continents are a stew of ancient continental fragments that have been rather well mixed through a long history of continental collisions and divisions. We have an abundance of detailed information on the geology of our continents, but because of their age and complexity, we are presently lacking a good comprehensive picture of their history. This is in contrast to the ocean bottoms, for which we have relatively little detailed geological information, but for which we have a good overall picture of their rather short and simple history.

From very basic information, we can construct models that provide us with very general features of the earth's evolution, although these models cannot provide the detailed history of any particular continent. From such models, we believe that heat production inside the earth was originally about four times its present value. Consequently, the young lithosphere was hotter, thinner, more flexible, and more buoyant than it presently is. As the heat production diminished, the lithosphere cooled, became less buoyant, and subduction zones began to form. Plateaus and other fragments of volcanically produced continental materials began to aggregate at these convergent plate boundaries forming larger blocks of continents. Plate motion caused repeated collision and division of these continental blocks, which is continuing today.

The actual details of these processes cannot be determined from models, but rather must be gleaned from the records held in the present continents. The older the continental materials, the greater the number and extent of the transformations they have undergone. This means that the most recent history is most easily deciphered.

We are gradually forming a fairly good picture of the history of the continents over the most recent 10% of earth history (Figure 3.26). Prior to the existence of Pangea, what is now the east coast of North America then belonged to the continent of Africa. About 400 million years ago, the continents collided, forming Pangea. As Africa collided with North America, the Appalachian Mountains were formed, just as the gigantic Himalayas are being pushed up by the collision of India with Asia today. Since that time, erosion has made the Appalachians into a gentle, rolling range, which is a far cry from the formidable rough range they must once have been.

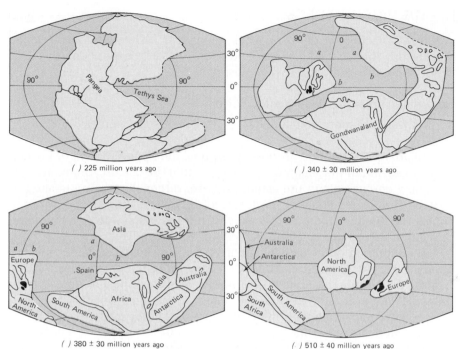

( ) 225 million years ago

( ) 340 ± 30 million years ago

( ) 380 ± 30 million years ago

( ) 510 ± 40 million years ago

FIGURE 3.26 *Sketch of our interpretation of the motion of the continents prior to the formation of Pangea. It is based on paleomagnetic data from the continents, so only orientation and latitudes may be inferred directly. (From Peter J. Wyllie, The Way the Earth Works, John Wiley & Sons, 1976. Based on results of A. G. Smith, J. C. Briden, and G. Drewry, 1973, in "Organisms and Continents through Time," editor N. F. Hughs, Palaeontology, Special Paper 12.)*

Later, about 190 million years ago, a ridge appeared, which started separating the continents and reopening the Atlantic Ocean. However, it formed to the east of the pre-Pangea continental margins, and so as Africa departed, it left North America with the piece of land to the east of the Appalachians.

If you look at a map of the earth, you find it covered with clues to previous and present tectonic activity. Why are the Ural Mountains in such a conspicuously straight line? What is going on in the Red Sea and the Gulf of Aden, with their peculiar geometry and an oceanic ridge running into them? Why the Rocky Mountains? Why the geysers and hot springs of Yellowstone National Park? We must remember that with the rapid rates of erosion and sedimentation, features more than a billion years old will be completely disguised, leveled, and buried, and their history will be very difficult to decipher. But even the history of the last billion years should keep us busy for a while.

H.   THE DRIVING FORCE

Although the details of the earth's geological history still need to be worked out, the fact that it has indeed been very active is undeniable. This observation raises the following questions:

1.   What causes this geological activity?

2.   Why are we more active than the other terrestrial planets?

In answer to the first of these questions, the apparent lack of tension or compression in the oceanic crust suggest that the oceanic plates are riding along on currents in the asthenosphere, rather than being pushed or pulled. Further evidence of the forces of this circulation in the asthenosphere is the elevation of the ridge and the apparently "voluntary" descent of the oceanic plate in the subduction zone.

If circulation of the asthenosphere drives the surface plates, then we need to explain this circulation. We think the circulation is caused by heating from below, similar to the motion of water in a pot heated from below. The warm fluid rises and spreads out in some regions, and then the cooler surface fluids fall in others, completing the cycle.

In searching for an interior source that could provide enough heat to drive this convection in the asthenosphere, scientists have come up with a number of possibilities. However, most think that the primary heat source is the decay of radioactive elements deep within the earth.

The answer to the second question is a little more straightforward. The reason that the earth is geologically more active than other terrestrial planets is a result of its greater size. You may have noticed that soup in a teaspoon cools more quickly than soup in a bowl. Smaller objects have more surface area in comparison to their volume, and can therefore cool more quickly.[6]

Planets having larger volumes have correspondingly more radioactive materials within them, but they have greater difficulty cooling off than do smaller planets. The radioactively produced heat tends to stay pent up inside, and so the interiors of larger terrestrial planets should be hotter. Being the largest of the terrestrial planets, we should expect the earth's interior to be hottest, and therefore it should be the most active geologically. It is.

I.   SUMMARY

Alfred Wegener was an early proponent of continental drift, and provided evidence, although it wasn't enough to sway the bulk of scientific opinion. However, several pieces of evidence arose during the 1950s that caused

---

[6]*If you take any object and break it, you expose new surface areas without changing the total volume at all. The finer the pieces you break it into, the more surface is exposed. In this way you can see that smaller objects tend to have more surface area in comparison to their volume.*

some reconsideration of Wegener's ideas. These included paleomagnetic evidence that the continents have moved, including evidence for rotations and changes in latitude. Also the mapping of the oceanic ridge system, particularly the symmetry of the Mid-Atlantic Ridge, and the discovery of the youth of oceanic crust, forced reconsideration of our ideas on drifting continents.

In 1960, H. H. Hess described many of the main components of our present ideas on plate tectonics in what he called "geopoetry." He suggested convection cells in the mantle, driving the lithosphere away from the ridge and eventually into subduction zones identified with the trenches. Confirmation came from paleomagnetic studies of the ocean bottom, revealing magnetic strip anomalies, resulting in speculation by Matthews and Vine regarding reversals of the earth's magnetic field. Confirmation of this came in studies of ancient lava flows and sedimentary deposits.

Present plate boundaries and motions can be determined from earthquakes. Along active divergent boundaries are found subterranean magma chambers and surface volcanism and hydrothermal vents. Along a lateral boundary one plate is slipping sideways relative to the other. Convergent boundaries involving oceanic plates are identified by ocean trenches, where plates are subsiding, and volcanism associated with distillation of some of the more volatile materials from the descending plate. Convergence of two continental plates results in mountain building.

Evidence of the earth's ancient history can be found on the continents only, because there are no ancient ocean bottoms. Ancient continental materials have undergone a great deal of transformation, so their histories are difficult to decipher.

The thermal energy that drives the earth's plates is thought to be provided mainly by radioactivity within the earth. The reason the earth is more active geologically than other terrestrial planets is due to its greater size. Larger objects tend to cool more inefficiently, so the earth's interior remains hotter than interiors of smaller planets.

QUESTIONS FOR CHAPTER 3

1. Explain some of the arguments used by Alfred Wegener in support of the idea of continental drift.

2. Briefly explain how a study of fossils might be used to date the separation of continents.

3. What was the general attitude toward Wegener's idea?

4. What is "paleomagnetism"?

5. What is ultimately responsible for the magnetic properties of minerals?

6. What is an electromagnet? How can a tiny atom be one?

7. What is the "Curie point"? Is it the same as the freezing point for a mineral? Explain.

8. What does the adjective "plutonic" mean? (Look it up in a dictionary, if necessary.)

9. Explain how igneous rock retains a record of the earth's magnetic field from the time it formed.

10. Are the magnetic records of ancient rocks identical to those of more recent rocks? What does that mean?

11. What is a path of "polar wandering"?

12. Did the magnetic pole really wander, or did the continents? How do we know? (It turns out, actually, that the poles do wander a bit, but not very much.)

13. How can you tell if a region of a continent has rotated since its formation?

14. What is the "dip" of the earth's magnetic field?

15. How can you tell at what latitude a certain mass of rock was formed? How can we use this to trace the north–south motion of continents?

16. Discuss the size and extent of the oceanic ridge system. If given a map of the world, could you sketch roughly where this entire system would be found?

17. In what ocean does the ridge run exactly down the center?

18. Do you suppose the erosion of the oceanic ridge system would be faster or slower than that of a continental mountain range? Why?

19. Where on the earth's surface do we find that heat flowing up from beneath is exceptionally high?

20. Briefly summarize the developments of the 1950s that indicated the earth's surface was in motion.

21. Explain why there is convection in a pot of water on the stove. Would chips of wood in this pot be carried toward or away from regions where hot water was rising? Why?

22. Why is there an oceanic ridge according to the "geopoetry" of H. H. Hess? Why the rift valley? How is new oceanic crust generated? Why is the ridge relatively hot?

23. According to the ideas of Professor Hess, why are there ocean trenches and what is happening there?

24. What are "magnetic anomalies," and what causes them?

25. Describe the pattern of magnetization on the bottom of the northeast Pacific.

26. What did F. J. Vine and D. H. Matthews suggest as the cause of the strange "strip anomaly" pattern of the ocean bottom?

27. Explain how studying the lava from successive volcanic eruptions

can be used to identify and date reversals of the earth's magnetic field.

28. Explain how dating the polarity reversals helps us determine the rate of spreading from the oceanic ridge.

29. Why can't lava flows be used to date reversals more than a few million years old?

30. How did an individual mineral crystal in the sediment originally become magnetized?

31. If the individual sediment particles were magnetized eons before they became sediment, how is it that a layer of sediment carries a record of the earth's magnetic field at the time of deposit?

32. In dating ancient reversals of the earth's magnetic field, what advantage is there in using sediment columns? What disadvantage?

33. What techniques may sometimes be used to try to establish the *absolute* age of a layer in a sediment core?

34. Can earthquakes be used to study ancient geological activities, or only what is happening today?

35. Are all earthquakes along plate boundaries? If not, why not?

36. What are the three general types of plate boundaries?

37. In what particular regions of the oceanic ridge do earthquakes most frequently occur? Why? Be particularly explicit along the fracture zones.

38. What happens when two continents collide?

39. In studying the earthquakes in subduction zones, how are the depths of the foci related to the distance from the trench? Why?

40. Why are volcanoes often located near ocean trenches?

41. Explain how our atmosphere, ocean, and continental materials are created in subduction zones. Why is continental crust lighter than oceanic crust?

42. What is the "ring of fire"? Why is it?

43. Why is older oceanic crust more difficult to study than young oceanic crust?

44. No clues to earth history of more than a few hundred million years ago will be found in the ocean bottom. Why?

45. If the Atlantic Ocean widened at an average rate of 4 cm per year for 200 million years, how many kilometers wide would it be today? Is that about right?

46. How did the Appalachians form? How did the Americas acquire what is now the East Coast of the United States?

47. Why is the oceanic ridge system elevated above the normal level of the ocean floor? Give two reasons.

48. Describe the magma chambers beneath the rift valley in active oceanic ridges.

49. What are hydrothermal vents and what causes them?

50. How does the flow from hydrothermal vents compare with that from the world's rivers? How does the hydrothermal vents' influence on ocean chemistry compare with that of the world's rivers?

51. We know of some colonies of living organisms on earth that do not rely on sunlight as their ultimate source of energy. Where are these colonies found and what is their source of energy for life?

52. What evidence indicates that oceanic plates may descend "voluntarily" in subduction zones rather than being forced down by the approaching plate?

53. How are mountain ranges created?

54. What happens when an ocean plateau enters a subduction zone?

55. Why is it probably impossible to find surface rocks older than 3.8 billion years?

56. How did the early lithosphere compare with today's?

57. Did the early earth have subduction zones? If not, why not?

58. What causes the earth's geological activity?

59. Why is the earth more active geologically than other terrestrial planets?

*60. If the convection in the mantle is driven by heating from below, what are some possible sources of this heat? Can you think of some not listed in the chapter?

*61. Look at a geological map of the earth, and list some peculiar features you think must testify to some earlier geological activity.

*62. One of the problems now confronting geophysicists is trying to determine whether continental drift is continuous or intermittent. That is, do the continents creep along a little every day, by small jumps every few years, by larger jumps every few centuries, and so on. How would you suggest we should go about trying to answer this question?

*63. It is possible that motion of plates of the lithosphere could drive the convection in the asthenosphere, rather than vice versa. If this were true, what do you think could be causing the lithosphere plates to move in the first place?

*64. If the interior heat source really is primarily the decay of radioactive nuclei, why should the rate of heat production be less now than earlier in earth history?

*65. Estimate how long it will take Los Angeles to become a suburb of San Francisco, using typical rates of plate motion and the fact that they are presently separated by 700 km.

*66. A billion years from now, what evidence will there be that people once lived here?

*67. Right now, no one can figure out how the Rocky Mountains were formed, as they don't seem to be attributable to collisions between continents. How do you think they might have formed?

SUGGESTIONS FOR FURTHER READING

1. Robert S. Dietz, "San Andreas: An Oceanic Fault that Came Ashore," *Sea Frontiers*, **23,** No. 5 (1977), p. 258.

2. Kenneth J. Hsü, "When the Mediterranean Dried Up," *Scientific American* (Dec. 1972).

3. Peter A. Rona, "Plate Tectonics and Mineral Resources," *Scientific American* (July 1973).

4. F. G. Walton Smith, "Baja: Yesterday, Today and Tomorrow," *Sea Frontiers*, **27,** No. 4 (1981), p. 194.

5. Peter J. Wyllie, *The Way the Earth Works*, Wiley, New York, 1976.

4
THE OCEAN BOTTOM

A section of "The Floor of the Oceans" by Bruce C. Heezen and Marie Tharp.

Our familiarity with earth surface features has been fairly restricted to the 29% of the earth not covered with water. Until recently, the remaining 71% has been entirely unknown to us. In fact, we are still more familiar with the detailed surface features of the moon than we are with those of our own seafloor.

In the previous chapter we learned that the rock underlying the oceans has a different origin, evolution, and even a different composition than that of continental rocks. Furthermore, the processes working to modify the underwater landscape are completely different from those at work on land. It should be no surprise, then, that the ocean bottom displays a spectrum of features quite different from those familiar to us in our terrestrial environment.

A. OVERVIEW

Going seaward from shore, the first submerged region is called the "continental shelf." It is really an extension of the continent that just happens to be submerged at this particular point in the earth's history. It hasn't always been submerged, and undoubtedly will be above water from time to time in the future in response to minor changes in sea level. The continental shelf is generally quite flat, having an average slope of only about 0.1°, being slightly steeper near shore and slightly flatter over its outer portions.

Shelf widths vary greatly, but the average continental shelf extends to a distance of about 75 km from shore, and reaches to a depth of about 135 m. This is very shallow by oceanic standards. The top of a 30-story skyscraper would just barely be submerged at this depth.

The shelf ends abruptly at the "shelf break," where there is a distinct change in slope. (See Figure 4.1.) Beyond the shelf break is the "continental slope," which plummets downward at an average angle of 4.3° to a depth of 3 or 4 km, typically. A slope of 4.3° is comparable to the slope of the steepest parts of a modern highway entering or leaving a continental mountain range.

At the base of the slope is debris that has been washed down the slope from the continent and deposited at the bottom. This wedge of deposited sediment is called the "continental rise." Often it takes the form of large alluvial fans, similar to those we see where mountain valleys empty onto the surrounding plateaus (Figure 4.2). The continental rise usually has only a slight slope, and it is often difficult to tell where the rise stops and the deep ocean basin begins.

A typical ocean basin would then extend for several thousand kilometers until it ran into the edge of the oceanic ridge system. The area covered by the oceanic ridge system is not as large as that of the basins, but is only slightly less than the surface area of the continents (Table 4.1); 29% of the earth's surface is covered by continents and 23% by the **87**

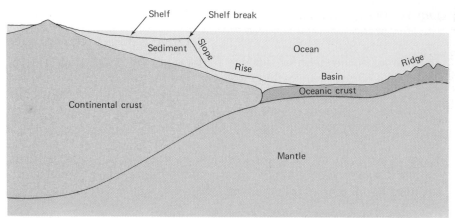

FIGURE 4.1   *Profile of a typical ocean bottom (with a great deal of vertical exaggeration). Going out from shore are successively the shelf, slope, rise, basin, and eventually the oceanic ridge. The dividing line between the shelf and slope is called the "shelf break."*

FIGURE 4.2   *Some large alluvial fans are seen at the base of this mountain.*

ridge system. The average ocean bottom would also have numerous small oceanic plateaus covering about 5% of its area altogether, and a still smaller percent of its area invested in trenches, and volcanic islands and ridges.

TABLE 4.1   Oceanic Regions

Region	Percent of World Area	Percent of Ocean Area	Percent of Ocean Water	Avg. Slope	Avg. Width (km)
Shelf	6	9	0.2	0.1°	75
Slope	4	6	3	4.3°	50
Rise	4	5	5	0.2°	40
Basin	30	42	53	—	—
Oceanic ridge	23	33	33	0.2°	1700
Trenches	1	2	4	3.0°	100
Volcanic areas	2	3	2	—	—

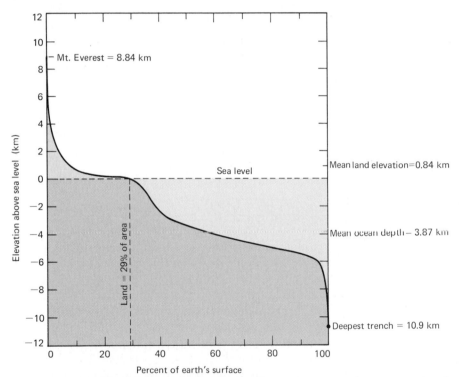

FIGURE 4.3   *Plot of the percent of the earth's surface area above a certain elevation vs. the elevation (the "hypsographic curve"). (From Sverdrup, Johnson, and Flemming, The Oceans, Prentice-Hall © 1942, renewed 1970.)*

The distribution of earth surface as a function of altitude is shown in Figure 4.3. This is a plot of the percentage of crustal surface above a certain elevation vs. the elevation. It is seen that more than half of the world's solid surface is in the range between 3 and 6 km below sea level, and about 20% at elevations between sea level and 1 km above. Together, these two regions involve more than 70% of the world's area. This means that relatively little of the earth's surface is found at other elevations: high

FIGURE 4.4 *If you put a pile of sand in a tub of water and then make waves, the waves eat away at the sand, forming a "seacliff" at the exposed face of the sand pile, and a terrace below water level.*

in continental mountains, low elevations in ocean trenches, or at intermediate elevations between sea level and 3 km depth.

As we saw in the preceding chapter, this rather surprising observation that little of the earth's surface lies at intermediate elevations, is a reflection of the two distinct types of crustal materials, one characteristic of oceanic crust, and one of the continents. Their distinctive compositions and structures cause them to "float" at different levels on the asthenosphere below, resulting in the two distinct ranges of surface elevations.

The average depth of the oceans is 3.87 km. If we could bulldoze the continents into the oceans, and smooth off the earth's entire solid surface, then the ocean would cover the entire earth to a depth of 2.69 km everywhere. This uniform depth to which any given material would cover a smooth spherical earth is called its "sphere depth."

B. THE SHELF

The continental shelf is most accessible to us, both because of its proximity to land and because of the shallowness of the water covering it. We have been fishing the continental shelves for millenia, and we've included in our diets organisms residing on the bottom as well as inhabitants of the waters above. In more recent years we have been developing the technology to remove oil and mineral resources from the shelf as well.

The primary cause of the continental shelf can be easily demonstrated in a large tub or puddle containing water of a few inches depth, as illustrated in Figure 4.4. If you put a pile of sand in the middle and then make some small waves, you find that the waves cut away at the exposed portions of the sand pile as they crash into it, leaving a terrace just below water level. If you observe carefully, you will see that there are two related processes that contribute to the creation of this terrace:

1. A terrace is left when the upper exposed portions of the sand pile are removed.
2. As the eroded sand is deposited in deeper water, the terrace is extended.

That is, both the erosion of exposed materials, and their deposition in deeper waters, are important in the excavation of the terraces.

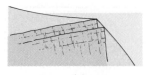  

(a) *(b)* *(c)*

FIGURE 4.5 *Sometimes the continental shelves are caused by sediment filling in behind natural dams, such as upturned basement rock (a), underwater volcanoes (b), or coral (c).*

On continental margins, waves do the same thing. Of the various kinds of waves that reach our shores, those that carry the most destructive energy are the large storm waves. Although the incessant smaller waves play a role in beach erosion and the shaping of shallow beach features, the more infrequent heavy storm waves actually perform the major task of removing exposed continental materials and depositing them in deeper waters.

Beneath these large storm waves, water motion to a depth of about 100 m is sufficient to transport sandy sediments, so we say that they "reach" to a depth of about 100 m. Consequently, the terraces excavated by these waves are typically at a depth of about 100 m, and are as predictable a feature on our continental margins as waves. Wherever there are waves, you'll find a continental shelf.

There are additional mechanisms that also contribute to the formation and maintenance of continental shelves in certain regions. One of these is the occurrence of natural dams near continents, behind which sediment is trapped, filling up the enclosed basins. These dams could have a variety of forms, such as upturned basement rock, a volcanic ridge, salt domes, or coral reefs. (See Figure 4.5.)

As we saw in the preceding chapter, the subsiding Pacific Plate creates a band of volcanic activity just east of the Asian continent. Sediment from the continent fills in behind this barrier and causes a large continental shelf. The shelf under the South China Sea has this origin, and it is one of the world's largest. In other places, the upward bending of basement rock just seaward of a continent can also make a natural dam. For example, the continental shelf off the east coast of the United States from the Carolinas northward is due to sediment fill behind this type of natural dam. Coral can also form a dam, although because coral is a shallow-water organism, it could only grow where the water was shallow to begin with. In some cases it has grown when the water was at a lower level than presently, and now is drowned and dead, but still a good dam. Since coral prefers exposure to waves, it grows at the outer limit of the shallow water, permitting sediment to fill in behind. Folds or domes in salt deposits sometimes provide similar natural dams, as is the case in some parts of the Gulf of Mexico.

In addition to natural dams, rivers can be shelf proliferators. Sediment is carried out onto the shelf and deposited beyond the river mouth,

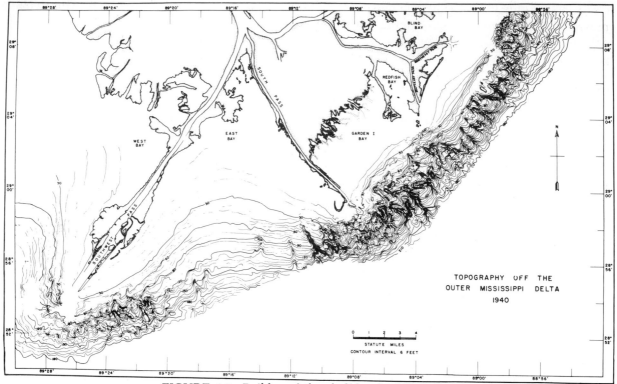

FIGURE 4.6 *Buildup of the shelf by the Mississippi River. Notice the gullies in the slope. (Courtesy Francis P. Shepard, Scripps Institution of Oceanography.)*

as in Figure 4.6. It turns out, however, that this direct excavation by rivers does not account for much extension of continental shelves when compared to other processes.

The sediment found on the shelf can be predicted from the climate of the nearby land mass. In humid rainy regions, the shelf tends to have very fine "muddy" sediments, and in cold harsh climates, coarser rock and gravel sediments are common. Coral and skeletal fragments are common in some warm balmy regions, and although sand is found almost everywhere, it is most abundant in temperate climates. You may wish to speculate over possible reasons for these correlations between sediments and climates. If you are not completely convinced by your own arguments, you may be comforted to know that oceanographers are not completely in agreement over the reasons either.

Along any one continental shelf, we might expect to find a gradation of sediments, with coarsest sediments appearing near the shore, and progressively finer sediments found farther out. This would simply reflect the tendency of finer sediments to be held in suspension longer, being carried farther before settling out. The actual distribution of sediments on the shelf, however, is much more complex than this, which tells us that other processes must also influence sediment distribution.

We know that large storm waves excavate the shelves, and so they must play a large role in rearranging the sediments. In addition, considerable excavation has occurred during recent ice ages. Ice sheets have left gouges and moraines on the shelves at higher latitudes quite similar to the scars left on the nearby land masses. In addition, some water normally in the oceans is removed in forming these ice sheets. The resulting drop in sea level exposes the shelves to subaerial erosion. Old stream beds and eroded canyons are common shelf features. Remnants of old beaches are found in many places on the shelf, corresponding to times when the water was lower. In some places, evidence of these "relic beaches" has been found down on the continental slope, 50 m or more below the shelf break, indicating that for some time during an ice age the entire shelves and even part of the slope have been exposed as part of the continents proper.

C.   THE SLOPE

C.1   General Features

The continental slope marks the division between the continental and oceanic provinces of the earth's crust. An average slope of 4.3° may not seem very decisive, but when plotted on a contour map as in Figure 4.7, or with some vertical exaggeration as in Figure 4.8, its significance becomes clear. The slope is really the edge of the ocean basin. Although the shelves have roughly 9% of the ocean area and are colored blue on maps, they are extremely shallow by comparison and have little influence on the major portions of the oceans. The real oceans begin at the slope, not at the beach.

There seems to be a tendency for the continental slopes to be slightly steeper on the "leading edges" than on the "trailing edges" of continents as they drift across the globe.[1] This may be related to the tendency to find convergent plate boundaries and subduction zones on the leading edges, and to the buildup of sediment deposits at the base of the rise on trailing edges. (These sediment accumulations seem to be absent on many leading edges.)

Along some slopes, the downward grade is interrupted by terraces. An example is the Miami Terrace illustrated in Figure 4.9. This particular terrace is thought to be carved by the flow of the main branch of the Gulf Stream through this region. Terraces may also result from faulting on the slope, as illustrated in Figure 4.10.

C.2   Submarine Canyons and Other Scars

There are some huge scars on the shelves and slopes. Understandably, some of these have aroused our curiosities and have received a great deal of attention.

---

[1]*See the discussion of continental drift in the preceding chapter.*

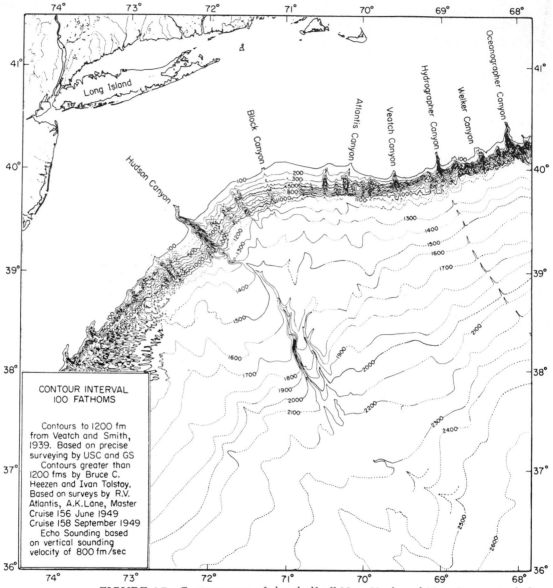

FIGURE 4.7 *Contour map of the shelf off New York and New Jersey, including the Hudson Submarine Canyon.*

The most impressive are the submarine canyons. Some of these have relief comparable to that of the Grand Canyon of the Colorado, and some are even bigger. Examples are the Monterey, Carmel, and Scripps canyons, illustrated in Figures 4.11, 4.12, and 4.13.

Most, but not all, of the submarine canyons appear directly off river valleys on the adjoining continent. Some submarine canyons cut almost completely across the continental shelf, and one canyon even extends into the mouth of the Congo River.

FIGURE 4.8 *Compare this physiographic diagram to the contour map of FIGURE 4.7. (A portion of the Physiographic Diagram of the North Atlantic by Bruce C. Heezen and Marie Tharp, 1968. Published by The Geological Society of America. Reproduced by permission. Copyright © 1968 by Marie Tharp.)*

The explanation of these curious features eluded us for many years. We were quite sure that some sort of current must flow through them. Ripple marks (Figure 4.14) were common on the canyon floors, indicating downward flow through them. Alluvial fans were found where the canyons emptied onto the ocean basin below, and remains of shallow-water organisms were found among these sediments. These things definitely indicated current flow down the canyons, and the fact that the sediment on the canyon floor was unsorted indicated that these currents must be fairly strong and turbulent. Yet, whenever we put instruments on the canyon floor to measure the current, we were not able to detect any.

You can see that for a while it seemed like nature was fooling us, somehow turning off the current whenever we put instruments in place to measure them. We concluded that the currents must be sporadic rather than continuous. In support of this idea, we have observed the sequential breaking of telephone lines laid across submarine canyons. These have indicated the currents flow in occasional violent underwater avalanches. From the timing of the cable breaks we know these flows can be remarkably fast—up to 50 km per hour may be possible.

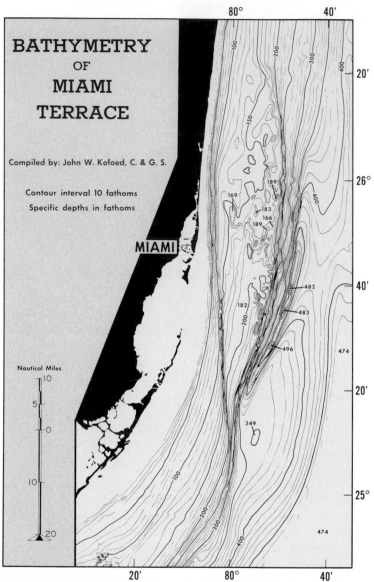

FIGURE 4.9 The Miami Terrace. (Kofoed and Malloy, Southeastern Geology, **6,** No. 3 (1965).)

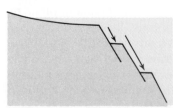

FIGURE 4.10 Terraces on slopes may be due to slipping along faults.

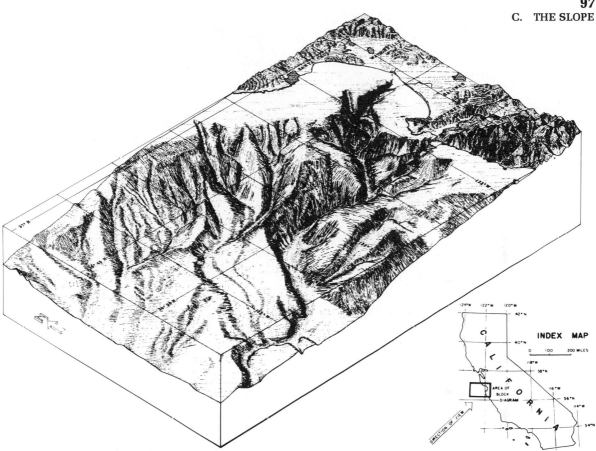

FIGURE 4.11   *The Montery and Carmel Submarine Canyons off the central California coast. (Diagram by Tau Rho Alpha, USGS.)*

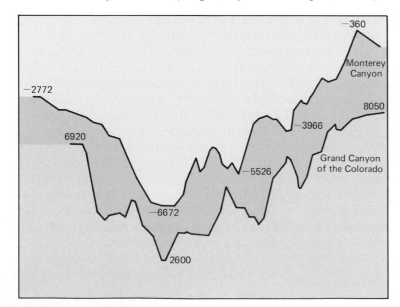

FIGURE 4.12   *A comparison of the profiles of the Grand Canyon of the Colorado River and the Monterey Submarine Canyon, showing them to have similar relief. Elevations relative to sea level are given in feet. Vertical exaggeration is 5X. (After F. P. Shepard, Submarine Geology, 3rd ed., Harper and Row, 1973.)*

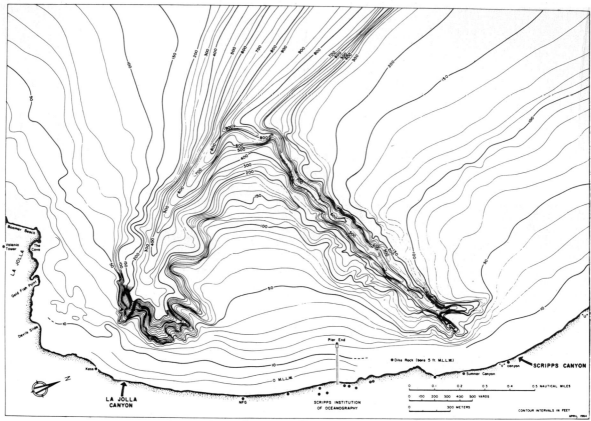

FIGURE 4.13 *(Below) La Jolla Submarine Canyon, (From F. P. Shepard, Submarine Geology, 3rd ed., Harper and Row, 1973.)*

We've noticed that all active canyons have large amounts of sediment flowing into canyon heads. One important source of sediment is rivers, which accounts for the canyons off river mouths. Those canyons not off river mouths receive their sediment from the "longshore transport." That is the transport of sediment along the coast by longshore currents. As evidence of this longshore transport of sediment, the beaches just "upstream" from a canyon head tend to be wide and sandy, whereas "downstream" beaches tend to be narrow and barren, indicating much of the sand has flowed into the canyon head rather than on down the beach.

We believe that this sediment collects in the canyon head until there is an "avalanche," and it is carried down through the canyon in a murky, underwater landslide that we call a "turbidity current." The turbidity currents can go for hundreds of kilometers down through the canyons and far out onto the ocean floor below. This indicates that there must be relatively little dissipation of energy through friction with the bottom.

Although this explains how submarine canyons are maintained and further eroded or extended, it does not explain how they formed in the

FIGURE 4.14 *Some of the evidence that submarine canyons are caused by occasional turbidity flows of the sediment that piles up in the head. (a) Ripples on the canyon floor indicate downward current flow. (b) Unsorted sediment indicates occasional turbulent flow rather than continuous smooth flow. (c) Sediment flows into the head of the canyon. (Photos by R. F. Dill, U.S. Navy.)*

first place. That is, in order for sediment to collect in the canyon head, there must be a canyon there first.

There are a variety of possible initial causes for submarine canyons. But the fact that so many are associated with the mouths of major rivers points to erosion by rivers and streams as a major cause. This could not be accomplished with the sea at its present level, because the fresh water from rivers and streams flows out along the surface of the ocean rather than beneath it. But there were times during the ice ages when sea level was lower and the shelves were exposed to direct erosion by the rivers and streams. This was undoubtedly the origin of many of the canyons that today are further eroded by turbidity currents. Since most submarine canyons today are considerably deeper than the level of sea level during ice ages, it is clear that considerably further erosion by turbidity currents must have taken place in those initiated by subaerial erosion.

Some submarine canyons have rather flattened bottoms, being more U-shaped in cross section that V-shaped. These tend to be found off large

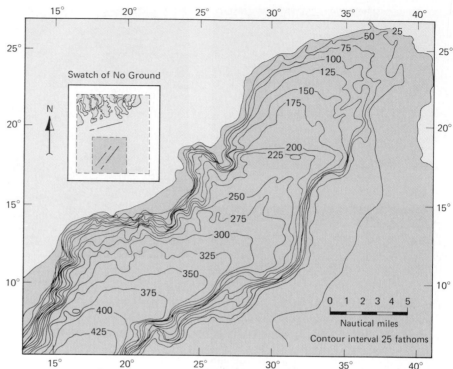

FIGURE 4.15 *The Swatch of No Ground is a delta-front trough that starts on the shelf off the Ganges Delta. (From F. P. Shepard,* Submarine Geology, *3rd ed., Harper and Row, 1973.)*

river deltas, such as the Ganges delta (Figure 4.15), and are called "delta-front troughs." Another common blemish on the slope are the "slope gullies," as shown in Figure 4.16. These seem to be especially common in regions of rapid buildup of loose sediment on the slope, such as off the mouth of the Mississippi River (Figure 4.6), and are probably due to slumping of the not very well packed area.

D. THE RISE

At the base of the slope, extending out onto the ocean basin, is a wedge of sediment known as the "continental rise." A large component of the sediment in this wedge arises from the deposits of turbidity currents. These deposits, whether on the rise or on the ocean basin below, are called "turbidites." Fan valleys, as in Figure 4.17, are often found cut across the rise, particularly as extensions of submarine canyons in the continental margins above. Often these fan valleys have leveed sides, as do rivers on land which flood their banks, and deposit their sediment on the adjoining flood plains. This suggests to us that turbidity currents are to the slope what

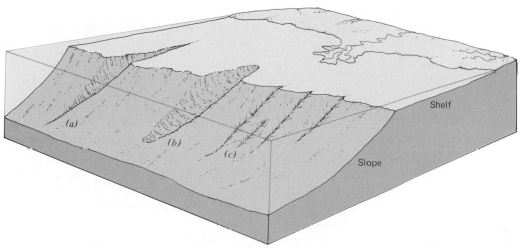

FIGURE 4.16   *Blemishes on the continental slope: (a) submarine canyons ("V"-shaped in cross-section), (b) delta-front troughs ("U"-shaped in cross section), and (c) slope gullies.*

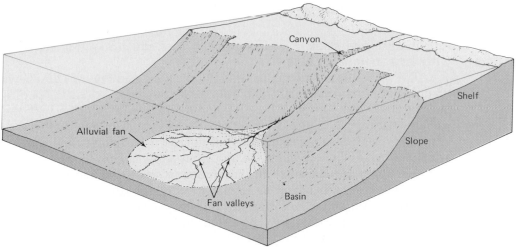

FIGURE 4.17   *Sediment carried down through canyons by turbidity currents is deposited at the base in fans. The fan valleys are current-cut channels across the fan.*

streams are to the edge of continental mountain ranges. They fall down from the highlands into the valleys below via canyons and ravines on the side of the slope, and cut their way through the fans of deposited sediment at the bottom. In flows of extra large volume they flood their banks, and the consequent slower flow causes the sediment to drop out and make levees.

Aside from finer sediment and gentler slope, these deposits must resemble the alluvial fans at the bases of tall mountains. If the water were removed and one could stand on the ocean bottom and look up at the con-

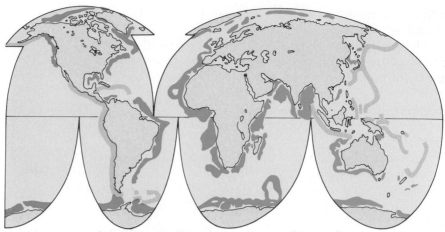

FIGURE 4.18 *The world distribution of continental rises, showing their scarcity on the margins of the Pacific. The thick shaded lines indicate the locations of the trenches. The lack of continental rises on these margins suggests "voluntary" rather than "forced" subsidence of the oceanic crust. (From K. O. Emery, Oil and Gas Journ.,* **67**, *No. 19 (1969) pp. 231–243.)*

tinental rise and slope, it would probably look much like the edge of a very large mountain range.

The world distribution of continental rises, seen in Figure 4.18, shows that they are found primarily off older, stable coastlines. Along the perimeter of the Pacific Ocean, where the continents are in collision with the oceanic crustal plates, the continental rises are not to be found. Presumably, some sediment still makes its way across the shelf, which is often quite narrow, and down the slope. But what happens to it then?

It is apparent that much of these missing sediments must be disappearing into the earth's interior along with the underthrust oceanic plate in a subduction zone. This suggests that the oceanic plate may not be *forced* under, but rather may be going under "voluntarily." If it were actually undergoing *forced* collision, then we would expect the sediments to be scraped off by the leading edge of the approaching continental plate, creating a bulldozed pileup of sediments where the trenches now appear (Figure 3.21). The fact that this does not happen to the extent that would be expected is one of the reasons we believe that in most cases the oceanic crust is going under "voluntarily," before actual collision, carrying its sediment cover with it.

E.   OCEAN BASINS

Just as the alluvial fans at the base of continental mountain ranges gradually blend into the valley floors below, so do continental rises gradually blend into the ocean basin floors below. Where one ends and the other begins is difficult to determine, and so it is often rather arbitrarily determined.

The ocean basins cover considerably more of the earth's surface than do the continents, and they display a considerable variety of features. The underlying rock is born on the crest of the oceanic ridge and spreads outward from there. Consequently, the underlying rock is youngest near the ridge, and older as you get farther and farther away from the ridge. Older basement rock has had more time to collect overlying sediments, so the sediment cover is thinnest near the ridge, and thicker as you go away from the ridge. Also, increasing proximity to the continental sediment sources tends to make the sediment cover thicken away from the ridge. Of course, the sediment cover tends to hide underlying relief features, so ocean basin topography tends to be most rugged near the oceanic ridge, and considerably less rugged near the continents.

The myriads of small hills on the ocean floor are called "abyssal hills" or "sneaknolls." They are the most abundant kind of topographic feature on the earth. Mountains that rise more than 900 meters above the surrounding seafloor are called "seamounts." Depressions due to crustal faults, called "fault valleys," are sometimes found. The deeper regions of the basins are referred to as as "deeps."

Near their continental margins, ocean basins sometimes have exceptionally flat regions, called "abyssal plains." These are probably due largely to sediment deposited by turbidity currents flowing out onto the basin, which cover up whatever relief features were there originally. An extreme example of this is the abyssal plain off the Washington coast. The oceanic ridge there is near shore and blocks the flow of sediment out to greater depths. Consequently, the sediment from the Columbia River is filling in this basin; it is already considerably shallower than the basin to the west of the ridge.

There are several other pieces of evidence that help substantiate the belief that the abyssal plains are caused by the sediment deposit of turbidity currents. First, among the sediment are found the skeletons of shallow-water organisms, indicating they were washed down with sediment from the continental shelf. Also, the deposits indicate "graded bedding." That is, the deposits come in layers, coarser materials being at the bottom of one layer, with successively finer sediments on top. (See Figure 4.19.) Then begins another layer with the same ordering of sediments, and so on. This would be expected if each layer came from a turbulent flow, with the coarser materials settling out first and the finest sediment last. Another piece of evidence is that the abyssal plains are flat rather than rolling. If the sediment settled out from above, then the sediment layers should be bumpy, displaying topographical features beneath, much like a blanket of snow over low shrubs. However, in an avalanche, the snow surface is smoothed out, just as are the abyssal plains. Finally, the preponderance of abyssal plains near the continents suggests the sediment is washed down from them.

One yet unsolved riddle of the ocean basins is the existence of a few

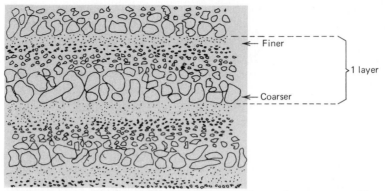

FIGURE 4.19 *Graded bedding. In any one layer the coarser sediment is found at the bottom and the finer sediment near the top. Then the pattern repeats itself in the next layer above.*

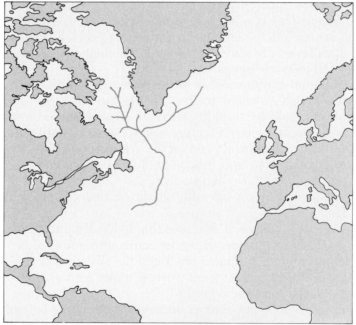

FIGURE 4.20 *A deep ocean channel extends along the north-western Atlantic Ocean bottom. It is rectangular in cross-section, resembling a large river bed.*

deep-sea channels. These resemble river valleys, and appear to be extensions of fan valleys along the abyssal plains. A map of the Mid-Atlantic Channel is given in Figure 4.20. It is typically one or two hundred meters deep, fairly steep-walled, and roughly 2 km wide. It is probably the result of current flow, but it is still not clear what these currents should be. Perhaps they are turbidity currents, perhaps some high-density highly saline water currents, or perhaps something completely unexpected.

F. THE OCEANIC RIDGE SYSTEM

In the previous chapter we saw how the discovery and study of the oceanic ridge system was central to the development of our ideas on plate tectonics. There we learned of its role in the production of new oceanic crust and in continental drift.

It is far too prominent a feature to be overlooked in a chapter that presents a description of the ocean bottom. But to avoid redundancy, we will confine our attention here to a geographical description and omit any discussion of its evolution or its role in evolutionary processes.

Its total length is about 65,000 km, which is slightly more than 1 ½ times the circumference of the earth. It averages over a thousand kilometers in width, being wider where it is more active. The elevation of the ridge crest is typically 1 to 2 km above that of the sea floor, but in some areas it may rise much higher than this and may even reach above sea level. Iceland in the North Atlantic and the Tristan da Cunha group of islands in the South Atlantic are exposed regions of the Mid-Atlantic Ridge. It is mostly confined to the ocean, but a branch of the system does invade the eastern part of the African continent.

The most rugged topography is found near the ridge crest (See Figure 4.21.), which runs centrally along the ridge system and is frequently displaced laterally along transform faults, as discussed in the previous chapter. The central rift valley ranges from 12 to 48 km in width, and is typically 0.5 to 1.5 km deep.

Bordering the rift valley on either side are the very rugged "rift mountains." The ridge crest is volcanically quite active, and is riddled with faults running both parallel to and perpendicular to the ridge axis. Going away from the ridge crest, the terrain becomes less rugged with decreasing elevation, and has increasing sediment cover corresponding to its greater age.

G. SMALLER FEATURES

G.1 Plateaus

There are numerous local elevated plateaus, rising typically 1 or 2 km above the surrounding seafloor (Figure 3.23). The crustal materials beneath these plateaus are thickened, as is necessary for them to gain the necessary buoyancy from the asthenosphere, and they seem to have composition characteristic of continental crust. Some of these are undoubtedly fragments that have broken off of existing continents. Others are probably the result of some local volcanism, that haven't yet had a chance to aggregate onto larger continental masses.

G.2 Ocean Trenches

As we saw in the preceding chapter, nearly all ocean trenches are found ringing the perimeter of the Pacific, identifying the location of convergent

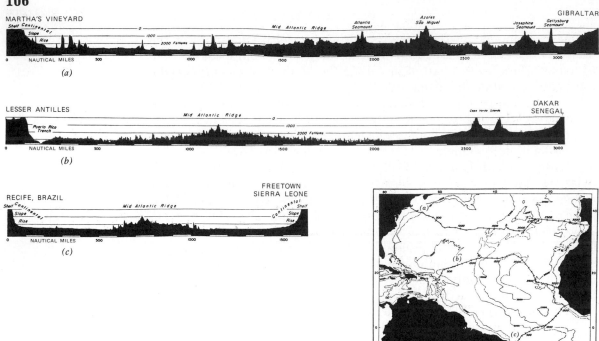

FIGURE 4.21 *Three transatlantic topographic profiles plotted at a vertical exaggeration of 40:1. Horizontal distances are indicated in nautical miles, and vertical distances are indicated by 1000-fathom depth intervals. (From Special Paper 65, "The Floor of the Oceans: I, by Tharp and Ewing," 1959. Published by The Geological Society of America. Reproduced by permission of Marie Tharp.)*

plate boundaries. The Java Trench in the Indian Ocean and the Puerto Rico Trench in the Atlantic are the result of small adjacent plates of lithosphere moving along with the Pacific Ocean bottom, thereby transferring the zones of convergence to the other sides of these two small plates.

Typically, trenches are thousands of kilometers long, fifty to one hundred kilometers wide, and four or five kilometers deeper than the neighboring ocean basin. A rough rule of thumb is that ocean basins are 4 or 5 km deep, oceanic ridges are half as deep, and trenches are twice as deep. Of course, there is a great deal of variation in these figures.

G.3 Volcanic Features

Most of the earth's volcanic activity is associated with subduction zones, and most of the remainder is located along the oceanic ridge system. But there are other isolated areas of volcanic activity as well.

The Pacific Ocean has numerous chains of volcanic islands. The Hawaiian Islands, for example, are at the southeastern end of a long chain of volcanic activity extending thousands of kilometers to the northwest, most of which no longer reach the ocean surface. Many other island

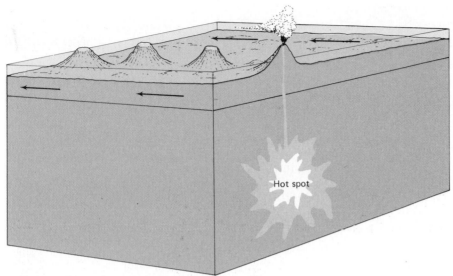

FIGURE 4.22 *We believe that chains of volcanic activity are caused by the passage of the lithosphere over some sort of hot spot anchored in the mantle below. A chain of ancient volcanoes traces the path that the lithosphere has followed over the hot spot.*

chains can also be found on maps, such as those in French Polynesia, the Line Islands, or the Marshall Islands.

A clue to the cause of these chains comes from dating the rocks of the islands. It is found that there is a tendency along any one chain for the volcanoes to get older to the northwest and younger to the southeast. In fact, the difference in age is typically a year for every few centimeters separating the two volcanoes. This is comparable to the rates of spreading from an oceanic ridge, and suggests the oceanic crust is sliding over some sort of stationary "hot spot" or "plume" in the mantle, which occasionally shoots up a volcano on the crustal surface passing overhead. (See Figure 4.22.)

We are quite curious about what could cause these local hot spots. This concern has generated an interesting variety of theories. Which, if any, are correct remains yet to be seen. Regardless of their cause, if we assume that they are somehow anchored in the mantle, then any one chain of volcanic activity leaves a record of the motion of that crustal plate over the mantle. From the dates and separations of the various volcanoes, we know the speed of the plate during any period, and from changes in the direction of the line of activity, we can infer changes in direction of the plate motion. Other techniques we have studied only tell us of the motion of one crustal plate relative to another, but not relative to the mantle beneath.

This means of determining the motion of crustal plates depends on the assumption that the hot spots are anchored in the mantle and are not themselves moving about. This assumption seems justified by the obser-

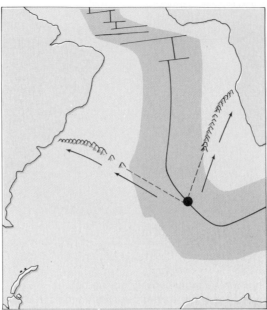

FIGURE 4.23   *A theory is that volcanic ridges on both sides of the South Atlantic may have been caused by a one-time "hot spot" under the ridge. As the crust separated and moved away from the ridge, the volcanic record of this motion remained. On the west is the "Rio Grande Rise," and on the east side is the "Walvis Ridge."*

vation that volcanic island chains in the Pacific tend to be in a northwest-southeast line. If the various hot spots were moving rather than the crustal plate, then the lines of islands should be in random directions, corresponding to the motion of their particular hot spots.

These volcanic chains are not restricted to the Pacific. Two such chains in the South Atlantic are illustrated in Figure 4.23. The volcanic activity on the Walvis Ridge has been so dense that it blocks the flow of bottom water on the east side of the Atlantic. These two chains give us some idea of the motions of Africa and South America over the mantle for the last 150 million years.

In any one chain of islands in the Pacific, the youngest ones are at the southeastern end. We can predict the next Hawaiian island will form to the southeast of present day Hawaii, for example. Toward the northwest along a chain, more and more of the volcanoes are submerged. This can be understood in terms of isostatic readjustment as follows.

Putting a massive volcano atop a relatively thin plate of oceanic crust must cause this plate to slowly subside until buoyant equilibrium is reobtained. Consequently, volcanoes that once were islands may now be found well below sea level. These "tablemounts" (or "guyots") are common submarine features, and are illustrated in Figure 4.25. They have flat tops due to subaerial and wave erosion, and they have subsequently subsided to depths sometimes more than a kilometer below sea level. In

FIGURE 4.24 *The new-born island of Surtsey, near Iceland in the North Atlantic. (June 18, 1964)*

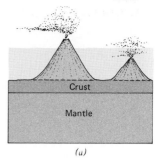

(a)

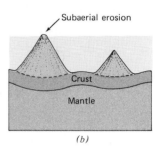

(b)

warmer climates where coral can grow, an atoll may form as the volcano subsides, providing there are no rapid variations in sea level that would drown the coral. Perhaps this will be the eventual fate of the present Hawaiian Islands.

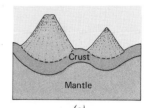

(c)

FIGURE 4.25

(a) *Volcanoes.* (b) *Those protruding through the ocean surface are subjected to subaerial and wave erosion, flattening the tops.* (c) *Heavy volcanoes on thin crust subside until isostacy is reobtained, leaving flat-topped undersea mountains, called "table-mounts" or "guyots."*

G.4 Coral Reefs and Atolls

Coral reefs are frequently found in warm shallow waters. They are rigid, porous, wave-resistant structures made primarily of the interwoven skeletons of generations of corals, with the skeletons of other members of the reef community making smaller but still significant contributions.

Corals require warm salty water, which means that most coral reefs are found in shallow tropical waters, away from the mouths of large rivers. The basis of the food chain on a reef is the food synthesized by a small algae living symbiotically with the coral. Therefore, living reefs are found only in shallow clear water, where there is sufficient sunlight for photosynthesis by this type of algae. The coral itself is a type of animal that gains its sustenance by filtering organic particles from the water, and so it does best where there is a great deal of water motion that constantly brings in fresh supplies of these suspended nutriments.

In tropical regions, trade winds blow the warm surface waters west-

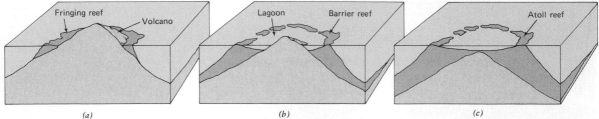

FIGURE 4.26 (a) *A coral fringing reef begins growing at the water's edge.* (b) *As the volcano subsides, coral grows upward at a matching rate, leaving a lagoon between the barrier reef and the volcanic island.* (c) *Eventually, the volcano sinks entirely below sea level, leaving only a ring-shaped atoll reef with a central lagoon.*

ward across the oceans, which means that the warm, salty, clear surface waters tend to accumulate along the western ocean margins. Consequently, we find the majority of coral reefs located in western tropical regions of the oceans, in shallow waters, far away from sources of freshwater run off from continents, and where there is a great deal of water motion due to waves or currents.

When conditions are right, "fringing reefs" first appear in the shallow waters bordering a land mass. As the reef matures, it tends to grow outward away from the land mass, because it grows best on the outer edge, where the wave action and water motion is greatest, and where it is farthest from the fresh-water runoff from the land. As it grows outward, it leaves a lagoon behind, and in this more mature stage it is called a "barrier reef." The accumulating skeletal debris extends the platform, enabling continued seaward growth of the barrier reef, with a resulting widening of the lagoon.

Sometimes, small ring-shaped reefs are found growing in what would otherwise be deep ocean waters. These are called "atolls," or "atoll reefs." The coral atolls of the South Pacific were among the things studied by Charles Darwin in his voyage on the *H. M. S. Beagle* in 1831 to 1836, and he postulated what has been subsequently shown to be the correct explanation for their formation. In balmy climates coral grows rapidly. As a volcanic island subsides a coral fringing reef starts growing around its flanks (Figure 4.26). As it continues to subside, the reef, now a barrier reef, grows upward at a matching rate, leaving a lagoon between it and the remaining emerged part of the volcano. Finally, after the entire volcano has subsided below sea level, only the reef remains, and we call it an "atoll."

Although coral can grow fast enough in balmy climates to keep pace with subsiding volcanoes, it cannot survive the rapid water level changes during the waxing and waning of ice ages. Consequently, living coral reefs today are less than 6000 years old, as this is the length of time that sea level has been relatively stable. However, one can find many drowned older reefs left over from previous times, and many present reefs are growing on the relics of previous ones.

FIGURE 4.27 *Atolls, looking southeast across Tuamotu Archipelago in the South Pacific, as seen from Apollo 7.*

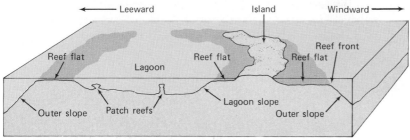

FIGURE 4.28 *Cross section of an atoll.*

Figure 4.27 is a high altitude photo of some atolls, showing the bright rings of shallow water. A schematic cross section of an atoll is shown in Figure 4.28. Coral grows best where exposed to the heaviest wave activity, as the churning waters ensure its constant supply of fresh nutriments. Consequently, the heaviest growth is found on the windward side of the atoll. In the interior regions of the atoll one finds less living coral and more coralline debris. The depth of the central lagoon is typically 50 m. "Patch reefs" sometimes grow up from the floor of the central lagoon, indicating that there may be enough nutriments there for the spotty growth of some species of coral, but not enough for the entire lagoon floor to fill in (Figure 4.29).

(a)

(b)

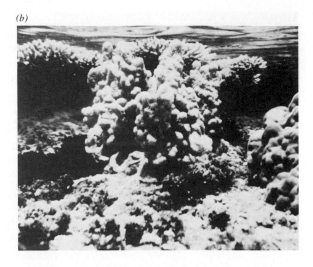

(c)

FIGURE 4.29
*Coral growth in atolls in the Marshall Islands. (a)
Aerial photo of the northwestern quadrant of a
reef flat. Notice the grooves in the coral growth
along the seaward edge, and notice the deeper
water of the central lagoon (lower right corner). (b)
Underwater photo of some species of coral on the
reef flat. (c) Patch reefs.*

H. DISTINCTIVE FEATURES OF
EACH OCEAN

A summary of the interesting and distinctive features of each of the oceans
will credit the Pacific, in addition to its size, with a preponderance of the
world's trenches, volcanic islands, and atolls. The East Pacific Rise does
not appear centrally located; rather it runs into the North American continent through the Gulf of Baja California (Figure 3.5), which consequently
is widening at a rate of some 6 cm/yr. The trench system extends around
most of the Pacific border, with a band of volcanic activity just beyond
them. The volcanic islands often come in long chains, indicating motion
of the oceanic crust over "hot spots" in the mantle below.

The Atlantic is characterized by its symmetry. The Mid-Atlantic
Ridge runs exactly down the center of the ocean, with its long S-shape
reflecting the same shape of the continental edges on both sides. The

Atlantic has wide, stable continental shelves, being on the trailing edge of each continent. The only trench system appearing in the Atlantic (off the east end of the Caribbean) is probably a carryover from the Pacific. As no trench appears off the West Coast of America (see Figure 3.20) opposite the Caribbean, its floor is probably moving with the Pacific Plate, moving eastwardly from the East Pacific Rise, and disappearing down this trench.

The Indian Ocean has an upside down "Y" in its ridge system. (See Figure 3.5.) The northern portion of the ridge runs into the Red Sea, and is currently separating the continent of Africa from Asia. The gigantic Bengal Fan off the Ganges River seems to be filling in the Bay of Bengal. It has numerous fan channels down its length (Figure 4.30). Finally, there is the mysterious "Ninetyeast Ridge" running south from the Bay of Bengal. (See Figure 3.5.) Some think it is an extinct remnant of an oceanic ridge system. Others think it may be caused by upbuckling of the crustal plate, being under some compressional duress.

I. THE MEDITERRANEAN SEAS

The mediterranean seas are considerably smaller and somewhat shallower than the oceans. Therefore, they tend to be influenced much more by the nearby land masses. It is seen in Table 4.3 that the Arctic Ocean is the largest of these. It has a curious system of narrow, parallel ridges and basins. The Gulf of Mexico has a fairly flat, inactive basin, but the floor of the neighboring Caribbean Sea is, in contrast, rough and active, probably due to its eastwardly motion along with the Pacific Plate, which results in collision with the Atlantic Plate. The Mediterranan and Black seas have several shallow sills, cutting off the flow of dense bottom water. The Black Sea is almost completely landlocked, having as its outlet only the very shallow Dardanelles and Bosphorus Straits. The flow along the bottom of the Mediterranean is blocked by sills of less than 300-m depth in the Strait of Sicily and the Gibraltar Sill. It has several narrow ridges and trenches in the eastern half, and nice flat basins in the western half, indicating it may be compressed in the east (Africa getting closer to Europe) and stretched in the west (Africa getting farther away). The mediterranean system, extending from the Bering Sea down to the South China Sea, is a consequence of the Pacific trench and resulting island arc system, which restricts communication of these waters with the open ocean.

J. SUMMARY

The continental shelf is the seaward extension of the continents, and the water over the shelf is quite shallow in comparison to other oceanic regions. Some shelves are created as terrigenous sediments fill in behind natural dams, and in some areas rivers build up the shelves by depositing their sediment load. But in all areas, waves play a major role in shelf excavation.

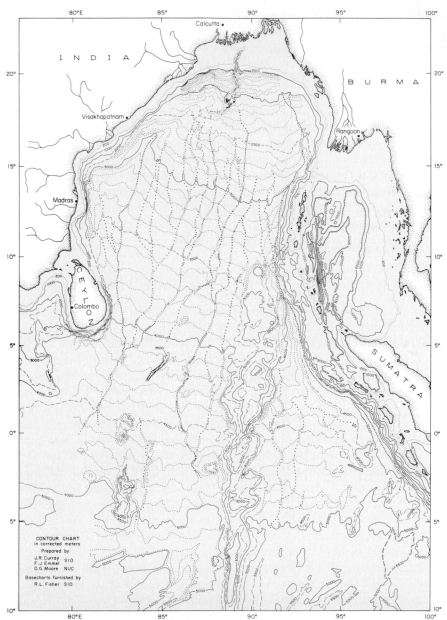

FIGURE 4.30 *The gigantic Bengal Fan seems to be slowly filling in the Bay of Bengal. (J. R. Curray and D. G. Moore, Geol. Soc. Am. Bulletin, **82** (1971). Chart revised 1975.)*

The continental slope separates the continental and oceanic provinces of the earth's surface. Some gigantic submarine canyons are found cut into the continental shelf and slope. The flow of turbidity currents maintains and extends submarine canyons, although turbidity currents alone cannot account for their origin.

TABLE 4.2 Areas of the Oceans

Ocean	Area (millions of km²)
Pacific	180
Atlantic	107
Indian	74

TABLE 4.3 Areas of Mediterranean Seas

Sea	Area (millions of km²)
Arctic Ocean	9.5
(Group along East Asian Coast)	6.0
Caribbean plus Gulf of Mexico	4.4
Mediterranean and Black seas	3.0

The wedge of deposited sediment at the base of the slope is called the continental rise. A large component of these sediments are found in the alluvial fan deposits where submarine canyons empty out onto the ocean bottom.

The ocean basins extend from the rise out to the oceanic ridge. Some extremely flat regions, called "abyssal plains," are found near the continental margins, and they are caused by the deposit of sediments carried in turbidity current flows. The ocean basins cover 30% of the earth's surface and have a variety of topographical features.

The oceanic ridge system covers only slightly less of the earth's surface than do the continents. It is quite active geologically. The most rugged parts are found nearest the ridge crest.

Away from subduction zones and oceanic ridges, the volcanic activity of the ocean bottom frequently comes in chains. We think that the explanation of these chains involves the passage of the lithosphere over isolated "hot spots" anchored in the mantle. Isostatic readjustment causes these volcanoes to sink gradually after their formation. Some that once were above sea level subsequently sank below the ocean's surface, and we identify them today as tablemounts. In warm, balmy climates coral may grow as these volcanoes sink, giving rise to atolls.

QUESTIONS FOR CHAPTER 4

1. Going seaward from land, what is the first region of the ocean bottom encountered? What is its average width, and roughly how deep does it go?

2. What is the "shelf break"?

3. Do you suppose the continental slope has valleys eroded down it like the mountain in Figure 4.2? If so, what might cause this erosion?

4. What feature in continental mountain ranges does the "continental rise" correspond to?

5. Roughly how deep are the ocean basins?

6. How does the area of the oceanic ridge system compare to the area of the continents?

7. If you smoothed off the earth's surface with a gigantic bulldozer so that the ocean covered the entire earth to a uniform depth, how deep would it be?

8. How does the sediment on the shelf correspond to the climate on the nearby land mass?

9. If all the sediment on the shelf were to remain where it was first deposited, how would you expect the sediment texture to vary with distance from land? Is this the way it really is? Why not?

10. What evidence is there that sometime(s) the entire continental shelves have been above sea level?

11. How might "natural dams" sometimes account for continental shelves? How might some of these "natural dams" be created?

12. What plays the biggest and most universal role in the excavation of shelves? Can you think of a "bathtub experiment" you could perform to substantiate this theory?

13. What is the average slope of the slope?

14. How does the slope on a trailing edge of a continent tend to differ from that on a collision edge?

15. What might cause terraces on the continental slopes?

16. How does the relief of some large submarine canyons compare to that of the Grand Canyon of the Colorado River?

17. Discuss some of the evidence for the conclusion that the submarine canyons are eroded and maintained by "turbidity current" flows.

18. What evidence is there that there is relatively little friction in the flow of turbidity currents?

19. Why couldn't the flow of fresh water from today's rivers account for submarine canyons?

20. Why couldn't the submarine canyons be completely caused by erosion when the shelves were exposed during an ice age?

21. What are "delta-front troughs"? What are "slope gullies"?

22. How would you expect the sediment on the rise to compare to that on the shelf above? Defend your answer.

23. How are levees sometimes formed on the sides of fan valleys?

24. Are continental rises more prevalent on trailing or collision margins of continents? Can you think of any possible explanation for this?

25. What is the difference between an "abyssal hill" and a "seamount"?

26. What is a "deep"?

27. What is "graded bedding"? Why does graded bedding in abyssal plain deposits imply they are created by turbidity currents?

28. What other evidence is there that abyssal plains are a result of turbidity current deposits (called "turbidites")?

29. Roughly what is the width and breadth of a deep-sea channel?

30. What is the rift valley? Where is it found? Describe it.

31. Briefly describe the relief of the oceanic ridge.

32. Why do the volcanic islands in the Pacific occur in long chains?

33. Briefly describe how a chain of volcanic activity might be used to trace the motion of a crustal plate over the mantle.

34. Which direction from Hawaii will the next island in the chain form? For every year that passes between now and then, about how much farther in this direction will the new island be?

35. What will be the eventual fate of the present island of Hawaii?

36. How are tablemounts formed?

37. How are atolls formed? Who first proposed this theory?

38. Why is it that any coral reef that is alive today probably is not over 6000 years old? Is it possible that some of these living reefs have grown on the relics of ancient ones?

39. On which part of an atoll would the coralline growth be hardiest?

40. Which ocean has most of the trenches and the most volcanic activity?

41. Which ocean has a mid-ocean ridge running exactly down its center?

42. What are some of the distinctive features of the Indian Ocean?

43. Why are mediterranean seas influenced more by climatic changes on nearby land masses than are the oceans?

44. Where are the shallow sills in the Black and Mediterranean seas?

45. What geological process is responsible for the mediterranean seas along the East Asian coast?

*46. An examination of Figure 3.20 shows that a trench appears at the eastern end of the Caribbean Sea. This indicates collision of oceanic crust with oceanic crust, as opposed to collision of oceanic and continental crustal plates elsewhere around the perimeter of the Pacific. Could this somehow explain the roughness of the bottom topography of the Caribbean as compared to the Gulf of Mexico, for example? How?

*47. Why do you suppose the continental shelf tends to have coarser sediments in colder climates?

*48. Why do you suppose delta-front troughs are more rounded than submarine canyons?

*49. Do you suppose the deep ocean bottom experiences more erosion or less erosion than the continents? Why?

*50. Do you suppose the Grand Canyon could have been a submarine canyon once? Why or why not?

SUGGESTIONS FOR FURTHER READING

1. Robert S. Dietz, "Iceland: Where the Mid-Ocean Ridge Bares Its Back," *Sea Frontiers*, **22,** No. 1 (1976), p. 9.

2. K. O. Emery, "The Continental Shelves," *Scientific American* (Sept. 1969).

3. Bruce C. Heezen and Charles D. Hollister, *The Face of the Deep*, Oxford University Press, New York, 1971.

4. J. R. Hertzler and W. B. Bryan, "The Floor of the Mid-Atlantic Rift," *Scientific American* (Aug. 1975).

5. Francis P. Shepard, "Coral Reefs of Moorea," *Sea Frontiers*, **22,** No. 6 (1976), p. 361.

6. Kathleen Mark, "Coral Reefs, Seamounts, and Guyots: Darwin's Double Theory," *Sea Frontiers*, **22,** No. 3 (1976), p. 143.

7. Alan C. Paulson, "The Coral Atoll," *Sea Frontiers*, **26,** No. 1 (1980), p. 36.

8. Francis P. Shepard, *Geological Oceanography*, Chapter 9, Crane, Russak and Co., New York, 1977.

9. Francis P. Shepard, "Submarine Canyons of the Pacific," *Sea Frontiers*, **21,** No. 1 (1975), p. 3.

10. J. H. McD. Whitaker (ed.) *Submarine Canyons and Deep-Sea Fans*, Dowden, Hutchinson and Ross, New York, 1976.

5
SEDIMENT MATERIALS

Sediments are one of the features of the earth that are most familiar to us in our everyday lives (Figure 5.1). In our early youth we enjoy spending summer days outside, playing in the dirt. "Please don't track in the dirt!" is a familiar parental expression. As we become older, we still enjoy the sediments, albeit in a slightly different manner. For example, favorite leisure diversions include caring for lawns or planting gardens. In fact, we complain about "concrete cities," which bury this life-giving material, the soil, under concrete buildings, sidewalks, and streets.

Sediments are hard to avoid. In fact, we must go to rugged continental mountain ranges in order to find significant areas of rock outcroppings not covered by this substance. Most sediments we encounter were once deposited in a shallow-water marine environment. The various tectonic motions of the earth have carried most portions of our present continents below sea level at one time or another. Even our mountain ranges display large regions of sedimentary rock formations indicative of former periods of submergence in shallow coastal waters. We now know that collision between continents or continental fragments is the aspect of tectonic processes responsible for the uplifting of these submerged coastal deposits into magnificent mountain ranges.

In the oceanic realm, we are familiar with the sediments on the beaches, which makes them such pleasant recreation areas. But oceanic sediments are not restricted to the beaches as are we. Sediments cover the ocean floor at all depths, although the composition of the sediments may vary from one region to another. In fact, if we wanted to find the small areas of ocean bottom not covered by sediment, we would have to look to rugged mountainous regions just as we would on land.

We can divide the ocean floor into two rather distinct regions according to the processes through which sediment accumulates. In the waters of the continental margins, sedimentation is generally quite heavy and mostly from continental or "terrigenous" sources. On the other hand, away from the land masses in the deep waters of the ocean basins and of the oceanic ridge system, sediments accumulate much more slowly and from different, less prolific sources.

We are developing increasing interest in ocean sediments for the resources they hold (Figure 5.2). As we deplete our terrestrial resources, we must turn to lower grade ores that are more difficult and more costly to mine. This, coupled with progress in ocean technology, makes ocean resources increasingly attractive. We are already removing large amounts of petroleum and natural gas from the sediments of the continental margins, and we are beginning to utilize ocean sediments for other resources as well.

But scientists would be interested in the ocean's sediments even if they had no obvious promise of economic value. This is simply because they are a result of some of nature's processes. If we want to learn more

121

FIGURE 5.1 *Without the sediment cover, continents would be barren, and scenes like these would not be.*

about nature, we should try to understand these processes.

The sediments hold many secrets that we wish to explore. In this chapter and the next we will examine many of these, and the techniques used in their study. In order to study the sediments, we need to have good representative samples to work with. For this reason, considerable effort is put into finding the best ways of retrieving samples from the ocean bottom. When analyzing a deposit of sediments, we wish to learn how they got there, where they originally came from, and what kinds of changes or transformations occurred between original source and eventual deposit. We may wish to know the conditions under which the original deposit formed, and what changes in the sediment's environment have occurred since then. Such information is gleaned from the properties of the sediment sample, such as its texture and its mineral composition.

FIGURE 5.2 *A semi-submersible mobile drilling platform, for drilling on the outer continental shelf.*

In this chapter we survey the various sources of sediments, we examine the methods used by oceanographers in retrieving samples for study, and we study the spectrum of properties displayed by the various kinds of sediments. In the next chapter we will study in more detail the various processes through which sedimentary deposits are created.

A. SEDIMENT SOURCES

The sediments of the ocean bottom display a wide variety of origins, but the largest single source of sediment is the erosion of the continents. Seventy-five percent of all ocean sediments come from this source, which is a result of the comparatively harsh environmental conditions here. Continents have to endure wide ranges in temperature, the percolation of fresh water, the freezing of this water in cracks and crevices, desiccation from sunlight and heat, and the probing organs and chemicals of plants and animals in search of sustenance.

The vast majority of these "terrigenous" sediments are deposited on the continental margins (Figure 5.3), which means that rates of sedimentation along the continental margins are quite a bit greater than sedimentation rates farther out to sea. Although there are large geographical vari-

FIGURE 5.3 *Aerial photo of the mouth of the Fraser River in British Columbia, showing the load of terrigenous sediments being fed into the Pacific Ocean.*

ations in rates of sediment accumulation, a typical number would be 20 cm of sediment per thousand years on the continental shelf, and only 0.2 cm of sediment per thousand years on the deep ocean floor away from continental margins. Most terrigenous sediments that reach the deep ocean floor are found near the continental margins in the turbidite deposits of abyssal plains, as discussed in Chapter 4.

Some terrigenous sediment is extremely fine, and can stay in suspension long enough to be carried by some current far out to sea before settling out. But this is a relatively small amount, so that most of the sediment of the deep ocean is dominated by other, less productive sources. Although terrigenous sediments are 75% of all sediments, they only dominate 20% of the area of the ocean bottom (Table 5.1), being very thick near the continents, but much less prominent farther out to sea.

A surprisingly large fraction of deep-sea sediments of terrigenous origin have been carried out to sea by the wind. These "eolian" sediments

TABLE 5.1   Source and Distribution of Sediments

Type	Source	Where Dominant	Approximate Percent of Bottom Area Covered
Terrigenous	Continental erosion	Continental margins, abyssal plains, and very high latitudes	20
Biogenic	Skeletons of organisms	Much of deep ocean bottom, especially at high temperate latitudes, and at depths less than 5 km in temperate and tropical latitudes	50
Clays	Miscellaneous. Fine inorganic particles carried by winds or currents, underwater volcanic eruptions, etc.	Deep ocean bottom, deeper than 5 km in temperate and tropical latitudes	30
Hydrogenous	Dissolved minerals	Virtually nowhere	<1
Cosmogenous	Outer space	Nowhere	0

are estimated to make up around 10% of all deep-sea sediments. This is not evidence of dirty air, but rather tells us that sedimentation from other sources is also slow.

Since the weathering and erosion of continental rock is such an important source, you might expect the weathering and erosion of the ocean bottom also to be important. However, the deep-sea environment is very stable, not subject to the extremes endured by the continent. So the weathering of the oceanic crust proceeds very slowly. Furthermore, most areas are protected from erosion by the overlying layers of sediment. Aside from some localized volcanic activity, the sea floor itself makes very little contribution to its sediment cover.

The largest single source of deep ocean sediments are the skeletal remains of microscopic plants and animals. These are produced mostly in surface waters, and then rain down on the ocean bottom as the organisms die, where they join sediment from other sources. If these "biogenic" sediments constitute more than 30% of the total, then the sediment is called an "ooze."

As you know, there is considerable biological activity over continental margins. But the sedimentation from terrigenous sources is so heavy that these biogenic sediments are usually very dilute on the continental margins. There are some localized areas near reefs, however, where biogenic sediments do dominate.

Over the deep oceans, biological activity is generally less than over the continental margins. However, dilution from other sediment sources is much less, so biogenic sediments may dominate large areas of the deep ocean floor.

There are other large regions of the ocean bottom where the skeletal debris dissolves as fast as it reaches the bottom. In these areas, clays of various types tend to dominate the sediments, due to the default by biogenic sediments.

Another source of sediments is the materials dissolved in seawater itself. These sediments, which are precipitated directly from dissolved materials, are called "hydrogenous," or "authigenic."

Interaction with seawater changes the character of many sediments. Organic skeletal debris is often dissolved. Fine clay particles from terrigenous sources may be noticeably altered during their suspension in the water and during their residence on or in the bottom sediments. Parts of the particles may be dissolved away, and others may be enhanced by precipitates from seawater. Additional waters of hydration may be given to some of the minerals.

In many large areas of the ocean, the clays are red or brown, indicating that they were laid down in an oxidizing environment. Their color is attributable to the oxidation of the iron. In other smaller areas we find green, blue, or black muds, indicating that they were deposited in an oxygen-deficient environment, and probably contain some organic matter that has not been able to decay completely due to lack of oxygen.

An interesting but very minor class of sediments is those whose origin is extraterrestrial. Countless meteoroids are floating about in the cosmos. As the earth travels about the sun and through the galaxy, it runs into these meteoroids. A few are large enough to heat up the air as they streak through the sky, causing the visible streaks known as "meteors" or "falling stars." By far, most are microscopic and slowly float down to earth through the atmosphere like tiny feathers. The earth acquires thousands of tons of material per day via this mechanism, and, of course, most of it lands in the ocean. However, the ocean is extremely large and even a few thousand tons per day spread out over the entire ocean bottom would be insignificant compared to the other forms of sedimentation. Therefore, it may be said that although cosmogenous sediments are found everywhere, they are important nowhere.

B. COLLECTING SEDIMENT SAMPLES

To study the sediments, we must have representative samples to work with. Most sediment samples are retrieved from aboard a surface ship, and we are usually separated from the ocean bottom by several kilometers of water. This extra obstacle means we must take extra precautions.

We want a sampler that is not likely to malfunction, as we do not wish to waste precious cruise time with a sampler that comes up empty. For this reason, simple mechanical devices are preferred over more complicated ones.

FIGURE 5.4   *Hanging to the left, a grab sampler. Hanging to the right, a thermo probe.*

Although sometimes we are content to get just anything we can, we frequently would like to ensure that our sample be representative of the bottom sediments. For this reason we would like our device to distort the sample as little as possible when it is taken, and then we would like our device to protect this sample from being washed or otherwise adulterated in any way by the sea during its journey back up to the ship. For coarse, rocky sediments taken by a dredge, this may not be a concern, but for samples of finer sediments, it is.

Sampling of surface sediments is most frequently done with some form of a dredge or with a grab sampler (Figure 5.4). Dredges are especially useful for retrieving nodules, rocks, gravel or other coarse sediments. Grab samplers have jaws that close when they reach bottom, and they are more useful for finer sediments. Usually some means is provided for protecting the sediments from being washed during the trip back to the ship.

"Corers" are used to sample a vertical column of sediments. Corers usually resemble very large darts hanging from a cable (Figure 5.5). The "needle" of the dart is a long hollow tube, several centimeters in diameter, called the "barrel." The corer may be weighted with several hundred kilograms of weights to help push the barrel into the bottom sediments. The

FIGURE 5.5 *Gravity corer.*

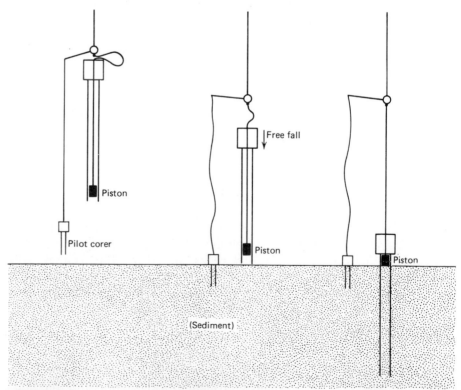

FIGURE 5.6 *The mechanics of a piston corer.*

barrel is carefully designed to minimize distortion as it plunges into the sediments, but clearly there will be some. In studying sediment cores, oceanographers scrape off the outer skin where the distortion will be greatest. The barrels usually contain a liner, so the samples may be removed along with the liner to minimize distortion during removal.

The invention of the "piston corer" (Figure 5.6) by Kullenberg enabled us to obtain much longer cores with minimal distortion. In this device, a corer with a very long barrel is lowered along with a small "pilot corer," which precedes it. When the pilot corer strikes bottom, the main corer is released from the cable (although still tied to a loose retrieving line), and accelerates toward the bottom. A piston within the barrel stops when it is even with the pilot corer, and the main corer continues its plunge. This piston acts like the piston in a hypodermic syringe. As it is drawn up into the falling barrel, it creates a vacuum that helps to draw the sediment sample up into the barrel, and with minimal distortion. The pilot corer gets a sample of the surface sediments, which are the ones most distorted by the piston corer. Core samples up to 20 m long have been retrieved using piston corers, whereas gravity corers usually sample only the upper meter or so of sediment. An improved type of piston corer, called a "hydraulic piston corer," is able to obtain even longer sediment cores.

New drilling techniques enable us to retrieve samples from very deep within the sediment, although these arrive at the surface in a drilling slurry rather than in firm, stratified layers.

C. TEXTURE

The sediments found in various regions of the ocean floor can display a large variety of properties, both physical and chemical. To describe the grain size of a sediment, we use a scale instituted by William Wentworth in 1922. The largest sediments are called boulders, followed by cobbles, pebbles, sands, silts, and clays, in order of decreasing size. The exact grain sizes implied by each of these terms is given in Table 5.2, and it corresponds quite well with common usage. (See Figure 5.7.)

Other descriptive terms used for describing a sediment are its "porosity," its "permeability," and its packing. Porosity refers to the amount of the volume not occupied by particles. Permeability refers to the ability of water to flow through the sediment. One might think that things that are more porous are also more permeable. This is not always true, however. For example, a piece of styrofoam and a kitchen sponge are both very porous, yet the styrofoam will not soak up water whereas the sponge will, indicating that the sponge is permeable and the styrofoam isn't. Clay deposits are usually quite porous but also impermeable, as the fine flaky clay particles have a very large surface area in proportion to their volume and provide a large amount of surface friction, which impedes the flow of water.

The packing of the sediment affects its porosity and permeability as

TABLE 5.2 Classification of Beach Material According to Diameter

	Classification	Sediment Diameter (mm)
Boulder		
		$256 = 2^8$
Cobble		$128 = 2^7$
		$64 = 2^6$
		$32 = 2^5$
Pebble		$16 = 2^4$
		$8 = 2^3$
		$4 = 2^2$
Granule		$2 = 2^1$
Sand	very coarse	$1 = 2^0$
	coarse	$\frac{1}{2} = 2^{-1}$
	medium	$\frac{1}{4} = 2^{-2}$
	fine	$\frac{1}{8} = 2^{-3}$
	very fine	
		$\frac{1}{16} = 2^{-4}$
Silt	coarse	$\frac{1}{16} = 2^{-4}$
	medium	$\frac{1}{64} = 2^{-6}$
	fine	$\frac{1}{128} = 2^{-7}$
	very fine	
		$\frac{1}{256} = 2^{-8}$
Clay	coarse	$\frac{1}{512} = 2^{-9}$
	medium	$\frac{1}{1024} = 2^{-10}$
	fine	$\frac{1}{2048} = 2^{-11}$
	very fine	
		$\frac{1}{4096} = 2^{-12}$
Colloid		

COARSE MEDIUM FINE

FIGURE 5.7 Some sedimentary rocks of different textures.

well as its resistance to erosion. Even if the sediment were composed of uniform spheres, it could pack in different ways, as illustrated in Figure 5.8. Poorly sorted sediments are generally less porous than well sorted ones because the finer grains tend to fill in the holes between the larger ones. This can be demonstrated with a drinking glass and marbles. After filling a drinking glass full of marbles, you will find that you can dump another third of a cup of sand into this glass without removing a single marble. The sand just goes into the voids between the marbles. You can make the mixture even less porous by putting in some finer-grained substance, such as talcum powder, to fill in the voids between the sand grains. The original glass of marbles accepted a lot more material without growing at all in volume.

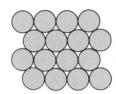

FIGURE 5.8
Two different ways of stacking marbles in two dimensions. The stacking on the above is clearly more porous than the one below.

D. MINERAL CONTENT

Two materials having similar texture may still be quite different (Figure 5.9). For example, sugar, table salt, and some beach sands all have similar textures. They have roughly the same grain size, similar porosity, and show similar permeability to the flow of fluids in which they're insoluble. Yet clearly, they are quite different materials. Even among beach sands of similar texture, there are wide ranges in composition. The ground-up skeletal debris of the white Florida beaches feel quite similar to the yellow quartz sands of California beaches.

There are two parts to the description of the composition of a sedi-

FIGURE 5.9 *Photo of four materials having similar textures, but quite different chemical compositions. (top left) A fine-grained quartz beach sand (SiO_2). (top right) A beach sand made up of skeletal debris of marine organisms ($CaCO_3$). (bottom left) Table salt (NaCl). (bottom right) Cane sugar ($C_{12}H_{22}O_{11}$).*

ment. One is its chemical composition, which tells us the various kinds of atoms that are in it and the relative abundances of these atoms. The other is the crystal structure, which tells us how the groups of atoms are arranged. The combination of chemical composition along with crystal structure defines a "mineral."

D.1 Chemical Composition

Molecules are groups of atoms that stick together and share some of their electrons. Normally, one part of the molecule is more generous with its electrons and the other part is greedy, accepting more than its share. The generous part carries a positive charge, as it is missing the negatively charged electrons that it donated to the other part of the molecule. This positively charged part is called the "cation." The other part of the molecule is negatively charged due to its excess of electrons, and is called the "anion." The charges of the two kinds of ions must be thought of as a time-averaged effect. The shared electrons are very lively and spend some time everywhere in the molecule. But on the average, they will spend more time near the anion than the cation.

An analysis of the materials of the earth's crust would show that almost all of the wide variety of crustal materials is made up of various combinations of just eight elements. These eight most abundant elements of the earth's crust, and their relative abundances by weight, are listed in Table 5.3.

In this chapter, we are interested in sedimentation, a process occurring right on the surface of the crust and not throughout its entirety. For this reason we add to the above list two more materials that are concentrated on the earth's surface and play an important role in sedimentation, although their presence at greater depths in the crust is not as significant. One is the element called "carbon," and the other is the compound called "water." Both are concentrated on the earth's surface through the volcanic processes described in Chapter 3, which continually distill materials at greater depths, sending volatiles up to the surface.

The carbon generally arrives as carbon dioxide gas, which frequently gets reworked through photosynthesis and the organisms that inhabit the earth's surface before finding its way into the sediments. The water, of

TABLE 5.3   Relative Abundances of Various Elements in the Earth's Crust

Oxygen	47.0%
Silicon	28.0%
Aluminum	8.0%
Iron	5.0%
Calcium	3.6%
Sodium	2.8%
Potassium	2.6%
Magnesium	2.1%
(All others together 0.9%)	

course, provides the medium in which most sedimentation occurs, but is mentioned here because the water molecule is frequently incorporated into the solid materials of the earth's surface as "water of hydration." That is, a molecule of some mineral will attach itself to one or more water molecules to make a larger "hydrated" molecule. For the next few paragraphs, however, we will concern ourselves with constructing some nonhydrated molecules, knowing that the actual minerals found in the earth may have varying amounts of water of hydration, depending on the environment in which they were formed.

So we add carbon to the previous list and then think about what kinds of molecules might be formed from these elements. What would make good cations and what would make good anions? Metals are very generous with their electrons, as is revealed by their ability to conduct electricity well. Consequently, we should consider the metals to be our cation prospects. These are listed along with the customary notation for indicating how many electrons they are willing to give up.

Aluminum	Al^{+++}
Iron	$Fe^{++ \text{ or } +++}$
Calcium	Ca^{++}
Sodium	Na^{+}
Potassium	K^{+}
Magnesium	Mg^{++}

The anions, then, must be somehow formed by carbon, silicon, and oxygen, as these are left over from our list. It turns out that these like to occur in groups called "radicals." The most common of these cation radicals is made of one silicon and four oxygens (SiO_4^{----}), called "silicate." However, on the surface of the earth, where carbon is concentrated, the "carbonate" (CO_3^{--}) radical is frequently found, being composed of one carbon and three oxygens. Many other anions are found, including oxides (O^{--}), sulfates (SO_4^{---}), phosphates (PO_4^{--}), and others. Although many of these are important, they are less common than are the silicates and carbonates, so we won't refer to them any more in this chapter.

Faced with a list of six available cations and two anions, a student can start inventing minerals (Figure 5.10). For example, the silicate likes to take on four electrons and the magnesium likes to donate two, so one possible mineral, would be $2(Mg^{++}) + (SiO_4^{---})$, or Mg_2SiO_4. Similarly, it is seen that four sodiums would each give up one electron, so a possible mineral would be $4(Na^{+}) + (SiO_4^{----})$, or Na_4SiO_4. Becoming slightly more sophisticated, one might replace one of the above sodiums with a potassium, getting $3(Na^{+}) + (K^{-}) + (SiO_4^{----})$, or Na_3KSiO_4. Adding to the complexity, one might notice that four aluminums give up a total of 12 electrons, which are acceptable by three silicate radicals, so another possible mineral is $4(Al^{+++}) + 3(SiO_4^{----})$, or $Al_4(SiO_4)_3$. Following similar reasoning, one can construct possible molecules having the carbonate radical. It isn't long before the student begins to realize that starting from a list of six cations and only two anions, a very large list of minerals is pos-

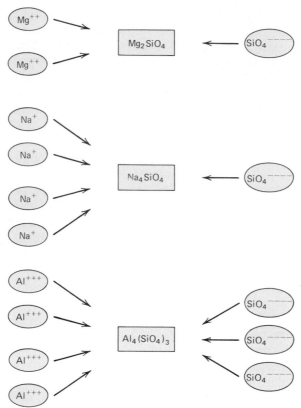

FIGURE 5.10 *Atoms combine in such a way that the total number of electrons donated by the cation equals the total number accepted by the anion.*

sible. For example, $AlCa_2Na_3K_2Fe(SiO_4)_2(CO_3)_3$ is a complicated but not impossible molecule.

The complexity becomes even greater when one considers the addition of water of hydration. The aluminum silicate molecules of the above paragraph, for example, might each choose to adopt two or three waters of hydration. Then we would have $Al_4(SiO_4)_3 \cdot 2H_2O$ or $Al_4(SiO_4)_3 \cdot 3H_2O$. Suppose half of the molecules had two waters of hydration, and half had three. Then they would have 2½ on the average: $Al_4(SiO_4)_3 \cdot 2\frac{1}{2}H_2O$.

It is seen that different minerals can exhibit large variations in chemical composition. For simplicity, all the various silicate minerals are placed in two broad categories according to the amount of iron and magnesium they contain, as these two elements have a large effect on the physical characteristics of the mineral. The minerals containing relatively large amounts of these two elements are called "ferromagnesian" minerals. They tend to be dark in color and slightly denser than "nonferromagnesian" minerals, which contain relatively little iron and magnesium and are light colored. Examples of ferromagnesian minerals are the black, glassy crystals of hornblende or augite, and black, flaky biotite, sometimes

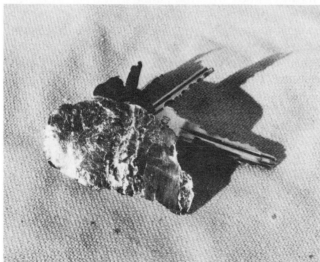

FIGURE 5.11 *(above) The darker rocks of the back row are some common ferromagnesian minerals. They are from left to right: biotite (or "black mica"), hornblende, olivene, pyroxene, and obsidian (or "black glass"). The front row shows some nonferromagnesian minerals. (Notice their lighter colors.) They are from left to right: plagioclase feldspar (whitish-translucent), potassium feldspar (pinkish), quartz (transparent). (below) The piece of biotite is propped up to reflect the sunlight and display shiny layered structure.*

called "black mica," or the greenish-gray olivene. Examples of nonferromagnesian minerals are the clear crystals of quartz, white feldspars, and the golden translucent muscovite, or "mica." (See Figure 5.11.)

D.2 Crystal Structure

A single molecule is typically less than a millionth of a millimeter across, far too small to be visible even with the most powerful optical microscopes. Consequently, the materials of even the very finest of fine sediments must be made of countless molecules grouped together. Sometimes the molecules are stuck together fairly randomly, such as the bodies of people crowding Times Square on New Year's Eve. These solids are called "amorphous." If there is some recurring order in the arrangement of the groups of atoms, such as the order of bricks in a brick wall or the cells in a honeycomb, then the solid is said to be "crystalline."

We can imagine building a three-dimensional crystal by placing identical groups of atoms at each corner of each brick in a neatly stacked pile of bricks. When the bricks are removed, the result is a periodic arrangement of these atoms in three dimensions. If the size of the bricks was very small—perhaps a millionth of a millimeter in length—then one might have a real crystal.

The imaginary periodic framework on which the atomic groups in a crystal are placed is called the "lattice." If the groups are centered on the corners of the imaginary tiny cubes, then the lattice is "cubic." If the lattice resembles a tiny honeycomb, then it is called "hexagonal." If the fundamental building block is an imaginary tiny brick, then it is called "orthorhombic." These are the most common crystalline structures, but there is one other whose basic unit resembles a distorted brick, having sides that are parallelograms rather than rectangles.

Strictly speaking, it is incorrect to refer to individual molecules within a crystal, as it is often impossible to tell which "molecule" a particular atom belongs to. Sodium chloride (table salt), for example, has a cubic structure with one Na-Cl group positioned at each corner of the imaginary tiny cubes. (See Figure 5.12.) Each chlorine atom finds the six nearest sodium atoms all equidistant from it. Likewise, each sodium atom finds the six nearest chlorine atoms all equidistant. Which belongs to which? Strictly speaking, the entire crystal is one gigantic molecule.

The interaction of neighboring groups within a crystal has an influence on its chemical composition. For example, the stacking of the silicate tetrahedra in a crystal affects the ratio of silicon to oxygen atoms, which is not always 1 to 4. As an example, consider quartz, which is made entirely of silicate tetrahedra, with no metallic cations present at all. The tetrahedra are stacked in such a way, however, that each oxygen is shared jointly by two neighboring tetrahedra. (See Figure 5.12b.) This means that each tetrahedron has four half-owned oxygen atoms, or equivalently, two oxygens. So the formula for quartz is SiO_2.

Another example of this is seen in the comparison of the minerals olivine and pyroxene. They are both a mixture of magnesium and iron silicates, olivene being made up of (Mg,Fe) and (SiO_4), and pyroxene being composed of (Mg,Fe) and (SiO_3). Notice that the silicon to oxygen ratio in pyroxene is 1:3 rather than 1:4 as in the olivine. This is because the silicate tetrahedra in pyroxene tend to form long chains, illustrated in Figure

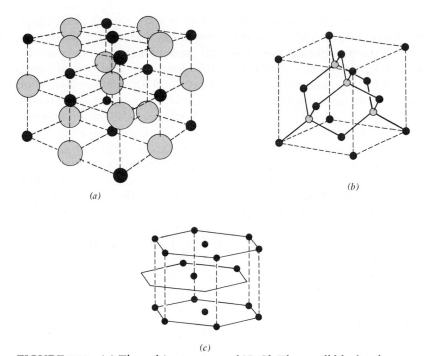

(a)

(b)

(c)

FIGURE 5.12   (a) *The cubic structure of NaCl. The small black spheres represent*
Na^+ and the large spheres represent Cl^-. Notice the six nearest
neighbors of the central chlorine are all sodium, and all equidistant
from it. (b) The cubic structure of quartz (SiO_2). The open spheres
represent silicon and the dark spheres represent oxygen. Notice
that each oxygen is attached to two different silicon atoms. (c) One
possible hexagonal structure.

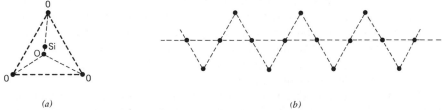

(a)   (b)

FIGURE 5.13   (a) *The silicate tetrahedron. (b) Sketch showing just the bases of the*
tetrahedra in a pyroxene chain. Each tetrahedron shares two oxy-
gens with neighboring tetrahedra.

5.13. In each tetrahedron, two of the four oxygens are shared with neighboring tetrahedra. Two wholly owned oxygens and two half-owned oxygens make only three oxygens altogether.

A material of a certain chemical composition might be able to have
several different cyrstalline structures. If it does, then the less stable forms
will erode more quickly, dissolve more quickly, or if left alone, they will
eventually transform into a more stable form. When window glass is first

poured, for example, it is amorphous. But over the years it crystallizes, becoming quite brittle. This is because the thermal motion of the molecules jiggles them around until they fall into a more stable periodic arrangement. The process is similar to "racking up" pool balls in an amorphous pile in the triangle and then shaking them. They quickly fall into a nice orderly arrangement, filling the triangle.

The particular crystalline structure that will be most stable depends not only on the chemical composition of the material but also on its environment. A certain mineral may crystallize deep within the earth, where high temperatures and pressures may favor one crystalline form. If this mineral is subsequently brought to the earth's surface, it may find that a different crystalline structure is favored there, due to the lower temperatures and pressures. Consequently, it would slowly transform into the new arrangement.

As an example, calcium carbonate is found both as an orthorhombic mineral called "aragonite" and as a hexagonal mineral·called "calcite." Under the prevailing conditions of the earth's surface, the hexagonal form is a little more stable. For this reason aragonite slowly transforms into calcite, and we find that all ancient calcium carbonate deposits are in the calcite form.

E. SUMMARY

The single most productive source of ocean sediments is the weathering of the continents; but these terrigenous sediments dominate only those regions of the ocean bottom near the continental margins. In deeper waters and away from land masses, sediments accumulate much more slowly and are dominated by biogenic sediments or clays.

The fact that we must retrieve sediment samples from rather large distances requires special care, especially with finer sediments. A variety of techniques are employed, including the use of dredges, grab samplers, and various types of corers. Samples from very deep beneath the ocean floor can be obtained through drilling.

We characterize sediments according to texture and mineral content. Sediment grain size is usually classified according to the Wentworth scale. Porosity, permeability, and packing are also characteristics of sediment texture.

The sediment's mineral content is determined by its chemical composition and its crystal structure. Most sediments are formed from various combinations of relatively few chemical elements; just eight chemical elements account for over 99% of all crustal materials. Portions of molecules that readily give up electrons are called "cations." The metals aluminum, iron, calcium, sodium, potassium, and magnesium are common cations in sediment materials. The portions of molecules that readily accept extra electrons are called "anions." The most common anions are the silicate $(S_iO_4^{----})$ and carbonate (CO_3^{--}) radicals. Mineral materials may also include water of hydration in addition to these other atomic groupings.

A mineral's crystal structure is the periodic framework, or "lattice," that describes the arrangement of the atomic groups. Different crystal structures may be stable in different environments.

QUESTIONS FOR CHAPTER 5

1. Can you think of any place in your neighborhood where the underlying crustal rocks are not covered by sediments?

2. Where on the ocean bottom might you find exposed crustal rock not buried beneath sediments?

3. Why are ocean bottom sediments receiving increasing attention as a source for various minerals?

4. What are some of the things we'd like to learn about sediment deposits?

5. What is the largest single source of ocean sediments? Why? What fraction of all sediments come from this source?

6. What are typical rates of sedimentation on the continental shelf? On the deep ocean floor?

7. Most terrigenous sediments on the deep ocean floor are found where?

8. In addition to turbidity currents, how else may terrigenous sediments reach the deep ocean floor?

9. Is the weathering of oceanic crust an important source of deep-sea sediments? Why?

10. What is the largest single source of deep-sea sediments?

11. What is an ooze?

12. Is the relative scarcity of biogenic sediments on the continental shelf evidence of low biological activity there? Explain.

13. Why don't biogenic sediments dominate the deep ocean sediments everywhere beyond the continental margins?

14. What are "hydrogenous" or "authigenic" sediments?

15. Discuss some ways in which seawater may alter the sediments from their original form. Why would such alteration in clay particles be greater than that in coarser-grained sediments?

16. Is most meteoritic material received in the form of macroscopic or of microscopic particles? Why don't they "burn up" as they fall through our atmosphere?

17. Why do we prefer simplicity in the mechanical design of sediment samplers?

18. How do dredges and grab samplers work? For what purpose would each best be used?

19. Describe a gravity corer.

20. Describe how a piston corer works.

21. List the major classes of sediment according to decreasing grain size. What would be a typical grain size for each class?

22. What is the difference between porosity and permeability? Are they often correlated?

23. Briefly discuss how the packing of the sediment may affect its porosity.

24. What are the two parts of the description of the composition of a sediment, in addition to its texture?

25. Atoms stick together to form molecules by sharing what?

26. What is a cation? What is an anion? What charge does each carry, and why?

27. What are the eight most abundant elements in the earth's crust, listed in order of decreasing abundance?

28. What two other materials are quite abundant on the surface? Which of these is not an "element"? Why are they concentrated on the earth's surface?

29. In what form does most of the carbon arrive at the earth's surface?

30. What is a "hydrated molecule"?

31. Suppose a certain element is a good electrical conductor. Why do we conclude, then, that the atoms of this element would make good cations?

32. What are some possible anion radicals? Which two of these are most common?

33. What does the "2½" in "$Al_4(SiO_4)_3 \cdot 2\frac{1}{2}H_2O$" mean? What does each of the subscripts, "4," "4," "3," and "2" in this formula mean?

34. What is the difference between "ferromagnesian" and "nonferromagnesian" minerals? How is this difference in composition reflected in their physical properties?

35. What is the difference between an amorphous solid and a crystalline solid?

36. What is the difference between cubic, orthorhombic, and hexagonal crystal structures?

37. Why is it that one cannot generally speak of individual molecules within a crystal?

38. If quartz is made up entirely of silicate tetrahedra, why is its chemical composition given as SiO_2 rather than SiO_4?

39. Why does glass become brittle with age?

40. Occasionally, mantle material can be found on the earth's surface. But its crystalline form is quite different from what it was when it was deep within the earth. Why?

*41. Invent six possible molecules, other than those mentioned in the text, made from the list of common cations combining with the silicate or the carbonate anion radical.

*42. Why are the continents blanketed by sediments?

*43. Some portions of continents were never covered by water but nonetheless are covered by sediments. Where did those sediments come from?

*44. According to Table 5.2, oxygen is far more abundant in the earth's crust than are carbon and nitrogen, which were of similar abundances in the earth's primordial materials. Can you think of any reason for this apparent lack of carbon and nitrogen relative to oxygen in the crust?

*45. Why do you suppose snowflakes have 6-sided symmetry?

*46. Look at some soil under a magnifying glass and describe what you see.

SUGGESTIONS FOR FURTHER READING:

1. John C. Fine, "Exploring the Ocean Bottom in Manned Submersibles," *Sea Frontiers,* **24,** No. 6 (1978), p. 327.

2. Francis P. Shepard, *Submarine Geology,* 3rd ed., Harper and Row, New York, 1973.

6

SEDIMENT PROCESSES AND DISTRIBUTIONS

A large gravity corer.

In this chapter we examine in more detail the various types of sediments and the processes through which they are formed. Regarding the terrigenous sediments of the continental margins, we wish to find out what they start out as. How are they removed from the continental rocks, and how are they transported to the ocean? What happens to them en route, and what determines where they are finally deposited?

Why do biogenic deposits dominate in deep ocean regions where biological productivity is relatively small, and why do they form in some regions and not others? Where would you expect to find biogenic sediments in shallow coastal waters and why? Under what conditions do materials dissolved in seawater precipitate out and form hydrogenous sediments? Of what value are they? Regarding the very tiny fraction of sediments having cosmic origin, what information does their presence give us, and what additional questions do they pose?

These are just a few of the questions we hope to answer in this chapter. We will concentrate on understanding the fundamental processes involved in sedimentation, and with this understanding it is easy to explain the actual observed distribution.

6
SEDIMENT PROCESSES AND DISTRIBUTIONS

A. TERRIGENOUS SEDIMENTS

A.1   The Skin of the Continents

The rocks of the earth's surface can be put into three classes (Figure 6.1). The first is igneous rocks, which are derived from volcanism and plutonism—hot magma flowing into fissures in the crust and sometimes flowing out onto the surface. The second is sedimentary rocks, which are manufactured through the compaction and lithification of sediments generally laid down in aquatic environments. (They could, in some cases, be laid down by the wind on a desert, for example, but generally an aquatic environment is necessary for cementing the sediments together.) The final class is metamorphic rocks, which have been subjected to sufficient temperatures and/or pressures to have significantly altered them from their original form without actually having melted. Although such environments are generally found only deep within the crust, metamorphic rocks are not rare on the surface, due to upheavals and other contortions endured by the continental regions during their long history.

Having digested Chapter 3 on plate tectonics, the student may remember that continental crust is born in the volcanism associated with subduction zones. Since the continental crust is originally and primarily volcanic or plutonic, the reader might conclude that terrigenous sediments are primarily weathered igneous rock. This, is false, however.

The reason is that the entire continental crust is not exposed to erosion, just the surface. And nearly 70% of the continental surface is   **143**

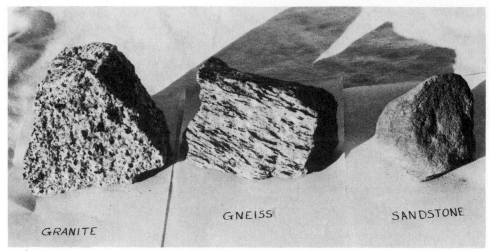

FIGURE 6.1 *One example from each of the three classes of rocks. Granite is igneous, gneiss is metamorphic, and sandstone is sedimentary. You might be able to tell from their light colors that the above three examples are all of predominantly nonferromagnesian minerals. The particular metamorphic rock type shown above is formed from granite, which has been subjected to high temperatures and pressures.*

blanketed by sedimentary rock, a testimony to the ups and downs of tectonic activity that brings most of the continent below water at some time in its history. From the heights of the Rocky Mountains to the depths of the Grand Canyon, we find the sedimentary rock covering the continent surprisingly prevalent and surprisingly thick (Figure 6.2). Thus, the greater part of terrigenous sediments are derived from the erosion of sedimentary rocks rather than igneous or metamorphic ones. We refer to the "sedimentary cycle," as the same minerals seem to go from sediment to sedimentary rock and back to sediment again, along with the ups and downs of the various parts of the continental surface.

But if we trace the sediments in the cycle back far enough, it is likely that they began as igneous rocks of the continental crust. The reasoning behind this conclusion follows from the observation that most of the sedimentary rock formations of the continental surfaces were generated in shallow waters, such as those of inland seas or continental margins. Since shallow water and nearshore sediments are predominantly terrigenous, then the cycle involved predominantly terrigenous sediments. The ultimate source of terrigenous material is in the volcanic birth of the continental crust. For this reason, we include in this chapter on sediments a quick study of the birth and subsequent weathering of igneous rocks.

A.2 Sediments Derived from Igneous Rock

Igneous rocks are born from hot, molten magma. As the magma cools, various minerals begin to crystallize. Minerals whose chemical "building

FIGURE 6.2 *The sediments derived from the weathering of this ancient sedimentary rock formation find their way into rivers and streams which carry them to the ocean, where they will join in the formation of new sedimentary deposits. (Why are the above strata so contorted?)*

blocks" bind most strongly begin to crystallize first. The more weakly bound the mineral is, the cooler the temperature at which it begins to crystallize from the magma. Crystals grow as the appropriate molecular groups attach themselves to the sides of the existing crystals.

The slower the magma cools, the more time the various molecular groups will have to migrate in search of an appropriate crystal of similar chemical composition. For this reason, when rocks cool more slowly the crystals tend to grow larger. A similar effect is noticed in the crystallization of water on snowy days. If the temperature is near the freezing point the snowflakes grow slowly, becoming quite large and heavy. In much colder weather, when the moisture condenses more quickly, the snowflakes tend to be small and powdery.

If the magma has poured out onto the earth's surface, it cools quickly. Such rocks have small grain size and are called "extrusive." When the magma intrudes into tunnels and cracks in the crust without reaching the earth's surface, it generally cools much more slowly in its insulated environment and the resulting "intrusive" rocks display larger grain size. As an example, rhyolite and granite have the same mineral content. However, granite has much larger crystals, indicating it cooled much more slowly. (See Figure 6.3.)

Those igneous rocks having relatively large amounts of ferromagnesian minerals are called "simatic," indicating the larger than normal abundance of iron and magnesium silicates. Igneous rocks that are relatively deficient in ferromagnesian minerals are called "sialic," this term presumably reflecting that when iron and magnesium are lacking there may be a slightly larger than normal abundance of aluminum silicates. The variations in the chemical content of these two types of rocks is gen-

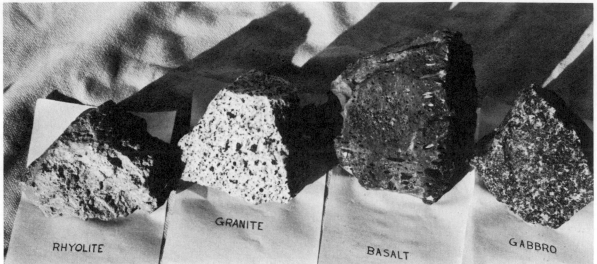

FIGURE 6.3 *The two light-colored rocks are sialic, and the two darker ones are simatic. The granite has a much larger grain size than the rhyolite, indicating it cooled much more slowly. Similarly, the gabbro has a much larger grain size than basalt, indicating it cooled much more slowly.*

erally only a few percentage points from the world average, listed in Table 5.3. The abundance of iron might be 7 or 8% in a simatic rock and down around 1 or 2% in a sialic rock. Similarly, magnesium might amount to 3 or 4% in a simatic rock and be below 1% in a sialic one.

Oceanic crust is characteristically simatic, and continental crustal rock is characteristically sialic. This pattern is in agreement with our understanding of plate tectonics. If continental crustal material is formed through the distillation of oceanic crust in subduction zones, then we might expect the distillation product to be depleted of the denser iron and magnesium silicates.

The two characteristics of igneous rocks we have discussed here are their grain size and their overall mineral content. Both are related to their birth and influence the sediments derived from them. Whether they cooled rapidly or slowly, and whether they were born in the oceanic ridge or in a subduction zone, are important factors in determining these characteristics. Figure 6.4 classifies various types of igneous rocks according to these two characteristics.

Our interest in igneous rocks is that the conditions under which they form affect the sediments that are eventually derived from them. Some of the minerals in the original igneous rock are quite unstable when exposed and weather quite quickly. Feldspars are examples of such easily weathered minerals, which are rapidly broken down into fine clays. Others, such as quartz, weather slowly, and will for a long time retain a grain size characteristic of that of their igneous birth. These coarser grains will tend to concentrate in harsh, high-energy environments, such as beaches. The

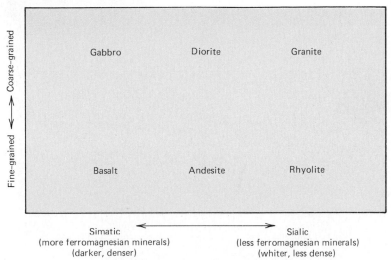

FIGURE 6.4 *Classification of some igneous rocks according to mineral content and grain size.*

finer clays will deposit only in quieter environments, such as in protected embayments or sloughs, or farther out to sea (See Figure 6.5.)

Grain size is not the only basis by which particles become sorted during their travels. Before reaching its eventual resting place, a particle gets picked up and redeposited several times. Minor differences in densities between sediments of similar grain size can cause them to eventually come to rest in different environments. You may have noticed that the color of the beach changes in times of changing wave activity, or from one area of the beach to another. The cause of this is that different minerals and different grain sizes tend to be deposited in different environments.

As a result of this sorting, the mineral content of the sediment in any one locale is usually considerably more homogeneous than that of the original igneous rock from which it was derived. Different parts of that original igneous rock have ended up in different environments. Studying the mineral content of the sediment in one particular area will give us little information about the mineral content of the original rock.

A.3 From Rock to Sediment

Whether the original rock is igneous, metamorphic, or sedimentary, the stages that lead to its eventual fate as sediment on some ocean bottom are fourfold. First there is the weathering of the rock, which causes parts of it to lose its structure and rigidity. Then there is the erosion of these particles from their original position in the rock. Then these particles must be transported across part of the continent and into some sea. Finally, there is the deposition of these particles on the sea bottom.

These sediments resulting from the breakdown of rock are called "clastic" or "detrital" sediments. Most clastic sediments are terrigenous, since rocks exposed on the continental surface experience a much harsher

(a)

FIGURE 6.5 *Sediments of high and low energy environments. (a) The exposed Main coast is a high energy environment, displaying a correspondingly coarse texture in its beach sediments. (b) and (c) A slough, shown here at high and low tides, has very protected and quiet waters, resulting in the deposit of fine-grained muds.*

environment and weather much more quickly than do rocks on the ocean bottom.

If the sediments are to become sedimentary rock, then this "lithification" becomes the fifth process involved in the transformation of the original rock material. It involves first squeezing out a lot of water via the weight of overlying layers of sediment, and then cementing together neighboring grains of sediment. The cementing could come from the precipitation of something previously contained in the seawater, such as calcium carbonate. Alternatively, the cement could come from the minerals of the sediment grains themselves. The water remaining in the sediment is confined by the overlying sediment layers. It slowly dissolves parts of the sediment grains and becomes saturated in these minerals. Once it is near saturation it will have to precipitate the minerals as fast as it dissolves them. The "pressure points," or the points where two neighboring grains press together (Figure 6.6), are weakest due to the stress and are dissolved most readily by the water. So the minerals are dissolved at the pressure points and deposited in the voids, and in this manner the grains become cemented together.

This last step in the sedimentary rock cycle is described first, because understanding it helps us understand the first step—the weathering of surface rocks. All rocks, be they igneous, metamorphic, or sedimentary, weather because they find themselves in a considerably different environment that that in which they were formed. In their original environ-

(b)

(c)

ment they were stable, but in their new environment they are not. Their degree of instability is reflected in the rate at which they weather.

The sedimentary rocks, for example, were born under heavy pressure of overlying layers of sediments, which tended to squeeze out excess water. The water that remained was nearly saturated in many minerals, which served as the precipitated cement bonding the particles together. When exposed at the surface, these rocks experience quite different conditions. Relieved of the pressure, they tend to soak up water rather than expelling it. Furthermore, the water they encounter is usually fresh and not saturated in minerals such as those forming the bonding cement between neighboring particles. As a result, material is dissolved and the rock falls apart.

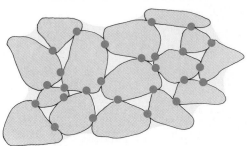

FIGURE 6.6 *Magnified cross-sectional view of sediment grains in some deep sediment layer. The dark spots indicate the points of contact between neighboring sediment grains. Here, the stress due to the weight of the overlying sediment layers is greatest. Consequently, the sediment grains are weakened here, and the entrapped water dissolves them at these pressure points, and deposits the dissolved minerals in the voids.*

Of course, some sedimentary rock weathers more quickly than others. The long, parallel ridges of the Appalachians are due to layers of sedimentary rock tilted on edge. The more quickly weathering layers have been washed away and now form the valleys, leaving the more stubborn layers to form the ridges. Similarly the series of cliffs down the walls of the Grand Canyon (Figure 6.7) testify to layers of sedimentary rocks that weather at different rates. When a more quickly weathering layer washes out from under a tougher layer, the tougher layer breaks off, forming a cliff.

Exposed igneous and metamorphic rocks also find themselves in an environment quite different from that in which they were born. Their birth environment was one of high temperatures and/or pressures, free of the daily and seasonal temperature fluctuations, and free of attack from water, winds, sandstorms, various living organisms, and so on. As we have seen, some minerals are considerably less stable than others when exposed to subaerial weathering and erosion. Some will be broken down into very fine clay particles, or may even be dissolved before reaching the ocean. Others will still maintain a grain size commensurate with that in the original rock.

The next two or three phases in the sedimentation process run together. It is difficult to tell where erosion stops and transportation begins. As generally used, erosion refers to the removal of a particle from its original site, and transportation refers to the carrying of this particle to some new site. Transportation is generally carried out by running water, but sometimes is carried out by gravity alone, wind, animals, ice, and so on.

The erosion–transportation–deposition cycle goes on many times before a particle reaches its final resting place in the ocean sediment. Swifter currents are required to transport larger grains of sediment. This is depicted in the graph of Figure 6.8a, which is a plot of the current speed below which a sediment particle will no longer be transported, as a function of the size of the particle.

FIGURE 6.7 *The terraced cliffs of the Grand Canyon testify to the different rates of weathering of different strata of sedimentary rock.*

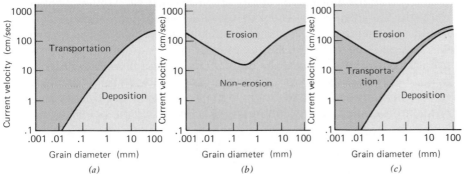

(a)   *(b)*   *(c)*

FIGURE 6.8 *Plot of current velocity vs. sediment size for deposition, transportation, and erosion of sediments. (a) For any point in the grey region, that current velocity is insufficient to keep that sediment type in suspension. (b) For any point in the colored region, the current velocity is sufficiently swift to erode and carry away an existing sediment deposit of that size. (c) For any point in the area marked "transportation", that sediment size will be transported by that current velocity without settling out. However, the current velocity is insufficient to erode away already existing deposits of that grain size. [F. Hjulström, Recent Marine Sediments, Am. Assoc. Pet. Geol., 1939.)*

Just as Figure 6.8*a* is a plot of deposition as a function of current speed, one can make a similar plot of erosion as a function of current speed. One would suspect that it would take larger currents to erode larger particles, but that is only half right. It turns out that sands are the most easily eroded deposits. The smaller particles erode with more difficulty due to their packing. The fine clays, for example, are very tiny, flaky particles with lots of surface area relative to their volume. Due to this large

FIGURE 6.9 *Diagram illustrating sediment deposits of increasing maturities. (a) Very immature—unsorted and very angular. (b) More mature—better sorting and less angular. (c) Very mature—the grains are all about the same size and are quite rounded.*

surface area they cling together well, forming a slick surface quite impermeable to the water flowing above. This is illustrated in Figure 6.8b.

Clays are transported more easily than sands, which are transported more easily than cobbles and boulders. Because of this, the farther sediments are carried from their origin, the more well sorted they tend to become, the clays taking the lead and the cobbles falling behind, for instance. Furthermore, the farther sediment is transported, the more rounded the grains become. Quartz sands, for example, are quite angular near their source, but as they bump and grind over each other during their long trip the corners get knocked off and they become rounded. It has become customary to refer to the "maturity" of a sediment based on these two criteria—the degree of sorting and the degree of roundness. (See Figure 6.9.) The alluvial fans at the base of continental mountain ranges contain a heterogeneous mixture of rock fragments and debris, characteristic of very immature sediment. On the other hand, beach sand is an example of mature sediment, being continually winnowed by the pounding waves.

**A.4 Deposition on the Continental
       Margin**

Once it reaches the ocean, the sediment settles out in the quieter waters there. Those who have been to the beach know that the ocean there is anything but "quiet." However, most of the ocean's water is neither near the surf zone nor near the surface, and is indeed much quieter than the streams and rivers that transported the sediment seaward.

We have seen that different types of sediments tend to settle in different environments. Coarse sediments tend to be concentrated in high-energy environments, such as on beaches or where currents are swift. Finer sediments collect in low-energy environments, such as farther out on the shelf away from swift currents and breaking waves, or in protected embayments. Materials in turbidite deposits of the continental rise tend to be a heterogeneous mixture of whatever collected in the canyon head above, although there is some vertical separation of sediment types within layers, as described previously.

You might think that the shelf sediments should be pretty much the same everywhere, since the continental crustal materials are similar the

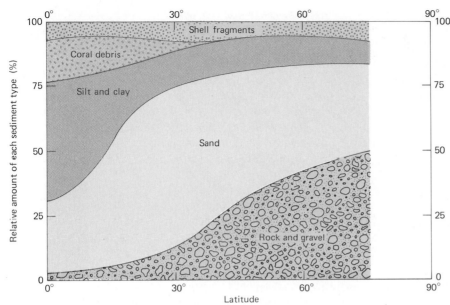

FIGURE 6.10 *Relative abundance of various sediment types on the shelf, as a function of latitude. (M. O. Hayes, Marine Geology, **5**, No. 2 (1967).)*

world over. A study of the sediments of the inner shelf, however, shows considerable variation with latitude (Figure 6.10). Although sand is prevalent everywhere, more rock and gravel are found at higher latitudes, and more silt and clay sized particles are found near the equator.

The coarse sediment at higher latitudes reflects glacial activity. Whereas rivers and streams would preferentially transport the finer-grained sediments, glaciers aren't so discriminating, taking cobbles, boulders, and all.

The preponderance of fine particles in equatorial regions is more perplexing, however. Perhaps it is a consequence of the large rainfall and generally wetter environment. If one lists the rivers of the world that carry the most sediment, almost all of them are near or within the tropics. Maybe the heavy precipitation in the tropics tends to give rise to large river valleys with slow-flowing waters near their large mouths. Perhaps the sand is deposited in their deltas and only the finest sediments stay suspended all the way to the ocean.

As a rule, the continental margins are slowly sinking. This is due in part to the weight of the accumulating sediments, which tend to shove the basement rock down into the mantle. This makes more room for future layers of sediment on top. Also, there is evidence of widespread downwarping of underlying rock due to tectonic forces on continental margins. Sedimentary deposits on the continental shelf are frequently several kilometers deep, every layer of which was originally deposited in shallow water. Similarly, the sediments on the continental rise are frequently thicker than one would suspect from external appearances. The materials of the Appalachian Mountains, for example, were once in a sinking con-

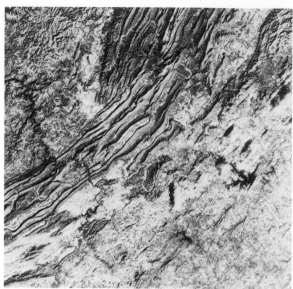

FIGURE 6.11 *Aerial photo of the Appalachian Mountains in Virginia, showing the characteristic long parallel ridges.*

tinental margin. They have been compressed, folded, and brought to the surface by subsequent tectonic activity. The erosion of different exposed layers at different rates has yielded the long parallel ridges characteristic of this range (Figure 6.11).

It is an idiosyncracy of modern times that most of the deposition now is not occurring on the continental shelves but rather in estuaries. These estuaries were formed at the end of the last ice age, as the melting back of the ice sheets caused the ocean water to rise. As it rose it flooded the river valleys that cut through the continents and across the once-exposed continental shelves. In this way, the rising seawater created "drowned river valleys" and bays at the mouths of rivers and streams (Figure 6.12). These estuaries are where much of today's transported sediments first encounter quiet water and settle out. Consequently, the outer shelves are presently accumulating little or no sediment, and frequently display exposed "relic" sediments from previous times.

B. BIOGENIC SEDIMENTS

With the exception of localized reef environments, the rapidly accumulating sediments from terrigenous sources dominate the deposits on the continental margins. Relatively few terrigenous sediments make it out to the deep ocean areas, however, so far away from land we find biogenic sediments in increasing proportion. In this section we will first examine the biogenic sediments of deep ocean waters, and then we will return to shallow water for a look at reefs.

FIGURE 6.12   *A section of the North Carolina coast, displaying a network of drowned river valleys and a long, thin barrier beach.*

B.1   Tests

Biogenic sediments are primarily the skeletal debris of marine organisms (Figure 6.13). Since the overwhelming majority of marine life is microscopic, it is the skeletons, or "tests," of these microscopic plants and animals that make up the bulk of biogenic sediments. The softer portions of the organisms generally decay and are returned to the ocean as nutrients, or as bits of organic nutriments for other animals.

The tests of some microorganisms are made of silica (SiO_2), whereas the tests of others are of calcium carbonate ($CaCO_3$). As sediments, the former are referred to as "siliceous," and the latter are known as "calcareous." The most common siliceous tests come from diatoms (plant) and radiolarians (animal). The most common calcareous tests are those of coccolithophores (plant) and foraminifera (animal). The skeletons of vertebrates are largely calcium phosphate, but are only a very minor component of biogenic sediments.

The formation of a biogenic sedimentary deposit depends on more than the biological productivity of the surface waters. In some cases the tests are dissolved into the water before having a chance to accumulate. In other cases the tests are diluted by heavy sedimentation from other sources. Coastal waters, for example, are often very active biologically. Yet a handful of beach sand often contains very little skeletal debris. The reason is that the sedimentation from terrigenous sources is very heavy and dilutes the biogenic sediments.

The ocean is everywhere undersaturated with silica, and is in many regions undersaturated with calcium carbonate as well. This means that when an organism dies, its skeleton may dissolve before reaching the bot-

(a)

(b)

FIGURE 6.13 *Tests of some microscopic organisms photographed through the Scripps Institution of Oceanography's scanning electron microscope. (a) The silica skeletons of radiolarians, about 0.2 mm in diameter. (b) Tests of radiolarians (coarse mesh), foraminifers, and sponge spicules. The needle-like sponge spicule in the central part of the photo is about 0.4 mm long.*

tom. Even if it does reach the bottom it may still dissolve after it gets there, unless it gets covered over with other sediment first.

As we have seen, a deposit that is more than 30% biogenic is called an ooze. But to be more specific, oceanographers may refer to "siliceous" or "calcareous" oozes, or they may refer to the type of organism that furnished the bulk of the tests, such as a "diatomaceous ooze" or a "foraminiferous ooze." Because of heavy dilution from terrigenous sediments, oozes are not normally found near land in spite of the higher biological productivity there.

It is the balance of the three processes of production, dilution, and destruction that determines whether an ooze will be formed. This turns out to be quite restrictive. The formation of a siliceous ooze, for example, requires that the surface productivity be large enough that the tests reach bottom at a rate faster than they are being dissolved. If they are to become an ooze, then they must also be in a region protected from dilution by calcareous, terrigenous, or other sediments, which might accumulate more rapidly than the tests.

Calcareous oozes are more plentiful not because calcareous skeletons are produced in greater abundance, but rather because there are large regions of the ocean that are saturated with calcium carbonate, so the skeletons won't dissolve. The ability of the water to dissolve calcium carbonate tests depends on the acidity of the water, which for the oceans is determined by the amount of dissolved carbon dioxide contained in the water.

The characteristics of a mass of seawater, such as its temperature and dissolved gases, are determined when it is at the surface and interacting with the atmosphere. Once it sinks below the surface layer, these characteristics don't change very much. The coldest, deepest waters were formed at high latitudes. Since cooler water dissolves CO_2 better (as soft drink and beer drinkers should know), cooler water masses tend to be more acid, and therefore tend to dissolve calcium carbonate better.

In equatorial and temperate climates, the surface waters tend to be relatively warm and saturated with calcium carbonate. As one goes deeper, however, the water tends to get cooler, more acid, and is able to hold more calcium carbonate. Beyond a certain depth, called the "calcium carbonate compensation depth," calcium carbonate particles dissolve as fast as they are deposited, so no calcareous oozes can accumulate. The exact depth beneath which this happens varies from place to place in the oceans, depending on the acidity of the deep water and on the rate of production of calcareous skeletons in the surface waters. It is generally around 4 to 5 km deep in equatorial and temperate latitudes.

Consequently, we could expect to find calcareous oozes where the ocean bottom is located above the compensation depth, providing biological productivity is high and dilution from other sediment sources is small. Much of the Mid-Atlantic Ridge fits these conditions, for example, and so we would expect to find calcareous oozes there. (See Figure 6.14.)

B.2   Reefs

The biogenic oozes cover vast expanses of some areas of the ocean bottom, but seem rather unimportant in the lives of us surface-dwelling creatures. The opposite is true of the biogenic reefs. They are a very small part of

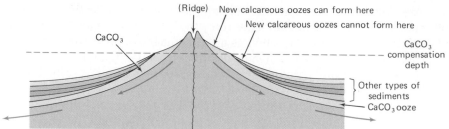

FIGURE 6.14 *Calcareous oozes can form above the calcium carbonate compensation depth, as, for example, on the oceanic ridge, as shown here. The CaCO₃ that exists below the compensation depth was deposited when high on the ridge. Then it spread away from the ridge, being carried along by the crustal rocks beneath it. Having been covered by other sediment layers, it may be protected from dissolution by the sea water at these greater depths.*

FIGURE 6.15 *The unique roots of the mangroves allow them to grow in shallow waters, trapping sediments and forming mangrove reefs.*

ocean bottom sediment, but since they are found only in shallow water every single little living reef must be identified as a potential navigational hazard.

Although coral reefs are most frequently mentioned, there are many noncoral reefs (Figure 6.15) and many nonreef corals. The roots of mangrove trees trap sediment, making small islands near the shore. There are oysters that grow on the cemented skeletons of previous generations. Similarly, there are some worms who secrete calcareous skeletal tubes in which they live and die, forming the basis on which future generations can attach their skeletal tubes and further the growth of the reef. Also, there are algae that have a sticky coating that traps sediment floating by, forming reefs of these stuck-together sediments.

The above noncoral reefs are generally small (usually less than a few meters across) and form in the intertidal zone near shore. Coral reefs (Figure 6.16), however, can be more than a thousand kilometers long and can grow in waters that are otherwise more than a hundred meters deep. The Great Barrier Reef along the eastern coast of Australia is an example of one of these enormous reefs.

The reef-building corals grow best where wave activity is greatest. This means that they tend to grow out into the waves, which makes the slope of the seaward edge of the reef quite steep, 30° inclines being typical. The seaward slope is covered with various corals near the surface. In deeper waters, the corals become more sparse and of more delicate varieties, and below 50 m in depth the slope consists primarily of coralline debris.

Behind the seaward "algal ridge" is a "reef flat." This is a shelf of coralline debris, usually at a depth of a few meters below low tide level. However, the course of the waves is changed in coming over the algal ridge, and so they sometimes pile the debris up in small islands, or "cays." These islands frequently appear and disappear with the falling and rising of the tides. The wave patterns are frequently seasonal, as are many of these islands. In some cases, the debris is piled above high tide level, and the island has a chance of becoming a more permanent feature. The evaporation of ocean spray falling on this debris leaves behind the ocean salts. These salts can cement the debris together, forming "beachrock," a more rigid island material. (See Figure 6.17.)

To the leeward side of the reef flat is frequently a lagoon. In the case of atolls, described in Chapter 4, the lagoon is around 50 m deep. In most reefs, however, 25 m or less is more typical. The lagoon is a much more protected environment than that enjoyed by the coralline algae of the algal ridge, but some species of coral flourish in this environment. They frequently grow upward from the bottom, forming "pinnacle" or "patch" reefs a meter or so across at the top. The sediments covering the reef floor are generally much finer than those of the reef flat above, clays and silts being typical. The cause is a combination of digestion and erosion. The lagoon is an environment enjoyed by parrot fish and many bottom-dwelling organisms that ingest and digest the coral debris for any remaining nutritional value it may have.

The leeward side of the lagoon may be bounded by a leeward reef flat, which is similar to the windward reef flat, but smaller and less vigorous in coralline growth.

B.3 Summary

To summarize the biogenic sediments, we will summarize where we would expect to find them. The coral reefs are a very minor part of biogenic sediments, but they occur in shallow water where people frequently encounter them. Because they enjoy warm, clear, salty water, they would tend to be located in the warm waters of the western tropical regions of each ocean, and away from large land masses, which supply too much fresh water runoff and sediments. The calcareous oozes would be found

FIGURE 6.16 *Various species of coral found on coral reefs.*

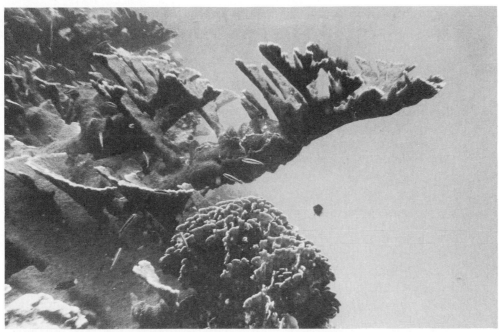

FIGURE 6.16 (continued)

FIGURE 6.17 *Blocks of beach rock have been dislodged by waves on this island on Bikini Atoll. Beach rock is formed by the evaporation of ocean spray, which leaves a residue of dissolved materials, cementing together the grains of sand.*

in deep ocean water, above the compensation depth and away from large land masses (which would otherwise dilute them with their large terrigenous sediment supply). They would not be found at all at higher latitudes where the water is cold at all depths. The siliceous oozes would be found where the biological productivity is high enough to exceed the rate of dissolution of the tests in seawater. Furthermore, they must be protected from dilution from terrigenous or other biogenic sources. This means that they would be found away from the continents and in regions in which the calcareous oozes do not form—namely, either at high latitudes or below the calcium carbonate compensation depth.

C. HYDROGENOUS SEDIMENTS

Aside from the fine clays, which become modified during their long suspension in seawater, hydrogenous sediments form a very minor part of all ocean sediments. They are beginning to receive a great deal of attention, however, due to the economic value of the minerals contained in some of them. Since the overall chemical content of the oceans does not change with time, the rate at which the various minerals precipitate out of solution and enter the sea floor sediments must be the same as the rate at which these minerals enter the seawater solution through rivers, streams, groundwater flow, and dissolution.

Salt is an important hydrogenous sediment that has been mined for millenia. In some areas we find nodular deposits enriched in manganese, iron, copper, and several other metals. These are called "manganese nodules," and often resemble large acorns, each being typically a few centimeters across (Figure 6.18). These are beginning to be mined commercially by dredging them from the ocean bottom. Other areas are accumulating deposits of phosphorite, a phosphorous rich mineral. The question of what causes a certain mineral to be deposited in one area and not another is currently under investigation. It is found, for instance, that fields of manganese nodules are often accompanied by a community of mud-eating worms and show evidence of organic action on nodule surfaces. Perhaps the formation of these nodules is triggered somehow by biological processes.

Some hydrogenous deposits are formed by the evaporation of large amounts of fresh water from the seawater solution, which causes the remaining water to become supersaturated in some minerals, and so they precipitate out. Deposits formed in this manner are called "evaporites." (See Figure 6.19.) The preferred conditions for evaporites to be deposited would be in arid, sunny regions where evaporation is high. Furthermore, if the water mass is isolated from circulation with the ocean or from continental fresh water run-off, then the evaporated water would not be replaced and the evaporites will form more readily. Lagoons and inland seas would fulfill this requirement.

It turns out that lagoons and shallow seas were much more prevalent

FIGURE 6.18 *(above) Manganese nodules on the floor of the South Pacific. (below) Unloading a dredge filled with nodules.*

in times past, as is witnessed by the large amount of sedimentary rocks covering the present continents. Today, evaporites are forming only in very small areas, such as the Gulf of California, the Red Sea, or the Persian Gulf.

As the water evaporates from a confined body of seawater, the order in which minerals are precipitated depends on the order in which they

FIGURE 6.19 *These exposed sedimentary strata are made of the evaporites calcite ("limestone") and dolomite.*

reach their saturation value. The first to be precipitated would be the carbonates, particularly calcite (one crystalline form of $CaCO_3$) and dolomite $(CaMg(CO_3)_2)$. Although the carbonates are far less abundant in seawater than some other minerals, seawater doesn't hold them very well, and so they reach their saturation value first. After the carbonates would come some sulfates, particularly gypsum $(CaSO_4 \cdot 2H_2O)$. When all but about 10% of the water is gone, common rock salt (NaCl) will start precipitating, and last to precipitate would be potassium and magnesium salts. After all the water is gone, it would be found that about ½% of the residue is carbonates, 3% is gypsum, and most of the rest would be rock salt.

The ocean is so nearly saturated with calcium carbonate that relatively minor changes in the water environment can cause it to precipitate. An example of this is the formation of oolite sands (Figure 6.20). When deeper ocean water is brought to the surface, both evaporation and the loss of dissolved CO_2 through the heating and photosynthesis can cause the precipitation of calcium carbonate. On the Bahama Banks, for example, the tidal currents keep bringing in fresh supplies of deeper water to be heated as it flows over the shallow banks, and the calcium carbonate precipitates out onto small shell fragments. These currents also keep the particles rolling, so that new layers may deposit on the old in a spherical concentric pattern resembling a tiny onion bulb. The particles continue to grown in this manner until they reach a size that the tidal currents can no longer budge.

D. COSMOGENOUS SEDIMENTS

The meteoritic sediments of the ocean bottom are extremely minor compared to other sedimentary sources. Nonetheless, they pose two interest-

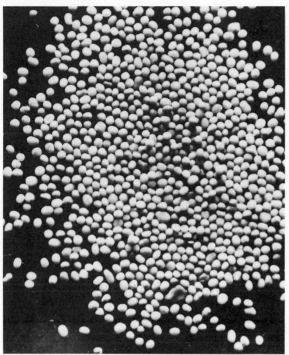

FIGURE 6.20 *Some grains of an oolite sand, under magnification.*

ing problems to scientists—one involving their origin and one involving their fate.

The meteorites that are recovered in sediments are each made up of either rocky minerals, such as those of the earth's mantle, or they are metallic, such as the earth's core (Figure 6.21). A combination of the two is the exception rather than the rule. How did the differentiation of the meteoritic minerals occur out in space? The answer is left to astronomers.

The other problem involves correlating the amount of meteoritic materials that land on the ocean surface with the amount that reaches the bottom and is incorporated into the sediment. Cosmogenous materials have joined the ocean bottom sediment at a rate of several thousand tons per year. Yet it is estimated that the rate at which meteoritic material strikes the earth's surface is several thousand tons per *day*. Could the estimates be that far off? Or is our meteoritic input greater now than it has been in times past? Or does much of the meteoritic material disappear somehow between the time it touches the ocean surface and the time it reaches the bottom?

It is presently thought that the last suggestion may be right. Most of the meteoritic input from outer space is in the form of tiny micrometeorites. Yet it is the smallest micrometeorites that are conspicuously missing from the ocean bottom sediment. They would take the longest to settle out of suspension in the water. This, plus their small size, makes one suspect that they are being dissolved before reaching bottom.

(a)

(b)

(c)

FIGURE 6.21 (a) *A charred stony meteorite.* (b) *A charred metallic meteorite. The holes were probably softer materials that were burned out on the trip through our atmosphere.* (c) *A metallic meteorite that has been cut and etched to reveal the crystallization of the metals inside. This testifies to the great age of this meteorite, as metals crystallize extremely slowly.*

E. DEPOSITION AND DISTRIBUTION OF DEEP OCEAN SEDIMENTS

On the oceanic ridge the sediment cover is thin because the crust there is young and hasn't had much time to collect sediments. Going away from the ridge, we encounter thicker sediments (Figure 6.22), reflecting the increased age of the underlying crustal rocks. Using a typical rate of sedimentation for the deep ocean bottom of 0.2 cm per thousand years, plus a typical rate of spreading for the crust from the ridge of 3 cm per year, we can calculate that the sediment cover should thicken as we go away from the ridge at a rate of roughly 1 m of sediment per 15 km distance. This means that the sediment covering the ridge would be typically 10 or

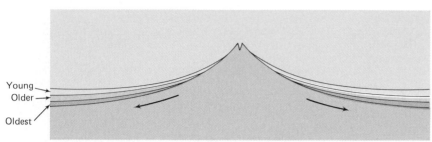

Young
Older
Oldest

FIGURE 6.22   *The sediment cover thickens going away from the ridge. In any sediment column, the most recent sediments will be on top, and the oldest on the bottom. Near the ridge, only the most recent sediments are present. Going away from the ridge, the older basement rock supports increasingly older layers of sediment.*

20 m thick, whereas the deep ocean sediment nearer the continent might be closer to 200 m thick.

On the continental margins, the sediment cover is much heavier than this, being typically several kilometers thick. This reflects both the greater age of the underlying continental crust as well as a higher rate of sedimentation.

The sediments of the ocean bottom tend to be much more homogeneous (horizontally) than those on the continental margins (Figure 6.23). The ocean bottom environment is quite well insulated from the environmental caprice of the surface, and so changes are very slow in distance and time. One sediment type may extend for several hundred kilometers in any direction. During a period of 20,000 years, for example, there would be no appreciable change in any ocean bottom environment, yet much would happen on the continental margins. An ice age could occur, for example, and expose much of the shelf to subaerial and glacial excavation. During a period of 20,000 years countless storm waves would have stirred up the sediments on the continental shelf, the erosion of surface materials from the land would probably have exposed different types of materials as a sediment source, and there would have been countless changes in the weather, which affects sedimentation from continental sources.

Due to the slow rate of sedimentation on the ocean bottom a sample taken of the upper few meters of sediment may encompass several million years of ocean bottom history. It was once thought that one could look as far back into the geological history of the earth as one wished, simply by looking deep enough in the ocean bottom sediments. Unfortunately, we now know that any part of the ocean bottom lasts at most a few hundred million years before diving into a subduction zone somewhere. So clues to the geological history of the earth from earlier periods will have to be found somewhere else.

Although most terrigenous sediments are deposited on the continental margins, the remainder do make a sizable contribution to the deep ocean

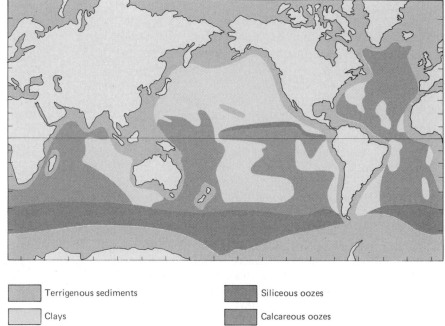

Terrigenous sediments

Clays

Siliceous oozes

Calcareous oozes

FIGURE 6.23 *The distribution of various sediment types on the ocean bottom. Terrigenous sediments dominate the continental margins and the very high latitudes where they may be rafted by ice. Siliceous oozes are dominant where surface waters are cooler. The clays identify the deepest parts of the ocean basins in temperate and tropical latitudes.*

sediments. They are particularly concentrated in the turbidite deposits of the abyssal plains bordering the continental margins.

Terrigenous sediments are also dominant at higher latitudes, above about 65° North or South. There are two reasons for this. One is the default of biogenic sources. Due to ice cover and meager sunlight, the biological productivity is rather low, yielding relatively little biogenic skeletal debris. The second reason is that ice in these waters tends to raft terrigenous sediments out to sea, where they either slowly sink through the ice or are dropped as the ice melts. The ocean bottom contains rocks and gravel among the sediments that could not have made it so far from land without the help of ice.

Rafting occurs at lower latitudes too, but by other mechanisms. Some terrigenous sediments are carried out to sea in the hold-fasts of plants that have been torn loose from the bottom during a storm. Others are carried to sea by animals that used them as gastroliths. But in general, rafted sediments are not an important component of the deep-sea sediments in latitudes lower than 60° or 65° North and South.

Biogenic sediments are quite prominent in intermediate and lower latitudes. Organisms with calcareous skeletons tend to prefer warmer waters. The production of siliceous tests seems to be most vigorous in cooler waters, such as those around 50° to 60° North and South latitudes, or near the equator where cooler deep water comes to the surface off the western margins of continents and is blown westwardly across the ocean surface by equatorial trade winds. Consequently, the ocean bottom in these areas tends to be rich in siliceous oozes, although they are missing in the narrow North Atlantic due to dilution from terrigenous sources.

Elsewhere in the moderate and lower latitudes, calcareous oozes would predominate the deep ocean sediments if they weren't soluble in the colder, more acidic deep waters. They are abundant among the sediments above the carbonate compensation depth, such as along the oceanic ridge system and in other elevated areas, but the deeper parts of the ocean in these latitudes are dominated by clays.

For all the oceans combined, about 48% of the deep ocean sediments are calcareous oozes, 14% are siliceous oozes, and 38% are "abyssal clays," which include turbidite deposits as well as clays of the deep ocean basins. There are some predictable variations in these figures from ocean to ocean. The Atlantic and Indian oceans have a large fraction of their bottom involved in oceanic ridges and other waters shallower than the calcium carbonate compensation depth. Therefore, these two oceans tend to have more than their share of calcareous oozes, whereas the deeper Pacific has more abyssal clays. The Indian and Pacific oceans have more siliceous oozes than does the Atlantic, because they have more cold bottom water. In the case of the Pacific, this is due to more of its bottom being below the calcium carbonate compensation depth, whereas in the Indian Ocean it is due to a large fraction of it being located at cold southern latitudes.

Both the hydrogenous mineral deposits and the biogenic coral reefs have attracted human attention to an extent that far exceeds their actual abundance as an ocean sediment. Cosmogenous sediments are nowhere significant, although they present some interesting questions.

F. SUMMARY

Terrigenous sediments are most voluminous, although they are deposited mostly on continental margins. Since most of the land is covered by sedimentary rock, most terrigenous sediments are derived from weathered sedimentary rock, although most of these minerals have their ultimate origin in the volcanic and plutonic birth of the continents. Of the various minerals in igneous rock, some weather quickly and some do not, which means that from any one igneous crustal rock there are derived a wide variety of sediment types and textures, which will eventually end up in a corresponding variety of sedimentary environments.

Rocks weather because they are in an environment quite different from that of their birth, in which they were stable. As the sediment travels from its source, it becomes increasingly well sorted, and the individual grains become more rounded. At this particular time in history, most terrigenous sediment is depositing in estuaries rather than on the continental shelves.

The major components of deep ocean sediments are biogenic. Although biological activity over the shelves is greater, dilution by terrigenous sediments prevents biological debris from being the major sediment component there. In some regions of the deep ocean bottom, where dissolution of the microscopic tests proceeds as rapidly as they are deposited, inorganic clays predominate. In the abyssal plains and in polar regions, terrigenous sediments predominate. But over a large part of the ocean bottom, especially at high temperate latitudes for siliceous oozes and above the calcium carbonate compensation depth for calcareous oozes, we find these oozes dominate.

Reefs are a minor form of sediment, but quite significant to us as they are navigational hazards. Some hydrogenous sediments are being mined for their economic value. Evaporites are not forming as much today as they have in ancient times when shallow seas were more prevalent. Cosmogenous sediments are not a significant component, although there are some interesting questions regarding them.

QUESTIONS FOR CHAPTER 6

1. What are the three major classes of rocks? How are rocks in each category formed?

2. If the temperatures and pressures required to produce metamorphic rocks are normally found only deep within the crust, why aren't metamorphic rocks rare here on the surface?

3. Since the continental crust is primarily igneous rock, why aren't most terrigenous sediments derived from weathered igneous rocks?

4. Roughly how much of the continental surface is covered with sedimentary rock?

5. What is the "sedimentary cycle"? What is the ultimate source for most of the material in this cycle?

6. As magma cools, do the most strongly bound or the more weakly bound minerals crystallize first?

7. Why is it that the individual mineral crystals grow larger when the magma cools slower?

8. Do extrusive or intrusive igneous rocks have larger grain size? Why?

9. What is the difference between simatic and sialic igneous rocks? Which are characteristic of the oceanic crust? Why? (Refer to the "distillation" process in subduction zones, described in Chapter 3.)

10. Classify gabbro, basalt, rhyolite, and granite according to their mineral content (sialic or simatic) and grain size.

11. Did the coarse quartz beach sands probably ultimately come from extrusive or intrusive igneous rocks? Explain.

12. Briefly explain why it is that different minerals from the very same igneous rock will probably end up in quite different sedimentary environments. Why is it that most beach sands are primarily quartz and lack the feldspars that were present in the original igneous rock?

13. Briefly explain why it is that the mineral content and texture of the sediments in any locale are usually much more homogeneous than were those of the original igneous rock from which they came.

14. What are the four processes involved in making a sedimentary deposit from continental rocks?

15. Why are most clastic sediments terrigenous rather than being derived from the oceanic crust?

16. In the lithification of sediments to form sedimentary rocks, what two processes occur?

17. When the water trapped in the sediment layers becomes saturated in the various sediment materials, then the rate of dissolution of these materials must equal the rate of precipitation. (The water cannot become too supersaturated.) From where are these materials preferentially dissolved, and where are they preferentially deposited?

18. Briefly discuss some of the aspects of the environment for an exposed surface rock that may differ from the environment in which it first formed.

19. Erosion and transportation generally involve water. What other vehicles might accomplish these tasks?

20. How is current speed related to the maximum size of a sediment particle that may be transported?

21. Why are swifter currents required to erode clay deposits than the coarser sand deposits?

22. How is the sorting of a sediment deposit related to the distance from the source of sediment?

23. What two criteria characterize a "mature sediment"? Explain why

24. How is the texture of the sediment on the continental margin related to the energy of the environment? Is this true for turbidite deposits?

25. How does the composition of shelf sediments vary with latitude? Can you explain this pattern?

26. What two processes have caused considerable sinking of continental margins?

27. On what part of the continental margin is most sedimentation occurring today? Why?

28. What are "tests"?

29. What are the two most common materials that tests are made of?

30. Are vertebrate skeletons a significant portion of biogenic sediments?

31. Why is it that biogenic sediments are relatively rare in coastal waters in spite of the great amount of biological activity there?

32. Suppose a region has high biological productivity and small dilution by other sediment types. What else influences whether a biogenic sediment deposit will form here?

33. Calcareous oozes are more abundant than siliceous oozes. Does this mean that organisms with calcareous skeletons are more numerous in the ocean? Why or why not?

34. Does CO_2 make water more or less acid?

35. Why is deeper water usually more acid? Where did all the CO_2 come from?

36. What is the calcium carbonate compensation depth? Why is it related to the temperature of the water?

37. What kind of sedimentary deposit would be favored on an oceanic ridge in temperate or tropical latitudes? Why? (There are two reasons.) Would the same be true at high latitudes? Why? (One reason.)

38. Why is it that reefs seem disproportionately important to us when compared to their actual small abundances among the sediments?

39. Give an example of a reef that is not made of corals.

40. What kind of environment is required by reef corals?

41. Why don't coral reefs grow close to continents?

42. Why is there more coral growth on the western margins of the oceans than on the eastern margins? (Look on a map for current direction. How is this related to water temperature?)

43. Why is the seaward slope of a coral reef so steep?

44. What is a "cay"? How does "beachrock" form?

45. Why are the sediments on the reef floor so fine?

46. Why are reef portions on the leeward side of a lagoon less well developed than those on the windward side?

47. Summarize the conditions required for the formation of calcareous oozes and for the formation of siliceous oozes. Where might these various conditions be met?

48. How do we know that the rates of formation of hydrogenous sediments are commensurate with the rates at which these various minerals are entering into solution from other sources?

49. What are evaporites?

50. Why is it that evaporites formed much more frequently in ancient times than today?

51. As water evaporates from seawater, in what order would the following materials precipitate out? rock salt, dolomite, calcite, gypsum?

52. Which of the above evaporite deposits would be thickest after all the water was gone? Which would be the next? Which would be the bottom layer of the deposit?

53. Discuss how oolite sands are created on the Bahama Banks.

54. Why is it strange that most meteoritic material is either stony or metallic and not a combination of the two? If a planet like the earth exploded, would its fragments display this same differentiation?

55. At what rate does the earth receive meteoritic material? At what rate has meteoritic material collected on the ocean bottom? What is thought to be the explanation for the big discrepancy between these two numbers?

56. Do larger or smaller meteoritic grain sizes seem most depleted from the sediment? How does this support your explanation in the above question?

57. Why does sediment cover characteristically thicken with increasing distance from the rift valley?

58. If the ocean crust is spreading from the ridge at 4 cm per year, how far along the ocean bottom away from the rift valley must we be (roughly) if the sediment is 10 m thick? 100 m thick?

59. Why are the sediments on the deep ocean bottom more homogeneous than those on the shelf, as a rule?

60. If a core of sediment 5 m long is taken from the ocean bottom, roughly how long a period in history does this core represent? What if the same core length were taken on the continental shelf?

61. Why do terrigenous sediments dominate the deep ocean sediments at very high latitudes?

62. How is the temperature of the surface water related to the type of ooze produced?

63. In what regions are siliceous oozes the dominant sediment? In what regions are calcareous oozes the dominant sediment?

64. Discuss the relative abundances of calcareous oozes, siliceous oozes, and abyssal clays in all the oceans combined. Explain deviations from the average figure for individual oceans.

*65. Can you think of any mechanisms for rafting sediments besides those discussed in the text?

*66. Most California beaches are composed nearly entirely of quartz (SiO_2) sands. Yet quartz was not the major component of the original crustal materials from which these sands were derived. Can you suggest any possible reason for this apparent discrepancy?

*67. It is difficult to classify some clay deposits as either terrigenous or hydrogenous. Why might this distinction be more difficult to make for clay deposits than for those of larger grain sizes?

*68. In the chapter we learned that most terrigenous sediments are derived from weathered sedimentary rock, and also that most terrigenous sediments ultimately have igneous origin. How can these two statements both be true?

*69. How do you suppose we might determine the ages of different sediment layers in the Grand Canyon or in the Appalachian Mountains?

*70. Worms burrowing 10 cm into deep ocean sediments mix up about how many years of earth's historical records?

*71. We learned that coarse deposits tend to be found in high-energy environments, and fine-grained deposits tend to be found in low-energy environments. Does that mean that coarse sediments do *not* deposit in low-energy environments? Explain.

*72. Look at the distribution of various sediment types in the Pacific Ocean in Figure 6.24. Discuss why each type is found where it is.

SUGGESTIONS FOR FURTHER READING

1. Brent K. Dugolinsky, "Mystery of Manganese Nodules," *Sea Frontiers,* **25,** No. 6 (1979), p. 364.

2. Alexander P. Lisitzin, *Sedimentation in the World Ocean,* Society of Economic Paleontologists and Mineralogists Special Publication no. 17, 1972.

3. G. Ross Heath, "Deep-Sea Manganese Nodules," *Oceanus,* **21,** No. 1 (1978), p. 60.

4. Turekian, Cochran, and DeMaster, "Bioturbation in Deep-Sea Deposits: Rates and Consequences," *Oceanus,* **21,** No. 1 (1978), p. 34.

7
PROPERTIES OF OCEAN WAVES

Among the most soothing experiences known to human beings are lonely, quiet strolls along the beach, or cruises at sea, where the rhythmic sounds and motions of the waves soothe our nerves, and our appreciation of their timelessness gives us a better perspective for viewing our own small problems, which we often blow so far out of proportion.

We are most familiar with those waves, one or two meters high, which are created by some winds or storms at sea, and end their lives after a few days by crashing on the shore at our feet. Waves come in all sizes, however, from the tiny ripples that appear and disappear at the caprice of the gusty winds, to the large tides that span an entire ocean, and are as predictable as is tomorrow. Different types of waves are generated by different forces, but the motion of the water beneath all waves is similar.

7

PROPERTIES OF OCEAN WAVES

A. TERMINOLOGY

The discussion of waves requires a certain amount of specialized jargon, so we begin this chapter by defining a few terms and symbols used by oceanographers and sailors; some of these are illustrated in Figure 7.1. The top of a wave is called its "crest," the low area between two waves is the "trough." The "wave height" (H) is the vertical distance between the crest of one wave and the lowest point in the neighboring trough. That can be measured, for example, by watching how far up and down the water moves on a piling as waves pass by. The "amplitude" is half the wave height, and indicates the maximum displacement of the water surface from its equilibrium position, called the "still-water level." The "wavelength" (L) is the horizontal distance between a point on one wave and the corresponding point on the neighboring wave, for example, from one crest to the next. The time it takes for two successive crests to pass a stationary point, such as a piling on a pier, is called the "period." There is, of course, a relationship between the wavelength and the period—longer waves having longer periods. We will also be using the term "wave steepness," defined as the ratio of the wave height to wavelength, H/L. In deep water no wave may have a steepness greater than $\frac{1}{7}$ without breaking, although, on occasion, a wave in shallow water may sustain a steepness slightly greater than this. This is roughly equivalent to saying that a wave cannot sustain a crest angle (Figure 7.2) more peaked than 120° without breaking. In describing the roughness of the sea, "significant wave height" (H_3) is often used, which is the average height of the highest $\frac{1}{3}$ of the waves. Sometimes the height of the highest $\frac{1}{10}$ of the waves is given, being indicated by the symbol, "H_{10}."

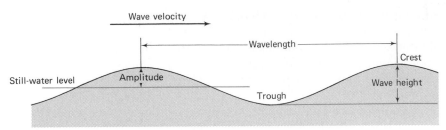

FIGURE 7.1 *Illustration of some of the wave terminology.*

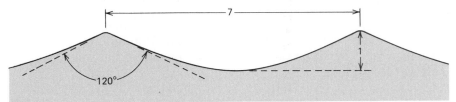

FIGURE 7.2 *A wave cannot sustain a crest angle less than 120° without breaking. Alternatively, a wave may not have a steepness (H/L) greater than ⅐ (except occasionally in very shallow water) without breaking.*

B. ORBITAL MOTION

Now we call on the experience of those readers of this text who have been in the surf. You probably noticed that as a wave passed under you, you were carried up and forward, and after it passed, down and back. Anyone who fishes has probably noticed that this is also the motion of a bobber when a wave passes under it. In fact, any floating object moves in a circular orbit as a wave passes, ending up where it starts from. The same is true of objects floating below the surface. You will notice that a piece of kelp, or a jellyfish, or your own body floating under the surface will also undergo a circular motion as a wave passes above.

The orbits of a few representative water molecules are shown in Figure 7.3. There are two things in particular to be noticed in this figure. First, as the water molecules progress clockwise in their orbits, the wave moves to the right (dotted line). Second, the size of the circular orbits decreases at greater depths. A rule of thumb is that at depth increments of ⅑ the wavelength, the orbit sizes are halved. Therefore, at a depth of ½ the wavelength, the orbits are relatively small. Consequently, if the ocean bottom were deeper than ½ the wavelength, it would have little effect on the wave. Likewise, the wave would have very little effect on the ocean bottom. Hence, when waves are in water deeper than half their wavelength, they are called "deep water waves," as the effects of the ocean bottom can be ignored. Notice that whether or not a wave is a deep water wave depends on its wavelength; in any given area some waves may be deep water waves, and others not.

For real waves, there is actually a slight amount of net forward transport of the water near the surface. As illustrated in Figure 7.4, the orbits do not quite close on each other. There is a corresponding return of the

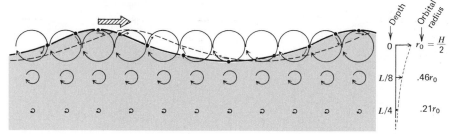

FIGURE 7.3 *Diagram illustrating the progression of the wave as the water mole-cules travel around their respective orbitals. The orbits of just a few out of the myriads of water molecules have been sketched here. The dashed wave position is after each molecule has gone ⅛ of a revolu-tion in its orbital from the solid-line wave position. The diagram on the right illustrates how the radius of the orbitals gets smaller with depth. At the surface, the radius is obviously half the wave height. At a depth of ⅛ of the wavelength (L/8), the radius is slightly less than half that at the surface, and at a depth of ¼ the wavelength, the orbital radius has been reduced by 0.21.*

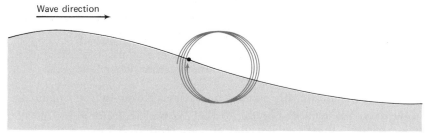

FIGURE 7.4 *In real waves it is found that there is a slight net forward displace-ment of the molecules in each orbit. The effect is so slight, however, that we will ignore it, considering the orbits to be stationary closed circles.*

water at depth. If the waveforms were purely sinusoidal, then the orbits would be closed circles. However, ideal water waves, being driven by the combination of gravity and inertia, are "trochoidal." The two forms are very nearly the same, especially for waves of low amplitudes. In any case, the actual orbits are very nearly complete circles; the net transport is so slight that we will ignore it from now on.

Some readers may have already noticed what is illustrated in Figure 7.5. When the molecules are at the tops of their orbits (i.e., when a wave crest is overhead), the vertical spacing of the molecules is more spread out than when they are at the bottom of their orbits (trough overhead). This would seem to imply that the water under a wave crest is less dense than normal, and the water under a trough more dense than normal, judging from the relative vertical spacings of the molecules indicated in the figure. The fault with this reasoning is that we have considered only the vertical spacings, and we have ignored the horizontal spacings. As illustrated in Figure 7.6, it turns out that when the molecules would be vertically dis-

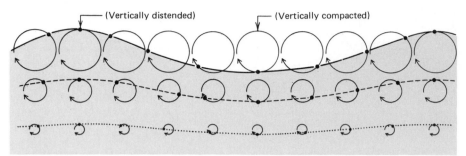

FIGURE 7.5 *The vertical spacings of a few representative molecules are closer together under the troughs and farther apart under the crests.*

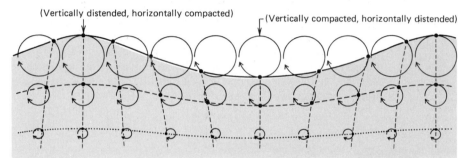

FIGURE 7.6 *But the horizontal spacings are the other way around, being farther apart under the troughs and closer together under the crests.*

tended, they are horizontally compacted, thus filling in the would-be voids. Likewise it is seen in Figure 7.6 that when the molecules are vertically compacted, they are horizontally distended. Hence, the water remains a constant density. (For demonstrative purposes, the cart has been put before the horse in this paragraph. It is actually the uniform density of water that causes the characteristic circular orbits under wave motion rather than vice versa.)

C. WAVE PROPAGATION

C.1 Superposition and Dispersion

The next experience that the reader is asked to recall is that of throwing a stone into a quiet pond, or bathtub, or mud puddle, for example. The splash creates a hodgepodge of short, choppy waves; yet, as you follow the leading waves toward shore, they become long and smooth. If you wait for the short and choppy ones to arrive at shore, you wait in vain. What happens to the short, choppy ones, and where do the long smooth ones come from?

There are two processes contributing to the decreased choppiness of the waves. One process, which is more important in puddles than in the ocean, is that waves in puddles tend to propagate outward in concentric rings. As each ring grows, the wave energy is spread out over an increas-

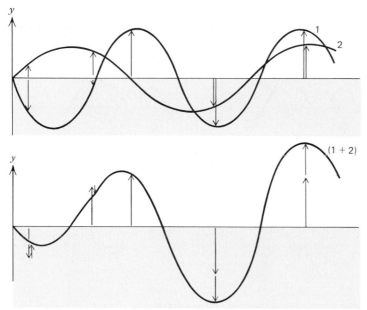

FIGURE 7.7   *In the superposition of waves, the resulting displacement is the sum of the individual displacements.*

ing area. This spreading out of the wave energy causes reduction of the wave height, just as the thickness of a rubber band must diminish as it is stretched. This process is not as important in the oceans as in puddles, because the cause of the disturbance in the oceans is frequently the wind, which generates waves that travel predominantly in one direction (the wind direction) rather than in all directions.

The second process causing the decreased choppiness of waves in puddles is the most important process in oceans. This is that the original disturbance generates many different waves at the same time and the same place. Short waves, long waves, and medium-sized waves all sit atop each other.

As illustrated in Figure 7.7, two waves added together create a combined wave that tends to be higher and more erratic than either of its two component waves. As shown in this figure, when several waves are combined, the resulting wave displacement at any point is simply the sum of the displacements of the individual waves. This is called the principle of "superposition." It applies to all waves, including seismic waves, sound waves, waves on a string, light waves, and so on, as well as ocean waves. The superposition of several low smooth sinusoidal waves to give one large and jagged "sawtooth" wave is illustrated in Figure 7.8.

In deep water, waves of different wavelength travel at different speeds. Those of longer wavelength travel faster. This means that after waves have been generated, they tend to sort themselves out as they leave the site of the disturbance. Longer waves take the lead, and shorter ones fall behind. The original waves were quite choppy because they were a superposition of many different kinds of waves all sitting atop each other.

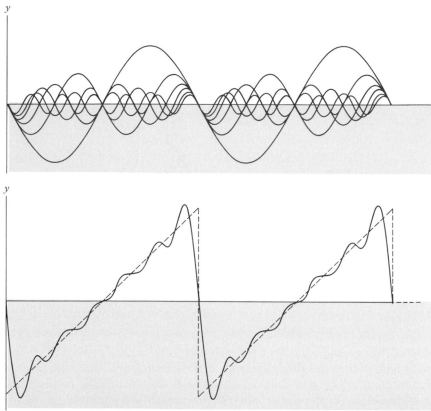

FIGURE 7.8 *The superposition of several low smooth waves can yield one big jagged one. (From Halliday and Resnick, Fundamentals of Physics, 2nd ed., John Wiley, 1981.)*

But as they travel away from the disturbance, they sort themselves out into nice orderly smooth wave trains, with the longer waves in front and the shorter ones behind. This orderly separation of the various components of the original wave group is referred to as "dispersion."

C.2 Wave Speed

The speed with which waves propagate can be calculated from fundamental considerations. We know that gravity is the driving force. Once a disturbance has been created, the water in the crest of one wave is at a higher elevation than water in the neighboring trough. Because of gravity, the water in the crest gets pulled downward, forcing upward displacement of the nearby trough. Because of its inertia, the water in both regions "overshoots" the equilibrium level, just as the inertia of a mass hanging on a spring causes it to continually overshoot the equilibrium level as it bobs up and down. The former crest continues right on down, passing the still-water level and forming a trough before its downward motion is finally stopped. Likewise water in the neighboring trough continues its upward motion past the still-water level, and becomes a new crest before its upward motion is stopped. At any one point, the water surface oscillates up and down past the still-water level, becoming alternately crest and trough.

This physical knowledge can be transformed into a simple mathematical expression, or "equation of motion" for the water. The possible solutions to this equation are constrained by our knowledge that the water on the bottom cannot move up and down through the seafloor, but rather may only move laterally along this boundary. This equation of motion along with this constraint, or "boundary condition," can be solved fairly easily to find a theoretical description of the motion of the water at any depth. When we compare the results of this theoretical, or mathematical, model to actual measurements, we find they agree, so we believe we basically understand how waves propagate.

Among other things, we get from this model a fairly complicated formula[1] that tells us how fast various kinds of waves travel. We find that, in general, the speed with which ocean waves travel depends on two things: the wavelength of the wave, and the depth of the water, measured to the still-water level. In particular, we find that:

1. Waves of longer wavelength travel faster.
2. Waves travel faster in deeper water.

As we have seen, water beneath passing waves undergoes circular orbital motion. The size of these orbitals decreases with depth beneath the wave. The bottom interferes with this orbital motion, but at sufficiently great depths, the orbital motion is so small that the bottom's influence is negligible. We find that in water depths greater than ½ the wavelength, we can ignore the influence of the bottom altogether, and the waves travel essentially as if the bottom were infinitely deep.

Waves in water deeper than ½ their wavelengths are called "deep water waves." For them, the speed at which they travel across the water (i.e., their "wave speed") depends only on their wavelength, with longer waves traveling faster. Notice that whether a wave is a "deep water" wave depends *both* on the depth of the water *and* the wavelength. At the same place, some waves may be deep water waves and others not, depending on their wavelengths. For example, in 1 meter of water, waves of wavelengths shorter than 2 meters would be deep water waves, whereas those with wavelengths longer than 2 meters would not.

As deep water waves head in toward regions where the water is shal-

---

[1]*The exact solution is* $v_w = [(gL/2\pi) \tanh (2\pi d/L)]^{1/2}$, *where* v_w = *wave speed*, L = *wavelength*, d = *water depth, and g is the acceleration of gravity* (g = 9.8 m/s^2). *For deep water waves* (d/L $\geq$ ½) *this becomes* $v_w = [gL/2\pi]^{1/2}$, *and for shallow-water waves* (d/L $\leq$ ⅟₂₀) *this becomes* $v_w = [gd]^{1/2}$. *If metric (SI) units were used, then these are:*

$$\text{For deep water waves:} \quad v_w = 1.25 \sqrt{L}$$
$$\text{For shallow water waves:} \quad v_w = 3.13 \sqrt{d}$$

where **L** *and* **d** *are in meters, and* v_w *in meters per second.*

lower, they begin to "feel" the ocean bottom, which slows them down. Waves of longest wavelength reach the deepest[2] and so they strike bottom first as the water shoals. Since they reach deepest and strike bottom first, they are slowed down more than are waves of shorter wavelengths. Although waves of longest wavelengths travel fastest in deep water, they slow down the most as they approach shallow water. When the water is sufficiently shallow, they have been slowed down to the point where they travel the same speed as all other waves. That is, if the water becomes sufficiently shallow, all waves travel at the same speed, irrespective of their wavelength. The shallower the water, the slower they travel.

It is found that when the water is shallower than about $\frac{1}{20}$ the wavelength, the wave speed depends only on the water depth; the effect of the wavelength is negligible. Such waves are called "shallow water waves." Notice that whether a wave is a "shallow water" wave depends *both* on the water depth *and* on the wavelength. In water that is one meter deep, for instance, waves having wavelengths greater than 20 m would be shallow water waves, but those of wavelengths less than 20 m would not.

In summary, we have seen that wave speed tends to depend both on the wavelength and on the water depth. Longer waves tend to travel faster, and any wave tends to travel slower in shallower water. However, if the water is sufficiently deep, the wave speed is essentially independent of the water depth, and if the water is sufficiently shallow, the wave speed is essentially independent of wavelength. In particular:

1. If the depth is greater than $\frac{1}{2}$ the wavelength, the wave speed depends only on wavelength (not on depth), with longer waves traveling faster.
2. If the depth is less than $\frac{1}{20}$ the wavelength, the wave speed depends only on water depth (not on wavelength), with waves traveling slower in shallower water.

Waves whose wavelengths lie in between these two extremes are called "intermediate water waves." For example, if the water depth was 1 m, waves shorter than 2 m would be deep water waves, those longer than 20 m would be shallow water waves, and those between 2 m and 20 m in wavelength would be intermediate water waves. For them, their wave speed depends on both wavelength and water depth.

In deep ocean water, waves of all wavelengths will be present. Coming in toward shore, the longer waves will touch bottom first, dissipating their energy sooner. Consequently, the waves washing up on our beaches will have relatively smaller amounts of the long wavelength components compared to the waves at sea.

C.3 Capillary Waves

The results presented above are applicable to "gravity waves"—those that move due to the force of gravity and the inertia of the moving water. The

---

[2]*That is, the water motion extends deepest beneath them.*

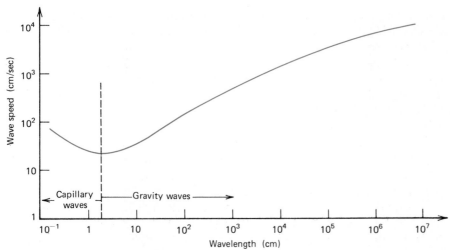

FIGURE 7.9   *Plot of wave speed vs. wavelength for deep water waves. The slowest waves are those of length 1.73 cm, and they travel only 23.1 cm/sec.*

very tiny waves that result from the forces of wind friction and surface tension are called "capillary waves." The wind generates these small waves, and the surface tension provides the restoring force necessary to cause them to propagate, much as the tension in a drum head or a stretched plastic sheet causes waves to travel across these surfaces. You can make these by blowing across the surface of a glass of water. For them, the shorter wavelengths travel faster. The minimum wave speed turns out to be for waves of 1.73 cm wavelength, so this is often used as the division between the two kinds of waves. A plot of wave speed vs. wavelength for both kinds of waves in deep water is seen in Figure 7.9.

D.   WAVE GROUPS

Another thing some students may have noticed about a group of waves going across a pond is that if you watch any particular wave in the group, it won't make it all the way across, even though the group does. For example, watch the front wave in a group, such as in the wake of a ship, as illustrated in Figure 7.10. You can follow it for a while, but then it diminishes and disappears. What was the second wave is now the first, so now you watch it for a while, and then it disappears too, leaving what was once the third wave in the group now in front, and so on. A similar phenomenon happens at the rear of a group of waves, where new waves keep forming. The waves in the group travel faster than the group itself, springing up at the rear, running up to the front, and disappearing, as illustrated in Figure 7.11.

A physical explanation of this phenomenon is as follows. Whatever wave is at the front of the group expends its energy into setting the newly encountered water in motion—that is, giving the water molecules their

FIGURE 7.10 *The wake of a ship provides a way to study the behavior of waves in a group. Evolution in time can be determined in this still photo by noting that points in the wake further from the ship have been under way longer than points near the ship. Following any one wave in the above group, you can see that it is born (point closest to the ship) at the rear of the wave group, and it dies (point farthest from the ship) at the front of the group.*

circular orbits. This saps the energy out of the lead wave, and so it dies. Stopping the orbital motions at the rear of the group yields extra energy, which appears as new wave forming at the rear.[3]

The group behavior can also be explained mathematically, but this is too complicated and unenlightening for this book. Suffice it to say that a wave group can be formed by the superposition of an infinite number of infinitely long wave trains. In deep water, where the different wavelengths travel at different speeds, the group velocity can be predicted to be ½ of the individual wave speeds. In shallow water, where all wavelengths travel the same speed, the group travels with the same speed as the individual waves, with minor modification due to wave attenuation through loss of energy to the bottom. In practice, it is found that the mathematicians are right; the deep water group velocity is ½ of the velocity of the individual waves in the group.

Another demonstration of the thesis that the wave group results from the superposition of many waves of different wavelengths, comes from tracking a wave group from some storm center. Near the storm center, the group is choppy and may be rather short, perhaps taking a day to pass an

[3]*Although this explanation is quite simple, and probably even right, it should be pointed out that it must be modified to explain shallow water behavior.*

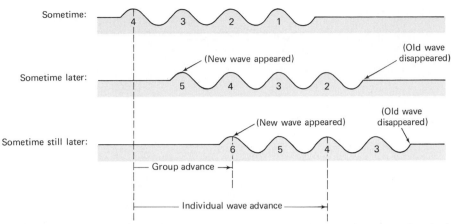

(1st day)

(2nd day)

(3rd day)

(4th day)

FIGURE 7.11 *In deep water, the group moves at half the speed of the individual waves.*

FIGURE 7.12
View from above of a wave group crossing an ocean on four successive days. As the group progresses, the long wavelength components (being faster) take the lead, and the slower, shorter components fall behind.

island. But as it progresses across the ocean, it becomes longer and longer; the faster, longer wavelength components of the wave group take the lead, and the slower, shorter wavelength components fall behind. (See Figure 7.12.) When the storm waves finally strike shore somewhere, oceanographers can tell how far out to sea the storm center was, simply by measuring how far the longer waves have gotten out in front of the shorter ones. In oceanographer's jargon, this is called measuring the "dispersion" of the wave group.

Because of dispersion, waves from a storm center sort themselves out as they cross the ocean into long, parallel, uniform waves we call "swell." Because longest wavelengths travel fastest, we'll notice that the long-period swell arrives at our shore first. On successive days, we'll notice that the period of the swell becomes shorter and shorter.

E.   CHANGING DIRECTIONS

In this section we study how waves going initially in one direction may be deflected toward some other direction. The three different ways this may be accomplished are called reflection, diffraction, and refraction.

E.1   Reflection

What happens to the orbital motion of water molecules when a wave strikes a wall? The molecules up against the wall cannot go in circles, as that would carry them through the wall half the time and away from the wall the other half (Figure 7.13a). The water molecules can only move along the wall. To solve the dilemma, the wave must be reflected from the wall. In Figure 7.13b it is seen that the superposition of a counter clockwise orbital motion, due to the incident wave moving to the left, upon the clockwise orbital motion, due to the reflected wave moving toward the right, yields a linear displacement of the water molecule along a line par-

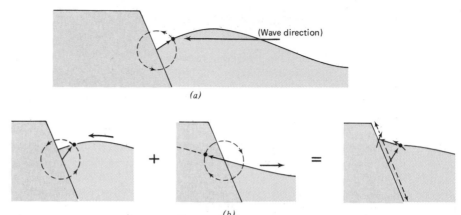

FIGURE 7.13   (a) *When a wave runs into a wall, the normal orbital motion would
require those water molecules at the edge to penetrate the wall half
of the time, and vacate the wall the other half. (b) However, if the
wave is reflected, then the superposition of the orbital motions of
the water molecules for incident and reflected waves yields linear
displacement along the wall.*

allel to the wall. Consequently, the water molecules neither have to penetrate nor vacate the wall, provided that the waves are reflected. So they
are.

We have seen that the orbital motion extends to some depth below
the surface, so to completely reflect a wave, the reflecting surface must
extend appropriately deep. If the reflecting surface only subtends a fraction of this depth, then only part of the wave will be reflected, the rest
being transmitted. There have been cases of fishing boats breaking up in
heavy seas, which resulted from the interference of storm waves with
those being partially reflected off an underwater escarpment.

The sides of swimming pools are straight and smooth, and generally
make good wave reflectors. On the other hand, breakwaters are often very
rough and full of holes and bumps. They hardly conform to the smooth
reflecting surface of Figure 7.13. So, although they do partially reflect
incoming waves back out to sea, they also simply absorb much of the incident wave energy, the rising water from an incident wave being caught
among the nooks and crannies and slowly trickling back down to sea level.

E.2   Diffraction

Every point on a wave crest is elevated above still-water level. Gravity
provides the restoring force that tends to pull these elevated portions of
the water surface back down. Indeed, in the absence of winds and other
wave generating forces, gravity would ensure that the sea surface would
be perfectly smooth, with no elevations or depressions at all.

Imagine we could isolate a single small portion of a single wave crest
above an otherwise smooth water surface. If we could somehow pull the
water surface up at one point and then let go, what would happen? Grav-

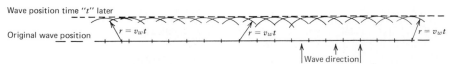

Wave position time "*t*" later

Original wave position

$r = v_w t$ $r = v_w t$ $r = v_w t$

|Wave direction|

FIGURE 7.14   *Huygens' Principle for a long straight wave traveling at velocity* v_w. *Each point on the wave crest acts as a point source for future waves. After time, t, the wavelet from each point source will have traveled distance,* $v_w t$. *Hence, where all these arcs of radius,* $v_w t$, *coincide (i.e., where they constructively interfere) will yield the new position of the wave.*

ity would, of course, pull it back downward immediately, and this disturbance would generate a ring of waves traveling outwards from the point of the disturbance.

This was the line of reasoning followed by a Dutch physicist named Christian Huygens in 1678 when he proposed that every point along any wave crest can be considered as a point source for new wavelets propagating radially outward from that point. This concept has become known as "Huygen's Principle."

As an application of this idea, suppose we know the original position of a wave crest, and wish to know where it will be located at some time (*t*) later. Each point on the wave crest acts as a point source for new wavelets traveling radially outwards, and after time *t* they will each have traveled a total distance (*r*) given by the product of wave velocity (v_w times time (*t*), $r = v_w t$. As illustrated in Figure 7.14, the superposition of all these wavelets from all these point sources is quite complex. At most points there will be the overlapping of many crests and many troughs, and the superposition of the crests and troughs of all these wavelets yields little or no net displacement of the sea surface. But at the foremost edge of all these wavelet arcs, at a distance $r = v_w t$ ahead of the position of the original crest, there will be only crests and no troughs from these wavelets, so they combine to form a large wave crest in this region. Therefore, that is the new position of the wave crest.[4]

As we saw in the above paragraph, application of the ideas in Huygen's Principle implies that a long straight wave crest will continue to propagate in the forward direction as a long straight wave crest. But if a wave is truncated so that only a small portion of it is allowed to enter a region (Figures 7.15 and 7.16), then it will fan out at the ends, propagating radially outwards from the point sources near the ends of the truncated waves. Whenever waves encounter obstacles or modifications whose dimensions are comparable to the wavelength of the waves, the spreading of the waves into the regions behind the obstacles is quite pronounced. It

---

[4]*There is a similar constructive interference from the wavelets at a distance* r = v_wt *behind the original wave. Therefore, you may wonder why the wave doesn't move in both directions from its initial position. It would, if the original wave crest were stationary. But if the original wave disturbance were moving, then the superposition of the various wavelets generated when the original wave was in various positions would constructively interfere only in the forward direction, and not in the backward direction.*

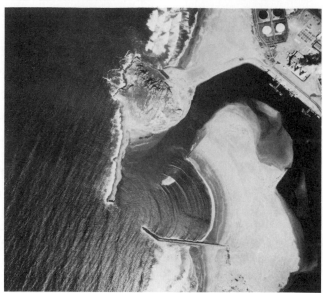

FIGURE 7.15   *Aerial photo of waves coming through the breakwater at the mouth of Morro Bay, California.*

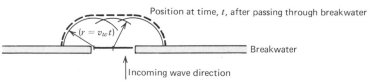

FIGURE 7.16   *Waves coming through a gap in a breakwater fan out in compliance with Huygens' Principle.*

is referred to as "diffraction." It happens to all types of waves, such as sound, light, radio waves, and ocean waves. Among other things, it means that boats anchored behind breakwaters are not completely protected.

E.3   Refraction

Besides diffraction, waves can be bent by another process, called "refraction," which is caused by variation in the water·depth. When at the beach, we see the waves coming in nearly directly towards shore, with wave crests parallel to each other and nearly parallel to the shoreline. Out to sea the waves are often quite chaotic, due to the superposition of many waves going many directions. But they change character, coming ashore very orderly, and almost parallel. Why?

The reason for this behavior is that waves tend to go slower in shallower water. As is true for all types of waves in all media, ocean waves bend towards where they go slowest. When an ocean wave comes in toward shore from any direction, the part of the wave that reaches shallower water first will go slower than that part still out in deep water. This means that the deeper water part will catch up with the shallower water

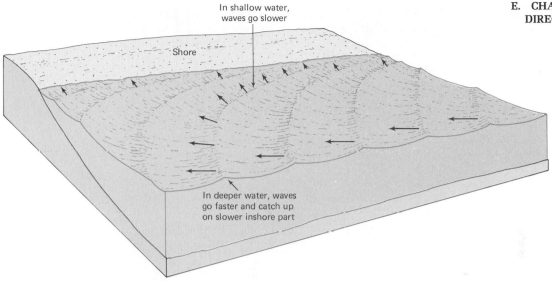

In shallow water,
waves go slower

Shore

In deeper water, waves
go faster and catch up
on slower inshore part

FIGURE 7.17   *Refraction of waves coming in toward shore.*

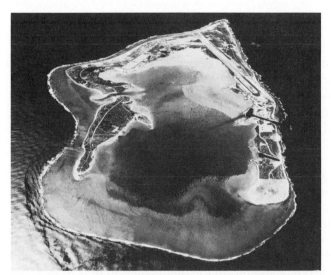

FIGURE 7.18   *Aerial photo showing the refraction of waves towards shallower*
waters at Wake Island, an atoll in the Pacific.

part until the whole wave is almost parallel to shore. This process is illustrated schematically in Figure 7.17. In the aerial photograph of Figure 7.18,
one can see this refraction process occurring. If you are marooned on a
"desert island," you can walk completely around the island and always
find waves coming in toward shore, regardless of what direction the waves
are moving farther out to sea. (See Figure 7.19.) However, the wave height
of the waves refracted completely around and coming in from the leeward
side will be less than that on the windward side. Often over coral reefs, a

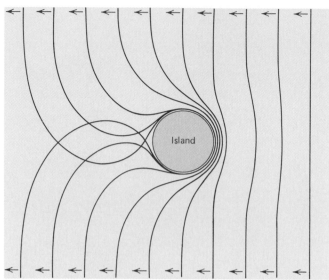

FIGURE 7.19 *Refraction of waves around a small island. The slower speeds in shallow water cause them to bend, and observers on shore anywhere would see the waves coming in toward them, even if they were on the leeward side of the island.*

large fraction of the incident wave is reflected back out to sea off the steep outer edge; the remaining part of the wave is refracted while passing over the reef. Skilled sailors look out for this pattern as a warning of an approaching reef (Figure 7.18).

The refraction of waves coming into shore is one of the processes responsible for straightening shorelines. Old salts have always recognized that on jagged shorelines the "points draw the waves." The coves between the points are generally rather quiescent. The reason is that incoming waves meet shallow water off the points first; consequently, they bend in toward the points. In Figure 7.20 the waves out to sea are divided into equal segments by the dashed lines, and so the dashed lines trace equal amounts of wave energy as they come into shore. It is seen that because the waves reaching shallow water are refracted in toward the points, the energy concentration is much greater on the points than in the coves. Consequently, on a headland one often finds high waves crashing into the point, with lots of noise, frothing, and spectacular jets of water shooting up into the air, even though the coves are quiescent and the sea itself may not be very rough. Lighthouses are often built out on points, and so the best tales of rough seas can often be told by lighthouse keepers.

An interesting case of wave refraction occurred off Long Beach, California, in April, 1930. Large breakers of a period of 20 to 30 sec (rather long-period swell) damaged part of the tip of the Long Beach breakwater, removing stones weighing from 4 to 20 tons. What was strange about this incident was that it was very calm at sea, according to reports from gambling ships anchored just past the three-mile limit. The nearby San Pedro

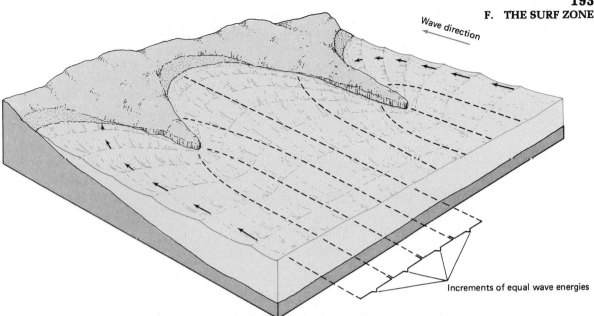

Wave direction

Increments of equal wave energies

FIGURE 7.20 *Incoming waves reach shallow water off the points first; therefore, they bend toward the points as they go. Dashed lines are along the direction of wave propagation (everywhere perpendicular to the crests), and have divided the incoming waves into equal energy increments. It is seen that the energy of the incoming waves is concentrated on the headlands, and the coves are relatively quiescent.*

breakwater experienced no breakers at all. Furthermore, lifeguards on the beaches reported no unusual wave activity. The breakwater had withstood many large storms before, but now on a calm day it incurred damage. This was puzzling.

The explanation was found in a hump in the underwater topography some 16 km to the south of the harbor. (See Figure 7.21.) Long-period swell, originating from some storm center in the Southern Hemisphere, was bent when passing over this hump, and much as a lens focuses light, this hump focused the incoming swell on the tip of the breakwater. Consequently, having height of only 60 cm at sea and a wavelength of longer than 1 km (hardly noticeable), this swell acquired waveheight of some 4 m at the tip of the breakwater, and did the damage.

F. THE SURF ZONE

F.1 Breakers

We now understand how waves line up and come in parallel to shore. But why do they break? Recalling that the wave steepness (H/L) must be greater than about ⅐ for a wave to become unstable and break, we conclude that as a wave comes in to shore, either its height must increase or

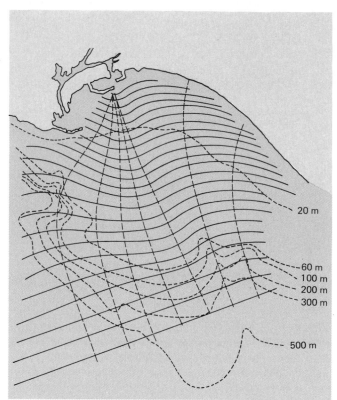

20 m

60 m
100 m
200 m
300 m

500 m

FIGURE 7.21 *Dashed lines are bottom contours, which caused the bending of the incoming swell. Solid lines are the wave crests, and the long-dashed lines indicate the direction of wave propagation (perpendicular to the wave crests). They also divide the waves into equal energy increments, indicating how the energy was concentrated on the tip of the breakwater. (From Waves and Beaches by Willard Bascom. Copyright © 1964 by Educational Services Inc. Used by permission of Doubleday and Company, Inc.)*

its length must decrease. Actually, both happen. The wavelength decreases because as waves come into shallower and shallower water, any one wave is usually in shallower water than the one behind it. Therefore it travels slower, and the one behind keeps catching up on it. With the wavelength decreasing, the water and the energy in a wave must be concentrated in a shorter and shorter region. Hence, it gets higher, and the crest angle nears the critical value of 120°. The proximity of the bottom causes once circular orbits to become elliptical, further adding to the wave's instability. In deep water, a molecule in the crest atop its orbit can count on support from the water immediately in front of it to complete its cycle. As the water shoals, however, there is not enough water immediately in front of the crest to be able to come all the way up to meet it. The crest appears to get ahead of its support, and it breaks (Figure 7.22).

The wave will break when the velocity of the water molecules in the

FIGURE 7.22   *Waves break when the crest gets ahead of its support.*

crest is greater than that of the wave beneath, such that the crest gets ahead of the wave. Theoretically, this should happen when the water depth, d, measured from the still-water level, is equal to the wave height, H. However, due to imperfections in real waves, it is found that they generally break a little before reaching water this shallow. Experimentalists find that the average wave breaks in water of depth equal to 1.3 times its wave height.

Waves will often break more than once before reaching shore, perhaps once over a bar offshore and again on the beach. When beaches are subjected to heavy wave activity, such as during winter storms, the waves will drag material from the beach and deposit it in longshore bars. This is the beach's defense mechanism, as the bar then makes the waves break prematurely, and thus partially protects the beach from the onslaught. One can sometimes look for the white water to detect the position of bars. (See Figure 7.23.)

The types of breakers are commonly placed in three classes, although in real surf some mixture of these are present. (See Figure 7.24.) In "spilling breakers," water from the crest tumbles down the face of the wave. These occur most frequently where the bottom slope is very gentle. Hence, they rise slowly, last a long time, and are most commonly used for surfing. The "plunging breakers" are more spectacular as the top curls over and falls in front of the wave. These often arise from the breaking of long-period swell, and usually where the bottom has slightly more slope than where spilling breakers are found. Finally, "surging breakers" are most common where long low waves meet a steep beach. The waves hardly having a chance to break at all before striking the beach.

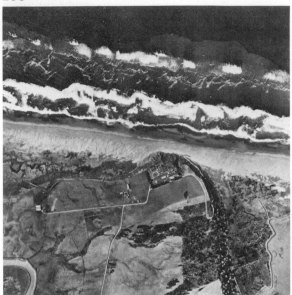 

FIGURE 7.23 *Waves may break several times before reaching shore. Left is an aerial photo taken at Table Bluff, California, showing waves breaking over longshore bars before reaching shore. The right photo shows waves breaking over a fringing reef at Guam. The island on the right is about one mile long and 0.1 mile wide.*

After a wave breaks, the foaming frothy remnant continues up the beach. This is called a "wave of translation," because the water actually moves up the beach with the wave. This is in contrast with its earlier form, where the water molecules underwent orbital motions and there was no net water transport with the wave. The velocity of a wave of translation is given by the same formula as other shallow-water waves, but the depth, d, is measured to the top of the wave rather than to the still-water level. Hence, one wave can ride on the back of another and overtake it, and big waves of translation often overtake little ones.

F.2 Surf Beat

Another wave phenomenon frequently observed by those at the beach is called "surf beat." As the waves come in, several in a row will be very high and wash high up on the beach. Then soon the waves are small again. Experienced surfers just lay on their boards and wait out these periods of small waves, because soon the high waves will be coming in again. The periodic rise and fall in wave height is due to interference between two or more different wave trains. Two wave trains going in the same direction may line up so that the crests of one lie in the troughs of another. Figure 7.25a shows that the superposition of these two yield no wave at all, and is called "destructive interference." However, when the two waves "constructively interfere"—that is, the crests of one lie on the crests of the other—the superposition of the two yield an even higher wave train (Fig-

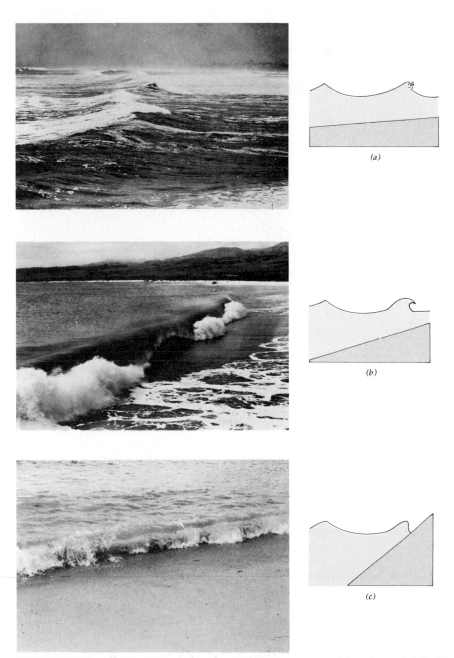

FIGURE 7.24 *Illustration of the three different types of breakers. (a) Spilling breakers are generally found where there is a gently sloping bottom. (b) Plunging breakers are found where the bottom slope is moderate. (c) Surging breakers are found where the bottom slope is so steep that the wave doesn't break until it is right at the shoreline.*

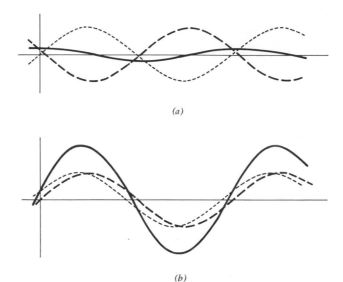

(a)

(b)

FIGURE 7.25 *Interference. The superposition of the two waves indicated by the dashed lines yields the wave indicated by the solid line. (a) Destructive interference. (b) Constructive interference.*

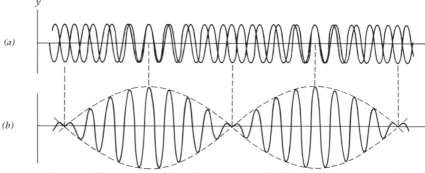

FIGURE 7.26 *The superposition of two waves of slightly different wavelengths* (a), *yields alternate regions of constructive and destructive interference, or "surf beat"* (b). *(From Halliday and Resnick,* Fundamentals of Physics, *2nd ed., John Wiley, 1981.)*

ure 7.25b). When two wave trains of slightly different wavelength are superimposed, as in Figure 7.26, there will be regions where they destructively interfere and regions where they constructively interfere. Thus, when more than one wave train (perhaps from different storm centers at sea) of differing wavelengths come to shore at the same time, the result will be alternate regions of constructive and destructive interference. This accounts for surf beat.

A particularly frustrating experience for the novice surfer is that the surf usually seems to be bigger elsewhere. Sitting on their boards, the surfers will notice that the surf is much bigger a few hundred yards just down the beach. So they lay down along their board and paddle furiously to get

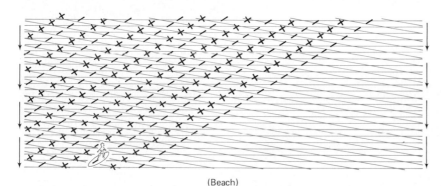

(Beach)

FIGURE 7.27   *Two wave trains coming in at slightly different angles. The pluses*
(+) indicate regions of constructive interference (crest on crest);
consequently, big surf. The minuses (−) indicate the regions of
destructive interference (crest on trough); consequently, small surf.
The surfer above is in a region of small surf, and will see big surf
just down the beach to either side. However, it is seen that she or
he won't have to move over to get it—it will come over to her or
him as the whole pattern moves shoreward with the waves.

to there, but when there, they find it quiet; the big surf has moved back to
where they were. This is more than the well known "grass is greener"
illusion of the mind's eye. Experienced surfers know that the surf beat
does often move along the beach. If the surf is bigger elsewhere, they just
wait a little while and it will come to them; they don't have to go to it.

The reason for this is that the two or more sets of waves, whose interference is creating the surf beat, usually will not be coming from the same
direction. Even in shallower water, where they are coming almost straight
in, there will be slight differences in their direction. Hence, although there
is constructive interference just down the beach from you, there may be
destructive interference where you are. As illustrated in Fig. 7.27, if you
just wait there, as the waves come in the big surf will soon move over to
where you are. The longshore component of surf beat belongs to a special
class of waves called "edge waves," which move parallel to the beach
rather than coming in perpendicular to it. Edge waves are generally
caused by some kind of interference, such as between two or more different wave trains, as discussed above, or between incoming waves and
those reflected back out off a steep beach.

Often when several large waves break in succession upon the beach,
the waves of translation carry the water up the beach faster than it can
wash back down. This stacking of the water on the beach may cause it to

FIGURE 7.28 *A small rip current can be spotted in this cove by the line of floating debris that is being carried seaward from the beach. Notice that the floating debris also causes a visible change in the nearby small surface waves.*

break back through the surf in certain places in strong down-beach rivers called "rip currents" (Figure 7.28). Swimmers caught in these currents should realize that they are both strong and narrow. Thus, swimming against them is futile, but swimming parallel to the beach just a few strokes should easily take them out of the rip, and then they can easily return to the beach.

G. SUMMARY

Water molecules undergo cyclical motion as waves pass, going up and forward as the crest passes over, and down and back beneath the trough. The size of the orbits decreases with depth beneath the wave.

When several different waves are superimposed, the net displacement of the water surface is just the sum of the displacements from the individual waves. The speed with which a wave travels depends on the length of the wave and the depth of the water. The longer the wave and the deeper the water, the faster it travels. When water is sufficiently deep, the wave speed depends only on the wavelength. When it is sufficiently shallow, it depends only on water depth. Capillary waves are small waves generated by the wind and driven by surface tension. For them, shorter waves travel faster.

After rough waves have been generated, the various components tend to begin sorting themselves out into ordered swell. Traveling faster, the longer wavelengths tend to take the lead and the shorter waves tend to fall behind in a process known as "dispersion." The individual waves in a group travel faster than the group itself, springing up at the rear and disappearing at the front. In deep water, the individual waves travel twice as fast as does the group itself.

There are three ways that waves can change their direction of travel. These are reflection, diffraction, and refraction. Reflection is the result of the water's inability to penetrate a solid barrier. Diffraction is explained by Huygen's Principle. Refraction is caused by slower wave speeds in shallower water, and is the reason waves come directly into shore, regardless of their direction of travel farther out to sea.

Waves break in shallow water when the crest gets out ahead of its support. Breaking waves are classified as spilling breakers, plunging breakers, or surging breakers. The frothy remnant of a broken wave is a wave of translation. Surfbeat is the result of the superposition of two or more wave trains of different wavelengths.

QUESTIONS FOR CHAPTER 7

1. What is the wave crest, trough, and wavelength?

2. How are the wave height and amplitude related?

3. What is the wave period?

4. What is wave steepness? What is the maximum steepness that a deep water wave can have?

5. What is "significant wave height"? What does the symbol "H_{10}" mean?

6. What is the motion of individual water molecules as a wave passes? How does this change with distance below the surface?

7. If the wave height is 2 m, what is the diameter of the orbits of the water molecules at the surface? At a depth of ⅑ the wavelength? At a depth of ⅔ the wavelength? ⅚ the wavelength?

8. Why is it that some waves in an area may be deep water waves while other waves in the very same area are not?

9. If wave forms were purely sinusoidal, would there be any net water transport with the waves?

10. Suppose you were to label certain water molecules and notice their positions. Describe what would happen to their vertical spacing and their horizontal spacing as a wave crest passes over. What happens when a trough passes over?

11. Suppose something generates a wave at a point, and this wave travels outward in all directions, forming a ring about the point source. How is it that conservation of energy demands the wave amplitude must decrease as the wave travels outward?

12. If waves of all wavelengths traveled the same speed, what would happen when one makes a splash in a quiet pond? (Remember conservation of energy arguments.) How is this different from what really does happen?

13. How fast would a deep water wave of 4 m wavelength travel? How about one of 9 m wavelength? How deep would the water have to be for each of these to be a "deep water wave"?

14. Does the speed of a wave depend on its height?

15. Why do you suppose it is that longer waves do *not* travel faster than shorter ones in shallow water? If they travel faster in deep water, shouldn't they travel faster in shallow water too?

16. If a wave is 20 m long, about how shallow must the water be before this can be considered a shallow water wave? How fast would it travel in water of this depth?

17. Suppose you were in water up to your waist. What range of wavelengths would be intermediate water waves in this water?

18. Why is it that the waves striking the beach are particularly depleted in the long wavelength components (compared to when they were at sea)?

19. What makes gravity waves travel? How does it do this? If there were no restoring force (e.g., gravity were shut off), what would happen when you make a wave?

20. What is the restoring force that drives capillary waves?

21. Do capillary waves of longer wavelengths travel faster or slower?

22. What happens to the individual waves in a wave group? What is the physical explanation for this?

23. How is the group velocity related to the wave velocity?

24. Explain the dispersion of a wave group as it travels across the ocean from a storm center.

25. How is it that one can tell the distance to the storm center by the dispersion of the waves?

26. What are the three different ways that the direction of travel of a wave may be changed?

27. Why is it that the water molecules near a wall cannot traverse circular orbits?

28. With the help of some sketches, indicate how the superposition of clockwise and counterclockwise orbital motions may yield linear displacements of the water molecules.

29. Must a wall extend all the way to the surface to reflect a wave? To completely reflect a wave?

30. Why are most breakwaters not particularly good wave reflectors?

31. What is Huygen's Principle, and what does it mean?

32. How does Huygen's Principle explain why waves entering through a gap in a breakwater will bend in behind the breakwater?

33. Why is it that waves always come almost straight toward shore, regardless of what direction they were headed when they were at sea?

34. At sea, many different wave trains headed in many different directions yield chaotic seas. How is this chaos removed as these waves come to shore?

35. Why do the points draw the waves? Use sketches to explain your answer.

36. If the heavy wave erosion occurs on the points, why does the erosional debris collect in pocket beaches in the coves?

37. On the day in 1930 when the Long Beach breakwater incurred the damage, what was the condition of the sea elsewhere in the area? What was the explanation for how all the wave energy could get concentrated at this one point?

38. As waves come into shore, why does the wavelength decrease? Why does the wave height increase?

39. Why does a wave break in shallow water?

40. How can one judge the depth of the water knowing the height of a wave when it breaks?

41. How do longshore bars help protect a beach from heavy wave activity?

42. What are the three general classes of breakers? How are they related to the slope of the bottom?

43. What is the difference between a "wave of translation" and an ordinary gravity wave?

44. If all waves travel the same speed in shallow water, why can one wave of translation overtake another?

45. What is constructive interference? Destructive interference?

46. What is surf beat? What causes it? Use a sketch to help you explain it.

47. Why is that the surf can be small at one point on the beach but large several hundred meters in either direction?

48. What are edge waves?

49. What is a rip current? What causes it? If you're a swimmer caught in one, how would it be best to get back to the beach?

*50. In a fashion similar to that of Figure 7.7, sketch two or three different sinusoidal waves, and then sketch the wave form that would result from the superposition of these.

*51. Can you think of a way of judging the height of a breaking wave at the beach without getting wet? (*Hint:* Perhaps you could find the elevation of the crest above still-water level, using the altitude of your eyeball when the crest appears to line up with the horizon. What is still-water level?)

*52. From an airplane, you can sometimes see waves reflected off of sheer rock cliffs but not from nearby breakwaters. Why do you suppose this is?

*53. What causes surface tension?

*54. How could boats in anchorage be protected by contouring the harbor floor rather than by building breakwaters? Why do you suppose this isn't done?

SUGGESTIONS FOR FURTHER READING

1. Willard Bascom, *Waves and Beaches*, Anchor/Doubleday, New York, 1964.

2. Gerhard Neumann and Willard J. Pierson, Jr., *Principles of Physical Oceanography*, Prentice-Hall, Englewood Cliffs, New Jersey, 1966.

3. John P. Robinson, Jr., "Superwaves of Southeast Africa," *Sea Frontiers*, **22**, No. 2 (1976), p. 106.

8
WAVES AND TIDES

We continue our study of ocean waves by looking a little more closely at the causes and characteristics of the various kinds of waves.

A. WIND DRIVEN WAVES

Most of the waves we observe are wind driven. Exactly how the wind interacts with the water is still being heavily researched, but some general features of the interaction are known. When you blow across a smooth water surface, such as in a bathtub or a glass of water, you notice that little "capillary waves" spring up quickly. These little capillary waves probably provide the primary source for energy transfer from the wind to the waves. Experienced sailors know that the presence of these little waves means the sea will be getting rougher. The small waves rapidly build up and break, giving some of their energy to the generation of larger waves. These larger waves build up and break, transferring some of their energy to still larger waves, and so on.

The longer the time that the wind blows, the larger will be the resulting waves. The buildup continues until their wavelength is so long that they are traveling nearly as fast as the wind. A diagram of how the wind helps build up the orbital motions of the water molecules is presented in Figure 8.1. Skimming the crests, it helps push the water forward in the top part of the orbital, and the eddies in the troughs help reinforce the backward motion of the bottom half of the orbital.

In addition to the duration of the wind, the size of the developing waves will also increase with the wind velocity and the distance over which the wind blows, which is called the "fetch." Figure 8.2 shows schematically how the smaller waves give way to larger ones as the sea develops. Table 8.1 lists the duration of the wind and the length of the fetch necessary for fully developed seas to form at different wind speeds. Table 8.2 gives the "Beaufort scale," which is the standard used for describing the state of the standard used for describing the state of the sea in ship logs

B. SEISMIC SEA WAVES

Another class of waves that are much less frequent, but carry large amounts of energy, are the seismic sea waves, or "tsunamis." At one time they were popularly called "tidal waves," but as they had nothing to do with the tides oceanographers looked for another word, coming up with the popular Japanese name, tsunami. Then it was found that "tsunami" means "tidal wave" in Japanese, so now there is a movement toward the use of "seismic sea waves" instead, as the energy of the wave originates in seismic activity in the earth's crust. Examples would be underwater landslides, or underwather slippage of the crust along a **207**

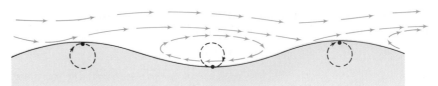

FIGURE 8.1 *The wind builds up waves by pushing the surface water in the crest farther and faster forward in its orbital motion. Also, eddies in the troughs reinforce the backward orbital motion there.*

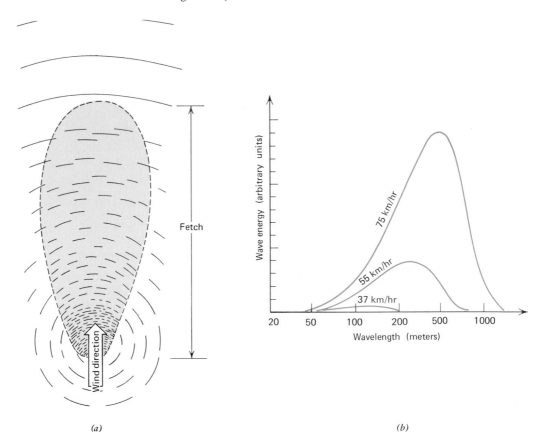

(a) *(b)*

FIGURE 8.2 (a) *As one goes downwind along the fetch, smaller waves give way to longer ones.* (b) *Plot of wave energy vs. wavelength for fully developed seas in winds of 75, 55, and 37 km/hr.*

fault, as illustrated in Figure 8.3. Eruptions of volcanoes, such as that of Krakatoa in 1883, also may cause tsunamis.

Typically, a seismic sea wave would have a wavelength of nearly 200 km, and would be less than 1 m high, hence not noticeable to ships at sea. This, of course, presents a problem in trying to track them. Noticing that waves of this wavelength are shallow-water waves even over the ocean basins, we can readily calculate that their wave velocity would be roughly 200 m per second, or 720 km per hour. The group velocity, however, might

TABLE 8.1 Conditions for Fully Developed Seas

Wind Velocity (km/hr)	Length of Fetch (km)	Wind Duration (hr)	Average Wave Height (m)	H_3 (m)	Wave Period (sec)
20	20	2.4	0.3	0.4	4
30	60	6	0.9	1.1	6
40	140	10	1.6	2.6	8
50	300	16	3	4.5	10
60	500	23	4.5	7	12
75	1200	42	9	14	16
90	2600	69	15	25	20

be somewhat less than this,[1] and that is what is important in predicting the arrival of seismic sea waves at some distant port.

The real problems in predicting seismic sea waves are even more severe than just finding their group velocity. First, we have to know whether they exist at all. With an earthquake, there may or may not be associated underwater crustal movement. Furthermore, we saw that ships at sea are no help, as the waves are too long and low to be noticed. Not until a seismic sea wave reaches the shallower coastal waters will it build up to more foreboding dimensions, and then it is too late. Furthermore, the refraction due to underwater topography may make its effects disastrous at some places and insignificant at others along the same ocean margin.

There is a fifty-fifty chance that when a tsunami strikes a coast, a trough will come in first. Curious bathers who might be lured out by the tidepools and flopping fish, will be in for another surprise in 5 or 10 more minutes when the first crest starts in. An example of this happened on April 1, 1946, when a fishing fleet in Half Moon Bay, California, found itself sitting dry on the bay floor, and then 10 minutes later it was floating above the roadway around the bay. Figure 8.4 shows some effects of seismic sea waves on this date.

Although recent history records frequent destructive seismic sea waves, one of the most destructive occurred on August 27, 1883, when Krakatoa erupted. The resulting seismic sea waves were felt in all oceans, even the North Atlantic. On the coasts of nearby Java and Sumatra, more than 36,000 people were drowned. Clearly, the destructive potential of seismic sea waves mandates that more effort should be put into early warning systems.

C. TIDES

C.1 The Astronomical Cause

Waves of even longer wavelength, carrying more energy, but which are very predictable, are the tides. They seldom can be considered destruc-

---

[1]For intermediate or shallow-water waves, it is difficult to predict their group velocity, except for idealized conditions, such as a perfectly smooth bottom, for example.

TABLE 8.2 The Beaufort Scale[a]

Beaufort Number	Wind Speed		Sailor's Description	Appearance of Sea
	km/hr	Knots		
0	under 1	under 1	calm	Sea like a mirror.
1	1 to 5	1 to 3	light air	Ripples; no foam crests.
2	6 to 11	4 to 6	light breeze	Small wavelets; crests have glassy appearance and aren't breaking.
3	12 to 19	7 to 10	gentle breeze	Large wavelets, crests begin to break; scattered whitecaps.
4	20 to 28	11 to 16	moderate breeze	Small waves, becoming longer; frequent whitecaps.
5	29 to 38	17 to 21	fresh breeze	Moderate waves, taking longer form; many whitecaps; some spray.
6	39 to 49	22 to 27	strong breeze	Large waves forming; whitecaps everywhere; more spray.
7	50 to 61	28 to 33	moderate gale	Sea heaps up; white foam from breaking waves begins to be blown in streaks.
8	62 to 74	34 to 40	fresh gale	Moderately high waves of greater length; edges of crests begin to break into spindrift; foam is blown into well-marked streaks.
9	75 to 88	41 to 47	strong gale	High waves; sea begins to roll; dense streaks of foam; spray may reduce visibility.
10	89 to 102	48 to 55	whole gale	Very high waves with overhanging crests; sea takes white appearance as foam is blown in very dense streaks; rolling is heavy, and visibility is reduced.
11	103 to 117	56 to 63	storm	Exceptionally high waves; sea covered with white foam patches; medium-sized ships are lost to view for long periods.
12 to 17	118 and up	64 and up	hurricane	Air filled with foam; sea completely white with driving spray; visibility greatly reduced.

[a]The Beaufort scale was initiated in 1806 by Admiral Sir Francis Beaufort of the British Navy. The numbers were related to the amount of sail a fully-rigged warship of his day could carry.

tive, however, because both coast and coastal dwellers are prepared for them, in contrast to the seismic sea waves. Tidal waves are caused by the gravitational attraction of other celestial bodies.

In Chapter 2 we saw that the gravitational force that one body exerts

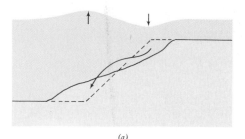 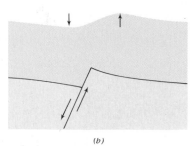

FIGURE 8.3 *Two possible causes for seismic sea waves are underwater landslides (a), and slippage along a fault (b).*

FIGURE 8.4 (a) and (b) *Scotch Cap Lighthouse, Unamak, Alaska, before and after the tsunami of April 1, 1946. (c) Pier at Hilo, Hawaii during the same tsunami.*

on another depends both on their masses, and on the distance between them. So if we look at the gravitational pull of various astronomical bodies on a certain point on the earth, we will have to consider the mass of the body and the distance to it. Of course, the earth itself pulls the most, and pulls itself into a rather spherical shape. But in this section we will be interested in other forces that try to distort this shape. Of these outside forces, by far the largest are due to the sun and the moon. The sun is important because it is massive, although relatively distant. The moon is

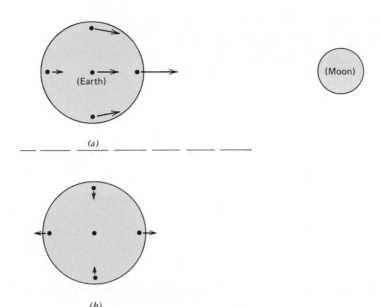

FIGURE 8.5 (a) *Acceleration of various parts of the earth toward the moon (not to scale). Regions closer to the moon are pulled harder.* (b) *Acceleration of various parts of earth relative to the earth's center.*

important because it is close, although relatively small. First, we examine the tidal force due to the moon.

The acceleration, a, of a certain mass, m, is directly proportional to the force, F, acting on it ($F = ma$). Considering the acceleration of various points on the earth toward the moon, we notice that since the gravitational force between two objects increases as their separation decreases, a unit mass on the side of the earth nearest the moon will be accelerated toward the moon more than a unit mass at the earth's center. And a unit mass at the far side of the earth gets accelerated even less. A good analogy is to imagine a sprint race between a rocket sled, an automobile, and a child on a tricycle. When given the signal, all three accelerate forward. Passengers in the automobile will see the rocket sled shoot out ahead, and the tricycle will be left behind.

As a result of the differential acceleration, there will be a tidal bulge on *both* sides of the earth (Figure 8.5). The side nearest the moon gets pulled out in front, *and* the side away from the moon gets left behind. Both bulges are the same size. Because of its greater mobility, the water gets stretched out more than the solid earth, although there are tidal bulges in both.

The student might ask that since we're being accelerated toward the moon, why don't we run into it? The answer is that we are moving sideways. When you twirl a ball on the end of a string, the string always pulls the ball toward your hand, yet since the ball is moving sideways, it never hits your hand.

Because the earth is 81 times more massive than the moon, the earth–

moon system spins like a very lopsided dumbbell. The earth and the moon each orbits a point on a line joining their centers, which is located 81 times closer to the center of the earth than the center of the moon. The force that keeps each in orbit about this point is the gravitational force of the other.

When an object is in orbit, portions farther from the center of rotation actually require *greater* force in order to keep them in orbit.[2] For example, portions of the earth farther from the moon should require a greater force. However, the force is provided by the moon's gravity, which *decreases* with distance rather than increasing. Therefore, there are actually *two* reasons for the tidal bulges: not only does the moon pull on the opposite sides of the earth differently, but the difference is opposite to that which is required.

The earth's daily rotation carries us through both tidal bulges each day. From our point of view, we say the tide "comes in" twice a day. From a broader perspective, however, it doesn't really "come in" at all; rather, we are carried into it by the earth's rotation.

The ideal moon tide should be about 70 cm high. This will vary slightly because the moon's orbit is an ellipse, so sometimes it is closer and the tides are higher than at other times. However, since the solid earth isn't really all that rigid, it has tides too, amounting to a bulge of some 26 cm. So the perfect moon tide should only be 44 cm high above the solid earth. It generally gets higher than this as it comes into shore. Since the wavelength of a tidal wave is halfway around the earth, even in the deepest ocean basin it is a "shallow water wave," and as it runs into shore it builds up like other waves do in shallower water. So among the things that determine how high the tide will actually be when it arrives at the shore is the underwater topography. Some bays can further amplify the tides by funneling them into narrow zones. An example is the Bay of Fundy, shown in Figure 8.6.

The sun, of course, does exactly the same thing as the moon, but the sun tides are only about 45% as high as those of the moon. When the sun and moon line up so that they're either on the same or opposite sides of the earth (new or full moon), their tides add together to make an extra high tide, called "spring tide," although it has nothing to do with the season. When they're at right angles to each other (so the moon is at first or third quarter), their two tides work against each other, as seen in Figure 8.7. Because the moon's tides dominate, there still will be a tide, but a rather low one, called "neap tide." Of course, most of the time will be spent somewhere between these extremes. The moon encircles the earth, going

---

[2]*The formula is $F = mw^2r$, where F is the required centripetal force in newtons, m is the mass in kg, w is the angular velocity in radians per second, and r is the radius of orbit in meters. An apparent complication arises from the fact that the barycenter, about which the earth–moon system rotates, actually lies within the earth. In doing the computations, however, it turns out that the position of the barycenter is irrelevant, since we are dealing with differences in the required forces, rather than their absolute values.*

FIGURE 8.6   *Photo taken at the Bay of Fundy as the water begins its reentry after low tide.*

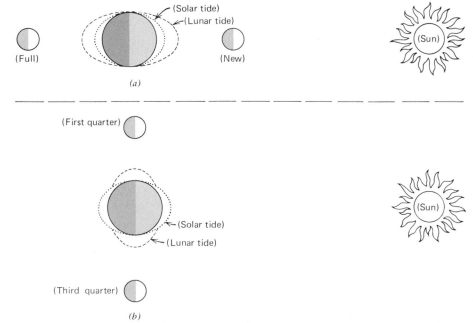

FIGURE 8.7   (a) *When the moon is new or full, the solar and lunar tides are in the same direction, giving the extra large "spring tides." (b) When the moon is at first or third quarter, the solar and lunar tides are at 90° to each other, yielding little net tides, called "neap tides."*

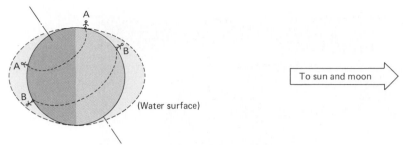

To sun and moon

(Water surface)

FIGURE 8.8 *Tidal pattern depends not only on the phase of the moon, but also on season and latitude (among other things). This sketch is for new (or full) moon in the northern winter. Observer A gets only one high tide per day (at midnight), and observer B will see two high tides per day (one at noon and one at midnight).*

through all its phases once every 29½ days, so during that period we should experience two spring tides, two neap tides, and be back where we started, tidewise.

C.2 Variations in Tidal Patterns

Since the earth spins around once every day under the tidal bulge, one would expect two high tides and two low tides per day. Actually, since the moon moves slightly from day to day, it takes 24 hr and 50 min to complete the tidal cycle. However, some places experience only one tide per day, and some places experience one for part of the year and two for part of the year. Still others experience one high high tide and one low high tide each day. How can this be?

The earth's axis is tilted by 23½° relative to the sun, and the moon's orbit varies between 18° and 29° from the plane of the earth's equator. As illustrated in Figure 8.8, this means that as observers spin with the earth, they will be carried through different parts of the two tidal bulges on opposite sides. Hence, they may be carried under the peak of the bulge on one side, and under the tail of the bulge on the other, or may miss it altogether. It can be seen that this will depend on latitude, season, and the phase of the moon. (Figure 8.8, for example, is drawn for winter in the Northern Hemisphere.)

It is reemphasized that bottom topography is also important. Since even in the deep ocean basins the tide is a shallow water wave, refraction due to the underwater topography is important in bending the tidal waves toward certain coastal regions and away from others. For instance, the Bay of Fundy gets tides in excess of 13 m. However, just to the south, Nantucket Island has tides of less than 0.5 m.

The wave velocity of a tidal wave is easily calculated from the shallow water wave speed formula to be about 720 km per hour. This is not fast enough to keep up with the sun and moon in their daily westerly trip across the sky, except at very high latitudes, so the tides often lag behind where they should be. Chesapeake Bay is so shallow, and the tidal wave

FIGURE 8.9   *Tidal bore on the Petitcodiac River, New Brunswick.*

velocity so slow, that although it is only 250 km long, two successive high tide crests can be found still traveling down its length at the same time.

In some places, a tidal wave stacks up quite spectacularly, the front running into shallow water and slowing down, perhaps also helped by traveling upstream against a current, and the crest of the tidal wave catches up with the trough. This creates a steep wave front called a "tidal bore." On the Tsien-Tang River in northern China, these reach heights of 3 m or so, and at the mouth of the Amazon River they have been reported as high as 8 m. (See Figure 8.9.)

Even if the water were deeper, the tides still couldn't really be considered stationary waves underneath which the earth spins once a day, because the continents get in the way. Due to the shapes of the basins and the relative shallowness of the water, the tides in some ways resemble forced oscillations, such as water sloshing around in a large bathtub. Collisions with the continents cause the tidal waves to bounce back and forth across the oceans, following circular paths due to the Coriolis deflection.[3] Just north of Antarctica, there are no land masses to interfere with the tidal wave's daily trek around the globe. Furthermore, at this latitude, it doesn't have to travel so far to get around the globe, so it does manage to keep up with the sun and moon. The tidal wave is quite pronounced, and as it passes by the various major oceans, it sends waves up into them from the south.

Due to all these complications, predicting coastal tides from theory

[3]*The curved paths are caused by the earth's rotation. See Chapter 13 for a further explanation of the Coriolis effect.*

would be impossibly complex, involving among other things a detailed knowledge of the relief of the entire ocean bottom. As a result, tide tables are produced via extrapolation from past data instead of from theory.

C.3   Tidal Jargon

In sailor's language, "high water" and "low water" refer to the time during the day when the tide is at its highest and lowest points, respectively. Along estuaries, "flood tide" refers to the time when the water is rising and coming into the estuary, and "ebb tide" refers to the water going out as the water level is falling. "High water slack" and "low water slack" refer to the times when water is neither entering nor leaving the estuary, occurring just after high and low tides, respectively. It is a good time for boats to come or go without having to battle strong currents at the mouth of the estuary. We have seen how various items like the relative position of the sun and moon, the latitude, time of year, topography of the ocean bottom, and so on can cause large variation in the tidal cycle. Sometimes it is such that the water will stay at a certain level for several hours during the tidal cycle, neither rising nor falling. This is referred to as a "stand" of the tide.

D.   INTERNAL WAVES

We are most familiar with waves on the surface of the ocean, propagating along the interface between air and water. But waves can propagate between any two fluids of different densities, such as between mercury and air, between oil and water, and between layers of ocean water of different densities. The greater the difference in density between the two fluids, the faster the waves will travel. For example, because mercury is so very dense compared to air, waves on top of mercury will travel very rapidly. On the other hand, water is not much denser than oil, so waves between these two fluids will move rather slowly. When two layers of ocean water have only slightly different densities, waves can form between these layers. These are called "internal waves," and they propagate extremely slowly.

Norwegian fishermen have complained that their boats get "stuck" in the "dead water" of fjords. This is caused by internal waves. To understand what actually happens one must understand that fjords are commonly hollowed out glacial valleys, with sills at the ocean ends where the glaciers terminated (Figure 8.10). Cold dense sea water fills the bottom of the fjord to the sill depth, and often the surface water is relatively light, since lots of fresh water streams run into the fjords. Consequently, there is an interface between water of different densities. The difference is only slight, however, and the waves between these two layers propagate very slowly. When the boats move slowly, they can set up waves between these two layers, which are not visible at the surface. Consequently, the boat becomes trapped, the energy going into making these internal waves, and

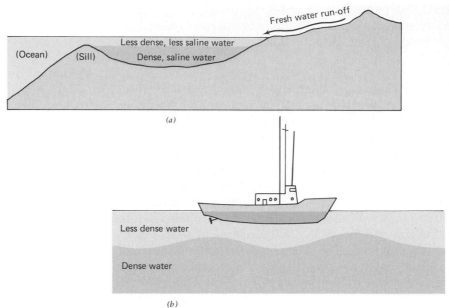

FIGURE 8.10 (a) *Cross section of a fjord. Due to fresh water run-off from land,
the surface water is light. Due to the sill at the mouth of the fjord,
the dense saline water remains trapped and fills the bottom.* (b)
*Slowly moving boats generate large internal waves, invisible from
the surface, but they can be detected by the boat's sluggish handling.*

the skipper can't understand why the engine is working so hard and the
boat still is barely moving.

One doesn't have to be in fjords to find internal waves; they are common at sea too. Sometimes when they are near the surface, long parallel
slicks can be seen, which are caused by the internal waves below. (See
Figure 8.11.) These slicks will generally be over the troughs. Because the
water above will fill the troughs and be thinner over the crests (Figure
8.12), the debris floating on the surface is carried toward the troughs. The
floating debris dampens the surface waves, and these regions of minimal
waves appear to us as "slicks."

E.   REDUCING WAVE ACTIVITY

In ancient times fishermen realized that they could reduce the roughness
of the sea around their boats by slowly leaking fish oils onto the water.
The explanation rests in the surface tension of liquids. Water has surface
tension, which can be demonstrated by floating a needle on the surface of
a glass of water. This acts like a rubber sheet stretched over the water
surface, which rounds off waves and helps prevent them from breaking
as soon as they would otherwise. A thin film of oil on the water also has
surface tension, and acts like an additional rubber sheet over the water

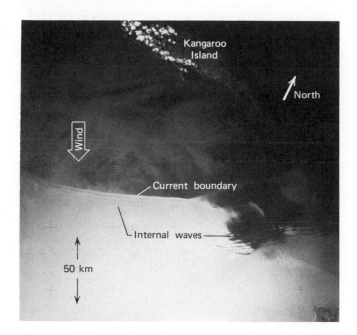

FIGURE 8.11   *(top) High-altitude photo of the ocean south of Australia. Internal waves can be identified by their enormous size and regularity. A current boundary is also clearly visible, as the wind creates differ- ent sea states where waters have different relative motion. (bottom) Low-altitude photo near the Main Coast. Internal waves are revealed through the presence of visible slicks on the surface. These slicks are caused by the accumulation of surface debris, some of which is visible in this photo.*

that helps round off the waves even more. Rounded waves have less steep sides for the wind to blow against, and they break less frequently.

In addition, waves may be calmed by putting anything in the water that helps sap the energy out of the waves. Ice or debris floating on the surface can accomplish this. Also mud or kelp in the water can help. In

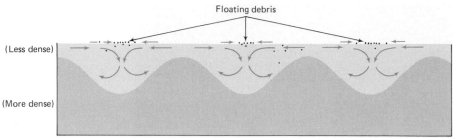

FIGURE 8.12 *Floating surface debris collects over the troughs in the internal waves, as the surface water must flow toward the troughs to fill them in. This debris dampens the wind ripples, making visible "slicks."*

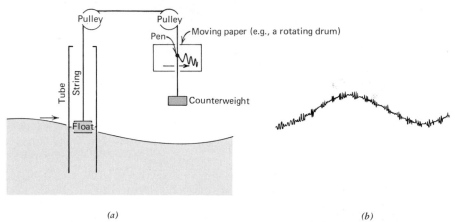

(a)   *(b)*

FIGURE 8.13 (a) *Basic design of a device for recording waves.* (b) *The record given by this device yields all changes in water level, both tidal changes and those due to smaller waves (and surf beat too).*

fact, artificial kelp is manufactured to help calm the waves where they are not desired, and where natural kelp doesn't grow.

F. WAVE MEASURING DEVICES

A wide variety of devices have been invented to measure and record waves. Electronic pressure sensors can be placed on the ocean bottom. Since pressure depends on the depth of the water, the passage of waves is indicated by changes in pressure. A mechanical wave recorder, which is an open-ended tube, is illustrated in Figure 8.13. As the water level rises and falls with passing waves, a float in the tube rises and falls as well. A string attached to the float moves a pen above the tube up and down on a sheet of paper, and the sheet of paper is driven sideways in time, so the pen records a wave pattern like that in Figure 8.13.

It is seen that this kind of gauge measures both the tides and the waves. If one wants to measure the tides alone, the lower end of the tube can be closed and a very small hole poked in it, so water can enter and

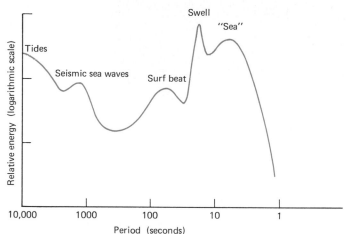

FIGURE 8.14   *If the tube had only a very small hole in the bottom end, then water
could enter and leave only very slowly. Thus the slow changes of
the tides would be recorded, but the more rapid changes of the
smaller waves would not.*

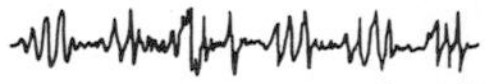

FIGURE 8.15   *The difference between the record of the tides alone and the record
of everything gives the record of the smaller waves alone.*

FIGURE 8.16   *The energy spectrum of ocean waves, depicting the relative
amounts of energy contained in each of the various kinds of waves
at any one time, on the average. (Data from Walter Munk, Scripps
Institution of Oceanography.)*

leave the tube only very slowly. Thus, during the quick passage of a
smaller wave, virtually no water has time to enter or leave the tube. But
during the six hours of rising tide the water can trickle in, and during the
six hours of falling tide it can trickle back out. The pen or electronic sensor
will then give only the pattern of the tide, illustrated in Figure 8.14.

To measure only the usual sea waves, and not the tide, involves one
tube of each kind. A pressure-sensitive electronic device is placed at the
bottom of each tube, and only the difference between the readings of each
device is recorded. Thus, the tide is subtracted from the total pattern, and
one is left with a record of the smaller waves only. (Figure 8.15.)

With wave measuring devices, one can attack the problem of finding
how much of the ocean's energy at any one time is stored in each of the
different kinds of waves. This will, of course, vary with the region and the
time of year, so one must take an average. The results are shown in Figure
8.16. It is seen that most energy is carried in the "sea" and swell, less in
the tides (much more per wave, but fewer waves), and still less in seismic
sea waves, capillary waves, and so on.

FIGURE 8.17 *A 40-ft Coast Guard vessel meets a breaking wave over the Columbia River bar.*

G. SHIP DESIGN

Ship design, like building design, depends on the size of the product. Small structures, such as dog houses, can be built rigidly. In the strongest earthquakes, the dog house will bounce around without breaking. Tall buildings, however, must be made flexible, lest a small earthquake or a strong gust of wind cause them to crack and tumble. Likewise, small boats are built rigidly, as they bounce around on the sea like a cork. Large ships, however, must be flexible to weather the pounding and twisting caused by their weight and the waves.

Contrary to what is commonly believed, the large waves of fully developed seas are not damaging to ships. Large waves have very long wavelengths compared to a ship's length, and the ship simply undergoes the orbital motion along with the water molecules, much like a cork would. Modern ships can easily accommodate whatever small amount of bending and twisting that is required, as they are built to be flexible. Often the plate metal sheets on the hull are built overlapping each other, much like scales on a fish, so they can slide over each other when the ship bends. They leak water, of course, but this water is often utilized in cooling the engine and bearings.

What is damaging to ships, however, are breaking waves. In a storm area where waves are developing, waves are also breaking. As a wave breaks, the water sliding down a wave front no longer is undulating in circular orbits; rather it can amount to a lot of mass moving at high speeds. This forceful mass can wipe the superstructure off a ship or break up the hull, particularly if it sweeps broadside to the ship. Hence, a captain will head the ship into the wind in a storm (Figure 8.17), as the bow can take

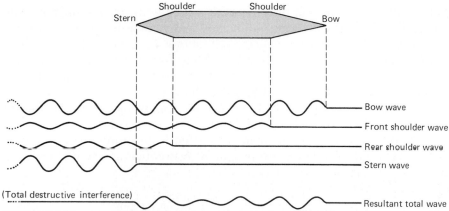

FIGURE 8.18 *At cruising speed, the waves produced at different points on a ship should destructively interfere, producing the smallest wake possible. The less energy wasted in the wake, the more efficient the ship.*

the beating of breaking waves better than any other part of the ship. Out of the storm center, it doesn't matter which way the ship is headed—no breaking waves, no damage.

Wave study is important in ship design from another point of view, too. A great deal of the energy that it takes to shove a ship through the water goes into making waves. The waves in the wake represent energy lost to the ship, as they carry away this energy to be dumped eventually when they break on distant shores. Consequently, the key to designing an efficient boat is to make a hull that will create as small a wake as possible at the ship's cruising speed. Waves are created primarily at four points along the side of a ship—at the bow, front shoulder, rear shoulder,and stern. (See Figure 8.18.) The idea is to get as much destructive interference as possible among the waves generated from these points so that the ship leaves as small a wake as possible. Less energy given to the wake means a more efficient ship.

H. SUMMARY

The size of wind driven waves increases with the velocity, fetch, and duration of the wind. If the duration and fetch are sufficiently large, then the waves will build up to a size that moves approximately as fast as the wind.

Seismic sea waves are generated by underwater seismic activity. It is difficult to predict their existence or arrival at any coast area. Their danger is their unpredictability.

The largest and most energetic of all waves are the tidal waves, which are generated by the gravitational forces of the moon and sun. The earth spins beneath the tidal bulges, generating two high tides and two low tides per day for most points in the ocean. There are many variations in the-pattern caused by the tilt of the earth's axis, the tilt of the moon's orbit, variations in underwater topography, and many other things. The tides are everywhere shallow-water waves, and cannot keep up with their astronomical generators because they travel too slowly. This, plus their confinement by the continents, makes them often resemble some sort of large oceanic seiche rather than an ideal tidal bulge.

Internal waves form at the interface between water layers of different densities. Surface wave activity is reduced by floating debris.

Breaking waves may damage ships, due to the force of the moving water mass. Efficient ships are designed to produce as small a wake as possible.

QUESTIONS FOR CHAPTER 8

1. If you're at sea in a small boat, what might you look for to tell if the waves will be getting rougher or calmer?

2. With the help of a sketch, show how the wind helps to reinforce the orbital motions of the water molecules in a wave.

3. The size to which wind driven waves develop depends on what three factors?

4. What is the Beaufort scale? What is the range of Beaufort numbers, and to what do a few representative numbers correspond?

5. What generates seismic sea waves? Give examples.

6. By what other names do people sometimes refer to seismic sea waves? Why are these names on their way out of usage?

7. What is a typical wavelength for a seismic sea wave? How deep would the ocean have to be for this not to be a shallow water wave?

8. How fast do seismic sea waves travel? Does the wave group travel this fast?

9. Discuss some of the problems involved in trying to forewarn people of impending seismic sea wave damage after underwater seismic activity has been detected.

10. Why do you suppose certain places frequently suffer damage from seismic sea waves and others nearby don't?

11. Seismic sea waves are frequently popularized as unannounced giant crests looming over unsuspecting coastal villages. Is this an accurate picture? What would be a more likely picture?

12. When did Krakatoa erupt? How far did the resulting seismic sea waves travel?

13. Why do you suppose seismic sea waves are more damaging during high tide than low tide?

14. If the tides are bigger, longer, and carry more energy than seismic sea waves, why aren't they more destructive?

15. What two astronomical bodies are the biggest generators of tidal forces on the earth? Why these and not others, like Venus or Mars, for example?

16. Why is there a tidal bulge on the side of the earth facing away from the moon? Why is there one on the side facing toward the moon?

17. If the earth's gravity were stronger, would the size of the tidal bulges be different?

18. If all parts of the earth are being accelerated toward the moon, why don't we run into it?

19. How big is the moon-caused tidal bulge in the solid earth? How much bigger than this would be the tidal bulges in an ideal ocean?

20. Why do tides get bigger as they approach shore?

21. How big are the sun's tides compared to those caused by the moon?

22. How is the phase of the moon related to the size of the net tidal bulge? What is its phase during neap tide? During spring tide?

23. With the help of a sketch, indicate why some coastal villages might experience only one high tide per day while others experience two.

24. Why does it take more than one full day to complete the tidal cycle?

25. Why is there such a big difference between the tides at the Bay of Fundy and those at Nantucket Island?

26. The equatorial circumference of the earth is 40,000 km. How fast would the tidal bulge have to travel to keep up with the sun and moon (roughly once around every 24 hours)? Does it travel this fast? Why not?

27. Why would the tidal bulge do better at keeping up with the sun and moon at higher latitudes?

28. What is a tidal bore?

29. Why do you suppose the tidal pattern in the ocean just north of Antarctica is quite pronounced and predictable? What ramifications does this have in the tidal patterns in the other oceans?

30. What are high water, low water, flood tide, ebb tide, and slack water?

31. What is a "stand" of the tide?

32. Can waves propagate between any two fluids, or must it be between water and air?

33. How are the densities of the two fluids related to the speed of waves propagating between them? If we had a denser atmosphere (like Venus), how would that change the speeds of waves on our oceans?

34. What are internal waves? Why do they go so slowly?

35. How can Norwegian boaters get their boats "stuck" in the water in fjords?

36. Do slicks appear over crests or over troughs in internal waves? Why?

37. How is it that floating debris makes a visible slick?

38. How does a thin oil film reduce wave activity?

39. What can be put into water to help reduce wave activity? Why is it easier to carry a soft drink that has crushed ice in it than one without ice?

40. How might you make a device to record the wave activity? What would you do to this if you wanted to record the tides only?

41. Suppose you were interested in finding how much of the ocean's total wave energy is carried in each type of wave. What class of waves carries the most energy? Second most? Third most? In which class does an individual wave have the most energy?

42. How is the size of a ship related to the flexibility it must have?

43. Why do large ships leak?

44. What kind of wave is damaging to ships and why?

*45. The sun's gravitational pull on the earth is about 200 times stronger than that of the moon. Why then are the moon-caused tides larger than those caused by the sun? (*Hint:* Does the tidal bulge reflect the actual gravitational force, or does it reflect the difference in the gravitational force from one side to the other?

*46. Why might ocean freighters last longer than those on the Great Lakes? (*Hint:* Perhaps a larger portion of the waves on the Great Lakes are not yet fully developed, due to the smaller dimensions and shorter "fetch" for the wind. What is it about waves not yet fully developed that might make them more damaging?)

*47. How is the energy efficiency of a ship related to the size of the wake it produces? Why is destructive interference important here?

*48. What tidal variation pattern would be observed at the North Pole?

*49. Would waves on fresh water or salt water travel faster? Why?

*50. Figure 8.18 makes it appear that total destructive interference can be accomplished among the waves produced at various points along a ship. Can it? Explain your answer.

SUGGESTIONS FOR FURTHER READING

1. William A. Anikouchine and Richard W. Sternberg, *The World Ocean,* Chapter 9, Prentice-Hall, Englewood Cliffs, New Jersey, 1973.

2. Willard Bascom, *Waves and Beaches*, Anchor/Doubleday, New York, 1964.

3. Alan P. Carr, "The Ever-Changing Sea Level," *Sea Frontiers*, **20,** No. 2 (1974), p. 77.

4. Edward P. Clancy, *The Tides*, Anchor/Doubleday, New York, 1969.

5. Kathleen Mark, "Earthquakes in Alaska," *Sea Frontiers*, **20,** No. 5 (1974), p. 274.

6. Michael J. Mooney, "Tragedy at Scotch Cap," *Sea Frontiers*, **21,** No. 2 (1975), p. 84.

7. Michael J. Mooney, "Tsunami!," *Sea Frontiers*, **26,** No. 3 (1980), p. 130.

8. George Pararas-Carayannis, "The International Tsunami Warning System," *Sea Frontiers*, **23,** No. 1 (1977), p. 20.

9. John P. Robinson, Jr., "New Foundland's Disaster of '29," *Sea Frontiers*, **22,** No. 1 (1976), p. 44.

9
BEACHES

Beaches are the most temporary of all the oceanic regions. Every time sea level changes slightly, as with the coming and going of ice ages, new beaches must form, and the old ones are abandoned, although the other oceanic regions remain unchanged. Even from day to day we notice small changes, and from season to season the changes are quite pronounced.

The character of a beach varies with position as well as with time. Just a few hundred meters from where we frolic or sunbathe in the warm soft sands, there may be a beach impassable to humans, or a beach where we must wear shoes to keep from cutting ourselves on sharp rock fragments.

Because we are terrestrial creatures, our most frequent contact with the ocean is at the beach, and so we are already familiar with some of the evidence of these on-going beach processes. But often we don't question what we see; we just accept it as the way things are. For instance, why is it that durable rocky outcroppings get broken down and washed to sea within a few decades, whereas the soft and easily moved sands of the beaches don't? This chapter explores the processes that cause this and other beach phenomena, which most of us have observed but never questioned.

9
BEACHES

A. COASTAL REGIONS

Coastal areas are often placed into two categories according to whether the coast is rising or falling relative to sea level. In regions of "rising topography," the land is rising relative to the sea level, and relict beaches are often found above the present one (Figure 9.1). This is true along most of the California coast. Similarly, much of northeastern Canada is now springing back up from the weight of the last glaciation. On the other hand, much of the East Coast of the United States is classified as "drowned topography," as relict beaches can be found under water out on the shelf, indicating that the water line has risen relative to the land (although it is presently rather stationary).

The actual line of contact between water surface and land is called the "shoreline." The entire active area is called the "beach." This is the region in which you might expect to find changes caused by the waves from month to month. It might characteristically extend up to an altitude of some 5 m or so above the high tide line, and perhaps some 10 m or more below the low tide line, according to this definition. We know that wave activity affects the shelf at depths much greater than this, but these changes are on the average slower and less predictable, only long period swell or storm waves reaching to this depth. Finally the "coast" encompasses the entire coastal area, and may in some places be many kilometers wide—for example where there are extensive dunes or salt marshes.

FIGURE 9.1 *Relict beaches above the present beach. (Photos by Anthony J. Buffa.)*

A cross section of a typical beach face is shown in Figure 9.2. Many will be backed up by a sea cliff, although some are not. Most will have a berm (the dry sand region where you sunbathe or have a picnic when you're at a beach) with one or more berm crests. The berm crest will indicate the highest point of normal wave activity. Behind it you can normally camp with reasonable assurance that you won't wake up soaked in the morning (Figure 9.3). From the berm crest, the beach face slopes down to the water line. Beyond the normal surf zone may be longshore bars, which serve as offshore reservoirs of beach sand (Figure 9.4). Sometimes these bars are so shallow that they are partially exposed at low tide, and sometimes large storm waves will make them at depths of 10 m or more. If and where we might expect to find bars will be better understood after we've learned how they are formed.

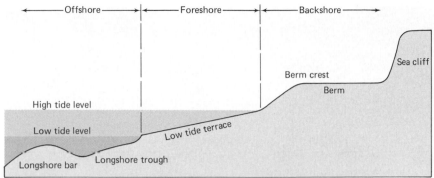

FIGURE 9.2 Cross section of a beach, showing various regions. (The details vary, depending, among other things, on whether the berm is growing, shrinking, or stable.)

FIGURE 9.3 High and low tides at Parrsboro, Nova Scotia. Clearly, the low tide terrace, revealed in the lower photo, would not be a safe place to spend the night.

FIGURE 9.4 *Aerial photo revealing some longshore bars.*

FIGURE 9.5 *A beach of dead seaweed.*

B.   BEACH MATERIALS

The beaches popular with most bathers are sandy, but sand is only one of a wide variety of materials from which beaches can be made. Rocks, shale, shingles, cobbles, mud, and gravel are frequent beach materials, and even beaches of dead seaweed and tin cans can be found (Figure 9.5). Even the sands have a wide variety according to their local ancestry. The white sand on some Florida beaches is due to ground up coral and skeletons of other marine organisms. Much of the yellow California sand comes ultimately from weathered granitic rocks, and grayish-green sands of the Pacific Northwest are derived from basalts. The ground up rocks of some volcanic islands yield a black beach sand.

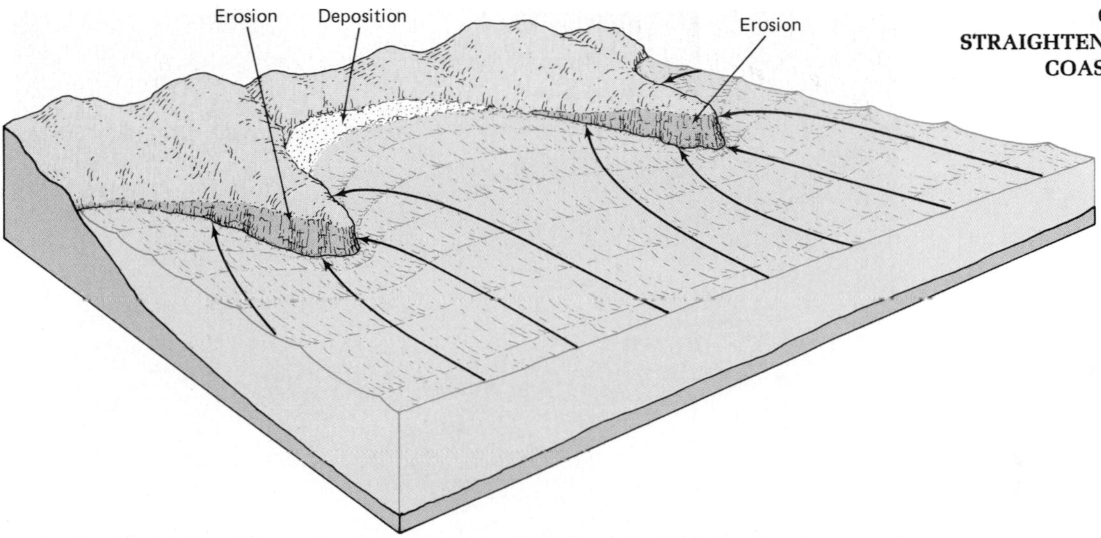

FIGURE 9.6 *Wave energy is concentrated on the headlands and diverted away from the coves. Consequently, the headlands erode, and the sediment deposits in the quiet coves, forming pocket beaches.*

C. THE STRAIGHTENING OF COASTLINES

To carry out the erosion of land and breaking down of large rocky cliffs into fine sediment, the waves have a multifaced attack. First, there is the hydraulic and pneumatic prying that comes when waves crash into cliffs, forcing water and entrapped air into the cracks, which gradually widens the cracks and pries the loosened rock out. Then, there are the rock fragments suspended in the wave when it crashes into land. These suspended fragments act as miniature chisels to chip out more fragments. Third, there is the abrasion of the suspended fragments among themselves, chipping and grinding each other down into finer and finer pieces. Also, rocks on the bottom are rolled and ground together by the crashing waves. Finally, we have seen that water is a good solvent of inorganic materials. It, along with the chemicals it carries, slowly corrode away the rocks.

As was seen in the previous chapter, on a jagged shoreline the wave energy concentrates on the points. Hence, these erosion processes progress faster on the points than in the coves between them. In fact, the quiescent water in the cove coaxes sediment out of suspension, and so the cove slowly fills in, as can be seen by the presence of pocket beaches in the coves. (See Figures 9.6 and 9.7.) These two processes both work to help straighten the coastline.

Another process that helps straighten coastlines is due to the "littoral" or "longshore" sediment transport, which will be studied more thoroughly later in this chapter. Suffice it to say for now that when wave crests do not come in exactly parallel to a beach (as they seldom do), there will be a net

FIGURE 9.7 *A cove between rocky headlands on the Oregon Coast.*

transport of sediment along the beach in the direction favored by the waves. When this sediment comes to a point on the coast, the rough water there will hold it in suspension past the point. But as it starts across the adjacent cove, it encounters more quiescent water, and settles out in a baymouth bar, as illustrated in Figure 9.8. This baymouth bar then further protects the cove from wave activity, and so any sediment entering the cove, either from the ocean or a stream, will even more readily settle out and fill in the cove behind the bar. While still in transition these regions become salt marshes or brackish estuaries, such as the sloughs so common along the California coast (see Figure 9.9). Protected from the bigger waves and predators, these regions are generally teeming with birds and smaller marine life.

D. IN AND OUT TRANSPORT OF SEDIMENT

D.1 The "In" Transport Mechanism

Before studying the littoral transport of sediment along a beach, we will focus on the in-and-out transport of sand up and down the beach face.[1]

Here, there are two opposing processes at work. The first process

---

[1]*Henceforth we will refer to all beach materials as "sand," whether they are or not, as the same processes affect them all.*

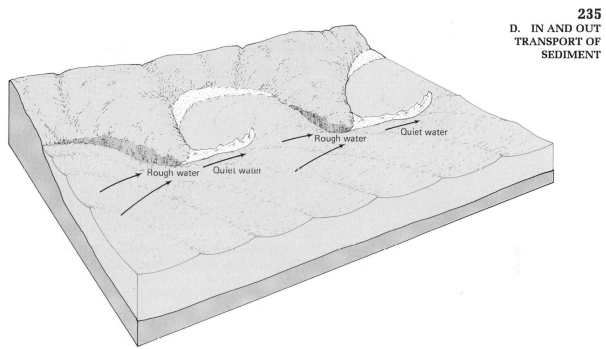

FIGURE 9.8 *Where there are longshore currents, the sediment will be kept in suspension in the rough water off the headlands, but will be deposited in baymouth bars in front of the quiet coves.*

FIGURE 9.9 *Slough at Morro Bay, California, with the tide out.*

tends to bring sand in toward the beach along the bottom, and is responsible for the sand staying on the beaches. When you think about it for a moment, you might wonder why the sand doesn't gradually work its way downhill out to sea and end up on the shelf somewhere, leaving us with barren rock in place of our beaches. The explanation involves the orbital motion of the water molecules below a wave. Sand particles stirred up on

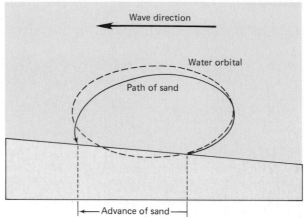

FIGURE 9.10 *Near bottom, the water molecules' orbits are flattened ellipses (much more flattened than shown in this drawing). When a wave passes overhead, the water at the bottom drags some sand grains with it as it progresses around its orbital. The sand, being denser, falls relative to the water molecules, and strikes the beach face well ahead of where it left it. This is aided by the ability of the water to penetrate the beach while the sand grains can't.*

the bottom by this motion will be carried up and forward with the water molecules, but as the orbital motion carries them down and back, they will tend to run into the bottom. Water can and does penetrate the bottom, but the sand particles cannot. Being heavier than water they will sink slightly relative to the water on the top part of the orbital motion, and so run into the bottom before making it all the way back (see Figure 9.10). Their friction with the bottom will then inhibit their backward motion.

D.2 The "Out" Transport Mechanism

The other process operates only in the surf zone, so although it sometimes carries sand grains back down the beach face, it doesn't take them out so far that they can't be retrieved by the wave motion described in the previous paragraph. When a wave breaks on a beach, a wave of translation is formed moving up the beach (the "swash"). This will often carry some sand up the beach with it. When the waves are small and far apart, much of the water in a wave of translation soaks into the unsaturated beach before washing back down the beach, depositing its sediment when it disappears into the beach. In the beachcomber's jargon, there will be a "swash" but little "backwash," the bulk of the water returning to sea through the beach rather than on top of it. Under these conditions, then, we'd expect the wave of translation to help transport sand up the beach, enlarging the berm.

However, when the waves are bigger and more frequent, the swash from one wave will find the beach still saturated with water from previous waves. Hence, the backwash will not be able to soak in, and will return

FIGURE 9.11   *Sand-laden backwash (left) washes under the next incoming wave (right).*

to sea along the beach surface. Those who frequent the beaches have noticed this film of returning water increases in speed as it flows down the beach face, becoming turbulent and so laden with sand that it can sometimes contain more sand than water. (See Figure 9.11.) This backwash washes under the next wave of translation, insulating it from the beach. Therefore, it minimizes the transport of sand up the beach by the next wave of translation, while continuing its own sand transport down the beach.

D.3   Seasonal Changes in Berm and Bar

With this understanding, we'd expect that these processes that work to transfer sand between berm and longshore bar should operate in the berm's favor during quieter periods, and fatten the bar at the expense of the berm when wave activity is greater. This is indeed the case. Figure 9.12 shows the same beach in the winter, when waves are rougher, and in the summertime, when the sea is generally calmer. The berm is significantly healthier in the summer time. An underwater photograph of the longshore bar would have shown the reverse relationship with the season. Figure 9.13 shows the profile of the Carmel beach for various months. Again it is seen that the larger winter waves remove sand from the berm, and the summer seas give it back. Lab experiments demonstrate that for very calm seas, with average wave steepness less than 0.025, the bar will disappear altogether. However, for average wave steepness greater than 0.03, there will always be bars.

Often during winter months, or other periods of heavy wave activity, two bars will form—one corresponding to high tide and one to low. It can be shown that for normal tidal fluctuations, the water level lingers near high tide and low tide levels, but goes very quickly from one to the other. For a sinusoidal tide fluctuation, dividing water level into equal thirds

FIGURE 9.12 *A beach at La Jolla, California. In winter (top) the sand is gone, and in summer (bottom) it returns.*

between high and low tide levels (see Figure 9.14), it is found that the water spends almost four times as long in either high third or low third as it does in the middle third. Hence, bars have time to form when near high or low tide, but do not have time to form when the tide is in transition between the two. Places with more complex tidal patterns may have more than two bars.

Severe storms can remove sand from the berm remarkably fast. During a storm off Delaware in 1962, many beaches retreated as much as 25

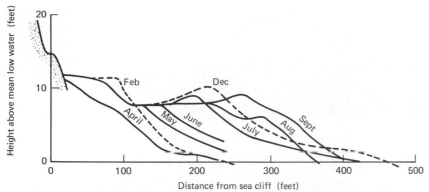

FIGURE 9.13 *Cross section of the beach at Carmel, California, showing the seasonal changes in the berm. (From Waves and Beaches by Willard Bascom. Copyright © 1964 by Educational Services, Inc. Used by permission of Doubleday and Company, Inc.)*

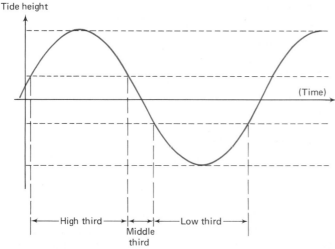

FIGURE 9.14 *Dividing the water level into three equal parts for an ideal sinusoidal tide, it is seen that much more time is spent either near high tide or near low tide than in transition between the two.*

m in two days. Hurricanes have been known to cause beaches to be cut back more than a half mile. The sand is deposited in storm bars at a depth determined probably by the size of the waves. Just how deep this can be is not known, but this could sometimes be in excess of 15 m.

D.4 The Slope of the Beach Face

The steepness of the beach face increases with both the amount of wave activity and with the coarseness of the beach materials. Regions of the beach being more protected from the wave onslaught generally have gentler slopes, as is illustrated for the particular case of Half Moon Bay (California) in Figure 9.15. The explanation of this lies in the ability of the

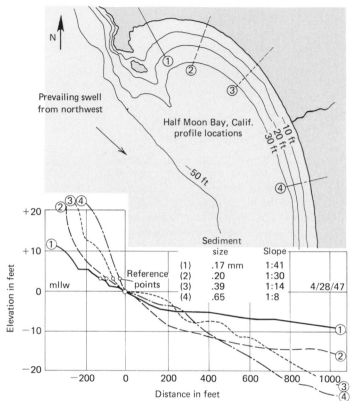

FIGURE 9.15 *Several beach profiles at Half Moon Bay, California. The more protected regions have gentler slopes and finer sediments. (From
Waves and Beaches by Willard Bascom. Copyright © 1964 by Educational Services, Inc. Used by permission Doubleday and Company, Inc.)*

waves to sweep the bottom sediment shoreward, as described in Section
D.1. Under smaller waves, the orbital sizes are smaller, so they don't reach
or sweep as deeply as do larger waves.

As for the dependence of the slope on the beach materials, it is noted
that the coarseness of the beach materials determines how much backwash there will be. The coarser the beach sediment, the more of the swash
will soak in, and the smaller the backwash will be. Since it is the backwash that tends to level the beach face, dragging materials from higher up
and depositing them lower down, coarser materials mean smaller backwash and steeper beach faces.

This demonstrates that the composition of the beach influences the
steepness of the beach face. But the reverse is also true. The steepness of
the beach face influences the composition of the beach materials. Suppose
one starts with a conglomeration of beach sediments of all sizes. What will
determine the eventual composition after it has been worked over by the
waves? Coarser sediment requires more energy for transport. Only in the
narrow surf zone of a steep beach will the wave energy be concentrated

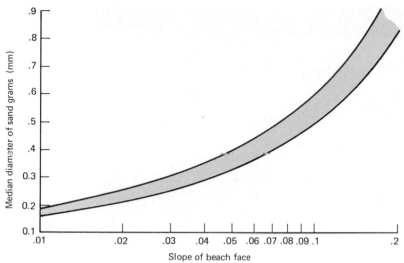

FIGURE 9.16   *Plot of grain size vs. slope of beach face for mature beaches.*

enough to move cobbles about. Finer sediments, being more easily kept in suspension, can be transported in the less violent, wide, gradually sloping surf zone. (See Figure 9.16.) On a beach of intermediate steepness, for example, the backwash might be sufficiently strong to remove the sand and deposit it in a longshore bar, but not sufficiently strong to remove the gravel from the beach face.

D.5   Minor Beach Features

Many other minor beach features due to the in-and-out motion of the water can be noticed by the novice beachcomber. The swash marks, illustrated in Figure 9.17, are created when the swash reaches its highest point on the beach before running back down. It deposits a little wall of sand grains it was bulldozing up the beach in front of it. As the tide recedes, successive swashes don't reach quite as high before depositing their bulldozed sand grains. The result looks like a giant fish scale pattern on the beach.

   On steeper beaches, where the backwash reaches higher velocities, it often carves little diamond-shaped valleys in the beach face, as illustrated in Figure 9.18. Sometimes when the tide retreats rapidly from beaches saturated with water, the water drains out of points on the beach face, like little springs, and runs down the beach face in branching streams, leaving little rills (Figure 9.19).

   When a wave washes up on dry sand, it may wet the top inch or so of the sand and seal off the air trapped beneath. The weight of the water may then push down on the beach, and the trapped air will sometimes bubble up, making little "pinholes" where the air came through. (See Figure 9.20.)

   An interesting and not completely explained feature of many beaches are the cusps, the crescent shaped indentations into the beach face. (See Figure 9.21.) These tend to form where the waves come pretty much

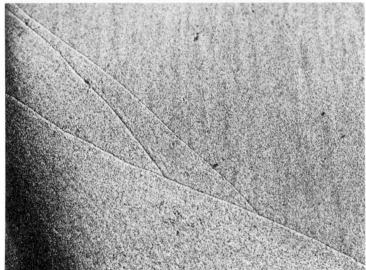

FIGURE 9.17 *Swash marks.*

FIGURE 9.18 *Backwash diamonds.*

FIGURE 9.19 *Rills.*

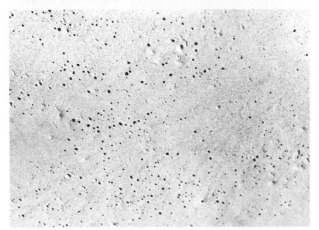

FIGURE 9.20 *Pinholes.*

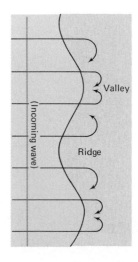

FIGURE 9.21 *(left) Photograph of beach cusps at Port San Luis, California. (right) The backwash tends to flow back out through the valleys. Hence, not only are the valleys repeatedly dredged by the backwash, but also the sediment is then deposited in front of the valleys, which helps to refract successive waves toward the valleys rather than toward the ridges.*

FIGURE 9.22   *Aerial photo of large cuspate beach features.*

straight into the beach, and tend to disappear when the waves come in on an angle, or when there are strong currents along the beach. They can vary in length from 0.5 to 100 m, the smaller ones associated with smaller waves and larger ones with larger waves. They seem to be caused by the water piling up on the beach, and then running back out to sea in the hollowed-out portion of the crescent, dragging sediment with it and maintaining the hollowed-out form. By depositing this sediment just in front of the hollowed-out portion, it probably helps future waves to refract into these areas rather than into the points. Cusps are a small-scale contradiction of the general thesis that waves work to straighten coastlines.

Cuspate beach features are also to be found on a much larger scale. The cuspate shape of the coast and barrier beaches on the East Coast (Figures 9.22 and 6.13) are good examples of this. There are currently three popular theories to explain these. One says they are due to erosion of headlands and formation of long baymouth bars, as discussed earlier. Another states they are overgrown longshore bars. A third indicates they are remnants of previous dunes that have been washed upward by the waves during the recent (last 20,000 years) rise in sea level. Probably each has validity for certain regions. Their cuspate shapes are probably reflections of eddies in the longshore current.

A bottom feature that waders notice are the ripple marks. There are actually two kinds. They both have a tendency to sort the sediment, so that coarser sediment is found at the top of a ripple and finer sediment in the valley. The oscillation ripples (Figure 9.23) are symmetrical, and are formed by the water's oscillatory orbital motion as a wave passes overhead. Current ripples are more asymmetrical, the more gentle sloping side facing into the current. They are usually a few centimeters long, but can attain a size of 15 m or more in channels with large currents. Although the sand transport is always in the direction of the current, laboratory experiments show that for current velocities between 0.66 and 0.76 m/sec, the ripple forms themselves move backward. (If this is confusing, notice that

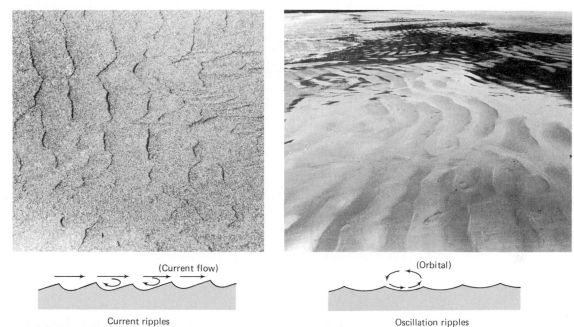

FIGURE 9.23 *Current ripples (left), and oscillation ripples (right).*

(Current flow)

(Orbital)

Current ripples

Oscillation ripples

FIGURE 9.24 *Sea caves and a sea arch.*

the body of a snake may move forward, while the form of the "S's" that its body makes remain stationary or moves slightly backward.) But for still faster currents (greater than 0.76 m/sec), the ripple marks disappear completely. The actual ripple marks found on the beach are usually caused by a mixture of various things, and so it is difficult to get information about either the waves or the currents from them.

Where there are sea cliffs, there are usually also a proliferation of sea caves, crevasses, and gullies (Figure 9.24). These have a variety of causes, aside from the fact that some materials are weaker and erode more

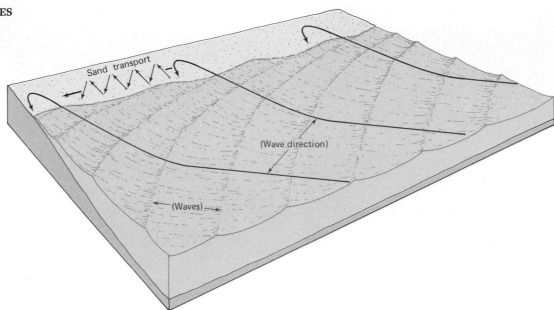

FIGURE 9.25 *Waves coming in at some angle to shore won't be quite parallel to shore yet when they break. Hence, the waves of translation will have a component of their motion parallel to the shore, and the sediment will follow a zig-zag path down the beach.*

quickly than others. The concentration of wind or wave energies in certain areas, or the routes that the ground water takes in its flow toward the ocean, also affect which regions will weather most quickly.

Sometimes when waves crash directly into the sea cliffs, or into the heavy rock debris at their bottom, the air that was being pushed in front of the breaking waves get forced up through cracks, causing spectacular jets of wind and spray known as "blowholes."

E. LONGSHORE TRANSPORT

E.1 The Transport Mechanisms

The second kind of sediment transport we are to study in this chapter is the longshore, or littoral, transport along the beaches. The first of two causes of the littoral transport is the swash–backwash motion of the waves hitting the beach on an angle. The swash of each wave moves up the beach on an angle, and the backwash comes back down, resulting in a net sideways transport of water and suspended sediment. (See Figure 9.25.) The second cause is other near-shore currents paralleling the beach. Of course, the two are related; the waves coming in on an angle often are caused by the wind blowing along the beach in that direction, and the wind often causes surface currents in the same direction.

When no longshore bars are present, the dominant method of littoral

FIGURE 9.26 *Aerial photo of Santa Barbara Harbor. The sand flow is from left to right on this photo. Notice that the sand has filled in the "upstream" side of the left-hand breakwater, and is now starting to fill in the harbor. (Photo from Mark Hurd Aerial Surveys.)*

transport is the sediment carried in the zig-zag path along the surf zone by the swash and backwash. This transport reaches a maximum when the incident wave crests make a 30° angle with the coastline. However, where there are bars, even more sediment is generally carried along them by the water currents than is carried along the surf zone by the swash–backwash.

E.2 When Humans Interfere

The volume of sand carried on this "littoral conveyor belt" is larger than one might expect. On the California coast, for example, typical values are 500 to 1000 cubic meters of sand passing a fixed point per day. This amounts to roughly fifty or a hundred dump trucks full. When one builds obstructions, such as breakwaters, the pile up of sand around the obstruction can be expected to be rapid. For example, the growth of the sandspit behind the Santa Barbara breakwater is shown in Figures 9.26 and 9.27.

Furthermore, no beach modification can be kept local. If you rob the ocean of its suspended sand at one point, the beaches downstream will pay it back. Figure 9.28 shows an aerial photo of Lake Worth Inlet, Florida, where a channel was built through the barrier beach. Notice how the beach sand is depleted downstream from the jetties. In the first 2 years after construction of the Santa Barbara Harbor breakwater, a downstream strip of sand cliffs extending for more than a mile was cut back by 100 or 200 ft. In Santa Monica, the breakwater was built paralleling the beach, hoping that this would allow the sand to drift through while protecting the harbor from the large waves from the sea. Unfortunately, in the calm sea behind the breakwater the drifting sand falls out of suspension and fills the harbor anyhow. (See Figure 9.29.) Downstream from the Santa Monica harbor, the beaches eroded, and a strip of valuable beachfront real estate one block wide was quickly lost to the sea. Consequently, these harbors must be dredged periodically, the sediment being placed on downstream

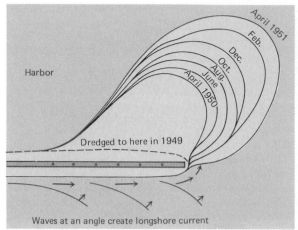

FIGURE 9.27 *Diagram showing the growth of the Santa Barbara sandspit in the years 1949 to 1951. (From Waves and Beaches by Willard Bascom. Copyright © 1964 by Educational Services, Inc. Used by permission of Doubleday and Company, Inc.)*

FIGURE 9.28 *Lake Worth Inlet, Florida. Sand flow is from right to left along the bottom of this photo.*

"feeder beaches." The ocean then carries the sediment on downstream from there, replenishing the lost sediment along the downstream beaches.

A study of California beaches shows there is compartmentalization of the sand flow down the beach. In each compartment the beach sediment

FIGURE 9.29 *Aerial photo of the Santa Monica Harbor. The breakwater parallels the shore. (The thing perpendicular to shore is a pier that is up on pilings, and not itself interfering with the sand flow.)*

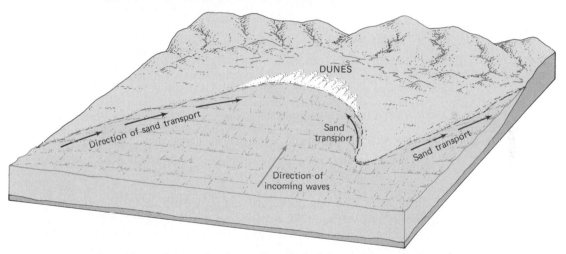

FIGURE 9.30 *Sometimes the predominant direction of incoming waves is such that a point on the coast can block the flow of sand, so it piles up in dunes. For example, this occurs frequently along the central California coast.*

is contributed by rivers and streams dumping into the ocean. Going south (downstream) along a compartment, one crosses more river mouths, and the beaches get wider. At the far downstream end of each compartment there is often a submarine canyon into which the sand gets dumped, disappearing permanently from the beaches. At the downstream end of some compartments, the sand piles up in dunes, as illustrated in Figure 9.30. Just south from one of these canyons or dune fields, then, the beaches are again narrow, with little access to sediment, widening farther downstream as more streams contribute, and so on. The major compartments are shown in Figure 9.31.

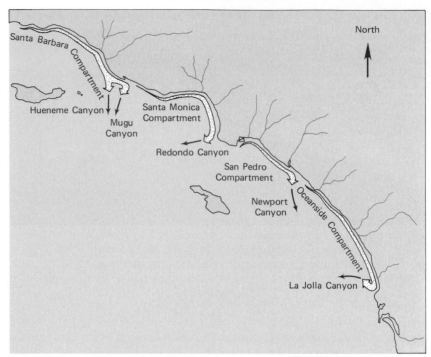

FIGURE 9.31 *The southern California coast, extending from Santa Barbara to the Mexico border. In each beach compartment, the beach will tend to be narrowest at the northern end. As the sand moves south along the shore, more is added from rivers, erosion of sea cliffs, and so on, and so the beach becomes wider. At the southern end of each compartment is a submarine canyon into which the sand is dumped, disappearing from our beaches forever.*

In the Los Angeles area, streams that used to carry sediment to the beaches are now dammed up for water reservoirs, and the remaining creek beds are lined with concrete to serve as storm drains. Robbed of this one-time sediment source, the downstream beaches are slowly eroding away. In one storm alone, 75 houses that were previously protected by wider beaches were damaged by waves. Downstream, the sand eventually gets dumped into the Newport Submarine Canyon, so one plan was to catch that sediment just before it reaches the canyon and periodically transport it back up to Los Angeles feeder beaches for recycling. Another was to periodically drain the reservoirs behind the dam, clean out the sediment deposited there in the quiet waters (which has to be done anyhow), and truck it down to feeder beaches on the ocean.

It seems that the best place to stop streams for water reservoirs, if necessary, would be the far downstream end of a compartment, where the arrested sediment would have just gone into a canyon anyhow, not contributing to miles of downstream beaches first. Likewise, the best place to build a harbor, if necessary, would probably be at the far upstream end of a compartment, where the sediment flow is small. The harbor won't fill so fast, and downstream beaches won't be robbed so fast. Another idea suggested is to build harbors just upstream from headlands, which interfere with the sandflow anyhow. This will tend to dampen the otherwise large fluctuations in downstream beach size due to the periodic dredging of the harbor.

Our studies of the beaches indicate that probably regardless of what we do, the ocean will eventually have its way. If it tries to modify a beach in a way we disapprove, we might temporarily postpone that modification, but we won't prevent it indefinitely.

F. SUMMARY

Beaches are the most temporary and variable of all oceanic regions, and can be made of a wide variety of materials.

The refraction of waves tends to concentrate the energy on the headlands, so erosion proceeds most rapidly there. Also, the longshore currents work to straighten shorelines by carrying sediments from the regions of high wave energy and depositing them in quiet coves. The formation of baymouth bars in this fashion further reduces wave activity in the coves, and hastens their filling in with sediment.

The orbital motion of the water near the bottom guarantees that sediment particles will be transported shoreward beneath incoming waves. This orbital motion is interrupted when the waves break in the surf zone, so whether the sediment will be carried farther up the beach face is determined by other factors. If wave activity is light (e.g., in the summer time), then the waves of translation will tend to carry the sediment on up the beach face before they soak into the beach, depositing their sediment load in the berm. In periods of heavy wave activity, however, the backwash will be strong enough to remove sediment from the berm and deposit it in longshore bars.

The slope of the beach face is correlated to the amount of wave activity, with smaller waves resulting in more gentle slopes. There is also a correlation between the slope of the beach and the texture of beach materials, with steeper beaches involving coarser materials. There are a variety of other beach features, such as pinholes, swash marks, rilles, ripples, cusps, and sea caves, reflecting a variety of beach processes.

When waves do not come directly into shore, there will be a net motion of water and water-carried sediments parallel to the shoreline in the surf zone. Wind driven longshore currents beyond the surf zone may also have significant longshore sediment transport. Human obstructions interfering with this littoral transport cause problems with no simple solutions.

The sand flow along beaches tends to be compartmentalized, with beaches being narrow at the upstream end and wide at the downstream end of each compartment. Compartments are usually terminated in either submarine canyons or dune fields.

1. Where would you find relict beaches in regions of rising topography? In regions of drowned topography? To which of these kinds of regions does much of the California coast belong? How about the Carolina coast?

2. What is the difference between the coast, the shoreline, and the beach?

3. What are the names of the various regions of the beach, and how are they identified? What is a berm? What is a longshore trough?

4. What different kinds of materials can you think of from which beaches could possible be made?

5. Why do beach sands display such wide variety in colors?

6. What are the five different processes mentioned that ocean waves use to transform rocky cliffs into fine sediment?

7. Why are there pocket beaches in coves?

8. How does the longshore transport work to straighten out coastlines?

9. Why would the presence of a baymouth bar tend to accelerate the filling in of the embayment?

10. What is a slough? How do you pronounce it?

11. Why doesn't the sand work its way downhill along the bottom and out to sea, leaving us with barren rock in place of our beaches?

12. How far "uphill" can the process referred to in the previous question bring the sand grains? (*Hint:* Does it work under waves of translation or only under normal gravity waves?)

13. Why does the berm grow in times of light wave activity?

14. Why does the berm shrink in times of heavy wave activity?

15. In periods of heavy wave activity, why does the sand removed from the berm only go out as far as the longshore bar? That is, what keeps it from continuing out beyond the surf zone?

16. Describe the expected seasonal changes in the berm and longshore bar.

17. How is the average wave steepness out to sea related to the presence of longshore bars in the surf zone?

18. Why might two longshore bars sometimes form?

19. Why is it that when you go to the beach, the odds are about four to one that you'll catch it either near high tide or low tide rather than somewhere in between?

20. Why is it that longshore bars created by severe storm waves are generally farther out in deeper water than most longshore bars?

21. How is the steepness of the beach face related to the amount of wave activity? What is the explanation for this?

22. How is the steepness of the beach face related to the grain size of the beach sediment? Why?

23. What causes swash marks?

24. What are rills?

25. How are pinholes created?

26. Describe how cusps are maintained by the flow of water in and out with each wave.

27. How does the grain size of the sediment differ at different points on a ripple? What might cause this segregation of sediment?

28. How can you tell if a certain set of ripples in the sediment were caused by currents or wave oscillations?

29. What is a blowhole?

30. Why would you always expect to find longshore currents when the waves come into the beach at an angle?

31. Describe the motion of the sediment particles involved in the longshore transport.

32. What is a typical rate at which sand travels along the California coast? If human beings interrupt this flow of sand, what effect does this have on the upstream and downstream beaches?

33. Why is a breakwater parallel to the beach not a satisfactory solution to the problem of letting the sand continue down the beach uninterrupted?

34. Describe the flow of sand within one beach compartment.

35. What is usually at the downstream end of a beach compartment on the California coast? What else might there be?

36. If you were traveling along the beach, what would you look for to indicate the presence of a submarine canyon head nearby?

37. If one dams up the rivers and creeks, what effect will this have on the downstream beaches, and why? How might you solve this problem? Would it be expensive?

38. If you had to dam up some rivers, near which end of the beach compartment should you do it, and why?

39. In which part of a beach compartment should harbors be located and why?

*40. In doing correlation experiments such as those referred to in Questions 21 and 22 above, it is often difficult to separate the cause from the effect. (For example, is the beach steep because the sediment is coarse, or is the sediment coarse because the beach is steep?) How would you design experiments to distinguish the cause from the effect in the correlations mentioned in Questions 21 and 22?

*41. It still isn't entirely known what kind of instability it is in the backwash that creates backwash diamonds. Do you have any ideas?

*42. Can you describe how the sediment would have to move in order for the ripple marks to move backward to the direction of current and sediment flow? That is, sediment must be eroded from what part of one ripple and deposited on what part of the next one?

*43. How many different reasons can you think of to explain why the erosion of a sea cliff might be irregular (caves, gullies, holes, pillars, etc.)?

*44. Explain why beach sands don't get washed away, and why continents don't eventually get washed away by the waves.

*45. We saw that the finer sediments are usually associated with low flat beaches. What do you suppose would become of a load of gravel or cobbles dumped on a low flat beach?

*46. Would you expect longshore bars to be more common for low flat beaches, or for steep beaches? Why?

*47. Why do you suppose it's more comfortable to sunbathe at the beach than in your backyard?

*48. How do you suppose you could detect a rip current without getting into the water?

*49. What do you suppose becomes of the dead seaweed that washes up on our beaches?

SUGGESTIONS FOR FURTHER READING

1. Willard Bascom, "Beaches," *Scientific American* (Aug. 1960).

2. Willard Bascom, *Waves and Beaches*, Anchor/Doubleday, New York, 1964.

3. Dolan, Hayden, and Lins, "Barrier Islands," *American Scientist*, **68** (1980), p. 16.

4. Kim Fulton, "Coastal Retreat," *Sea Frontiers*, **27**, No. 2 (1981), p. 82.

5. Haynes R. Mahoney, "Dune Busting: How Much Can Our Beaches Bear?," *Sea Frontiers*, **26**, No. 6 (1980), p. 322.

6. "The Coast," *Oceanus*, **23**, No. 4 (Winter 1980/81).

7. Francis P. Shepard, *Geological Oceanography*, Chapters 5, 6, and 7, Crane, Russak and Co., New York, 1977.

10
THE COMPONENTS OF SEAWATER— PART 1

The "Castle Geyser," Yellowstone Park.

From what we have been able to learn with our telescopes, space probes, and other sophisticated equipment, the earth is a very special place. Among the members of our Solar System, it is even quite unique. The most striking difference is the copious quantities of liquid water covering most of the earth's surface. This in itself is intriguing, for all astronomical bodies are born of the same universe and of the same universal materials. Why, then, are we so different?

Water is such a familiar substance to us earthlings that we tend to take it for granted. But water is truly exceptional by any standards. Just look at how strikingly different our planet is from the others! Without it, the earth would be barren: no clouds, no rivers and streams, no oceans, no life. Temperatures here would vary by hundreds of degrees between day and night, and from season to season. Birds, trees, flowers, people, and all the other complex life forms could not have developed here without it.

Life depends not only on the water itself, but also on its remarkable ability to dissolve other materials and hold them in solution. Many of the materials dissolved in seawater are vital to the functions of living organisms such as ourselves. Others are essential in some of the physical and chemical processes that are continually remodeling the earth's surface. No other substance ever invented could possibly have nearly the profound influence on physical, biological, chemical, or geological processes as does the water found on the earth's surface.

As you can guess, we are naturally quite curious to find out just what it is about water that gives it these remarkable properties. And we arc equally curious to find out just how these properties influence the various physical, chemical, and biological processes here that are so critical to our existence.

Such a study is carried out in this chapter and the next. We first examine the water itself to learn about its remarkable properties. Then we examine the materials held in solution and the important physical, chemical, and biological processes that they are involved in.

A. THE WATER

A.1 Where Did It Come From?

A brief look around us in space reveals that our planet is exceptional. (See Figures 10.1 and 10.2.) We have 1.4 billion cubic kilometers of water on our surface, and our neighbors have little. Venus and Mars have very little surface water at best, and none at all in the liquid form. Our moon has no surface water. Since we all presumably condensed from the same primordial cloud of interstellar gas and dust, then our constituent materials[1] should all be the same. But judging from surface appearances, this isn't the case. So what makes us so special? Where did our surface water come from?

---

[1]These exclude the very light gases of hydrogen and helium, which have escaped due to insufficient surface gravity to retain them at the ambient temperatures.

FIGURE 10.1   *Mosaic of photographs of Mercury taken by Mariner 10. The resulting composite picture is the surface of Mercury as seen from above
its south pole.*

FIGURE 10.2   *View of Jupiter from above its north temperate latitudes, taken by
Pioneer 11.*

It is relatively easy to establish where our surface water did *not* come
from. It did not come from the crust, as the rocks of the crust don't hold
enough water to account for the oceans. It did not come from our atmosphere, as the atmosphere could hold only enough water to raise sea level
by a few centimeters at best. Furthermore, present scientific evidence
indicates that originally we had no atmosphere at all.

If the ocean water did not come from the crust or atmosphere, then
what was its source? The belief now is that through the dynamics of the
earth's mantle described in Chapter 3, there has been considerable outgassing of our more volatile materials from the mantle through the crust
and onto the earth's surface. Measuring the present rates of emergence of
new water from sulfur hot springs, volcanoes, geysers, and so on proves

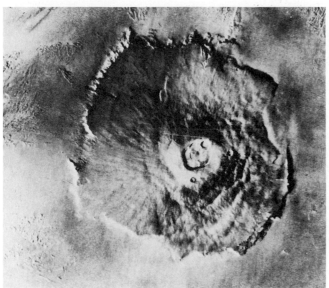

FIGURE 10.3 *Olympus Mons as viewed from above by Mariner 9. This gigantic volcano is more than 500 km in diameter at the base and is about 30 km high. The central caldera is about 70 km in diameter.*

to be quite difficult, but not impossible. The reason for the difficulty is that much of the water emerging from these sources is simply recycled surface water that has trickled down into cracks and fissures in the crust, and relatively little of it is water that is appearing on the surface for the first time.

The results of the measurements that have been made indicate that the present rate of outgassing is probably sufficient to account for the accumulation of the entire 1.4 billion cubic kilometers of surface water, providing the rates of outgassing have been roughly the same throughout the 4.6 billion years of history. Apparently, we gained most of our atmosphere the same way.

This theory provides only half the answer to the question posed, for although it explains where the earth's water came from, it does not explain why our neighbors in the Solar System should be different. We now believe that they are no different fundamentally, just a bit slower in their geological evolution. Because the moon, Venus, Mercury, and Mars are not as large as the earth, we would expect their interiors to be somewhat cooler (see Chapter 3, Section H). Cooler temperatures would imply less motion in their mantles, less surface volcanism, and therefore slower rates of outgassing. The discovery of volcanoes on the Martian surface by Mariner 9 (Figure 10.3), along with some ice at the Martian poles, has led to increased optimism that the "outgassing" theory is correct. The difference between nearby planets would then be the rate of outgassing, not their original chemical makeup.

It is the abundance of surface water that distinguishes the earth from its neighbors. It is the exceptional properties of water that make this distinction an attractive one. The water molecule has some very special prop-

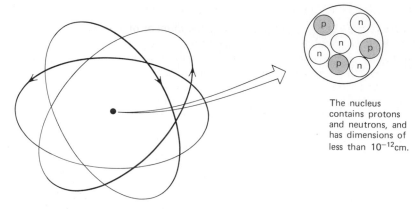

The nucleus
contains protons
and neutrons, and
has dimensions of
less than 10^{-12}cm.

The electrons in the "electron cloud"
have orbits with characteristic dimensions
of 10^{-8}cm. This is 10,000 times larger
than the nucleus.

FIGURE 10.4   *An atom is composed of an extremely small and dense, positively charged nucleus, with a much more nebulous cloud of electrons in orbit around it. Both protons and neutrons are nearly 2000 times more massive than electrons, so essentially all the mass of an atom is in its nucleus.*

erties not shared by other light molecules, such as methane, ammonia, hydrogen, or carbon dioxide, which are abundant in the atmospheres of some of our neighbors in the Solar System. In the following pages we will investigate these special properties of the water molecule.

A.2   Atomic Preliminaries

To appreciate the remarkable properties of water, we must understand the basic structure of molecules and their atomic constituents. The fundamental chemical building blocks of materials are the atoms. An atom is made up of a tiny, dense, positively charged nucleus with negatively charged electrons orbiting it, as illustrated in Figure 10.4. The electrostatic attraction between the two is what keeps the electrons in their orbits. Typically, the "electron cloud" of an atom is about 10^{-8} cm across, whereas the nucleus is more than ten thousand times smaller, being less than 10^{-12} cm across. Inside the nucleus are individual protons and neutrons. Each proton carries a positive charge, equal but opposite to that of a negatively charged electron. Neutrons have no charge, as the name implies. By itself, an atom is normally electrically neutral, so the number of electrons in the electron cloud equals the number of protons in the nucleus. Table 10.1 lists the atomic structures of the 13 lightest elements.

The simplest atom is hydrogen, being composed of just one proton with one electron in orbit around it. The shorthand notation for the element hydrogen is $_1H^1$, where the "H" stands for "hydrogen," the subscript indicates the number of protons in the nucleus (or equivalently, the number of electrons in the neutral atom), and the superscript indicates the total number of nucleons (neutrons plus protons) in the nucleus. Hydrogen is

TABLE 10.1   Nuclear Structure of the 13 Lightest Elements.

The chemistry of the element is determined by the number of electrons in the neutral atom, and this is the same as the number of protons in the nucleus. The number of neutrons in the nucleus has no bearing on the chemistry of the element, only on its mass. Nuclei of the same element, having different numbers of neutrons in the nucleus, are called "isotopes" of the element. Below are listed the relative abundances of the naturally-occurring isotopes for each of the first 13 elements.

Element	Number of Protons	Number of Neutrons	Nuclear Symbol	Fraction of that Element's Total Abundance
Hydrogen	1	0	$_1H^1$	99.9844%
	1	1	$_1H^2$	0.0156%
Helium	2	1	$_2He^3$	0.000134%
	2	2	$_2He^4$	99.999866%
Lithium	3	3	$_2Li^6$	7.4%
	3	4	$_2Li^7$	92.6%
Beryllium	4	5	$_4Be^9$	100%
Boron	5	5	$_5B^{10}$	18.83%
	5	6	$_5B^{11}$	81.17%
Carbon	6	6	$_6C^{12}$	98.892%
	6	7	$_6C^{13}$	1.108%
Nitrogen	7	7	$_7N^{14}$	99.64%
	7	8	$_7N^{15}$	0.36%
Oxygen	8	8	$_8O^{16}$	99.76%
	8	9	$_8O^{17}$	0.04%
	8	10	$_8O^{18}$	0.20%
Fluorine	9	10	$_9F^{19}$	100%
Neon	10	10	$_{10}Ne^{20}$	90.51%
	10	11	$_{10}Ne^{21}$	0.28%
	10	12	$_{10}Ne^{22}$	9.21%
Sodium	11	12	$_{11}Na^{23}$	100%
Magnesium	12	12	$_{12}Mg^{24}$	78.6%
	12	13	$_{12}Mg^{25}$	10.1%
	12	14	$_{12}Mg^{26}$	11.3%
Aluminum	13	14	$_{13}Al^{27}$	100%

by far the most abundant element in the universe, as you might guess from its simplicity.

The second simplest atom is also the second most abundant in the Universe. This is helium, $_2He^4$. It has two protons and two neutrons in the nucleus, and two electrons in the electron cloud. Together, hydrogen and helium seem to comprise more than 99% of all materials in the universe.

You might think that the third most abundant element would be the third simplest, lithium, which has three protons, followed by beryllium with four, boron with five, and so on, but that is not the case. It turns out

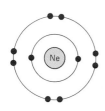

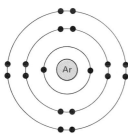

FIGURE 10.5
*Schematic illustration
of the preferred sym-
metric electronic con-
figurations, which are
those of the inert
gases. Here are heli-
um, neon, and argon,
which have one, two,
and three completed
electron "shells," re-
spectively. It should
be remembered, how-
ever, that the electrons
in real atoms are mov-
ing extremely quickly
at all times. This dia-
gram, therefore, is a
conceptual aid only,
and not an accurate
representation of re-
ality.*

TABLE 10.2   Estimated Relative Abundances of the 20 Most Common Elements
in the Universe, in Terms of the Number of Atoms of That Element
Per Million Hydrogen Atoms.

*Other than hydrogen and helium, the abundances of the elements are a reflection
of nuclear reactions in stars. It is seen that hydrogen and helium comprise the vast
majority of all materials, Presumably, the earth and inner planets are made of the
small residue left over when the hydrogen and helium escaped.*

Element	Atomic Number[a]	Atomic Weight[b]	Abundance Relative to 10^6 Hydrogen Atoms
Hydrogen (H)	1	1	1,000,000
Helium (He)	2	4	80,000
Oxygen (O)	8	16	690
Carbon (C)	6	12	420
Nitrogen (N)	7	14	87
Silicon (Si)	14	28	40
Neon (Ne)	10	20	37
Magnesium (Mg)	12	24	32
Iron (Fe)	26	56	25
Sulphur (S)	16	32	16
Aluminum (Al)	13	27	3.3
Calcium (Ca)	20	40	2.5
Nickel (Ni)	28	59	2.1
Sodium (Na)	11	23	1.9
Argon (Ar)	18	40	1.0
Chromium (Cr)	24	62	0.69
Phosphorus (P)	15	31	0.39
Manganese (Mn)	25	55	0.26
Chlorine (Cl)	17	35	0.22
Potassium (K)	19	39	0.12

[a]Atomic number is the number of protons in the nucleus.
[b]Atomic weight is the total number of nucleons in the nucleus, given here for the most common
isotope of each.

that the heavier elements are manufactured in the nuclear fusion reactions inside of stars, and their relative abundances reflect the details of
these complicated reactions. As a result, for instance, the sixth, seventh,
and eight elements, carbon ($_6C^{12}$), nitrogen ($_7N^{14}$), and oxygen ($_8O^{16}$) are all
more abundant than the third, fourth, and fifth elements. The relative
abundances of the 20 most common elements are listed in Table 10.2.

A.3   Molecules
The chemistry of an atom is determined by its electrons. Just as a ball will
roll downhill rather than up, the electrons of an atom also prefer to find
the state of lowest energy. Because of this, certain electron configurations
are preferred over others. It turns out that certian symmetric configurations involving one "shell" with two electrons, or an additional shell with
eight more electrons (10 altogether), or a third shell with still eight more
electrons (18 altogether), or additional larger completed "shells," are the
electronic arrangements most preferred by atoms. (See Figure 10.5.)

As an example of how this determines the chemistry of an atom, we notice that hydrogen, $_1H^1$, has only one electron. This is one short of the preferred arrangement involving two electrons. Consequently, hydrogen willingly combines with other atoms to share an electron in order that some of the time it may enjoy the preferred state of two electrons. For example, hydrogen will combine with another hydrogen atom, when possible, to form the H_2 molecule. (Here the subscript on the right indicates the number of atoms of hydrogen in one molecule.)

The second simplest atom, helium, already has its preferred arrangement of a completed two-electron shell. Consequently, it doesn't share electrons at all with other atoms, and so it doesn't interact chemically at all. In fact, even the interatomic forces between helium atoms themselves are so weak that helium doesn't even liquefy until it is cooled below 5.2K. Consequently, it is called an "inert gas." Other inert gases are neon ($_{10}Ne^{20}$), with its full complement of 10 electrons in two completed electron shells, argon ($_{18}Ar^{40}$), with its full complement of 18 electrons in three completed electron shells, and three other, much heavier and rarer inert gases, with four, five and six completed shells, respectively (krypton, xenon, and radon).

As final examples of how the electrons determine the chemistry of atoms, we look at carbon, nitrogen, and oxygen (Figures 10.6 and 10.7). Carbon ($_6C^{12}$) is four electrons short of the preferred arrangement of 10 electrons in two closed shells. It is also four electrons in excess of the two-electron shell arrangement. Consequently, it is quite eager to share four electrons with other atoms (we say, it has a "valence" of $+4$ or -4), which enables it to form the basis of quite complicated molecules, such as those in living tissues. As a particular example, one carbon with its valence of four, willingly combines with four hydrogens, each having a valence of one, to form the CH_4 molecule, called "methane." Because of the abundance of hydrogen, this is a fairly abundant molecule in the universe, although not nearly as abundant as H_2 or He. Similarly, nitrogen, $_7N^{14}$, is three short of the preferred arrangement of 10 electrons, and so it willingly combines with other atoms to share three electrons. Ammonia, NH_3, is a fairly common molecule in the universe. Finally, oxygen ($_8O^{16}$) needs two electrons to form a completed shell, and so it willingly shares two, forming water, H_2O. Notice that two oxygens each having a valence of two could combine with carbon, having a valence of four, forming carbon dioxide, CO_2. This would not be as abundant as CH_4 or H_2O though, because of the greater abundance of hydrogen in the universe.

It is particularly important for the chemistry of oceanography that electrons shared by two atoms are usually not shared equally. That is, they usually spend more than half the time near one of the atoms, and less than half the time near the other. As an example, examine the sodium chloride molecule, NaCl. Sodium, $_{11}Na^{23}$, has one more electron than is necessary for the preferred state of 10 electrons in two completed shells. So it eagerly gives up one electron, and isn't particularly eager to take it back. Chlorine ($_{17}Cl^{25}$), on the other hand, is one electron short of the preferred electronic configuration of 18 electrons in three completed shells. So it

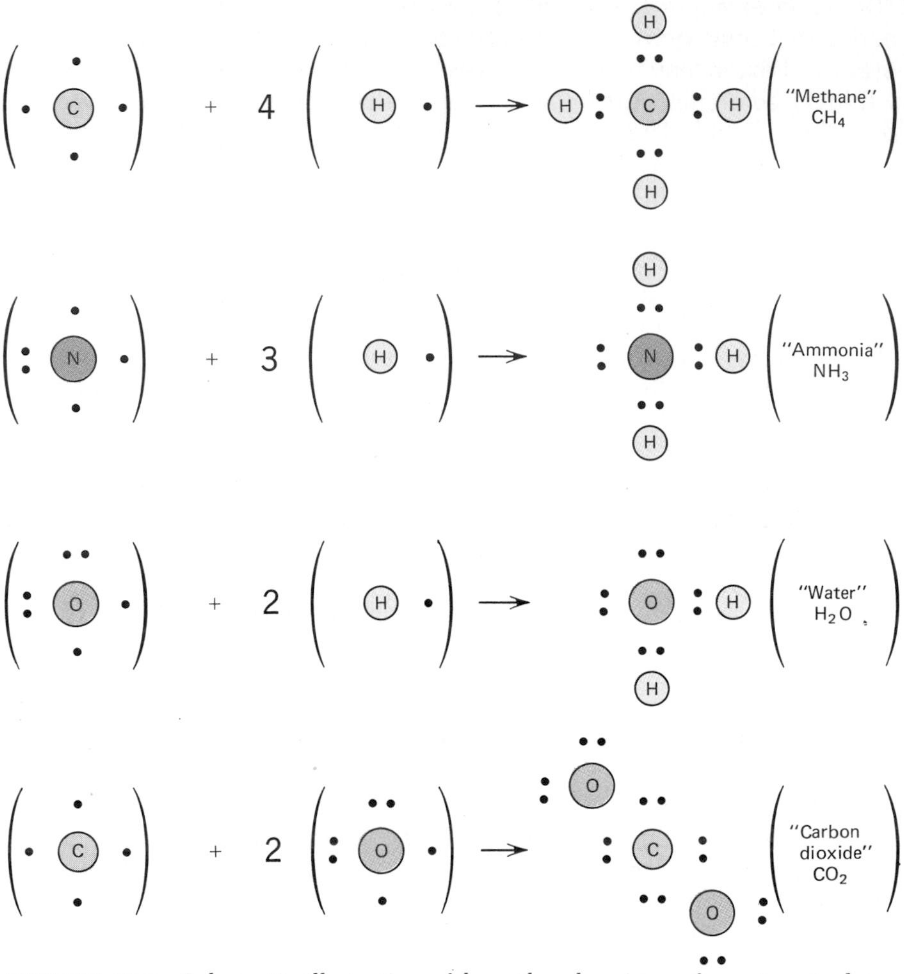

FIGURE 10.6 *Schematic illustration of how the chemistry of an atom is determined by the desire to have completed outer electron shell (i.e., the ideal gas electronic configuration). Carbon is four electrons short of the "neon" configuration with eight electrons in the outer shell. So it gladly shares its four with four others. Similarly, nitrogen shares three and oxygen two, for the same reason. Hydrogen is one electron short of the "helium" configuration with two electrons in the first shell. So it eagerly shares its one with one other.*

accepts one electron, but isn't particularly eager to give it back. Consequently, when these two get together, the "shared" electron spends about 94% of its time near the chlorine atom and only about 6% of its time near the sodium atom. When this salt dissolves in water, the chlorine atom usually takes the extra electron with it as the NaCl molecule breaks up, yielding Na^+ and CL^- "ions," rather than the neutral atoms.

FIGURE 10.7 *Three-dimensional illustration of the location of the hydrogen atoms in the methane, ammonia, and water molecules. (a) The methane molecule is quite symmetrical, with the hydrogens located at the four corners of a tetrahedron, and the carbon atom inside. (b) In the ammonia molecule, the hydrogens are located at three of the corners of a tetrahedron, and the nitrogen atom oscillates back and forth through the plane of the hydrogen atoms. This makes the ammonia molecule quite symmetrical, too. (c) The two hydrogens in the water molecule are located at two corners of the tetrahedron, and so the molecule is quite asymmetrical. This asymmetry is responsible for all the remarkable properties, as we shall see.*

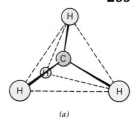

(a)

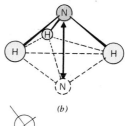

(b)

A.4 The Lighter Molecules of Atmospheres and Oceans

We have seen in earlier chapters that as a planet evolves, denser materials tend to sink deeper toward the core of the planet, and lighter materials tend to work their way toward the surface to form the crust, oceans, or atmosphere.

The very lightest materials would be hydrogen gas (H_2 molecules) and helium. But because they are so light, it is difficult for a planet to retain them in the atmosphere. These lightest and most volatile molecules move the quickest at any temperature. At the ambient temperature of the earth, for example, the molecules of these two materials quickly escape from the earth, as at any instant a fraction of them are traveling faster than the earth's escape velocity (Table 10.3). In order to retain these very light molecules, then, a planet must either be very cold (so the molecules move more slowly and don't exceed the escape velocity), or it must have a very strong gravity.

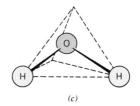

(c)

TABLE 10.3 Tabulation of the Average Speeds of Various Gas Molecules at Temperatures Typical of Our Present Environment (0°C).

Notice that heavier molecules move more slowly. Notice also that the motion is extremely fast (measured in km/sec).

Molecule	Molecular Weight	Average Speed, km/sec	Root Mean Square Speed, km/sec
H_2	2	1.70	1.84
He	4	1.20	1.30
CH_4	16	0.60	0.65
NH_3	17	0.58	0.63
H_2O	18	0.56	0.61
N_2	28	0.45	0.49
O_2	32	0.42	0.46
CO_2	44	0.36	0.39

TABLE 10.4 Molecular Weights of Some Lighter Molecules That Might Be
Found in Planetary Atmospheres, Using the Most Common Isotope
for Each Element.

A common mineral in ordinary crustal rocks is included for comparison.

Molecule	Constituents	Molecular Weight of Molecule
H_2	$2(H^1)$	2
He	$1(He^4)$	4
CH_4	$1(C^{12}) + 4(H^1)$	16
NH_3	$1(N^{14}) + 3(H^1)$	17
H_2O	$1(O^{16}) + 2(H^1)$	18
N_2	$2(N^{14})$	28
O_2	$2(O^{16})$	32
CO_2	$1(C^{12}) + 2(O^{16})$	44
$FeMgSiO_4$	$1(Fe^{56}) + 1(Mg^{24}) + 1(Si^{28}) + 4(O^{16})$	172

The inner planets of Mercury, Venus, Earth, Mars, and our moon are too close to the sun, and therefore, too warm to be able to hold onto gaseous hydrogen or helium. The reason that the outer planets, Jupiter, Saturn, Uranus, and Neptune, are so much larger than Earth is probably that larger distance from the sun, and consequent cooler temperatures, have enabled them to retain hydrogen and helium, which are so much more abundant than the heavier materials from which our planet is made. We are the "residue" that's left after the hydrogen and helium escapes.

The next group of moderately light materials would be those molecules composed of hydrogen plus one moderately light atom, such as carbon, nitrogen, and oxygen. These would be methane (CH_4), ammonia (NH_3), water (H_2O), and fragments of these, such as CH, NH, and so on. Still heavier would be molecules involving two or more moderately light atoms, such as O_2, N_2, CO_2, and so on. These also would tend to work their way into the surface oceans and atmospheres of planets. Still heavier molecules would involve heavier atoms, like those we studied in the chapter on sediments. The molecular weights of some common light molecules that might be found in planetary atmospheres are listed in Table 10.4.

You may wonder why H_2O tends to be a solid or liquid, whereas NH_3, CH_4, and even the heavier O_2, N_2, and CO_2 are gases on the earth's surface. This is just one reflection of the properties that make water an extremely unique material, as will be studied in the pages to follow. For now we'll just say that we owe our lives to these unique properties, and it is because of these unique properties that we have such an interest in the presence of water elsewhere in the Solar System. We know there is some water in the Venusian atmosphere, there is ice on Mars, and probably much more within the surface sediments, and there may be permafrost beneath the surface of the moon. Three of Jupiter's four largest satellites (two of these about the size of Mercury) have large amounts of surface water in frozen form. Clearly, oceanographers will be asked increasingly in the future to expand their studies beyond the earth.

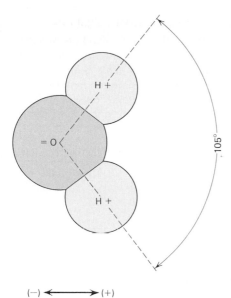

$(-)$ ◄————► $(+)$

FIGURE 10.8 *In the water molecule, the shared electrons are not shared equally. They spend more time in the vicinity of the oxygen atom and less time in the vicinity of the hydrogens. This makes the molecule electrically polarized, with the oxygen side carrying a negative charge, and the hydrogen region carrying a positive charge.*

A.5   The Polarized Water Molecule
The things that make water such a distinctive substance can be traced to its molecular structure. The shared electrons are not shared equally. Oxygen is a bit greedy for shared electrons, and hydrogen is more generous. Consequently, the shared electrons spend more of their time around the oxygen atom than in the neighborhood of the hydrogen atoms. This means that the region around the oxygen atom carries a net negative charge, and the hydrogen region is positively charged. Furthermore, both hydrogen atoms are located on the same side of the molecule rather than being at opposite ends, as illustrated in Figure 10.8. This means that the molecule is not symmetrical, but rather is shaped like a small boomerang. One side of the molecule carries a net negative charge, and the other side a net positive charge. Thus, the molecule is strongly electrically polarized, and to this we can attribute most of its remarkable properties.

A.6   Physical Properties of Water
One of these properties is that ice melts, and water boils at amazingly high temperatures for a molecule that is so light. The methane molecule is similar in weight to the water molecule, but methane melts at 89K and boils at 112K, whereas the melting and boiling points of water are 273K and 373K. Even molecules considerably heavier than water have lower melting and boiling points. Oxygen (O_2) and nitrogen (N_2) have nearly twice the molecular weight, but it can be seen from Table 10.5 that they melt

TABLE 10.5   The Melting and Boiling Points For Materials With Light Molecules.

At atmospheric pressure, heavier molecules move more slowly, which is reflected in higher melting and boiling points, usually. Notice that the melting and boiling points for water are extremely high for a molecule so light. This is because it is electrically polarized, and therefore "sticky." The two organic materials are included to show that for fairly symmetrical molecules, their masses must be roughly six times heavier in order to have similarly high melting and boiling points.

Material	Molecular Weight	Melting Point (K)	Boiling Point (K)
H_2	2	14	20
He	4	—	5.2
CH_4	16	90	111
H_2O	18	273	373
N_2	28	63	77
O_2	32	55	90
HCl	36	161	189
CO_2	44	(sublimes)	194
C_2H_5OH (ethyl alcohol)	46	161	351
C_8H_{18} (octane)	114	216	399

and boil at much lower temperatures. Carbon dioxide has nearly three times the molecular weight of water, yet it remains a solid only until 195K.

The reason for the elevated melting and boiling points of water is that the highly polarized water molecules tend to stick together. Consequently, when ice melts, it takes a good deal more energy and higher temperatures to shake the molecules apart from each other than it would for other substances of similar atomic weights. Similarly, when water is boiling, it takes more energy (higher temperatures) to shake water molecules loose from the liquid surfaces and into the gaseous state than it would if the water molecules were not so polarized and didn't stick together. It is clear that if the water molecule weren't so polarized, it would not be a liquid here at the ambient temperatures, and we wouldn't have any oceans.

As we saw in the above paragraph, because the water molecules are "sticky," it takes a great deal more heat energy to accomplish a change in water than it would to make the same change in other substances. We say that water has a very high "heat capacity." For example, much more heat must be removed to cool a cup of hot soup than to cool a cup of hot air, sand, or popcorn by the same amount.

Heat energy is measured in a unit called a "calorie," which is defined as the amount of heat required to raise the temperature of one gram of water by one degree Celcius. So the heat capacity per gram of water[2] is one calorie per gram per degree Celcius, and this is much higher than the heat capacity per gram of any other common substance.

Ice has a very high "latent heat of fusion," which means that once

[2]*Heat capacity per gram is called "specific heat."*

you bring it to the melting point, you must still add a lot of heat just to melt it. All this energy goes into shaking the molecules loose from their frozen, rigid, stuck-together structures without raising the temperature at all. (That is, the heat changes ice at 0°C to liquid water at 0°C.) Also, water has an exceptionally high "latent heat of vaporization," which is just the large amount of energy to shake the water molecules loose from the liquid surface and into the gaseous state, when the temperature is at the boiling point. These concepts will be quantified in Chapter 12, where we'll study the essential role that these remarkable thermal properties play in moderating our climate on earth. But for this chapter, it is important that you understand how the *cause* of these properties is traced to the polarization of the water molecule.

Another interesting property of water is that it freezes *over* instead of freezing *under*. As most materials freeze, the cooler, denser material sinks, and the frozen portions will first appear on the bottom rather than at the top. But in the case of water, it turns out that the solid phase, ice, is *less* dense than the liquid phase.

The reason for this can be traced back to the polarization of the water molecule. As the water cools, the thermal motion of the molecules lessens, and they begin sticking together in fairly rigid structures, with one positive hydrogen portion of one molecule attracted to the negative oxygen portion of the next. The angular structure of the individual molecules makes for an angular crystal structure, which is quite porous, resembling chicken wire in three dimensions (Figure 10.9). Consequently, this porous solid phase is less dense than the liquid phase, and so it floats on top.

As ice melts, these structures start breaking apart, and the pores disappear. Actually, even after the ice has melted, there is still some tendency for neighboring molecules to stick together, and, consequently, pure water doesn't reach its highest density until 4°C. Dissolved salts do have some effects on these properties and we'll see later, but in any case it is far different from the pattern of most materials, which simply get denser as they get colder in all their various phases.

For water, both the solid and the gaseous states are less dense than the liquid state. This means that it must expand when it freezes, and it must expand when it boils. Under heavier pressures, expansion becomes more difficult, and so the atmospheric pressure has an effect on the freezing and boiling points of water. Under heavier pressures these transitions become more difficult, and so you must cool the water further before it will freeze, and heat it hotter before it will vaporize. Under lower pressures the converse is true, expansion is easier, so water will freeze at higher temperatures and boil at lower temperatures (Table 10.6). In fact, if you hook a flask of water to a vacuum pump and reduce the pressure to 0.6% of one atmosphere, it will boil and freeze at the same time. Figure 10.10 displays a "phase diagram" for water. On it you can see at least qualitatively how the transitions between phases are effected by both temperature and pressure.

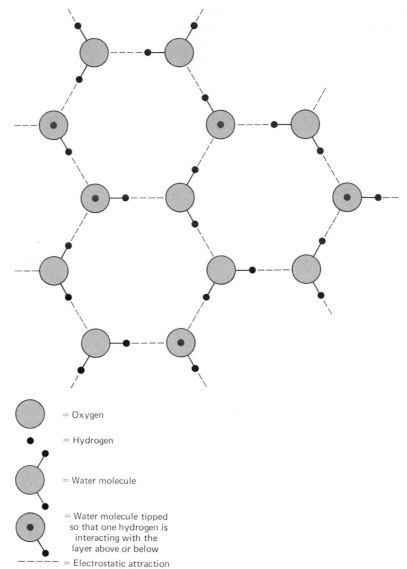

= Oxygen

= Hydrogen

= Water molecule

= Water molecule tipped
so that one hydrogen is
interacting with the
layer above or below

- - - - - = Electrostatic attraction

FIGURE 10.9   *Molecular view of one layer in an ice crystal. Actually, each "layer"
is not flat; alternate oxygens in the above diagram should be slightly
above or slightly below the plane of the paper. The porous, hexagonal structure is clearly evident, and is reflected in the fact that ice
floats and snowflakes have hexagonal symmetry.*

A.7   Solvent Properties of Water

Another important property of water that is attributed to its polarized molecule is that it is an extremely good solvent. It is one of nature's best universal solvents, and definitely the best one that occurs in appreciable
quantities. It is to this property of water, in particular, that we are

TABLE 10.6   The Boiling Point of Water as a Function of the Pressure.

Pressure (Atmospheres)	Boiling Point (°C)
0.90	97.1
1.00	100.0
1.10	102.9

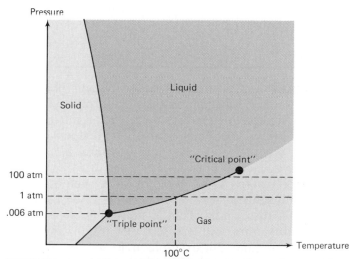

FIGURE 10.10   *Phase diagram for water (the scale is not linear). At higher pressures, the melting point gets lower, and the boiling point gets higher. Especially interesting is the "triple point," where water can freeze and boil at the same time. Also interesting is the "critical point," above which there is no clear noticeable difference between liquid and gaseous phases.*

indebted for life, and the rest of this chapter and the next will be devoted to studying it.

When a molecule of a foreign substance is introduced into water, the positively charged hydrogen ends of the water molecules will seek out the negatively charged portions of the foreign molecule, and the negatively charged end of the water molecules will pull at the positively charged regions of the intruder. Together, they'll tear the molecule loose from the parent material, and in many cases, they'll even tear the molecule itself apart as they dissolve it. The only materials which are not soluble in water are those having extremely even charge distributions, so that neither end of the polarized water molecule will find something to pull on. The water molecules are sort of like miniature, hungry piranha, and it is a rare material that they can't find some way of getting their teeth into.

One of these rare materials is oil. As the student knows, "oil and water don't mix," and this implies that oil molecules must have a very symmetric charge distribution. This is no accident, of course, as oils are

TABLE 10.7 Dissociation of the Most Common Ions in Seawater

Ion	Percent as a Free Ion
Positive ions	
Sodium (Na$^+$)	99
Magnesium (Mg^{++})	87
Calcium (Ca^{++})	91
Potassium (K$^+$)	99
Negative ions	
Chloride (Cl$^-$)	100
Sulfate (SO$_4^{--}$)	50
Bicarbonate (HCO$_3^-$)	67
Bromide (Br$^-$)	100

derived from organisms (present or past) that necessarily have produced tissue materials insoluble in water. Most known organisms rely on the solvent properties of water to transport needed materials to and into the cell and waste products out and away. But, of course, the organic tissues themselves must be insoluble, or else the cell itself would disappear along with the waste products.

The most soluble materials tend to be the ionic salts, such as NaCl, which tend to dissociate into charged ions when dissolved. Most salts in seawater are more than 90% dissociated. That is, more than 90% of the molecules will have been broken up, and less than 10% of the salts' molecules will be intact at any one time. For example, most sodium chloride in seawater will be found as separate Na$^+$ or Cl$^-$ ions. Relatively few NaCl molecules will be found intact. A few salts are less than 90% dissociated. For example, only about half of the sulfate appears as the dissociated SO$_4^{--}$ ion, and only about 87% of the Mg^{++} is completely dissociated. (See Table 10.7.)

When the water is holding as much of a certain material as it can hold, it is said to be "saturated" with that material. The amount of a material that a liter of water can dissolve before it becomes saturated depends on the balance between how strongly a molecule of that material is attracted by the water molecules and how strongly it is attracted by other molecules of its own kind. For example, as a salt block is first immersed in water, the salt will begin dissolving quickly. But as the solution becomes saltier, more and more water molecules will find themselves tied up with dissolved salt ions, and fewer and fewer water molecules will be available to attract new salt ions (Figure 10.11). Consequently, as the solution becomes saltier, the water's affinity for additional salt molecules becomes less and less. Eventually, a point is reached where a salt molecule is attracted equally strongly by the remaining salt block as by the water. At this point, as many salt molecules will crystallize out of the solution as are dissolved into the solution. The solution gets no saltier, and the salt block gets no smaller. The solution is "saturated."

The amount of a material that water can hold when saturated varies greatly from one material to the next. It also varies with the temperature

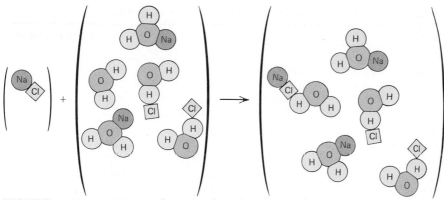

FIGURE 10.11 *As more and more salt molecules dissolve in the water, there are fewer and fewer water molecules left to attach to the salt ions. Eventually, attraction by the water molecules is reduced to a point where incoming salt ions are attracted as strongly by other salt ions as they are by the water, and they begin to crystallize out of the solution. When this happens, we say that the solution is "saturated."*

TABLE 10.8 Solubility of Various Salts in Water as a Function of the Temperature.

Units are kilograms of dissolved salt per kilogram of water.

| | Temperature (°C) | | | |
Salt	0	10	20	30
NaCl	0.357	0.358	0.360	0.363
MgSO$_4$	0.409	0.422	0.445	0.453
CaCl$_2$	0.595	0.650	0.745	1.020
K$_2$CO$_3$	1.050	1.080	1.110	1.140
Ca(HCO$_3$)$_2$	1.615	1.638	1.660	1.682

and pressure of the solution. Hotter temperatures mean more lively molecules, which generally increases the solubility of most salts (Table 10.8), as one might expect. However, we shall see that the solubility of gases *decreases* with increased water temperature.

Increased pressure increases the solubility of all materials. The reason for this is that the combined volume of the material and the water is less when the two are in solution than when they're not. That is, if you dissolve one cup of table salt in six cups of water you find the resulting solution has a volume less than the seven cups that it had originally. Increased pressure naturally makes the system favor decreased volume, which tends to put more of the material into solution.

Since opposite charges attract, dissolved salts tend to have positive parts congregating around the negative end of the water molecule and negative parts congregating around the positive end of the water mole-

TABLE 10.9   Abundances of the Various Elements in Seawater

Element		Concentration, Parts Per Billion	Element		Concentration, Parts Per Billion
Oxygen	O	857,000,000	Nickel	Ni	2
Hydrogen	H	108,000,000	Vanadium	V	2
Chlorine	Cl	19,000,000	Manganese	Mn	2
Sodium	Na	10,500,000	Titanium	Ti	1
Magnesium	Mg	1,350,000	Tin	Sn	0.8
Sulfur	S	890,000	Cesium	Cs	0.5
Calcium	Ca	400,000	Antimony	Sb	0.5
Potassium	K	380,000	Selenium	Se	0.4
Bromine	Br	65,000	Yttrium	Y	0.3
Carbon	C	28,000	Cadmium	Cd	0.1
Strontium	Sr	8,000	Tungsten	W	0.1
Boron	B	4,600	Cobalt	Co	0.1
Silicon	Si	3,000	Germanium	Ge	0.06
Fluorine	F	1,300	Chromium	Cr	0.05
Argon	A	600	Thorium	Th	0.05
Nitrogen[a]	N	500	Silver	Ag	0.04
Lithium	Li	170	Scandium	Sc	0.04
Rubidium	Rb	120	Lead	Pb	0.03
Phosphorus	P	70	Mercury	Hg	0.03
Iodine	I	60	Gallium	Ga	0.03
Barium	Ba	30	Bismuth	Bi	0.02
Indium	In	20	Niobium	Nb	0.01
Zinc	Zn	10	Lanthanum	La	0.01
Iron	Fe	10	Thallium	Tl	<0.01
Aluminum	Al	10	Gold	Au	0.004
Molybdenum	Mo	10	Cerium	Ce	0.005
Copper	Cu	3	Rare Earths		0.003–0.0005
Arsenic	As	3	Protoactinium	Pa	2×10^{-6}
Uranium	U	3	Radium	Ra	1×10^{-7}

[a]Nutrient nitrogen only; the dissolved gas is not included.

cule, neutralizing the charges at both ends. Therefore, dissolved salts have the effect of tempering the charge separation of the bare water molecule. Since this charge separation was responsible for the remarkable properties of water, it is no surprise that these remarkable properties are also tempered when salt is in solution. For example, at the 3.5% salinity of most ocean water the freezing point is lowered to $-1.9°C$. (Still it is remarkably high when compared to other materials of similar molecule weight.) Similarly, its heat capacity is reduced by a few percent.

When water vaporizes, the dissolved materials are left behind, except for the dissolved gases, of course. Similarly, when water freezes, the ice crystals show a definite preference toward molecules of their own kind, and the dissolved salts tend to be rejected. In fact, only about one third of the original salt is incorporated into the ice. Sea ice, then, is only about 1% salt, and would be fit to furnish drink for thirsty sailors (if necessary), even though seawater is not.

TABLE 10.10 Amounts of the Principal Salts and Ions in Seawater of Salinity 34.32‰[a]

Material	Grams per Kilogram of Seawater	Percent of Total Salt by Weight
Chloride (Cl^-)	18.980	55.04
Sodium (Na^+)	10.556	30.61
Sulfate (SO_4^{--})	2.649	7.68
Magnesium (MG^{++})	1.272	3.69
Calcium (Ca^{++})	0.400	1.16
Potassium ($K^|$)	0.380	1.10
Bicarbonate (HCO_3^-)	0.140	0.41
Bromide (Br^-)	0.065	0.19
Boric acid (H_3BO_3)	0.026	0.07
Strontium (Sr^{++})	0.013	0.04
Fluoride (F^-)	0.001	0.00

[a]From Sverdrup, Johnson, and Fleming, *The Oceans*, Prentice-Hall, Englewood Cliffs, New Jersey, copyright 1942, renewed 1970.

B. SOURCES OF THE DISSOLVED SALTS

Because water is such a good solvent, it is no surprise that just about everything imaginable can be found in seawater. (See Table 10.9.) Of the 92 stable elements found on earth, more than 80 have been found in seawater, and there is every reason to believe that the others are there too, albeit in sufficiently small quantities to have escaped detection so far. These observations tantalize the curiosity of the scientists. We must see if we can tell where these various materials came from, how they got into the water, and why they are present in the observed abundances.

B.1 Weathering of the Crust

It is often assumed that the salts in the ocean (Table 10.10) are there because the rains and subsequent water drainage leach them from the soil and carry them out to sea. However, in analyzing the soluble materials in river water (Table 10.11), we find them not in the same proportion as they are present in the sea. This is not all that alarming, as many materials undergo chemical, geological, or biological processes, which alter them or remove them from the sea. For example, potassium is used in biological processes, but if the dead organism decays completely, it will return to the sea. As another example, calcium carbonate is used in some animal skeletons, such as coral, which may remove it from the sea altogether. We have seen that it is expended in formation of some non-biogenic sediments as well. From considerations such as these, we must expect some differences between the relative abundances of dissolved materials in seawater and river water.

To determine whether a certain substance could have come from weathered crustal rock, we use the following equation.

TABLE 10.11 Average Amounts of the
Principal Salts and Ions in
River Water[a]

Material	Percent of Total Salt, by Weight
Carbonate (CO_3^{--})	35.15
Calcium (Ca^{++})	20.39
Sulfate (SO_4^{--})	12.14
Silica (SiO_2)	11.67
Sodium (Na^+)	5.79
Chloride (Cl^-)	5.68
Magnesium (Mg^{++})	3.41
Oxides (Fe_2O_3 and Al_2O_3)	2.75
Potassium (K^+)	2.12
Nitrate (NO_3^-)	0.90

[a]From C. W. Wolfe et al., *Earth and Space Sciences*, D. C. Heath, Boston, Mass., 1966.

$$\text{(content in original crystalline rock)}$$
$$= \text{(content now in oceans and atmosphere)}$$
$$+ \text{(content remaining in the sediment)}$$

That is, the materials that were previously in the original rock are either still in its weathered remains, or are now in the oceans and atmosphere. The last two parts of this equation are readily determined at the earth's surface, but to measure the content of the substance in the original crystalline rock, we must bring up samples from sufficiently deep in the earth's crust that they couldn't yet have been weathered.

Carrying out this analysis for the various substances in the oceans, we find, indeed, that there are many, such as calcium, magnesium, sodium, and potassium, whose abundance in the oceans is quite compatible with the available supply in the crust. In addition, there is another group of materials, such as gold, copper, or uranium, that are so rare that they could have come from anywhere, so no one worries about explaining their presence.

B.2 The Problem of the Excess Volatiles

However, there is a third and very important group, including chlorine, bromine, sulfur, nitrogen, carbon, and the water itself, that are present far too much in excess to be explained by the weathering of crustal rock (Table 10.12). They are called the "excess volatiles." As you might expect, their abundance has provided much food for the imagination.

For a while it was thought that they might have been held in a very heavy primordial atmosphere. This theory has fallen into disfavor, however, as a result of studies of the temperature of the primordial atmosphere. The way this early temperature is known is by looking at the

TABLE 10.12 Maximum Amounts of
Various Excess Volatiles
That Could Have Come
From Crustal Rocks

Material	Fraction That Could Have Come From the Crust
Water (H_2O)	1%
Carbon (C)	1%
Chlorine (Cl)	2%
Nitrogen (N)	1%
Sulfur (S)	20%

TABLE 10.13 The Average Molecular Velocities for the Various Inert Gases at
Various Temperatures

For comparison, the escape velocity from the earth is 11.2 km/sec.

Inert Gas	Molecular Weight	Average Speeds (km/sec)		
		T = 300K	T = 1200K	T = 4800K
Helium	4	1.26	2.52	5.04
Neon	20	0.56	1.12	2.24
Argon	40	0.40	0.80	1.60
Krypton	84	0.28	0.55	1.10
Xenon	131	0.22	0.44	0.88

amount of depletion of the various inert gases from our atmosphere, compared to their relative abundance elsewhere in the universe, as inferred
from stellar spectra, for example. Being chemically inert, they should have
been present in the early atmosphere, and subsequent chemical processes
would have had no effect on their abundances. The amount of depletion
of a particular inert gas depends on what fraction of its molecules have
exceeded the escape velocity from the earth. The speed of a molecule, in
turn, depends on its mass and on the temperature, lighter masses and
higher temperatures implying higher speeds (Table 10.13). The relationship between molecular speed, mass, and temperature is well known (Figure 10.12), so one can use the known molecular masses and measured
depletion of the various inert gases (Figure 10.13) to calculate the temperature of the early atmosphere.

It is found that the temperature of the primordial atmosphere was sufficiently high that it wouldn't have held much of anything. The molecular
velocities of virtually all the excess volatiles would have exceeded the
earth's escape velocity, and not much of an atmosphere would have
remained. Therefore, the excess volatiles now present must have come
from somewhere else.

B.3 The Solution of the Problem

It is now believed that the correct explanation of the excess volatiles must
be in the outgassing from the mantle. A check of the bubbles trapped in

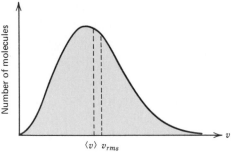

FIGURE 10.12   *Plot of the distribution of speeds among the molecules of a gas. The calibration of the axis depends on the temperature, but the shape of the distribution is the same for all temperatures. The "root mean square speed," v_{rms}, is slightly larger than the average speed. Notice that a small number of molecules will have speeds much greater than the average speed. This means that a gas might slowly escape the earth, even if the average speed is much smaller than the escape velocity.*

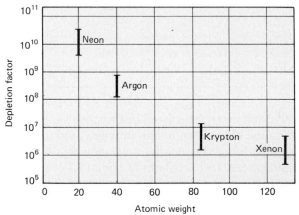

FIGURE 10.13   *Plot of "what once was, to what is now," for the various inert gases, showing the strongest depletion for the lightest atoms. From this, we can tell the temperature of the primordial atmosphere. (From Peter K. Weyl, Oceanography, John Wiley, 1970, and G. P. Kuiper, The Atmospheres of the Earth and Planets, pp. 258–266, Univ. of Chicago Press, 1952.)*

solidified igneous rock, formed deep in the crust where they couldn't have interacted with ground water or sediment, shows they contain the excess volatiles in about the correct ratios. An independent check comes from checking fallen meteorites for their water content. Assuming them to be made roughly of the same material as the earth, we deduce that about ½% of the mantle must be water. To explain the oceans, then, we only need to outgas less than 10% of the mantle's water content. This seems like a reasonable amount in view of the large amount of geological activity that the earth has undergone, and present-day measurements of volcanic outgassing support this judgment.

This volcanic outgassing is a natural consequence of plate tectonics. We have seen previously that our recent understanding of plate tectonics has tied together the explanations of many phenomena, which previously seemed puzzling and unrelated. Here we've added another to the list.

B.4 Chemical History

From our understanding of the origin of the excess volatiles, we can make two particularly interesting observations. One of these is the observation that the sodium and the chlorine in common table salt (NaCl) have had quite different histories. The chlorine is outgassed from the mantle with the other excess volatiles. Subsequently, the chlorine atoms from the Cl_2 molecule combine with metal atoms (mostly sodium) derived from the weathering of crustal materials. Undoubtedly, chlorine dissolved in the water has some effect in expediting the weathering, so it isn't entirely coincidental. Nonetheless, it seems interesting that the salt you put on your food has had an interesting and divergent origin. One half of it came as a gas from the mantle through a volcano, and the other half came out of weathered crustal rocks.

The other interesting observation is that the ocean must always have had roughly the same salinity it has today. That is, it isn't getting saltier with time. The chlorine has been coming out of the mantle along with the water all the time, and then combining with metals from the weathered crust. Consequently, salts have been entering the ocean in the same ratio with new water all the time.

Of course, there have undoubtedly been slight changes in the ocean's chemistry over time. We know, for example, that the accumulation of free oxygen in our atmosphere has been relatively recent, being released during photosynthesis. This increasing presence of oxygen over time is noticed when comparing ancient sediments to more modern ones. Biological activity has a more immediate effect on other aspects of the ocean's chemistry, and we will study some of these later.

C. APPROPRIATE UNITS FOR DESCRIBING CONCENTRATIONS

C.1 Types of Materials in Seawater

The dissolved substances are conveniently grouped into four broad categories:

1. major constituents
2. dissolved gases
3. nutrients
4. trace elements

The major constituents comprise about 99.7% of all dissolved materials, the remaining 0.3% belonging to the last three categories. This does not mean that the last three categories are unimportant, however. Indeed, life as we know it could not have developed and survived on this planet were

TABLE 10.14   Some Typical Values for the Concentrations of Some of the
Important Members of Each of the Four Categories of Seawater
Constituents, By Weight.

*Each is expressed in parts per thousand, parts per million, and parts per billion.
Clearly, parts per thousand is the most appropriate unit for the major constituents,
parts per million is most appropriate for dissolved gases and nutrients, and parts
per billion is most appropriate for trace elements.*

	Parts per Thousand	Parts per Million	Parts per Billion
Major Constituents			
Cl^-	19.3	19,300	19,300,000
Na^+	10.7	10,700	10,700,000
SO_4^{--}	2.7	2,700	2,700,000
Mg^{++}	1.3	1,300	1,300,000
Dissolved gases			
CO_2	0.09	90	90,000
N_2	0.014	14	14,000
O_2	0.005	5	5,000
Nutrients			
Si	0.003	3	3,000
N	0.0005	0.5	500
P	0.00007	0.07	70
Trace elements			
I	0.000060	0.060	60
Fe	0.000010	0.010	10
Mn	0.000002	0.002	2
Pb	0.00000003	0.00003	0.03
Hg	0.00000003	00.0003	0.03

it not for the dissolved gases, nutrients, and trace elements present in seawater. A great deal of scientific effort is directed toward studying these
materials, far in excess of what one might think from their small relative
abundances.

The concentrations of the dissolved gases, nutrients, and trace elements are greatly affected by biological organisms, the climate, and some
geochemical processes. This means that it is impossible to say what the
relative proportions of these various materials are, as their concentrations
vary so much from place to place, and from time to time. As a very rough
approximation, the dissolved gases are generally one or two hundred
times more abundant than the nutrients, and the nutrients, in turn, are
thousands of times more abundant than are trace elements (Table 10.14).
The four categories, then, are listed in order of decreasing relative
abundance.

In addition to what is in solution, seawater also contains some fine
particulate matter in suspension. Some of this particulate matter will
eventually dissolve, and some will eventually add to the sediment on the
ocean bottom. Organic particulate matter has other possible fates as well.
It may be consumed directly by organisms or it may decompose through

bacterial action and return to solution as nutrients, reusable by plants in photosynthesis.

The suspended particulate matter has a noticeable effect on the sea-water chemistry. The particles may exchange some of their cations or anions with those dissolved in the water. They also provide surfaces on which some dissolved materials may precipitate out of solution, and from which other materials may be dissolved into solution. Because of their small sizes, they provide a remarkably large surface area for interaction with the water. A teaspoon of fine clay sediments, for example, has over 10 m² of surface area.

C.2   The Major Constituents

Just as a baker would use different units of measure for the salt than for sugar or flour in his recipe, an oceanographer finds it appropriate to use different units for describing the concentrations of trace elements than for describing the dissolved gases or major salts. We will now introduce the units used to describe the concentration of the materials in the various categories, and compare them to each other. (Refer to Table 10.14.)

The major constituents are measured in terms of grams of dissolved salt per kilogram of seawater, or equivalently, parts per thousand. The symbol used for "parts per thousand" is "‰," which is an easily recognized extension of the more familiar symbol for parts per hundred, "%." It must be remembered that this definition is parts per thousand by *weight* and *not* by numbers of atoms or molecules. In Table 10.10, for example, it is seen that the chloride content of that particular sample is 18.98‰, whereas the sodium ion concentration is only 10.56‰. Does this mean that there are 1.8 times more chloride ions than sodium ions in seawater? No, it does not, because one chlorine atom is 1.6 times heavier than one sodium atom (the average atomic weight of chlorine is 35.5, whereas that of sodium is 23.0). You can see that the chlorine ion is somewhat more abundant than the sodium ion, but not by as much as their relative abundance by *weight* might imply.

C.3   The Dissolved Gases

The dissolved gases are measured in units of the number of milliliters of the gas which are dissolved in 1 l of water. This unit requires further embellishment, because the amount of material in 1 ml of a gas depends on its temperatures and pressure. Under heavy pressure or cold temperatures a milliliter of a certain gas contains a good deal more material than it would at lower pressures or higher temperatures. So the concentration of a dissolved gas is measured in terms of how many milliliters it would occupy if removed from a liter of water and placed under conditions of standard temperature and pressure (STP), defined as 0°C and 1 atmosphere of pressure.

Gases are very sparse compared to liquids, and so a milliliter of gas does not contain much material. For example, 1 ml of nitrogen (N_2) gas at STP has a mass of about 1.3×10^{-3} g, and 1 ml of CO_2 gas at STP has a

TABLE 10.15   The Three Common Units for Measuring the Concentrations of
Dissolved Gases

Gas	Milliliters per Liter	Parts per Million	Millimoles per Liter
Conversion from 1 ml/l			
CO_2	1	1.96	0.045
N_2	1	1.25	0.045
O_2	1	1.43	0.045
Typical concentrations			
CO_2	40	78	1.79
N_2	10	12.5	0.45
O_2	5	7.2	0.23

mass of about 2.0×10^{-3} g. Since 1 l of liquid water has a mass of 1000 g,
these units of milliliters per liter for gases dissolved in water are roughly
equivalent to parts per million by weight, varying from 1.3 parts per million for a milliliter of nitrogen to 2.0 parts per million for a milliliter of
carbon dioxide.

Another unit sometimes used to measure concentrations of dissolved
gases is millimoles per liter. One millimole of any gas at STP occupies a
volume of 22.4 ml, so the conversion between units of millimoles per liter
and milliliters per liter, is simply a factor of 22.4. Table 10.15 illustrates
these various units.

Having different units is not really as confusing as it might appear,
because we're usually interested in the relative concentrations of the various dissolved gases, and we don't really care how these compare with the
other dissolved materials. We know, for instance, that about 99.7% of all
dissolved materials are the major constitutents, and that most of the
remaining 0.3% are the dissolved gases in their various forms. What is
usually of interest to us is how the concentrations of the various gases
compare with each other, and how they vary with position in the ocean
and with time. What unit we use to make these comparisons is irrelevant,
as long as we are consistent.

We have seen that various processes affect the concentrations of the
gases. Biological processes can have a particularly pronounced effect on
oxygen concentrations, due to its short supply. As a result, oxygen concentrations in the ocean vary from less than 1 ml/l up to nearly 10 ml/l (or
roughly 0 to 14 parts per million), and carbon dioxide in its various forms
is usually present in abundances of around 45 to 55 ml/l (or about 90 to
110 parts per million). Dissolved nitrogen is about twice as abundant as
dissolved oxygen, although not as interesting. All other gases are in much
smaller concentrations. You can see that together the dissolved gases
amount to somewhat over 100 parts per million, or 0.1 parts per thousand,
by weight. Since the total salinity of seawater is typically 35 parts per thousand, you can see that the dissolved gases account for about 0.3% of the
total (Figure 10.14).

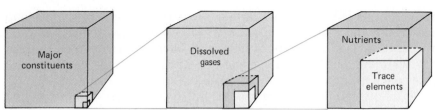

FIGURE 10.14 *Sketch illustrating the relative proportions of the major constituents, dissolved gases, nutrients, and trace elements among the materials dissolved in seawater.*

C.4 The Nutrients

The nutrients are equally important, but much less abundant than the dissolved gases. The nutrients of greatest concern, and those which are most heavily studied, are the elements nitrogen and phosphorus, which are contained in nutrient complexes usable by plants in photosynthesis. Dissolved nitrogen gas, for example, is *not* usable except by a few specialized nitrogen-fixing plants, and so it is not considered a nutrient. The usable nitrogen comes in the nutrient complexes of the nitrate ion (NO_3^-), the nitrite ion (NO_2^-), or occasionally the ammonium ion (NH_4^+), and these are far less abundant than dissolved nitrogen gas. A typical value for the concentration of nitrogen contained in these nutrient complexes is 0.5 parts per million by weight, with large fluctuation due to biological activity. Plants usually obtain their phosphorus through phosphates ($H_2PO_4^-$ or HPO_4^{--}). The concentration of phosporus in seawater is typically 0.07 parts per million, again with wide variations. Other items must be considered "nutrients" for certain organisms, such as the dissolved silica used in the skeletal material of certain organisms. In each case, the appropriate units for their concentrations in seawater are parts per million, or some fraction thereof.

C.5 Trace Elements

Trace elements are receiving a great deal of attention, because many of them have been found to have a crucial effect on life, especially more complex organisms with more complex and specialized systems (such as birds, people, porpoises, etc.). Sometimes this "crucial" effect is positive, in that an organism needs them to survive, and sometimes it is negative, in that an excess will kill or harm the organism. In any case we are giving them a great deal attention, largely motivated by the obvious self-interest. Their concentrations are frequently so small that it is a major accomplishment to detect certain trace elements at all, not to mention making a quantitative determination of their abundances. The trace elements are typically thousands of times less abundant than the nutrients, so units of parts per *billion* by weight (or smaller) are appropriate.

There are a few substances, such as lithium, rubidium, strontium, and fluorine, whose abundances are a little too low to be measured conveniently in units of parts per thousand, as are the major constituents, and whose abundances are a little too high to be measured conveniently in parts per billion, as are the trace elements. These are sometimes grouped along with the nutrients into a category called "minor constituents," whose concentrations are best expressed in terms of parts per million. In this book, however, we will not use this fifth category; rather, we will find room for everything in the four categories listed above, knowing there will be a few marginal or ambiguous assignments.

D.   SUMMARY

The earth is geologically more active than our neighbors in the Solar System, which accounts for the comparative abundance of light molecules forming our atmosphere and oceans. The electrical polarization of the water molecule gives this substance some remarkable properties to which life is indebted. This includes thermal properties unparalleled among light molecules, and an ability to dissolve a wide range of materials, among which are those essential to life.

None of the materials in the ocean could have come from a primordial atmosphere, and not all of them could have been derived from weathering of the earth's crust. The remaining materials are called the "excess volatiles," and must have originated deep within the earth.

Since the ocean contains such a wide range of materials in such a wide range of abundances, no one unit is appropriate to describe the concentrations of them all. This is one reason for dividing the various dissolved materials into four broad classes, each having its own set of units for describing concentrations. In order of decreasing abundance, these groups are: the major constituents, the dissolved gases, the nutrients, and the trace elements.

QUESTIONS FOR CHAPTER 10

1.   Who are our neighbors in the Solar System? How does their surface water compare with ours?

2.   How much water is there on the surface of the earth?

3.   One can demonstrate that in the beginning we had no surface water, so all water has come to the surface since then. How much new water must arrive on the surface each year (on the average) to account for the accumulation of this volume of water in the 4.6 billion years of earth history?

4.   Why should the constituent materials of the various planets be similar? Why do we ignore hydrogen and helium in this statement?

5. Why couldn't our present oceans be derived from water held in the crustal rocks? From water held in the atmosphere?

6. Where did our surface water come from, and how did it get from there to here?

7. Why don't the other nearby astronomical bodies (the moon, Mars, Venus, and Mercury) have an abundance of surface water too?

8. The "active" Martian volcanoes are very few, very large, and very ancient in origin compared to those on earth. Can you think of any implication this might give concerning crustal plate movement on the two planets?

9. Roughly what is the size of an atom? A nucleus?

10. What types of particles are found in atoms, and where are each found?

11. Describe the simplest and next-to-the-simplest atoms. What does the notation $_2He^4$ mean?

12. What are the third, fourth, and fifth most abundant elements in the universe? Why are they more abundant than lithium, $_3Li^7$?

13. What electronic arrangements are most preferred by atoms?

14. Why is hydrogen chemically active and helium not? Besides helium, what are some of the other elements that are chemically inactive?

15. Explain the chemistry of carbon, nitrogen, and oxygen in terms of electron sharing and completed electron shells.

16. Why is it that in the sodium chloride molecule the sodium usually is positively charged, and the chlorine half is negatively charged?

17. What two factors determine whether a planet is able to retain H_2 and He gases in its atmosphere, and why?

18. Are O_2 and N_2 molecules more massive than H_2O or NH_3? Which is heaviest?

19. Discuss the charge distribution in a water molecule.

20. Why should one expect that materials with lighter molecules should have lower melting or boiling points?

21. How do the melting and boiling points for water compare with those of other light molecules?

22. Qualitatively, how does the polarization of the water molecule account for the high melting point, latent heat of fusion, boiling point, latent heat of vaporization, and heat capacity of water.

23. Why is ice less dense than liquid water?

24. For pure water in the liquid phase, is colder water always denser? Explain. At what temperature is liquid water densest?

25. How does the pressure affect the freezing point and boiling point for water? Why?

26. Why is it that in order to be insoluble in water, a material must have molecules with extremely even charge distributions?

27. The balance of what two forces determines how much of a material can be held in solution when saturated?

28. On a molecular level, why is it that the ability of water to dissolve more of a material depends on how much material is already in solution?

29. On a molecular level, why would you expect a salt to be more soluble in water at higher temperature?

30. Why does increased pressure increase the solubility of materials?

31. Explain why we should expect the freezing point and heat capacity of salt water to be lower than that of fresh water. In view of this, how can you explain the fact that dissolved salts *raise* the boiling point?

32. If seawater is typically 3.5% salt, roughly what would you expect to be the salt content of sea ice?

33. The relative abundances of the various salts found in river water are quite different from their present relative abundances in the ocean. Does this mean that as these rivers continue to dump their salts into the ocean, the ocean's chemical makeup will change? Explain.

34. How do we determine whether a material now present in our ocean (or atmosphere) could possibly have come from weathered crustal rocks? What are some materials now in the ocean that could have had this origin?

35. What are some of the important materials in the ocean that could not possibly have come from weathered crustal rock? Why not? What are they called?

36. How is the average speed of a gas molecule related to the temperature? Its mass?

37. Explain how observation of inert gases in the atmosphere can tell us the temperature of the primordial atmosphere. Why do we look at inert gases rather than oxygen, nitrogen, and carbon dioxide, for example?

38. How do we know that the excess volatiles could not have come from a very heavy primordial atmosphere?

39. What evidence supports the idea that the excess volatiles were outgassed from the mantle?

40. Why do you suppose we examine fallen meteorites for insight into the mantle's composition rather than checking the mantle directly?

41. Discuss the history of the sodium and the chlorine in your table salt.

42. Why is it thought that the oceans have always had the same salinity?

43. What are the four categories of dissolved substances?

44. Roughly what fraction of the dissolved materials belong to the group of major constitutents? Typically, how many times more abundant than the trace elements are the nutrients? Typically, how many times more abundant than the nutrients are the dissolved gases?

45. What are some of the possible eventual fates for suspended particulate matter?

46. For equal volumes of sediment, finer sediments have greater total surface area. Explain why. Why do suspended sediments with greater surface area have a larger effect on the chemistry of the seawater?

47. What are the symbols for "parts per hundred" and "parts per thousands"?

48. In a certain sample of seawater, the chloride content is 18.0‰ and the sodium content is 10.0‰. Does this mean there are 1.8 times as many chlorine atoms as sodium atoms in the solution? Explain.

49. How does the amount of material in a milliliter of oxygen gas vary with the temperature and pressure? What temperature and pressure are taken to be the standard?

50. What is the size of a milliliter in terms of cubic centimeters? (Look this up in a dictionary if necessary.) Roughly how many grams of a typical gas are in a milliliter at STP?

51. One milliliter of CO_2 gas dissolved in one liter of water is how many parts per million by weight? Is a CO_2 molecule heavier or lighter than an O_2 molecule? An N_2 molecule?

52. If you were told that the concentration of CO_2 gas in a certain water sample was 2 "millimoles per liter," what would this concentration be in terms of "milliliters per liter" (ml/l)?

53. When are typical values in ml/l for the concentrations of dissolved O_2, N_2, and CO_2 gases in seawater? Convert one of these to units of parts per million.

54. Nitrogen that can be used as a nutrient by plants in photosynthesis usually comes in what complexes? How about phosphorus?

55. What are typical values for the concentrations of nutrient nitrogen and nutrient phosphorus in seawater?

*56. Is all the water coming up through a volcano or geyser emerging for the first time on the earth's surface? Can you think of any way of estimating what fraction of this water is recycled surface water?

*57. Give a possible explanation for why the planets beyond Mars are so much larger than the inner planets.

*58. Why do you suppose terrestrial organisms (such as ourselves) more often suffer from lack of certain trace elements than do marine organisms? Can you think of any advertisements you've seen, heard, or read recently that address our need for a trace element?

*59. Why do you suppose CO_2 is much rarer in the universe than CH_4 or H_2O?

*60. Why do you suppose the presence of salt in water should change its boiling point? Its freezing point?

*61. Why do you suppose some materials are more soluble in water than others?

*62. If methane (CH_4) and ammonia (NH_3) are nearly as abundant as water (H_2O) in the universe, why do you think they aren't as abundant as water on earth?

SUGGESTIONS FOR FURTHER READING

1. Drake, Imbrie, Knauss, and Turekian, *Oceanography*, Chapter 9, Holt, Rinehart and Winston, New York, 1978.

2. R. A. Home, *Marine Chemistry*, Wiley, New York, 1969.

3. Ferren MacIntyre, "Why the Sea Is Salt," *Scientific American* (Nov. 1980).

11
THE COMPONENTS OF SEAWATER— PART 2

In the preceding chapter we studied the various kinds of constituents in seawater. Our interest in them actually goes farther than just identifying what they are and explaining where they came from. Many of the constituents are involved in important physical, chemical, and biological processes. To some of these we owe our lives. Others have less crucial, but equally interesting, effects on our environment. In this chapter we examine a little more closely some of the processes affecting these dissolved constituents, and some of the processes through which they affect us and our environment.

11
THE COMPONENTS OF SEAWATER —PART 2

A. THE MAJOR CONSTITUENTS

A.1 Constant Ratios

In 1884 a British chemist named Dittmar did a very careful analysis of water samples collected on the Challenger Expedition 10 years earlier. His analysis of 77 samples collected at a wide variety of geographical locations (some from each major ocean) and various depths confirmed what had been indicated in earlier works as a remarkable property of the oceans. This is that all of the major constituents occur everywhere in the same relative proportions, although the relative amount of water in the mixture varies.

In areas of high precipitation or much fresh water discharge, the salinity will be low. For example, due to the large fresh water influx from nearby land masses, the suface water of the Baltic Sea is quite brackish, having a salinity of only about 10‰. On the other hand, the surface waters in regions where evaporation is high will be hypersaline, having well above the 34‰ to 36‰ normal range for ocean salinities. The Red Sea and Persian Gulf are examples of such hypersaline areas. Regardless of the salinity, when chemists analyze the dissolved materials alone, they find that 55.04% by weight is the chlorine ion, 30.61% is sodium, 7.68% sulfate, 3.69% magnesium, and so on. That is, regardless of the salinity of the sample, the ratios of the various salts among themselves do not change. (See Table 11.1.)

There are, of course, some coastal regions where pollution or river discharge (Figure 11.1) is sufficiently large to have a noticeable effect on the ratios, and so for these minor coastal areas the rule of constant ratios is violated.

The rule applies to the major constituents, but not to all materials. We know that the dissolved gases, nutrients, and trace elements show variations in their relative abundances due to biological, atmospheric, and geochemical processes. But these processes have little effect on the ocean's salinity, as the dissolved gases, nutrients, and trace elements together amount to much less than 1% of the total dissolved material.

291

TABLE 11.1   The Concentrations by Weight of the Major Constituents in a Water Sample of 35‰ Salinity and One of 15‰ Salinity.

The relative proportions of the various salts are the same for both samples

Salt		Abundance in Water of 35‰ Salinity (‰)	Percent of Total Salt	Abundance in Water of 15‰ Salinity (‰)	Percent of Total Salt
Cations					
Sodium	Na$^+$	10.71	**30.62**	4.59	**30.62**
Magnesium	Mg^{++}	1.29	**3.68**	0.552	**3.68**
Calcium	Ca^{++}	0.413	**1.18**	0.117	**1.18**
Potassium	K$^+$	0.385	**1.10**	0.165	**1.10**
Strontium	Sr^{++}	0.007	**.02**	0.003	**0.02**
Anions					
Chloride	Cl$^-$	19.27	**55.07**	8.26	**55.07**
Sulfate	SO$_4^{--}$	2.70	**7.72**	1.16	**7.72**
Bicarbonate	HCO$_3^-$	0.140	**.40**	0.060	**0.40**
Bromide	Br$^-$	0.067	**.19**	0.029	**0.19**
Fluoride	F$^-$	0.035	**.10**	0.015	**0.10**
Totals		35.00	**100.0**	15.00	**100.0**

FIGURE 11.1   *The Nile river delta. The law of constant relative proportions for the major constituents may be violated in some localized coastal areas such as this since the proportions of the salts in the river water is different from their proportions in seawater.*

A.2   Salinity Variations

The significant variations in salinity in the ocean are caused by *physical* processes, such as freezing, precipitation, or evaporation (Figure 11.2), which have the effect of adding or subtracting fresh water from the solution. Of course, physical processes may also have a minor effect on some of the dissolved constituents. For example, heating or evaporation tends to drive out some of the dissolved gases. Driving off dissolved carbon dioxide also has an effect on the dissolved calcium ion concentration. But these are minor effects. To a very good degree of approximation, we can say that changes in salinity of the oceans are caused by *physical* processes, which tend to change the amount of fresh water in the solution only. This is why the ratios of the concentrations of the various dissolved salts do not change, although the salinity does.

General patterns of salinity in the oceans can be guessed from our knowledge of physical processes. We would guess that near land, the surface salinity would tend to be lower due to fresh water runoff. We also know that climate varies with latitude, so we'd expect some effects of this on the surface salinity. Near the tropics, the sunshine evaporates water, and we'd expect high surface salinities there. This might be moderated along the equator because the afternoon showers return some of the fresh water to the ocean surface. At very high latitudes we'd expect low surface salinity, because there is relatively large precipitation and little sunshine. (See Figure 11.3.) Finally, in temperate latitudes we'd expect large seasonal variations in surface salinity due to seasonal climatic variations.

We can also predict the general variation of salinity with depth. Deep waters should be less saline, since they're formed in higher, colder latitudes, where the surface water is fresher. There should not be much change in this prediction until one gets near the surface, where the processes described in the previous paragraphs take effect. These surface effects should only extend as deeply as there is mixing of the surface waters, typically a few hundred meters. (See Figure 11.4.)

A.3 Mixing

The fact that the major salts always appear in the same ratios, whether the sample is taken in the Bering Sea or the Bay of Bengal, indicates that the oceans are well mixed. There are two types of processes at work to accomplish this mixing, one on a microscopic scale and one on a macroscopic scale. They are called "diffusion" and "advection," respectively.

Diffusion is due to the thermal motion of the dissolved material. This

(a)

(b)

(c)

FIGURE 11.2

Illustration of some important physical processes that cause changes in the salinity of seawater. (a) Precipitation adds fresh water to the mixture. (b) Evaporation removes fresh water from the mixture. (c) The formation of sea ice removes fresh water from the mixture.

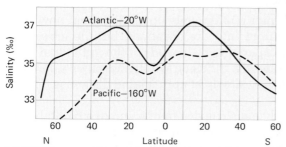

FIGURE 11.3 *Variation of surface salinity with latitude for the Atlantic and Pacific oceans during summer in the Northern Hemisphere. Can you explain the dip in the curves near the equator? (From Peter K. Weyl, Oceanography, John Wiley, 1970.)*

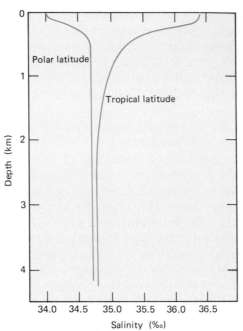

FIGURE 11.4 *Idealized plot of how we would expect salinity to vary with depth in the oceans. In reality, subsurface currents often bring waters formed at different latitudes and in different seasons, causing corresponding wiggles in the lines. In polar latitudes, the ice pack frequently reduces wave activity, so the surface waters don't mix very deeply.*

motion is surprisingly fast on a molecular scale, being several hundred meters per second for a typical dissolved ion. However, on a macroscopic scale, diffusion progresses very slowly, being typically a few millimeters per minute. The reason is that these fast moving ions keep bumping into things (e.g., water molecules), and so they keep changing directions, frequently ending up where they started. It is something like an army of fast, but blind, ants starting out from an ant hill to explore the world. Because they're blind, they stumble around, bumping into each other, going in circles, and so on, and most of their motion ends up getting them nowhere. Gradually, some will stumble out into the distances, but the army will spread at very slow speeds compared to the speeds of the individual ants, due to wasted motion. Consequently, diffusion will homogenize a solution on a small scale, but it works very slowly at appreciable distances. This is illustrated in Figure 11.5.

If you would very carefully put some dye in one end of a bathtub, you could notice it slowly diffuse away from the point of placement. You would grow impatient waiting for it to reach the other end, and so you might facilitate its mixing by introducing *advection*, for example, by stirring it with your finger, or blowing on the surface, or using any other means you might think of to get water currents to help carry the dye from one end toward the other.

(a)

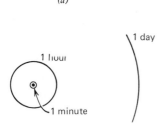

(b)

FIGURE 11.5   *Diffusion. (a) On a microscopic scale, the individual molecules move at very high speeds, typically several hundred meters per second. However, their direction is random, changing directions after each collision. The result is that their net progress is small. (b) Full-scale representation of the progress from a point source. The circles indicate the average distance from the source for the molecules after one minute, one hour, one day, and three weeks, respectively. After one year, the average molecule would be only 60 cm from the source.*

Although diffusion can account for the homogeneity of the ocean on a small scale, advection is needed to account for its mixing from one area to another, or from one ocean to the next.

Near the surface, mixing proceeds quickly, where the wind-induced waves and swift surface currents (with the help of seasonal thermal over-turns in termperate latitudes) insure that the surface waters of all oceans are thoroughly mixed within a few years.

The deeper currents proceed much more slowly, requiring hundreds or thousands of years to mix well. Deep water mixing must be measured indirectly, as experiments designed to measure it directly, such as following tracer materials, would outlast the lifetime of the experimenter.

A.4   Measuring Mixing Time for Deep Waters

One method used to estimate the mixing time for deeper water employs carbon-14 (C^{14}), and is illustrated in Figure 11.6. This radioactive isotope of carbon is produced through the bombardment of our atmosphere by cosmic rays, so a small fraction of the CO_2 present in our atmosphere is made from this heavier isotope of carbon. This isotope is unstable and decays with a half-life of 5600 years, but the constant influx of cosmic rays into our atmosphere insures that it gets replenished as rapidly as it decays.

Once removed from the atmosphere, however, the carbon-14 isotopes continue decaying, without replenishment, and so the relative amount of

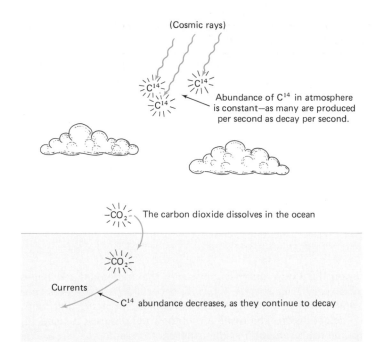

FIGURE 11.6 *Illustration of how we can tell how long it has been since a particular water mass has been at the surface by the depletion of C^{14} in the dissolved CO_2.*

C^{14} decreases with time. By measuring the amount of depletion of C^{14} relative to C^{12}, one can deduce how long it has been since the carbon was in the atmosphere. This is done, for example, to date fossils, which had incorporated carbon into their tissues from the CO_2 in the atmosphere while the organism was alive. It is also used in measuring the mixing of the deep water. By examining the depletion of C^{14} relative to C^{12} in the dissolved CO_2, we can tell how long it has been since that water mass was at the surface and in contact with the atmosphere. A typical value for the deep water in our oceans is several hundred to a thousand years.

B.   THE DISSOLVED GASES

B.1   Solubility of Gases in Seawater

An analysis of our atmosphere shows that dry air is about 78% nitrogen, 21% oxygen, and nearly 1% argon. Adding up these three you can see that all the other gases must amount to very little. Indeed, the next most abundant atmospheric gas, carbon dioxide, constitutes only about 0.03% of dry air. In the ocean, the relative abundances of these various dissolved gases is remarkably different, which can be seen in Table 11.2.

The ocean can and does hold considerably more CO_2 than either N_2 or O_2. In fact, even though it is not nearly saturated, the ocean contains 62 times more CO_2 than does the entire atmosphere. This is in contrast to

TABLE 11.2   Comparison of the Abundances of Various Gases in Seawater with
Their Abundances in the Atmosphere

Gas	Abundance in Dry Air (%)	Abundance in Seawater (ppm)	Ratio of Total Amount in the Oceans to Total Amount in the Atmosphere
N_2	78.0	12.0	0.004
O_2	21.0	7.0	0.01
CO_2	0.03	90.0	62.0

TABLE 11.3   Solubility of Gases in
Seawater Having Salinity of
36.1‰, as a Function of the
Temperature

Temperature (°C)	Solubility (ml/l at atmospheric pressure)		
	N_2	O_2	CO_2
0	14.05	7.97	8700.0
10	11.29	6.28	8030.0
20	9.42	5.15	7350.0
30	8.07	4.33	

dissolved O_2 and N_2, for which the oceanic content is only about 1% of
that which is in the atmosphere.

Table 11.3 shows the amount of each of these gases that a liter of seawater can hold when saturated in equilibrium with the atmosphere. Compared to O_2, N_2 would be nearly twice as abundant, and CO_2 would be
roughly 1000 times more abundant if the water were saturated. Colder
water holds the dissolved gases better, which is no surprise to those familiar with soft drinks.

The primary source for these dissolved gases is, of course, the ocean's
contact with our atmosphere. As a general rule, the deeper oceanic waters
contain more dissolved CO_2 than do the intermediate and surface waters.
The primary cause for this can be traced back to when the various water
masses were last at the surface. The deeper waters were formed at high
latitudes, where their colder temperatures encouraged them to dissolve
more carbon dioxide. They took this CO_2 with them to their present deep
ocean abode. The intermediate and surface water masses at temperate
and tropical latitudes were formed in warmer climates, and consequently
they dissolved less CO_2 during their formation. The concentrations of dissolved N_2 show a similar variation with depth.

B.2   Biological Effects

Biota also have a noticeable influence on the concentrations of dissolved
gases. This biological influence is especially pronounced in oxygen concentrations, due to the small amounts of this gas in the original water
masses. One can find oxygen concentrations varying from the anoxic

0 ml/l where respiration has depleted it, up to the supersaturated 10 ml/l, where photosynthesis is vigorous. The larger abundance of CO_2 insures that biological processes will cause smaller relative changes in its concentration than was the case for O_2. Usually, CO_2 concentrations will be found in the range of 45 to 55 ml/l.

Biological activity has very little effect on dissolved nitrogen gas. There are some nitrogen-fixing algae, which convert dissolved nitrogen into nutrients, and some denitrifying bacteria, which do the reverse. However, these are relatively rare organisms, and their overall effect is negligible. Typical values for nitrogen concentrations are around 10 ml/l, slightly below saturation. Argon is present only in very small concentrations, and being chemically inert, it is uninteresting anyhow.

Because of their involvement in life processes, there is predictable variation in O_2 and CO_2 concentrations with depth. Near the surface, photosynthesis causes a depletion of CO_2 and an enhancement of dissolved O_2. Below the photic zone, photosynthesis is impossible. Respiration depletes the oxygen and enhances the CO_2 concentration. Animal activity is especially vigorous in and just below the food-producing surface waters, so oxygen is particularly depleted just below the surface waters, where it isn't replenished by photosynthesis. If we plotted the concentrations of these gases as a function of depth, we expect to see an inverse correlation, O_2 being concentrated in surface waters and depleted at depth, and the reverse being true for CO_2. Within the deeper waters, CO_2 should be the most concentrated in the deepest water, because of the colder climate where it was formed. Oxygen should be most strongly depleted just below the photic zone, due to animal activity there. This is shown in Figure 11.7.

Occasionally, in deeper water where the O_2 is depleted, some microorganisms can survive by carrying out more sophisticated chemistry. They take the sulfate ion (SO_4^{--}), remove the O_2 from it, and discard the sulfur.

$$SO_4^{--} \rightarrow 2O_2 + S^{--}$$

They use the O_2 for respiration, and the sulfide ion (S^{--}) soon encounters a couple of hydrogen ions, making hydrogen sulfide gas (H_2S). This gas then rises to the surface, where it might encounter O_2 in the surface water, and the sulfate ion is remade.

$$2O_2 + H_2S \rightarrow 2H^+ + SO_4^{--}$$

Alternately, it might escape into the air, where one can readily recognize the "rotten egg" smell. Consequently, such "sulfate-reducing bacteria" in deeper water cause the depletion of the sulfate ion in deep water and its enrichment near the surface. Such a place in the Black Sea, where the long residence time of the deep water (about 2000 years) means that virtually all the oxygen is used up, and only these sulfate-reducing bacteria can survive.

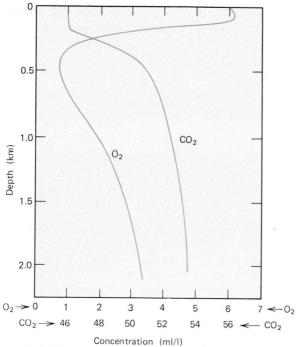

FIGURE 11.7 *Plot showing a typical variation of dissolved O_2 and CO_2 with depth.*

B.3   A Buffer for Atmospheric CO_2

Its ability to dissolve CO_2 makes the ocean a good buffer for the CO_2 in the atmosphere. Since the advent of the Industrial Revolution, we have been burning fossil fuels and dumping CO_2 into the atmosphere at increasing rates. The environmental impact of atmospheric CO_2 will be seen in more detail in Chapter 12. Carbon dioxide is opaque to much of the infrared radiation from the earth. As is illustrated in Figure 11.8, it allows the sun's radiation in, but it absorbs the earth's reradiation going back out and makes the atmosphere warm (the "greenhouse effect"). Since the beginning of the Industrial Revolution, both the CO_2 content and the mean temperature of the atmosphere have been slowly increasing. Fortunately, since the ocean absorbs much of the additional CO_2, the environmental impact of our releasing large quantities into the atmosphere isn't as bad as it would be otherwise.

During the years 1958 to 1979, the atmospheric CO_2 concentration was measured at a station as remote as possible from industrial pollution centers, near the summit of Mauna Loa on Hawaii. The results (Figure 11.9) shows a small gradual increase, about 0.2% gain per year. Notice the seasonal fluctuations due to plant life. During spring and summer the plants use up CO_2 in photosynthesis, producing O_2 and organic compounds, and the atmospheric CO_2 content falls. However, in fall and winter photosynthesis decreases (trees lose their leaves, etc.), and the plants burn organic compounds and O_2 to produce CO_2. Hence, the atmosphere's CO_2 content increases in these seasons.

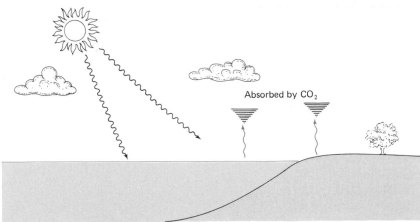

FIGURE 11.8 *Illustration of the effect of atmospheric CO_2 on our climate. CO_2 is transparent to the sun's rays, but it absorbs longer wavelength radiation, such as that given off by the earth. It lets sunshine in, but traps the earth's radiation going back out, like a blanket. The more CO_2, the thicker the "blanket."*

Many researchers have attempted to predict the environmental impact of our continued reliance on fossil fuels. Although the details of the various models differ, the resulting predictions are remarkably similar. Continued reliance on fossil fuels would mean an increase in atmospheric CO_2 by somewhere in the range between four and eight times its present abundance by the year 2200. This would probably cause an increase in ambient temperatures on earth of roughly 6°C. This is a major climatic change, and would necessitate large-scale shifts in human populations, mostly toward higher latitudes. For comparison, temperatures during the last ice age were only 5° to 10°C below the present norm. One particular concern of these researchers is whether climate changes are gradual, allowing for gradual population adjustments, or whether climatic changes come in sudden jumps, which would be more catastrophic.

B.4 Effect of CO_2 on the Chemistry of Seawater

When a carbon dioxide molecule (CO_2) dissolves in water, it combines with one water molecule (H_2O) to form the hydrated carbon dioxide molecule H_2CO_3. In this larger molecule, the hydrogen ions are less strongly bound than they were in the original water molecule (Figure 11.10), and so the hydrated carbon dioxide molecule tends to dissociate into a hydrogen ion and a bicarbonate ion ($H^+ + HCO_3^-$), and sometimes even releases both hydrogen ions, leaving the carbonate ion ($2H^+ + CO_3^{--}$). Under normal conditions in the oceans, over 90% of the dissolved CO_2 is found in the bicarbonate form (HCO_3^-), and relatively little in the nondissociated (H_2CO_3) or doubly dissociated (CO_3^{--}) form.

To describe the hydrogen ion concentration in a solution, scientists use the term "pH." In pure water, one out of 10^7 water molecules disso-

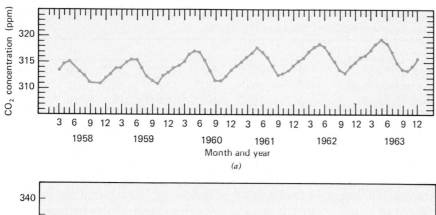

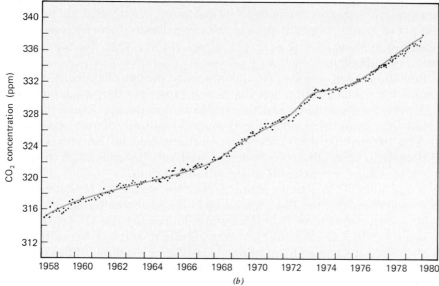

FIGURE 11.9 *Concentration of CO_2 in the atmosphere at Mauna Loa Observatory, Hawaii, measured in parts per million: (a) monthly average; (b) 12 month running average. (From J. C. Pales and C. D. Keeling, Journ. Geophys. Res.,* **70** *(1965). Data for (b) also from Keeling.)*

FIGURE 11.10 *When CO_2 dissolves in water to form H_2CO_3, the hydrogens are not as tightly bound in this larger molecule as they were in the original water molecule. So more hydrogens are shaken loose, resulting in more free hydrogen ions in the solution. That is, dissolving CO_2 in water makes it acidic.*

TABLE 11.4   Concentrations of Hydrogen Ions (H^+) and Hydroxide Ions (OH^-) in
Solutions of various pH Values

pH	1	3	5	7	9	11	13
H^+ concentration	10^{-1}	10^{-3}	10^{-5}	01^{-7}	10^{-9}	10^{-11}	10^{-13}
OH^- concentration	10^{-13}	10^{-11}	10^{-9}	10^{-7}	10^{-5}	10^{-3}	10^{-1}
product of the two	10^{-14}	10^{-14}	10^{-14}	10^{-14}	10^{-14}	10^{-14}	10^{-14}

ciates into $H^+ + OH^-$, so in pure water the H^+ ion concentration is one in 10^7. We say then that pure water has a "pH" of 7. When things are added to the water, the hydrogen ion concentration changes. If it increases, we say that the solution is "acidic." For example, if the H^+ ion concentration increases to one in 10^6 or one in 10^5, we say that the solution has a pH of 6 or 5, respectively. On the other hand, if the hydrogen ion concentration decreases (e.g., a pH of 9), we say that the solution is "basic," or "alkaline."

Both the H^+ ion and the OH^- ion are very reactive chemically, which accounts for the fact that when the two get together they stick together very tightly, forming the remarkably stable H_2O molecule. Recall that the thermal velocity of these molecules is several hundreds of meters per second. This means that they undergo several hundred billion very violent collisions with each other every second, and yet only one in 10^7 of these molecules will be in fragments at any instant of time.

If one adds H^+ ions to the solution, then the increased H^+ concentration means that it is more likely for an OH^- ion to encounter an H^+ ion to combine with and reform H_2O. Therefore, increasing the H^+ concentration in a solution depresses the OH^- concentration, and vice versa. In fact, the product of the two concentrations is 10^{-14}. For example, if the hydrogen ion concentration is increased from 10^{-7} (or, one part of 10^7) to 10^{-6}, then the OH^- ion concentration gets correspondingly depressed from 10^{-7} to 10^{-8}. (See Table 11.4.)

What this all means is that when the solution is acid (pH less than 7) the chemical properties are dominated by the behavior of the H^+ ion, and when the solution is basic (pH greater than 7) the chemical properties are dominated by the behavior of the OH^- ions. Other ions in solution do have some effect on the chemical properties, but since these are generally less reactive than H^+ or OH^-, their contributions are less pronounced.

You can make a solution more acid by adding, for example, hydrochloric acid (HCl) or sulfuric acid (H_2SO_4) to the water, which releases H^+ ions upon dissociation. Similarly, you could make it more basic (or "alkaline") by adding sodium hydroxide (NaOH), for example, which provides OH^- ions upon dissociation.

We can see that carbon dioxide dissolved in seawater reduces the pH. It combines with a water molecule to form "carbonic acid," H_2CO_3, which normally dissociates into a hydrogen ion, H^+, plus the bicarbonate ion, HCO_3^-. Because of the other things in solution, seawater usually remains basic, averaging a pH of about 7.8. This varies from around 7.5 in deep waters, where CO_2 is in greater abundance, to around 8.4 in warm surface (more CO_2) → (more H_2CO_3) → (more H^+) → (combine with CO_3^{--} to make HCO_3^-) (less CO_2) → (less H_2LCO_3) → (less H^+) → (HCO_3^- dissociates to make CO_3^{--})

FIGURE 11.11   *Illustration of why more dissolved CO_2 means less free carbonate (CO_3^{--}) ions in the solution. More dissolved CO_2 releases more H^+ ions, which combine with the free carbonate ions, reducing their numbers. Without sufficient free carbonates with which to combine, the calcium (Ca^{++}) remains in solution.*

waters, where warm temperatures and photosynthesis have reduced the abundance of dissolved CO_2.

B.5   The Formation of Calcium Carbonate Sediments

Among the dissolved salts found in seawater are the calcium (Ca^{++}) and carbonate (CO_3^{--}) ions. These two can combine to form the neutral $CaCO_3$ molecule, and precipitate out of solution, providing something else doesn't get to the CO_3^{--} ion first. The hydrogen ion, H^+, is very reactive, and when competing with Ca^{++} ions for CO_3^{--} ions, the H^+ ion usually wins out, forming the bicarbonate (HCO_3^-) ion, or even the carbonic acid molecule (H_2CO_3) when sufficient H^+ ions are present.

When the solution is acid, there are many H^+ ions present, and some of them combine with the carbonate ion, thereby preventing it from combining with the Ca^{++} ion and precipitating out of solution. When the solution becomes more basic, the removal of the H^+ ions frees up some of the CO_3^{--} ions, which can then combine with Ca^{++} and precipitate out. Consequently, we find that when seawater is slightly more acid than usual, the calcium carbonate tends to stay in solution, and when seawater is slightly more basic than usual, the calcium carbonate tends to precipitate out.

In the chapter on sediments, we saw that calcium carbonate sediments do not form in deeper waters. Furthermore, those that fall into deeper waters dissolve. We can now understand why. The increased quantities of CO_2 in deeper waters means that it is more acid, or equivalently has a higher H^+ ion concentration. As a result, any free CO_3^{--} ions floating around are more likely to find a H^+ ion to combine with and form bicarbonate, HCO_3^-. As a result, there will be fewer unattached CO_3^{--} ions around for Ca^{++} to combine with and form sediment. (See Figures 11.11 and 11.12.)

Even worse for the fate of $CaCO_3$ sediment already formed is that the very reactive H^+ ions will displace the Ca^{++} ion in this compound, and put the calcium back into solution.

$$H^+ + CaCO_3(\text{sediment}) \rightarrow Ca^{++} + HCO_3^-$$

Therefore, sediment already formed will go back into solution.

In shallower waters, the H^+ ion is less abundant (less dissolved CO_2 means fewer free H^+ ions), and so it cannot compete as successfully with

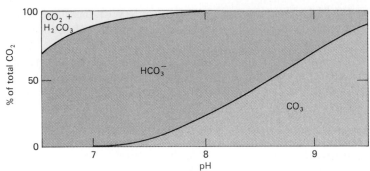

FIGURE 11.12 *Plot of the fraction of the dissolved CO_2 that is in the various carbonate forms as a function of pH. Normally, the pH of seawater is
around 7.8. It is seen that as the hydrogen ion concentration is
increased (lower pH), there is less and less of the free carbonate
(CO_3^{--}) form and more of the bicarbonate (HCO_3^-) form in the
solution.*

Ca^{++} ions. Therefore, calcium carbonate sediments form and survive better in shallow water.

Calcium is the only major cation whose concentration is near saturation. Because of the keen competition between Ca^{++} and H^+ for carbonate ions, anything that reduces the H^+ concentration at all will enable
some calcium carbonate to precipitate out (except, of course, in deep
water, which is not saturated with calcium). The most common way of
reducing the H^+ ion concentration is through the removal of dissolved
CO_2.

Consequently, anything that removes CO_2 from the water may cause
precipitation of calcium carbonate. The most common ways this is accomplished include heating of the water, either via extra sunshine, or through
the surfacing of deeper, cooler waters. Photosynthesis also removes CO_2
from the water. Those familiar with soft drinks may guess that reduced
pressures or agitation will also cause the release of dissolved CO_2. In the
oceans these mechanisms are encountered through lower atmospheric
pressures and waves.

Since these processes that remove CO_2 from solution and cause precipitation of calcium carbonate are so common, and since calcium is not
particularly abundant in the ocean in the first place (compared with
sodium, for example), you might guess that the average calcium ion does
not spend a very long time in solution. You would be right. The average
oceanic residence time for a calcium ion is only about 1 million years, by
far the shortest of any major cation.

In this section we have discussed calcium and calcium carbonate
extensively because of its importance as a sediment. Magnesium behaves
chemically similarly to calcium, and so it participates in this process as
well. However, Mg^{++} is not near saturation in the oceans as is Ca^{++}, and
so magnesium precipitates are not nearly as common as calcium precipitates, even though magnesium is more abundant in seawater. Nowhere do

we find pure magnesite deposits ($MgCO_3$) as we do pure calcite ($CaCO_3$). However, there are some "high-magnesium calcite" deposits where 10 to 20% of the cations are actually magnesium rather than calcium. The highest magnesium content is found in dolomite, which is about half and half. These, of course, had to be formed under conditions of high salinity, where the magnesium concentrations approached saturation.

C.   THE NUTRIENTS

As we have seen, marine organisms influence the chemistry of their environment, and so we will briefly outline that influence here. A more detailed treatment of this subject will be presented in Chapter 17, which is specifically devoted to marine biology.

The biggest effect of marine organisms is on the abundances of the dissolved gases oxygen and carbon dioxide. Between the two, oxygen is in the shortest supply in seawater, so it displays the largest fluctuations caused by biological processes. The nutrients also have large fluctuations in their concentrations. They are not consumed as much as are the gases, but they are in much shorter supply.

The term "nutrient" enjoys a vague definition. It is usually considered to be anything besides water and carbon dioxide that is needed by plants in the synthesis of organic matter or skeletal materials. The most universal of such materials are nitrogen and phosphorus in usable forms, but this particular definition may include many other materials as well, including many of the trace elements, which we discuss in a separate section. The usable nitrogen is generally present in the nitrate or nitrite radicals (NO_3^- or NO_2^-), but some ammonium (NH_4^+) is also present, largely from animal excretion. The usable phosphorus is generally present as a phosphate (PO_4^{---}), with one or two hydrogens attached (HPO_4^{--} or $H_2PO_4^-$).

Nutrient nitrogen is more abundant than the phosphorus, and is also in greater demand in the fabrication of organic tissues. On the average, nutrient nitrogen is approximately 16 times more abundant than nutrient phosphorus by number of atoms, but only seven times more abundant by weight. (Phosphorus has more than twice the atomic weight of nitrogen.) Typical concentrations would be 0.5 parts per million for nitrogen in nutrients, and 0.07 parts per million for phosphorus.

Nutrients are removed from the water to be incorporated into organic tissue during photosynthesis (or during synthesis of skeletal material), and are returned to the water when these organic tissues are oxydized during respiration. This respiration is performed either by the original plant, or by animals which have consumed the materials produced by the plant. Complete oxidization of the nitrogen or phosphorus back to its nutrient form generally requires its being cycled through more than one organism. This is especially true of the nitrogen in organic tissues, special bacteria being required for the final steps.

Since they are consumed during photosynthesis and released during respiration, we'd expect the variations in nutrient concentration to paral-

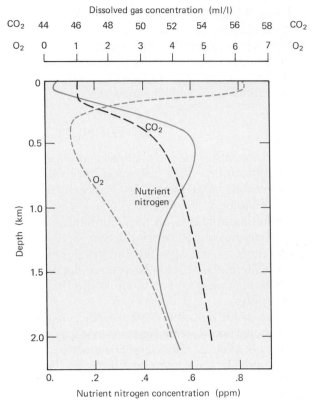

FIGURE 11.13 *Plot showing a typical variation in the concentration of dissolved
nutrient nitrogen with depth. The concentrations of O_2 and CO_2
are included for comparison.*

lel that of carbon dioxide, and be inversely correlated to oxygen concentrations (when plotted on appropriate scales, of course). This is shown in
Figure 11.13. Near the surface, carbon dioxide and nutrients will be
depleted, and oxygen enriched due to photosynthesis, with large seasonal
and geographical variations. Below the photic zone reside many animals,
feeding on organic matter produced above, and so carbon dioxide and
nutrients are returned to the water and oxygen is depleted. That is, in
these dark waters, photosynthesis cannot compensate for animal respiration as it does in surface waters. At greater depth, there is diminished animal activity, and this is displayed in somewhat increased oxygen concentrations and decreased nutrients.

D.   THE TRACE ELEMENTS

You might wonder how anything whose abundance must be measured in
parts per billion, or smaller, can have any significance at all. However,
there are other ways of discussing trace elements, which makes them
sound much more impressive. One of these ways is economical. In each

cubic kilometer of water there is over a million dollars' worth of gold in solution alone. For the entire ocean, this amounts to over a million billion (or, a quadrillion) dollars' worth at today's prices. The economic value of many other materials in the ocean are even greater. Naturally, this has generated much interest in methods of extracting these materials from seawater, but so far nothing has proven feasible for these rarer materials. On the other hand, nature has found a means of concentrating some of these materials in sediments in certain regions of the ocean's floor, and there is much interest in figuring out how nature does it.

The biological import of these trace elements was realized when early attempts to make artificial seawater for use in aquariums failed. Organisms died even when the various salts were carefully added to reproduce all known abundances in seawater. Clearly, some vital things were missing, even though their abundances in seawater were so small that they escaped detection.

Because of their importance, organisms have become quite proficient at hoarding and concentrating some trace elements in their own tissues. In fact, several trace elements are found mostly in organic complexes in seawater rather than as dissolved ions. This ability to hoard clearly has been important to the development and livelihood of the organisms needing these trace elements. But now, with the increased pollution of our environment, it works to our detriment, as organisms are concentrating some of these poisonous pollutants in their tissues as well. These tend to get increasingly concentrated at each step of the food ladder, and by the time the organic material reaches secondary carnivores, such as tuna fish or people, it may be unfit to eat.

E. MEASURING SALINITY

E.1 Definition of Salinity

So far, we have been discussing salinity in terms of the number of grams of material dissolved in a kilogram of seawater. Unfortunately, the formal definition must be made slightly more complicated, because of complications in the actual measurement.

To determine salinity, the student might pour one kilogram of seawater into a pan, and then boil it dry. The mass of the residue would be the mass of the salts dissolved in the original water, plus a little more. This additional mass is caused by some of the salts still holding on to some water of hydration. To drive the water from these hydrated salts would require additional heating. This, however, would cause some of the other salts to break down. Consequently, the actual fundamental determination of salinity must be a little more laborious, and to facilitate this more laborious work, a slightly different definition of salinity is used.

The salinity is defined to be the total mass in grams of all dissolved substances per kilogram of seawater, with all carbonate converted to oxide, all bromine and iodine replaced by chlorine, and all organic matter oxidized to 480°C. Since the four materials involved in the modified definition (carbonate, bromine, iodine, and organic matter) are relatively minor, the "salinity" differs only very slightly from the total mass of dissolved solids, the two being within ½% of each other.

E.2   Chemical Analysis

The fact that the relative abundances of the major salts is invariant greatly simplifies the chemical determination of salinity of a sample. For example, since 55.04% of all dissolved materials is chloride,

$$\frac{\text{all dissolved materials}}{\text{Chloride}} = \frac{100.0}{55.04} = 1.817$$

One need only to measure the chlorine content and multiply by 1.817 to determine the salinity.

The way it is commonly done is to measure the amount of all the halogens (fluorine, chlorine, bromine, and iodine) present, by replacing the bromine and iodine by chlorine and then chemically precipitating them out of solution. The halogen content of seawater is called the "chlorinity," which will be slightly larger than the chlorine content due to the presence of the other halogens (55.35% as opposed to 55.04%). Again, due to the constancy of the relative abundances of the dissolved materials,

$$\frac{\text{all dissolved materials}}{\text{all halogens}} = \frac{100.0}{55.35} = 1.80655$$

Therefore, one need measure only the "chlorinity," by titration with silver nitrate and then multiply it by 1.80655 to get the salinity.

E.3   Other Methods of Analysis

Any measurable property of salt water that varies with salinity may be used to determine salinity. The property most frequently used is electrical conductivity. It is interesting that solid salt by itself is electrically insulating, but when dissolved in water it becomes conducting. This is because the ions are fixed in position and cannot move in the solid salt, but in solution these charged ions gain mobility and can flow from one electrode toward the other.

The electrical conductivity also depends on the temperature of the sample. At higher temperatures, the thermal motion of the dissolved ions is greater. They are more mobile, but they also have higher rates of collision. Consequently, the electrical conductivity depends both on the temperature and the salinity of the sample, and in order to use electrical measurements to determine salinity, one must also measure the temperature of the sample. The electrical apparatus that performs these measurements is called a "salinometer." (See Figure 11.14.) This is now a much more popular technique for determining the salinity of a sample than the more laborious chemical analysis.

FIGURE 11.14   *A salinometer.*

Another property that varies with salinity is the speed of light in sea-water. The fact that light travels with different speeds in different materials causes it to bend (or "refract") when going from one material to another. Also, it takes longer to pass through one material than another of equal thickness. Instruments using this property to measure salinity are called "refractometers."

Sensitive electrical instruments, such as salinometers and refractometers, must frequently be recalibrated because the electrical properties of the sophisticated electrical circuits vary with temperature, barometric pressure, humidity, and so on. In order that all instruments may be calibrated the same, international standard seawater samples are used, which are produced and bottled by the International Sea Water Office in Copenhagen.

E.4   Making Measurements at Sea

Carrying out any analysis at sea is no trivial task. To make salinity measurements we must either bring up water samples from depth to be analyzed, or we must lower electrical probes on an electrical cable to the appropriate depths and make the electrical measurements *in situ.*

At any one "oceanographic station," measurements are made on samples at many different depths. If water samples are to be retrieved, sampling bottles must be attached to a weighted cable as it is lowered into the sea. If *in situ* measurements are to be made, the probes and electrical cable must be appropriately attached to the hoist cable as it is lowered. Being exposed, the ship is more influenced by the caprice of the winds than are the surface currents, and if the captain is not careful, the cable may be dragged under the ship, where the devices may be jarred or damaged. Usually, a small propeller under the bow (called a "bow thruster") is used to help keep the bow into the wind, and the captain tries to keep

FIGURE 11.15 *Retrieving water samples with a Nansen bottle (left) and a Niskin sampler (right).*

the ship headed in a way that the cable appears to be vertical, that is, he tries to keep the ship at rest relative to the surface currents. The hoist cable goes from a winch over a pully system held over the side of the ship. An oceanographer may attach the bottles or probes appropriately from a small "hero platform," which folds up to the railing when not in use. (See Figure 11.15.) The pitching and rolling of the ship frequently add to the challenge of this particular job.

A variety of different collectors are in use for samples that are to be retrieved. The Nansen bottle is an old standby and is still popular, although being metal, it causes some contamination. Larger, plastic-walled samplers are quite popular also, as they tend to contaminate the sample less. Scientific instrumentation is an art whose appreciation is geared toward simplicity. The simpler something is, the more likely it is to work properly. This is especially true in oceanography, where the laboratory provided us by nature is much more unruly than one we would have chosen for ourselves.

When water samples are retrieved from depth, some mechanism must be used to indicate from what depth the sample was taken. This may be quite different from the amount of line let out after the appropriate bottle was attached, as the weighted cable may not hang vertically at greater depths, even though it is vertical near the surface. Subsurface currents may drag the line horizontally. Also, the sampling bottles may be accidentally triggered prematurely, before they reached the desired depth.

Another trouble with retrieving a sample from depth is that as the sample rises from depth, the pressure on it is reduced, and so it expands and cools slightly. This adiabatic temperature change must be taken into

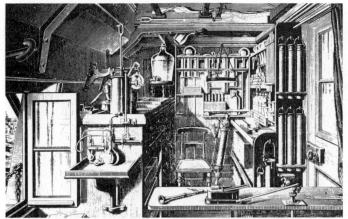

FIGURE 11.16   *Chemistry laboratory aboard the H. M. S. Challenger.*

consideration when one compares measurements made *in situ* to those made on retrieved water samples.

Although salinity determinations may be made *in situ*, clearly any detailed chemical analysis requires the collection of water samples. Contamination of the sample is an ever present problem. Atmospheric gases or other materials adhered to the walls of the sample bottle are one possible source of contamination. There are several techniques used to combat this problem. Some sample bottles are kept open all the way down, so they are continually flushed until reaching the appropriate depth. Larger samples dilute the effect of local contaminants, such as those from the bottle walls, but larger samples take up more space if they are to be stored aboard ship. Clearly, some compromise must be made.

Also, there is the likelihood of the sample's chemical nature changing due to organic activity. If a chemist wishes to study things that are noticeably affected by organic activity, such as nutrients, dissolved gases, or trace elements, then it is best that the measurement be made quickly before much distortion has taken place. If the sample is to be stored and worked on later ashore, then it should probably either be frozen, or it should be filtered through fine porcelain filters to remove organisms.

In any case, there is a clear advantage in performing any analysis quickly aboard ship, if possible (Figure 11.16). Such apparatus must be rugged as well as accurate. For any particular analysis, more than one measurement is made as a check on the precision of the result, and in an effort to eliminate errors due to malfunction, local contamination, and so on.

F.   RESIDENCE TIMES

F.1   First Entry and Permanent Removal

In this section we will be concerned with determining how long it is between the time a certain substance *first enters* the ocean and the time when it is *permanently removed*. This is called the "residence time" of

FIGURE 11.17 *As these formations erode and their minerals are washed to sea, can we really say that they are entering the ocean for the "first time?" Weren't they there hundreds of millions of years ago when these deposits were formed?*

the material. The phrases *first enters* and *permanently removed* are italicized because they require further amplification.

We consider, for example, that a substance is *permanently removed* when it joins the bottom sediment. But from our chapters on plate tectonics and on sediments, we are quite suspicious that nothing is really *permanently* removed from the oceans. Over periods of hundreds of millions of years, entire oceans disappear and others appear. What was once on the ocean bottom may now appear high on the continents as sedimentary deposits (Figure 11.17), or as volcanic material that has once disappeared with the ocean bottom into some trench and has returned to the surface through volcanoes. This material weathers, some of it dissolving and returning to the oceans. Is it fair to say, then, that this material was really *permanently removed* from the ocean? Doesn't much of it really get recycled? Similarly, isn't much of the material we think of *first entering* the ocean really recycled oceanic material from hundreds of millions, or billions, of years earlier?

Of course, the answer is "yes." But for this section we will consider anything that is removed from the oceans for hundreds of millions of years or longer to be *permanently removed*, and similarly, anything that is entering the ocean for the first time in hundreds of millions of years to be *first entering.* In short, although things do get recycled during these long geological time scales, we will ignore it here, having covered that in previous chapters.

F.2 Processes that Remove Salts from the Sea

Since the chemical makeup of the ocean has remained pretty much the same over the ages, each salt must be leaving the ocean at the same rate it is entering. How are the salts removed from the ocean? We do not know all the processes involved, even though we'd like to know them for both scientific and economic reasons.

In addition to direct precipitation via methods we've already discussed, we know that chemical interaction with suspended sediment is also an important process. The fine clays are especially important in these processes, because for equal volumes of sediment, the finest sediments have the largest surface area and remain in suspension longest. One of the most important salt-removal processes is adsorption onto the surface of the sediment. There is also exchange of cations and anions between the seawater and sediment. In cation exchange, more reactive cations in seawater, such as magnesium or potassium, tend to replace less reactive cations in the sediment. In anion exchange, the sediment tends to become enriched in phosphates and bromine. Trace elements, such as manganese, nickel, cobalt, zinc, copper, and the compound phosphorite (P_2O_5) are somehow adsorbed onto especially attractive surfaces, forming manganese and phosphorite nodules. Biological sedimentation is another important process removing some materials from the sea.

F.3 Recycling of Salts

Between their *first entry* into the ocean and the time they are *permanently removed*, most salts may undergo several "temporary" removals from the oceanic environment. There are several mechanisms which may cause this "temporary" removal, or "recycling," of a salt. When near the ocean, for example, we can smell the "salt air." The water has evaporated from ocean spray (Figure 11.18), leaving these tiny salt particles floating in the air, as our noses tell us. These may stay suspended for hours, days, or weeks before forming nuclei for the condensation of fog or rain droplets. These droplets return the salt either directly to the sea or to the ground, from which they return to the ocean with the drain-off from land.

Salts get recycled in other ways too. For example, the rising and falling of sea level with tides, seasons, or ice ages may lead to the formation of "salt flats," where water has been evaporated from confined shallow areas. These salts dissolve in the fresh-water run-off and return to sea, too, in times that are short compared to geological time scales. So these salts are recycled too.

F.4 The Problem of Residence Times

These various mechanisms causing the recycling of salts makes a determination of their residence times in the ocean a great deal more difficult.

FIGURE 11.18 *One way that salt is removed temporarily from the sea is through the evaporation of the water from sea spray, leaving microscopic salt particles in the air.*

The residence time of a salt is calculated by comparing how much of it is present in the ocean to the rate at which it is entering for the first time (or alternatively, to the rate at which it is permanently removed). For example, if it is found for a certain salt that $\frac{1}{50}$ of it enters each year and $\frac{1}{50}$ of it is permanently removed each year, then one molecule of that salt would last 50 years in the ocean on the average. That is, the turnover time, or "residence time," would be 50 years. Of course, no salt turns over this quickly, since the vastness of the oceanic reservoir guarantees that it would take a long time to replace any of the salts.

So in order to determine the residence time of a salt, we need to know both its present abundance in the ocean and the rate at which this salt is entering the ocean for the first time. The problem arises when we measure the rate of influx of this salt into the ocean via river water or other run-off from land. How do we know how much of this is entering the ocean for the first time and how much of it has simply been recycled? (Figure 11.19)

Fortunately, there is a straightforward answer to this question, and it hinges on the chloride ion in the river water. We know how much chlorine (an "excess volatile," remember) is entering our environment each year

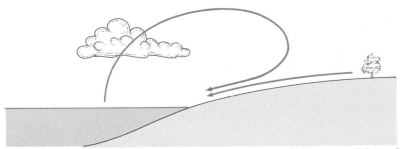

FIGURE 11.19 *Of the salts entering the ocean with the run-off from land, some are entering the ocean for the first time and some are recycled sea salt. How do you tell how much is recycled?*

TABLE 11.5 Illustration of How to Calculate How Much is Being Recycled and How Much is Entering the Ocean for the First Time, for Various Salts in a Sample of River Water.

Since the chlorine content is 11.4 mg, and since chlorine is 55.0% of sea salt, then there is a total of 20.73 mg of recycled sea salt in the sample.

Salt	Abundance in River Water Sample (mg)	Relative Abundance in Sea Salt	Amount That Is Recycled Sea Salt (mg)	Amount That Is Entering Ocean for First Time
Cl^-	11.4	0.550	$(.550) \times (20.73) = 11.4$	0
CO_3^{--}	70.4	0.0041	$(.0041) \times (20.73) = 0.1$	70.3
Ca^{++}	40.8	0.0116	$(.0116) \times (20.73) = 0.2$	40.6
SO_4^{--}	24.2	0.0.768	$(.0768) \times (20.73) = 1.6$	22.6
Na^+	11.6	0.306	$(.306) \times (20.73) = 6.5$	5.1
Mg^{++}	6.8	0.0269	$(.0269) \times (20.73) = 0.5$	6.3

through volcanic outgassing. This new chlorine only accounts for a negligible fraction of the chlorine in river water. Therefore, essentially all the chloride ion in river water is recylced sea salt.

Since all the major salts occur in constant ratio in sea salt, knowing how much chloride ion is recycled tells us how much of each of the other major salts is recycled sea salt as well. That is, since the concentrations of the various recycled salts occur in constant ratios, determining the concentration of any one of them determines the concentration of them all. Chlorine is especially well suited to set the scale for the recycled salts, since essentially all the chlorine in the land run-off is recycled from the sea.

Therefore, we measure the amount of a particular salt in the run-off from land, subtract from this the amount that has been recycled, and we are left with the amount of it entering the ocean for the first time. (See Table 11.5.) From this, we can determine its residence time, as described above.

The residence times for some of the dissolved salts are give in Table 11.6. These numbers are approximate. As we cannot test all the routes via which salts enter the oceans, our estimation of their rate of entry is a big extrapolation from limited data. The same sorts of extrapolation go into

TABLE 11.6 Approximate Residence Times For Various Materials in Seawater

Material	Residence Time (millions of years)
Cl	(∞)
Na	210
Mg	22
SO$_4$	11
K	10
Ca	1
CO$_3$	.11
H$_2$O	.044
Si	.035
Mn	.007
Fe	.0001
Al	.0001

estimates of rates of removal. Even our assumption that the ocean's chemistry remains constant is not strictly correct. For example, we know the oxygen content of our atmosphere and ocean have been increasing over time. We also know that climatic temperature changes, such as those during ice ages, have affected the amount of CO_2 (hence Ca^{++}) in our oceans, and this is reflected in the sediment record.

Accepting that the residence times in the table are approximate, we still see the indications that we'd expect. The water's residence time is very short due to its evaporation and subsequent precipitation over land. But among the major salts, the fastest turnover times are for CO_2 (mostly in the bicarbonate ion) and calcium, as we might expect from our discussion of this chemistry earlier in the chapter.

G. SUMMARY

With the exception of certain localized coastal areas, the major constituents of seawater are everywhere in the same relative proportions. This implies that variations in salinity are due to physical processes, which add or subtract fresh water from the solution, rather than to chemical processes that would add or substract certain salts. It also implies that the ocean is very well mixed. Time scales for the mixing of deeper waters are hundreds or thousands of years, and must be determined by indirect methods.

The relative abundances of dissolved gases are quite different from their relative abundances in the atmosphere. Carbon dioxide is much more abundant in the ocean than in the air, which makes the ocean a large buffer for the CO_2 released by our industrial society. Unlike the major constituents, the abundances of the dissolved gases vary widely, and biological organisms play a large role in these variations. Dissolved carbon dioxide affects the chemistry of seawater in a way that is particularly influential on the formation or dissolution of calcium carbonate sediments.

Biological organisms also have a large influence on the small but essential nutrient concentrations. The nutrients tend to be removed from seawater during photosynthesis and returned during respiration, as does carbon dioxide. Many trace elements are also essential to various organisms, and their evolved ability to hoard the more precious elements now threatens their existence as we pollute our environment.

The fact that the major constituents are in constant relative ratios simplifies the chemical measurement of salinity. Finding the concentration of just one of these salts determines them all. The salinity of a sample may be determined from its electrical or optical properties as well. There are advantages to doing the analysis either *in situ* or as soon as possible after the sample has been retrieved. Laboratories at sea, however, must endure some difficulties not encountered on land.

The average amount of time a certain salt spends in the ocean between the time it first enters and the time it is permanently incorporated into bottom sediment is called its "residence time." Determinations of residence times are complicated by the fact that most salts are "temporarily" recycled out and back into the ocean many times before their eventual permanent removal. Measurements of residence times reveal a very wide range among the various salts, with the carbonate ion having the shortest oceanic life expectancy among the major dissolved salts.

QUESTIONS FOR CHAPTER 11

1. Explain what is meant by the statement that dissolved salts are in constant ratios regardless of the salinity of the sample.

2. Suppose you were given two identical samples of seawater, and were asked to make one more saline and the other less saline, without changing the relative proportions of the salts in either sample. How would you do it?

3. What are the six most abundant ions dissolved in seawater, listed in order of decreasing relative abundance? (See Table 11.1.)

4. Where would you expect the rule of constant relative proportions for the salts to be violated, and why?

5. Why wouldn't you expect the rule of constant relative proportions to apply to dissolved gases, nutrients, and trace elements?

6. What two physical processes do you think have the biggest effect on the ocean's surface salinity?

7. What evidence is there that the oceans are well mixed?

8. If the individual molecules travel at speeds of hundreds of meters per second, why does diffusion progress at rates of only a few millimeters per minute?

9. If a person put cream in coffee but didn't stir it, would it ever get mixed? How?

10. Why is it that surface waters get mixed much more quickly than deeper waters?

11. Why are indirect measurements necessary for studying the mixing in deep waters? What methods can you think of for studying surface mixing?

12. If carbon-14 decays, why does the amount in our atmosphere stay constant?

13. Discuss how C^{14} can be used to measure mixing of deeper waters.

14. What are the abundances of O_2, N_2, and CO_2 in dry air?

15. How do the abundances of O_2, N_2, and CO_2 dissolved in seawater compare with their abundances in the atmosphere?

16. How do the solubilities of O_2, N_2, and CO_2 compare with each other? Does increased temperature make them more or less soluble?

17. Why have deeper waters more dissolved CO_2?

18. Why are the relative variations in dissolved O_2 concentration so much larger than the relative variations in dissolved CO_2? (That is, O_2 varies by $\pm 100\%$ from its average concentration, whereas CO_2 varies by only about $\pm 10\%$ from its average. Why?) Shouldn't organisms have a roughly equal effect on both?

19. How does biological activity affect concentrations of dissolved O_2 and CO_2?

20. Sketch a typical vertical profile of O_2 and CO_2 concentrations as a function of depth. Explain any variations exhibited that are caused by biota.

21. Explain how sulfate-reducing bacteria survive in anoxic waters.

22. What is the environmental impact of increased CO_2 in the atmosphere? How does the ocean moderate the environmental impact of CO_2 released from fossil fuels?

23. Explain the seasonal fluctuation in atmospheric CO_2 in Figure 11.9a. Explain the gradual overall increase in Figure 11.9b.

24. Continued reliance on fossil fuels for 200 more years will cause roughly what increase in earth temperatures? How does this compare with temperature changes during ice ages?

25. Describe what happens to a CO_2 molecule when it gets dissolved in water.

26. Lemon juice is quite acid, and soapy water is quite alkaline. From this, can you guess how the feel and taste of water are affected by the hydrogen ion concentration?

27. What is the concentration of OH^- in a solution of pH $=$ 5? What is the normal range in pH values of seawater?

28. Why is it that the OH^- ion concentration is inversely related to the H^+ ion concentration? That is, why should increasing the H^+ concentration cause the OH^- concentration to go down?

29. Explain why adding CO_2 to water reduces its pH.

30. Explain why Ca^{++} ions do not combine with CO_3^{--} ions in deep water to form sediments.

31. Explain why $CaCO_3$ dissolves in deep water. That is, what happens on an atomic scale?

32. Discuss some common processes that remove CO_2 from seawater. How does this affect the H^+ ion concentration? How does this facilitate the precipitation of calcium carbonate?

33. If Mg^{++} is much more plentiful in seawater than Ca^{++}, why does much more $CaCO_3$ precipitate out than $MgCO_3$?

34. Nutrient concentrations show much greater relative fluctuations than does the CO_2 concentration, even though the nutrients are used in far smaller quantities by biota. Explain why this is so.

35. What is a nutrient?

36. Nutrients are removed from seawater during what biological process? How are they returned to seawater?

37. Why should nutrient concentrations vary similarly to CO_2 concentrations, and oppositely to O_2 concentrations?

38. The oceans have 1.4×10^{18} metric tons of seawater. (1 metric ton $=$ 1000 kg and weighs about 2200 lb) How many metric tons are there of a substance whose concentration is one part per billion? One part per trillion?

39. Why is it difficult to produce artificial seawater for use in aquariums?

40. Discuss how the ability to concentrate and hoard certain trace elements may now work to the detriment of a species.

41. Why can't one measure salinity simply by weighing the residue left in a pan after boiling dry 1 l of seawater?

42. What is the formal definition of salinity? How close is this to simply the total mass of all dissolved solids per kilogram of seawater?

43. Explain how you can determine the salinity of a sample of seawater by measuring the concentration of only one substance.

44. Why is it that an electrically insulating salt becomes conducting when dissolved in water?

45. Would you think that electrical conductivity should increase or decrease at higher temperatures? Whichever your answer, defend it.

46. What two things must a salinometer measure to determine the salinity of a sample?

47. What property of seawater is used by a refractometer to measure salinity?

48. Why are international standard seawater samples produced?

49. What is a "bow thruster"? A "hero platform"? If the hoist cable hangs vertically, does it mean the ship is stationary? Explain.

50. Look up the verbs "pitch" and "roll" in a dictionary. How are they different (as applied to the motion of a ship)?

51. Why is there a bias toward simplicity in scientific apparatus and instrumentation?

52. Why can't you figure the depth at which a water sample was taken by the amount of line let out after the bottle was attached?

53. Why does the temperature of a sample change as it is brought up from depth? (Assume the sample bottle is insulated.)

54. How can you minimize the effects of microorganisms which alter the chemical makeup of a sample after it has been retrieved and brought aboard ship?

55. Discuss why it is difficult to say whether a certain salt entering the ocean is really entering for the first time. Could it have been there before, eons ago? Is there a similar problem in deciding whether something can get *permanently removed* from the ocean?

56. Under what conditions do we consider something to be *first entering* the oceans, or *permanently removed* from the oceans.

57. Discuss some of the processes we know of that *permanently* remove some materials from the ocean.

58. Why are the finest suspended sediments most significant in affecting the ocean's chemistry?

59. Discuss some of the processes which recycle salts out and back into the sea for relatively short times (short compared to geological time scales of hundreds of millions of years).

60. What is "residence time"? How do we determine the residence time of a salt?

61. How much of the chloride in river water is recycled?

62. Discuss how we can determine how much of each salt in river water is recycled sea salt, and how much of it is entering the ocean for the first time.

63. Of the major constituents, which has the longest residence time? Which has the shortest?

*64. Describe how you'd expect the salinity of surface water to vary with distance from land and with latitude.

*65. Draw an idealized rectangular ocean basin extending from Antarctica to the Arctic Circle, and bounded by continents on the east and west. Put numerals at many places on this ocean, indicating what you'd expect the surface salinity at those points to be. (Normal range is 34.0% to 36.0%)

*66. If the water molecule were not so stable (i.e., strongly bound together), would the pH of pure water be more or less than 7? Explain.

*67. List some compounds that you think would reduce the pH of a solution when dissolved in water. List some that you think would increase the pH of the solution.

*68. If the CO_3^{--} (carbonate) ions in the ocean arise from dissolved CO_2, which hydrates (H_2CO_3) and then dissociates (to HCO_3^- or CO_3^{--}), how is it that *removing* CO_2 from sea water can *increase* the CO_3^{--} concentration? (*Hint:* The distribution of the *remaining* dissolved CO_2 among the HCO_3^-, CO_3^{--} forms may depend on the hydrogen ion concentration.)

*69. What are some advantages and disadvantages you can think of for working with retrieved water samples, as opposed to making *in situ* measurements?

*70. Suppose the chlorine, sodium, and calcium content of a certain river was 10 parts per million apiece. How much of each is entering the ocean for the first time?

*71. In previous years we have assumed the dissolved gases to be in equilibrium with the gases of the atmosphere when deep water masses were first formed. Now we find they are not. In particular, there is proportionately less carbon-14 in the forming water masses than there is in the atmosphere. How would this recent discovery alter our estimation of the ages of deep water masses?

*72. According to Table 11.3, how many liters of CO_2 can be dissolved in one liter of water at 10°C?

*73. What would be the salinity of a sample of seawater if the Mg^{++} ion concentration was 1‰? If the Cl^- concentration was 24‰?

*74. Why do you suppose gases are more soluble in cold water, whereas solids are more soluble in warm water?

SUGGESTIONS FOR FURTHER READING

1. E. K. Duursma and R. Dawson (eds.), *Marine Organic Chemistry*, Elsevier, New York, 1981.

2. "Research Vessels," *Oceanus*, **25,** No. 1 (Spring 1982).

12
THE OCEAN AND OUR CLIMATE

Infrared imagery shows the warm Gulf Stream (dark areas) moving through colder waters (lighter areas) as it passes the east coast of the United States on its way toward Northern Europe.

Compared to other terrestrial planets, the earth's climate is extremely mild. This is due to the abundance of surface water and its remarkable thermal properties.

Both ocean and atmosphere are only a few kilometers thick but extend completely around the earth. This makes their relative thickness in comparison to their breadth comparable to that of this sheet of paper. They are as intimately interconnected as are two successive pages in this book when the book is closed. Clearly their behaviors are closely related.

Although roughly equal in thickness, the oceans are 90 times more massive than the atmosphere, so it is really the oceans that control the atmosphere, and not vice versa. The oceans cover only 71% of the earth's surface. Over the 29% covered by continents, the atmosphere displays a little more caprice, being temporarily removed from the dominant moderating influence of the oceans.

In this chapter we study the earth's climate. Since it is controlled by the oceans, we must understand the oceans in order to understand our climate.

12
THE OCEAN AND OUR CLIMATE

A. THE HYDROSPHERE

A.1 Water Reservoirs

The exterior of the earth is called the "hydrosphere." The hydrosphere has several types of water reservoirs, listed in Table 12.1. It may be somewhat surprising to us terrestrial beings, who rely so heavily on our water for our livelihood, that less than ½% of the water is stored in ground water and only a trace (.04%) in our beloved lakes and rivers. Furthermore, in spite of our sentiments during the rainy season, the atmosphere holds only a thousandth of a percent of the hydrosphere's water. If in one gigantic worldwide rainstorm all the atmosphere's water were to fall to the earth's surface, it would only amount to about 3 cm of rainfall. Practically all of the water (98%) is contained in the oceans, and most of the remainder is stored in the polar caps and other ice. Another way to visualize the relative amounts of water in the various reservoirs is to imagine that we could make the earth's surface perfectly smooth and spread out the water from each of these reservoirs evenly. Then the respective depths of the waters, called "sphere depths," would be those given in Table 12.1.

A.2 The Hydrologic Cycle

In spite of its low water content, the atmosphere serves as an important agent in the transfer of water from one reservoir to another. The cycling of water among the reservoirs is called the "hydrologic cycle," and is depicted schematically in Figure 12.1. The ocean loses water to the atmosphere via evaporation, but gains it back through precipitation, **323**

TABLE 12.1   The Amount of Water in the Various Reservoirs, in Terms of
Percent of Total (1.4 Billion km³), and in Terms of Sphere Depths

Reservoir	Percent of Total	Sphere Depth (m)
Oceans	97.96	2685
Polar caps and ice	1.64	45
Ground water	0.36	10
Rivers and lakes	0.04	1
Atmosphere	0.001	0.03

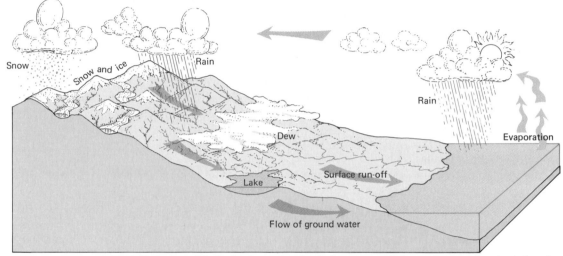

FIGURE 12.1   *Hydrologic cycle, depicting some of the major mechanisms for the
transfer of water between reservoirs.*

run-off from the land, and melting of ice. The atmosphere carries some of
the water from the oceans to the polar caps, where it is deposited as snow,
and it deposits water on the continents via rainfall, snow, dew, and so on.
Over the continents, the precipitation exceeds the evaporation, and so
some of the water must be returned to the oceans via the rivers and underground flow.

B.   THE ATMOSPHERIC RESERVOIR

B.1   The Atmospheric Water Content

Everyone is aware that the water in an open pan will slowly evaporate.
The warmer and drier the air, the faster the water will disappear. The
amount of water vapor that air can hold depends on the temperature of
the air, which increases roughly by a factor of two for every 10°C increase
in the temperature. (See Table 12.2.) At room temperature (about 20°C)
the atmosphere can be up to 2.31% water vapor. There are several terms
dealing with the water content of air. We call air "saturated" if it is hold-

TABLE 12.2 Water Vapor Content of Saturated Air as a Function of the Temperature

Temperature (°C)	0°	5°	10°	15°	20°	25°	30°
Water vapor content when saturated	.60%	.86%	1.21%	1.68%	2.31%	3.10%	4.17%

ing as much water as it can (i.e., under normal conditions, or in equilibrium with liquid water). For example, if air at 25°C is 3.1% water vapor, then it is saturated. If you were to cool this air down to 15°C, then it would be "supersaturated," and about half of the water would condense and precipitate out.[1] There are several terms for describing how much water there is in the air. "Absolute humidity" is simply the percentage of air that is water vapor. The "relative humidity" relates the actual water content to the maximum possible content. For instance, if air at 25°C were 3.1% water vapor, then the absolute humidity would be 3.1%, whereas the relative humidity would be 100%, as according to Table 12.2, that air would be saturated. If this same air held only half as much water as it could (e.g., 1.55% at 25°C), then the absolute humidity would be only 1.55%, and the relative humidity 50%. If air at 100% relative humidity is cooled, then it becomes supersaturated and the excess moisture precipitates. The "dew point" is that temperature at which the air will become saturated. For instance, if the water vapor content of a certain air mass is 1.68%, then the dew point is 15°C, as is seen from Table 12.2 that this is the temperature at which the air will be saturated. Notice that the dew point does not depend on the actual temperature of the air, only on its water content.

B.2 Latent Heat of Vaporization

It takes a lot of heat to evaporate water. We have seen that 1 calorie will raise the temperature of 1 g of water by 1°C. Therefore, it takes 100 calories to bring 1 g of water from its freezing point to boiling. But an additional 540 calories are required to actually evaporate the water. Thus, a pan of water on the stove readily comes to boiling, but takes a long time to boil dry. If it weren't for this we'd have difficulty boiling an egg or making soup. This is also why we perspire. The heat required to evaporate the water comes from our skins, cooling us off.

From a microscopic point of view, the following happens when water evaporates: Temperature is a measure of the average kinetic energy (energy of motion) of the molecules of a substance. At higher temperatures, the molecules are moving faster, on the average. Due to collisions between molecules, their motions are quite chaotic, and at any instant some will be moving faster than others. Some will be moving fast enough to burst free of the water surface and join the atmosphere as water vapor. These have "evaporated." At higher temperatures, more will be moving fast enough to evaporate, so the water evaporates faster at higher temperatures.

---

[1]*Actually, it could remain somewhat supersaturated if it did not come into contact with dust particles, ground, water droplets, or other condensation nuclei.*

Since the fastest moving molecules are the ones that evaporate, the ones left behind are slower moving, on the average, which means that the temperature of the remaining water is lower. This is why evaporation is a "cooling process," and why you feel cold when you step out of a shower. The fastest moving molecules join the atmosphere, taking their energy with them as "latent heat." They leave the slower moving ones behind so the remaining water is colder. If you wish to continue the evaporation, you must continually add heat to replace the energy removed by the evaporated molecules.

The heat given to water in evaporation is released again when it condenses. The condensing water vapor gives heat to our atmosphere during a rainstorm. So one way our atmosphere stores heat is in the latent heat of the evaporated water. At any one time relatively large quantities of heat are contained in the form of latent heat. As an example, a rainfall of 1 cm releases an amount of heat to the earth's surface and lower atmosphere that is equivalent to more than an entire day's sunshine. There is enough heat stored as latent heat in the earth's atmosphere right now to equal that received from the sun by the entire earth in four days.

It is sometimes convenient to divide the atmosphere's thermal energy into two components: "sensible heat" and "latent heat." Sensible heat is the thermal energy stored in the motion of the air molecules, and is reflected by the temperature. Latent heat is the energy stored in the evaporated water molecules, which is released when they condense into the liquid state.

Suppose we take a certain amount of water-saturated air and raise its temperature by 1°C. Some of the added heat would go into raising the temperature (sensible heat) and some into evaporating more water (latent heat). Although the required amount of sensible heat does not depend on the initial temperature of the air, the amount of latent heat does. For example, the same amount of sensible heat is required to raise the temperature from 10°C to 11°C as from 20°C to 21°C. But because the water content of saturated air increases exponentially with temperature, nearly doubling for every 10°C increase in temperature, we know that nearly twice as much water can evaporate when the temperature changes from 20° to 21°C as when the temperature changes from 10° to 11°C.

As the above example shows, at higher temperatures, relatively larger fractions of the added heat can go into evaporating water. (See Figure 12.2.) In fact, above 10°C, more added heat can be stored as latent heat than as sensible heat. This is quite remarkable, considering what a small fraction of the air is water vapor at these temperatures (Table 12.2).

B.3 Evaporation vs. Precipitation

Since the average water content of the atmosphere remains about the same from year to year, then the total precipitation in the world must equal the total evaporation. However, there are regional variations in this result. Over the oceans the evaporation exceeds precipitation, and over land the reverse is true. The excess of evaporation over precipitation

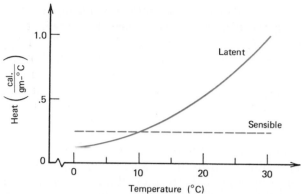

FIGURE 12.2 *Plot of the amount of heat required to raise the temperature of 1 g of air by 1°C (sensible heat), and of the amount of heat required to evaporate additional water in keeping it water saturated (latent heat) vs. temperature. Notice that above 10°C more heat goes into latent than into sensible heating.*

amounts to 7 cm per year for the oceans (worldwide average). Over land, precipitation exceeds evaporation by 17 cm per year, on the average. Since for every 7 km² of land area there are 17 km² of oceans, the excess water gained on land just balances the loss from the oceans, this amount being returned to the oceans via surface and underground run-off.

Seeing this discrepancy between the relative amounts of evaporation and precipitation may cause one to wonder what special talents the land has to be able to make the precipitation exceed evaporation. One reason is obvious. The land has less exposed water surface. The less water exposed to the atmosphere, the smaller will be the rate of evaporation.

In addition to lower rates of evaporation, the land also has two special talents for coaxing moisture out of the air. First it is higher. Air coming off the oceans must rise to get over the continent. At higher altitudes air encounters lower atmospheric pressures; consequently it expands and cools. At cooler temperatures air cannot hold so much moisture, so some condenses and falls on the continents. Good examples are provided by the Olympic Mountains in the state of Washington, and the Sierra Nevadas in California. In both these regions, winds come predominantly from the west. Air masses rising to pass over these mountian ranges deposit about ten times the annual rainfall as do the predominantly falling air masses on the other side.

The second special talent that the land has to help coax water from the air is its large daily and seasonal temperature fluctuations. A typical daily temperature fluctuation for the land surface is 15°C, whereas for the sea surface it would be only about 1°C. At its early morning low, the land can coax much moisture out of the atmosphere near it. Hence, fog appears primarily in the early morning. The same holds for the seasonal temperature changes. In winter, the warmer air over the warmer oceans holds more moisture. When it encounters the cooler land, it cools and precipitates. Even in the warmer seasons, moisture-laden air from the oceans

TABLE 12.3   Temperatures in °C for Various Canadian Cities of the Same
Latitude, Showing Role of Oceans in Moderating the Climate

	Victoria	Winnipeg	St. Johns
Mean January minimum (°C)	2	−22	−7
Mean July maximum (°C)	20	27	21

may survive the warm day over land, but not the cool night.

That the proximity of the ocean moderates the climate can be seen by comparing the seasonal temperature variations of cities at the same latitude, as in Table 12.3. Clearly the two coastal cities have more moderation of their seasons than does Winnipeg, which is on the Canadian Great Plains. The winds at this latitude come primarily from the west, so between the two coastal cities, the climate of Victoria (having the ocean on its west) should be somewhat more moderated than that of St. Johns. This is also confirmed by the data in the table.

Similar effects are noted elsewhere. At 40° N Lat., the surface of the Pacific Ocean changes at most by 8°C during the year (10° to 18°C), whereas at the same latitude the seasonal average temperature fluctuation on the interior of the Asian continent is from about −4°C to +34°C. Although the seasonal temperature fluctuations on the Asian continent are large compared to those on the ocean (38°C compared to 8°C), they are still quite small compared to the 120°C mean seasonal fluctuations on the Martian surface, where there are no oceans at all. (Daily temperature fluctuations there are also about 120°C.) All other things being equal, being a sea explorer should be a more comfortable job than being a land explorer.

B.4   Why Larger Thermal
Fluctuations Over Land?

There are several good reasons why the land surface temperatures fluctuate so much compared to those of the sea. One is that the heat capacity of water is greater than that of the crustal materials. For example, the heat capacity of sand is only half as great as that of an equal volume of water. Therefore, putting equal amounts of solar heat into equal volumes of sand and water, the sand should heat up by twice as much. Likewise, in cooling the two, subtracting equal amounts of heat from both, the temperature of the sand should diminish by twice as much as that of the water. Although this effect is obviously an important contribution to the relatively larger temperature fluctuations on land, the other causes have even greater effect.

Another reason for the greater temperature fluctuations on land is that the heating and cooling take place only at the surface, whereas in the oceans, solar radiation penetrates to greater depths. Most people know the difficulty in walking barefoot on a beach on a hot afternoon. Although the surface of the sand is so hot that you can only bear a few quick steps at a time, you can quickly cool your feet just by shoving them through the surface of the sand; just a few centimeters down it is refreshingly cool. Similarly, when your feet start freezing at a midnight beach party, you can

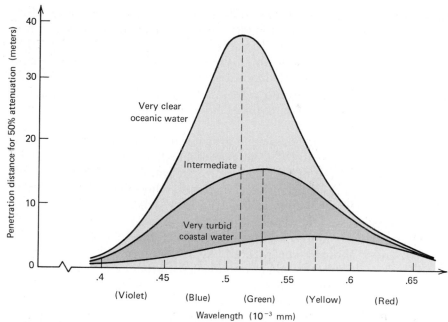

FIGURE 12.3 *Plot of the distance that light can travel before being diminished by 50% as a function of the wavelength, for ocean waters of various turbidities. Notice that very clear water tends to be bluer, and very turbid water tends to be yellower.*

quickly warm them up just by shoving them down into the sand below the surface. Virtually all the temperature fluctuations take place right at the surface; below the surface, the sand remains mild day and night. As a further example, building codes in many areas in the north-central states only require foundations of houses to be sunk one-half meter below the surface. Even in the longest, coldest winters, the ground won't freeze below that.

Therefore, a second reason why the land surface temperatures fluctuate so much is that the solar radiation is caught, stored, and released only in the upper few centimeters. In oceans, however, the penetration depth is much greater. Consequently, the solar heat is spread out over a much larger volume, causing the smaller temperature fluctuations.

The penetration length for light in water depends both on how murky the water is and on the wavelength of the light. For clear ocean water, blue-green penetrates the best, going nearly 40 m with only 50% attenuation. The extremes of the visible spectrum fare less well (Figure 12.3). Red and violet go only about 4 m with 50% attenuation. This explains why clear ocean water looks blue-green; light going any distance through it will have relatively more colors in the middle part of the spectrum remaining, and less red and violet. In murkier water, all colors are attenuated more quickly than in clear water, and the blue is reduced even more than the other colors. Consequently, murky water tends to look greenish-yellow. In addition to suspended sediments, the presence of phytoplankton and other organic matter tends to add to the yellowish tinge of the water. More details can be obtained from Figure 12.3. In any case, even in clear oceans, at sufficient depth all surface light is extinguished. Beyond 100 m depth, virtually no plant life can survive. That is not to say that at great depths the ocean is completely black, because some animals there provide their own lighting. But beyond a few hundred meters, solar lighting is missing, the existing light is feeble, and one cannot tell the difference between night and day.

Besides the larger heat capacity of water and the greater penetration depth of solar heating in the oceans, another effect that tends to reduce the thermal fluctuations of the ocean's surface, compared to those on land, is mixing. Even in the murkiest ocean waters, where only the upper few meters receive solar heating directly, the heat will mix as far as 100 m downward within an hour due to the wave motion.

Mixing is particularly important in redistributing the heat from the infrared. Water is opaque to infrared. Although not much infrared from the sun makes it down through our atmosphere, what does is absorbed by the upper few millimeters of water. Were it not for mixing, the surface skin of the oceans would be much hotter.

Finally, the evaporation of water from the ocean's surface removes latent heat, which also tends to keep the sea surface cool on sunny days.

C. WORLDWIDE TEMPERATURE CONTROLS

C.1 Heat Radiation

All bodies radiate heat. The amount of heat radiated by a body increases with its size and with its temperature.[2] Radiation comes in all wavelengths, from the short wavelength gamma rays through the long wavelength radio waves (Figure 12.4), and all wavelengths travel with the speed of light. Although all bodies radiate energy of all wavelengths, the region where most radiated energy is concentrated depends on the temperature of the body; the hotter it is, the shorter is the wavelength of maximum radiation. This relationship is expressed by the equation,

$$\lambda_{max} = \frac{2.9(mm \cdot K)}{T} \tag{12.1}$$

Putting the temperature of the sun's surface (about 5600K) into this equation, one finds that the wavelength of maximum solar radiation is in

[2]*The rate at which energy is emitted from each square meter of surface on a perfect radiator is given by σT^4, where T is the temperature in Kelvin, and σ is a constant, equal to 5.67×10^{-8} watts/m^2 · K^4. Notice that the rate of energy emission increases rapidly with the temperature. When the temperature is doubled, the rate of emission increases by $2^4 = 16$ times! Notice also that the rate of energy emission increases with the size of the object. Larger objects have larger surface areas (more square meters) from which this energy can be radiated.*

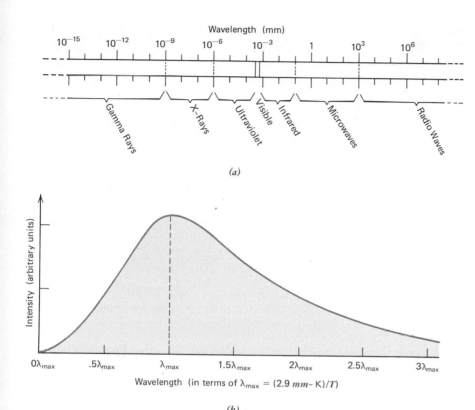

(a)

(b)

FIGURE 12.4 (a) *Electromagnetic radiation comes in all wavelengths, from the very long radio waves to the very short-wavelength gamma rays. Visible light encompasses only a small part of the total spectrum.* (b) *Plot of intensity of radiated energy vs. wavelength. Any body radiates at all wavelengths, but mostly near a wavelength determined by* $\lambda = 2.9/T$, *where* λ *is in mm and T in K. (Beware of the difference in the horizontal scales between (a) and (b). One is logarithmic and the other is linear.)*

the visible range—5.2×10^{-4} mm, which is yellow. Although the sun puts out most of its radiation in the visible range, some is emitted in the regions on both sides of the visible range. That is, small amounts of the solar radiation come in the ultraviolet and infrared wavelengths. Negligible amounts are in other wavelengths.

When the sun is directly overhead, the solar heat reaching a square centimeter of the outer atmosphere amounts to about 2 calories per minute.[3] Sometimes a unit called a "langley" is used, which is a calorie per square centimeter. Hence, the solar heat flux at the outer atmosphere would be about two langlies per minute. Not this much reaches the earth's surface, however, for several reasons. First, reflection and absorption by

---

[3]*Expressed in metric (SI) units, the rate at which solar energy reaches our outer atmosphere is 1370 watts per m². This is called the "solar constant" for the earth. The rate at which it reaches the earth's surface, averaged over all seasons, day and night, and over the entire earth, is about 175 watts/m².*

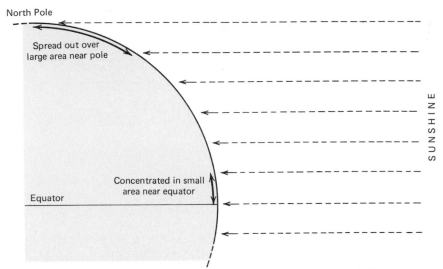

FIGURE 12.5 *At high latitudes the earth's surface receives much less heat per unit area than at lower latitudes, as equal quantities of incoming radiation must be distributed over a much greater area.*

clouds block about half the incoming radiation. So only about one langley per minute would reach the earth's surface, on the average, when the sun is directly overhead. But then, the sun can seldom be considered to be directly overhead—only in the tropics and only near midday. Away from the tropics the sunlight strikes the surface more obliquely, and so the solar heating is even less. (See Figure 12.5.) Furthermore, half the time will be nighttime, when there will be no sunshine at all. When averaged over the entire earth's surface and over a 24-hr period, the solar flux actually reaching the earth's surface averages to about 0.25 langley per minute. Since the earth's orbit is slightly elliptical, the varying distance from the sun during the year causes a 7% variation in this number during the year.

The earth is also radiating, but it is much cooler than the sun, averaging about 290K. Putting this temperature into the formula of Equation 12.1 yields that the earth radiates mostly in the infrared ($\lambda_{max} = 10^{-2}$mm). Since our eyes are sensitive only to the visible range, we can see the sun's radiation coming in, but not the earth's going back out. Since the earth is not getting hotter, it must (on the average) put back into space exactly as much heat as it receives from the sun.

C.2   The Greenhouse Effect

Our eyeballs tell us that the atmosphere is transparent to the visible wavelengths, but they do not tell us whether the atmosphere is transparent to the other wavelengths. It turns out that the ozone (O_3 molecules) in the upper layers of the atmosphere is opaque to some of the ultraviolet wavelengths, and that the H_2O and CO_2 molecules make the lower atmosphere opaque to a large portion of the infrared region (Figure 12.6). This doesn't have much effect on the total solar radiation coming in, as it is mostly visible. True, some of the ultraviolet and infrared will be screened out, but

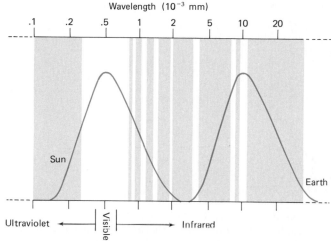

FIGURE 12.6 *Plot of intensity vs. wavelength for radiation received from the sun and emitted from the earth. Shaded areas are wavelengths for which the earth's atmosphere is opaque. Opacity in the ultraviolet is largely due to ozone (O_3), and opacity in the infrared is largely due to the CO_2 and H_2O in the atmosphere.*

they were only minor constituents of the sun's radiation anyhow. Most of it makes it through our atmosphere with no trouble, except for interference by clouds.

But trying to get the earth's radiation back out is another story. Since the atmosphere (due to the CO_2 and H_2O) is opaque to the infrared, earth radiation has real difficulty trying to get out. Any energy radiated from the ocean's surface stands a good chance of being quickly absorbed by a water or carbon dioxide molecule in the atmosphere before making it through. When this molecule reemits this absorbed energy, it could go in any direction, so there is a 50-50 chance that it will go back downward toward the ocean rather than up toward outer space. If it goes back to the ocean, then it must start its trip all over again. If it goes toward outer space, it still stands a good chance of being absorbed by some other water or carbon dioxide molecule higher up in the atmosphere. And so it goes. You can see that the presence of water and carbon dioxide make it difficult for infrared radiation, emitted from the earth's surface, to make it back out into outer space.

In this way, the atmosphere acts like a blanket which tends to trap the heat inside, keeping the earth's surface warmer. This is called the "greenhouse effect," because greenhouses work essentially the same way. The greenhouse glass is transparent to the incoming solar radiation, but opaque to the outgoing infrared, keeping it warm inside.

There is currently some concern over the rate at which we are dumping CO_2 into our atmosphere by burning our fossil fuels. More CO_2 makes the atmosphere more opaque to the infrared, which makes it warmer, which makes it hold more evaporated water, which makes it even more opaque to infrared and warmer still, and so on. Perhaps some salvation

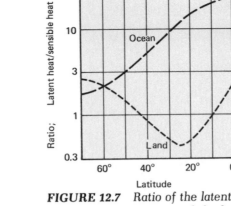

FIGURE 12.7 *Ratio of the latent to the sensible heating of the atmosphere by the oceans and the land as a function of the latitude (averaged over a year). (After W. D. Sellers,* Physical Climatology, *University of Chicago Press, 1965.)*

can come from increasing plant life, which through photosynthesis uses up CO_2 to make foods and O_2.

C.3 Heat Transfer Through the Lower Atmosphere

All the heat leaving the earth eventually exits the outer atmosphere in the form of electromagnetic radiation into space. We have seen that the earth's lower atmosphere provides a considerable barrier to the emigration of heat from the earth's surface. In this section, we examine the various mechanisms through which the heat can get through this barrier.

In studying the ocean's role in moderating the earth's climate, it is interesting to note that heat can go from the ocean to the air, but not vice versa. The ocean recieves most of its heat directly from the sun; the atmosphere, however, receives most of its heat from the ocean. We have seen the difficulty the ocean has in reradiating the heat through the atmosphere. In fact, it seems that the harder the ocean tries to radiate, the more difficulty it has. In the tropics the ocean is warmest, and would otherwise radiate more heat. But in the tropics, the air holds more water, which blocks the infrared radiation, and so the ocean actually succeeds in radiating less.

The largest means of getting heat back out through the atmosphere in the tropics is through latent heat (Figure 12.7). Evaporating water takes the latent heat from the ocean's surface and carries it high into the atmosphere, where it is released when it condenses in clouds. Up there the heat has less atmosphere left to go through to make it out into space. Furthermore, what atmosphere is left is thinner and less opaque. This type of heat transfer, however, is a one-way process. The evaporating water takes heat from the ocean, but the rainfall does not give it back. It is rather given off to the region of the atmosphere, where the condensation takes place.

Some heat exchange also occurs via conduction between ocean and atmosphere, but this is also a one-way process. To demonstrate this, we'll

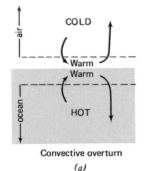

Convective overturn
(a)

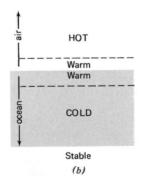

Stable
(b)

FIGURE 12.8

The conduction of heat between ocean and air is a one-way process. It can go from ocean to air, but not vice versa. (a) Convective overturn. (b) Stable.

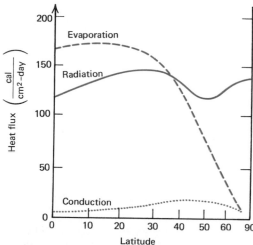

FIGURE 12.9 *Heat transfer from ocean to atmosphere by each of the three processes as a function of latitude, measured in terms of calories of heat released by each square centimeter of ocean surface per day (yearly average). (The horizontal scale is made so that equal distances represent equal amounts of the earth's area. At higher latitudes, the earth's area is smaller, and the scale is correspondingly compressed.) (After A. Defant, Physical Oceanography, vol. 1, Pergamon Press, 1961.)*

first consider cold air over a hot ocean and demonstrate that the heat will go up, and then we'll consider hot air over a cold ocean to demonstrate that it won't go down (Figure 12.8). In the first case, consider the strip of cold air just over the hot ocean. It'll warm up. Being warm, it will rise through the cold air above, bringing a new layer of cold air in contact with the ocean. Similarly, the surface of the ocean will cool, having lost heat to the air above, and will sink through the warmer ocean below, bringing more hot water to the surface to touch the cold air above and continue the cycle.

However, in the reversed case, a stable insulating layer will form between the hot air above and the cold ocean below. Consider the thin strip of hot air in contact with the cold ocean. It will be cooled. Being now cooler than the hot air above, it will stay below, and the air higher up will never be able to touch the cold ocean below. Similarly, the thin layer of water on top will be warmed by the hot air above, and consequently it will float atop the cooler, denser water beneath. So the hot air above remains insulated from the cold ocean below.

Of these three methods of getting heat from the ocean to the atmosphere, conduction is the smallest. The other two are about equal, latent heat being more important near the equator where the air is warmer and can hold more water vapor. Near the poles, direct radiation from the ocean is more important, as the moisture content of the air is smaller and inhibits the infrared radiation less. The relative amounts of heat transfer to the atmosphere by each of the three methods as a function of latitude is shown in Figure 12.9.

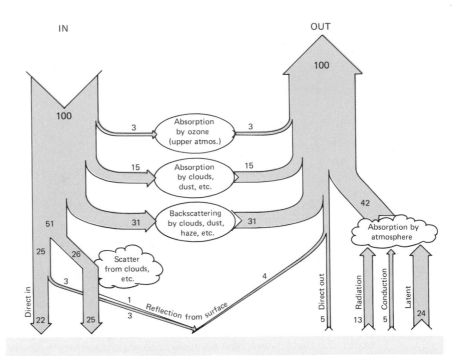

FIGURE 12.10   *Earth's heat balance, indicating the percentage of the incident radiation following each of the various courses; 47% of the incoming radiation is absorbed by the ocean surface.*

C.4   Detailed Heat Balance

Figure 12.10 illustrates schematically much of what has been said so far. Of the solar radiation incident on our atmosphere, some is reflected off the clouds, and a small fraction is in the ultraviolet or infrared to which our atmosphere is quite opaque. But still 51% of the incident radiation manages to make it to the ocean's surface, 4% being reflected ("seaglitter"), and 47% absorbed. This is then reradiated by the earth in the infrared. Most of it gets out by going first from the ocean to the atmosphere by one of the three processes described in the previous section, and then from the atmosphere into outer space. Only 5% is radiated directly from the ocean's surface into outer space without being hassled at all by the atmosphere. This is because there are some "infrared windows"—that is, there are some narrow hands of infrared radiation to which our atmosphere is transparent (Figure 12.5), although it is opaque to most.

Although we know that the total heat into the earth has to equal the total heat out, the actual measurement of heat going in both directions yields a surprise. Near the equator, the earth receives more heat than it gives off, and near the poles it returns more than it gets (Figure 12.11). Since the equator is not getting hotter, and the poles aren't getting colder, some heat (about 20%) that is received in the tropics must be transported poleward before being reemitted. How is this accomplished? We quickly look to our glamorous ocean currents, such as the Gulf Stream, which

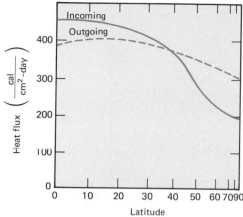

FIGURE 12.11 *Plot of heat received and reradiated by the earth as a function of latitude. Notice that in the tropics we receive more than we reradiate, and that in polar regions, the reverse is the case. (From Peter K. Weyl, Oceanography, An Introduction to the Marine Environment, Wiley, 1970.)*

carry only about 10% to 20% of the necessary amount of heat. Surprisingly enough, the major portion is carried as latent heat. In the warmer tropics, water evaporates from the oceans, but as air moves from lower latitudes to higher latitudes, it cools and cannot hold so much moisture, so the water condenses and precipitates. In this way, the latent heat is taken at lower latitudes and returned to the earth's atmosphere at higher latitudes. It is somewhat surprising that the atmosphere, being such a small reservoir of the hydrosphere, can play such a large role in distributing heat around the earth. It wouldn't be possible if the latent heat of vaporization of water weren't so large.

With this pattern in mind, if we were to plot evaporation minus precipitation as a function of latitude, we would expect it to be negative at high latitudes, where precipitation is larger, and positive at lower latitudes, where evaporation is larger. In Figure 12.12 it is seen that this is generally the case, with the exception of a dip in the curve near the equator. The cause of this dip is that the intense sunshine at the equator not only evaporates water but heats the air so much that it rises. Rising moist air means large rainfall. Consequently, it rains a lot at the equator, creating tropical forests and jungles on land. The dip, then, is due to the high amount of precipitation there. Except for this dip, though, the general pattern is as we predicted it should be, with evaporation dominating near the tropics, and precipitation dominating nearer the poles.

When seawater evaporates, it leaves the salt behind. Consequently, in regions where evaporation exceeds precipitation, one would expect to find the surface water more saline than normal. Similarly, the return of fresh water to the surface should reduce the salinity in regions where precipitation is larger. Therefore, the surface salinity of the oceans at the various latitudes should mimic the plot of evaporation minus precipitation. It is seen in Figure 12.12 that it does.

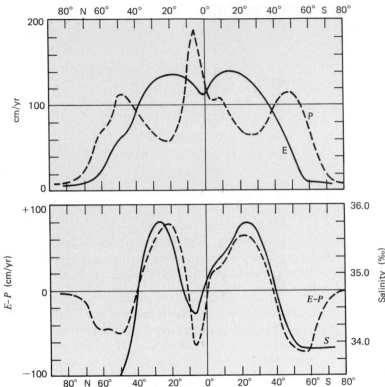

FIGURE 12.12 *(above) Plot of evaporation (E) and precipitation (P) as a function of latitude. (below) Plot of the difference, E–P, and the surface salinity (S) as a function of latitude. (After G. Wüst, Oberflächen- selzgehalt, Verdunstung und Niederschlag auf dem Weltmeere, Festschrift Norbert Krebs, Stuttgart, 1936, pp. 347–359.)*

D. THE EFFECT OF THE SALT ON THE THERMAL PROPERTIES OF WATER

The data on water presented in this chapter have been for fresh water. The presence of salt modifies these figures only slightly. For instance, tak- ing a salinity of 3.5%, which is about average for the oceans, the freezing point is depressed to $-1.9°C$, and the boiling point is raised slightly to $100.6°C$. The heat capacity is lowered by 4%, so it takes only 0.96 calories to raise the temperature of 1 g of seawater by $1°C$ instead of 1 calorie. Although the water evaporated from the sea into the atmosphere leaves all the salt behind, when seawater freezes, about 30% of the salt is retained and the remaining 70% ejected. This means that sea ice is about 1% salt. The difference between this and the fresh water ice formed on land is nonetheless sufficient that Captain James Cook could tell that the ice he encountered in the southern seas was formed on land.

FIGURE 12.13   *The nuclear submarine USS Seadragon at the North Pole, after surfacing through the ice cover. Notice that the ice is very thin in comparison to ice found in glaciers on land, for example.*

E.   ICE AND ICE AGES

We have seen that it takes 1 calorie to raise the temperature of 1 g of water by 1°C, and 540 calories to evaporate 1 g of water at 100°C. It also takes quite a bit of heat to melt 1 g of ice at 0°C. The "latent heat of fusion" of ice is 80 calories per gram. Thus it takes slightly less heat to melt ice at 0°C than it takes to raise the temperature of the water from freezing to boiling. To freeze ice, of course, the same amount of heat must be removed. Even in the coldest regions of the earth, the sea ice doesn't freeze very thick (Figure 12.13), one reason being it simply requires the removal of too much heat.

Ice is a good insulator, and it floats. Consequently, where the ocean is covered with ice, it ceases to warm up the atmosphere and temperatures can get very cold. Because of its good insulation properties, the ocean underneath the Arctic ice cap does not lose much heat to the atmosphere, so even in the coldest winters, the ice there seldom gets much thicker than about 2 m. One might wonder, then, from where these large icebergs come that threaten North Atlantic shipping lanes, and that sank the Titanic. These come from land, primarily from Greenland. We have seen that on land there is greater temperature fluctuation and more precipitation. Large glaciers form, slowly slide seaward from the coastal ranges, and break off in gigantic bergs. The largest bergs will float out to sea a ways before breaking away from the glacier. These flat "tabular" bergs can be several kilometers long and several hundred meters thick. Smaller irregular chips are called "splinter bergs" (Figure 12.14); one of these sank the Titanic.

Geophysicists have been able to correlate the ice ages with changes

FIGURE 12.14 *The International Ice Patrol keeps track of icebergs, such as this one that endanger North Atlantic shipping lanes. (The U.S. Coast Guard vessel in this photo is actually beyond the berg, and not in the lagoon, as it appears.)*

in the earth's orbit. These changes are caused by an interplay of many factors, including gravitational perturbations by other planets, the precession of the equinoxes, and the slight ellipticity of the earth's orbit. These small changes have sufficiently pronounced effects on the earth's climate to make the great ice sheets come and go.

What is also interesting is the apparent inertia of the ice ages. Once the ice sheets are here, they tend to stay here, and once gone, they tend to stay gone. The time spent in transition between the two stages is short in comparison. The reason for this lies in the absorption and reflection of incident solar radiation. We have seen that the oceans absorb most of the solar radiation incident on them, reflecting very little. For ice, the reverse is the case. With the help of the air bubbles trapped within, ice reflects about 80% of the incident radiation. Consequently, areas covered by ice stay cold due to reduced solar heating, and areas not covered by ice stay warmer and ice free.

The ice ages have had a big influence on sea level. Typically, they have caused a 100-m or more drop in the sea level, which exposed most of the continental shelves. Determining the eustatic[4] changes in the sea level is fairly difficult. One coastline may be rising while another is sinking. The measurement of sea level at some time past would yield different answers, depending on which coastline was used as a reference. Conse-

---

[4]*"Eustatic" changes are those that affect the sea level worldwide rather than just locally.*

quently, many measurements along many coasts must be made, and some sort of average must be used. Measurements of changes in sea level relative to land over longer time periods are even more difficult, because of the geological ups and downs of the earth's crust.

One of the complications involves isostatic adjustments of the crust due to the rising and falling of the water itself. As an example, consider suddenly removing the top 150 m of ocean water in a quick ice age. Then there would not be so much weight on the oceanic crust, which is pushing into the mantle. Consequently, the crust would be buoyed back up by the mantle. Since the crust is roughly three times as dense as the evacuated water was, it would rise only about 1 m for every 3 m of water removed. This would amount to 50 m of "rebound" in this example, where 150 m of water was removed. This means that when 150 m of water is removed, one would only measure a 100-m drop in sea level relative to land, provided that the isostatic readjustments of the crust were uniform and quick.

A further complication is that much of the water removed from the oceans is deposited in massive ice sheets on the continents at temperate and high latitudes. This additional mass shoves the continents down into the mantle. As an example, consider an ice sheet 3 km thick. The additional weight would shove the continent about 1 km farther down into the mantle below (again using a 3 to 1 ratio for the density of the crust to the density of ice). This is the case in Greenland even today, where most of the interior regions have their land surface actually below sea level. As the ice sheets came and went, not all portions of all continents adjusted at the same rate or by the same amount, further complicating our efforts to determine past changes in sea level relative to land.

In spite of such obstacles, attempts have been made to determine the recent history of sea level, and one of these is shown in Figure 12.15. Figure 12.15b amplifies the last portion of 12.15a—the rise of sea level accompanying the melting back of the most recent ice sheet.

At this particular point in time, the sea level seems to be relatively stable. A few places, such as Scotland, having been pushed down by the weight of the last ice sheet, have sprung back up faster and farther than the rising sea level. So in these areas one can find relict beaches from the previous ice age, not down on the continental shelf but actually above the present shoreline. However, in most places sea level is stable. There is still some melting back of the northern ice sheets, but there is also evidence of the south polar cap growing at a comparable rate, resulting in no net change to the oceans. (The student may find it interesting to speculate why a slight warming trend in the climate may cause the south polar cap to grow rather than shrink.)

F. SUMMARY

The earth's surface water is stored in a variety of reservoirs, the ocean being by far the largest of these. Although the atmosphere holds only a very tiny fraction of the total, it does play an important role in the transfer

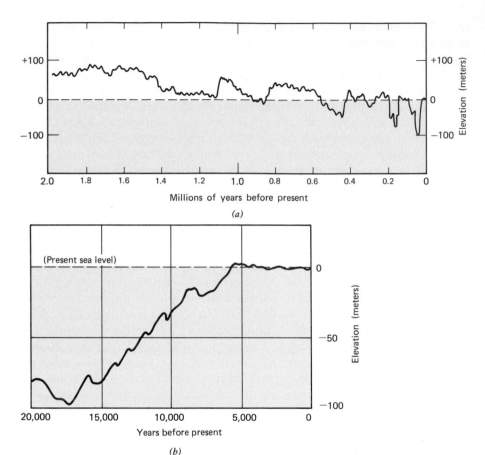

FIGURE 12.15 *Eustatic changes in sea level. (a) Over the last 2,000,000 years, the general trend has been downward, with ice ages causing big dips. (b) Over the last 20,000 years, there has been a rise in sea level, corresponding to the retreat of the last ice age. For the last 6000 years, sea level has been rather stable. (Courtesy of Rhodes W. Fairbridge, Columbia University.)*

of water between reservoirs, in what is called the "hydrologic cycle."

Water has a very high latent heat of varporization, which means that the very small amount of water in the atmosphere holds a surprising amount of heat energy. On the average, water joins the atmosphere through evaporation at the same rate it leaves through precipitation, although there are large geographical variations in the rates of these two processes. Over the land masses, precipitation exceeds evaporation. One reason is the smaller amount of water available for evaporation. But also the air masses must rise to pass over continents, and must endure much greater fluctuations in temperature there. The reasons that temperatures tend to be more stable over oceans than over land include the higher heat capacity of water, the distribution of heat within a thicker surface layer through sunlight penetration and mixing, and the conversion of some solar heat into latent heat of vaporization.

The atmosphere is fairly transparent to incoming sunlight, but opaque to outgoing earth radiation, giving rise to the "greenhouse effect." The primary means of transferring heat from the earth's surface through the earth's lower atmosphere in lower latitudes is as latent heat, because the water vapor content of the atmosphere blocks infrared radiation.

Climatic changes in past times have resulted in ice ages, which have had a pronounced effect on sea level.

One unique property of the earth is the abundance of surface water. The reason we are particularly interested in water, and not, for instance, silica or nitrogen, is that water has many very special properties that help regulate our environment. Among these is its existence in copious quantities in all three phases—solid, liquid, and gas—right here on the earth's surface. The mobility of its gas and liquid phases, as well as the thermal properties of all three phases, enable it to greatly moderate our climate.

QUESTIONS FOR CHAPTER 12

1. What is the hydrosphere?

2. What are the five major types of water reservoirs in the hydrosphere? What percentage of the water is held by each?

3. If all the water in the atmosphere were to come out in one worldwide rainstorm, how many inches (or centimeters) of rainfall would it amount to? How does this compare to the water in the oceans? The ground water?

4. Discuss as many processes as you can think of that are involved in the hydrologic cycle.

5. Suppose air at 25°C is saturated with water (100% relative humidity). What fraction of the air is water? What would be the answer if the temperature were 35°C? 15°C? 5°C? Does the answer change by roughly a factor of two for every 10°C?

6. What happens when saturated air is cooled?

7. What is the difference between absolute humidity and relative humidity?

8. What is the "dew point"?

9. What is the latent heat of vaporization of 1 g of water? How does this compare to the heat required to bring it from freezing to boiling?

10. How much rainfall is required to release as much latent heat as there is heat received in one day's sunshine?

11. What is the difference between "sensible heat" and "latent heat" in air?

12. The net evaporation from the oceans in only 7 cm of water per year,

whereas the net precipitation on the land is 17 cm per year. Is more water coming down than going up? Explain.

13. What are two special talents that the land has that the oceans don't have for coaxing moisture out of the air?

14. Why do you suppose that the western side of the Sierra Nevada is lush and forested, whereas the eastern side is a desert?

15. Why do you suppose fog appears during cooler hours of the day, and disappears during warmer hours?

16. Why is it that West Coast cities have more moderated climates than East Coast cities in the United States?

17. How do seasonal temperature fluctuations over the oceans compare with those over continents of the same latitude? How about those on Mars, which has no oceans at all?

18. Explain why materials with lower heat capacities would show greater daily temperature fluctuations.

19. Besides heat capacities of the materials, what other things promote the higher temperature fluctuations on land compared to those of the ocean surface?

20. What color of light penetrates clear seawater best? How far can it go with 50% attenuation? How do the answers to these two questions change as the water gets more turbid?

21. Why does clear water look blue, and murky water look green?

22. The temperature of the sun's surface is about 6000K. At what wavelength does it radiate the most energy? Visible light lies in the range from 4×10^{-4} mm (violet) to 7×10^{-4} (red). Does the wavelength you calculated lie in this range? About what color would you guess it is?

23. Your body and most of the things in our immediate environment have temperatures around 300K (give or take 20°). What is the wavelength of maximum energy radiated by these bodies? From Figure 12.4a, find what range of the electromagnetic spectrum this is in.

24. Would your body give off any radio waves at all? (Check Figure 12.4b) Any X-rays? Any visible light? If so, why can't you see yourself in a dark closet?

25. What is a calorie?

26. What is the solar heat flux at the outside of our atmosphere? What is the average solar heat flux reaching the surface of the earth? What are some of the factors contributing to the difference between these two numbers?

27. Which molecule in our atmosphere protects us from the sun's ultraviolet rays? Which molecules absorb much of the infrared radiation?

28. How is it that our atmosphere inhibits the outward flow of heat much more than the inward flow?

29. Why do you suppose this is called the "greenhouse effect"?

30. Children sledding on a snowy hill slide down the hill quickly, but take a long time going back up. Consequently, a snapshot would show many more children climbing up the hill than sliding down at any one time, although just as many go up as come down altogether. How does this parallel the flow of heat through our atmosphere?

31. What do you suppose is an "infrared window" in our atmosphere? (*Hint:* On Figure 12.6 do you see any infrared wavelengths to which our atmosphere is transparent?)

32. Why is it that the warmer waters of the tropics actually radiate *less* heat through the lower atmosphere than do cooler waters at higher latitudes?

33. Describe the primary mechanism by which heat is transferred up through the lower atmosphere in the tropics.

34. Why is it that heat may be conducted from a warmer ocean to a cooler atmosphere, but not from a warmer atmosphere to a cooler ocean?

35. Of the three basic ways of getting heat from the oceans through the lower atmosphere, which dominates at lower latitudes? Which dominates at higher latitudes? Why?

36. Of the solar radiation incident on the outer atmosphere, about how much of it makes it down to the earth's surface? How much of this comes through directly, and how much reaches the earth's surface only after scattering off the clouds, air molecules, and so on?

37. Of the radiation that strikes the ocean's surface, is most reflected or absorbed?

38. Of the radiation incident on the outer atmosphere, how much does not make it down to the earth's surface? What becomes of it?

39. Why is it that the rate at which heat leaves the earth must equal the rate at which it is received from the sun? Are the two rates equal at the equator? At the pole? What conclusion must be drawn from this observation?

40. What are the two primary mechanisms for the transport of heat poleward from tropical regions? Which of these has the larger effect?

41. Why is it that such a small amount of water in the atmosphere can transport so much heat?

42. Suppose you were going to go to a region and measure the amount of water precipitated in that region during the year, and the amount of water evaporated from that region during the year. Which of these two processes would dominate in tropical climates? In Arctic climates?

43. Why is there less evaporation at the equator than at latitudes 20° North and South? Is it related to the larger precipitation there?

44. How does the difference between evaporation and precipitation affect the salinity of the surface water in a region?

45. What happens to the salt when seawater freezes?

46. Is the heat capacity of salt water larger or smaller than that of fresh water? By what percent?

47. What is the latent heat of fusion for water? How does this compare to the latent heat of vaporization? To the heat required to raise 1 g of liquid water from freezing to boiling?

48. What is the thickest that one would normally expect the Artic Ice Cap to get? Why so thin?

49. Where do the large icebergs in the North Atlantic come from?

50. Why do the ice ages seem to display so much "inertia"? That is, why do the ice sheets tend to stay here once they're here, and to stay gone once they're gone?

51. Roughly how much water is removed from the ocean during a typical ice age (i.e., how much does sea level drop)?

52. What isostatic readjustment of the oceanic crust would accompany an ice age? Would there be any isostatic readjustment of the continental crust? If so, where and why?

53. Is sea level changing much now?

*54. If you put a pan of ice cold water on a stove and turned the burner on high, it would take some time before it started to boil, and longer for it to boil dry. How many times longer (roughly)?

*55. Why might it be that a slight warming trend in the earth's climate would cause the south polar cap to grow rather than shrink?

*56. Suppose 200 m of water were removed from the oceans and deposited in ice sheets on continents. Given that crustal materials are about three times denser than water, make the following estimates. What would be the amount of isostatic readjustment of the ocean bottom? What would be the net drop in sea level? If the ice sheet on one continental margin were 400 m thick, what would be the amount of downward isostatic readjustment for this continental margin?

*57. Why do you suppose clear nights are cooler than cloudy nights?

*58. Explain how changes in ice cover tend to amplify climatic variations. Would you think cloud cover would tend to amplify or moderate climatic variations, and why?

SUGGESTIONS FOR FURTHER READING

1. Robin Burton, "Icebreaking by Hovercraft," *Sea Frontiers,* **26,** No. 2 (1978).

2. Ferren MacIntyre, "The Top Millimeter of the Ocean," *Scientific American* (May 1974).

3. Albert Miller and Jack C. Thompson, *Elements of Meteorology,* 2nd ed., Merrill, Columbus, Ohio, 1975.

4. Gerhard Neumann and Willard J. Pierson, Jr., *Principles of Physical Oceanography,* Chapter 12, Prentice-Hall, Englewood Cliffs, New Jersey, 1966.

5. "Oceans and Climate," *Oceanus,* **21,** No. 4 (Fall 1978).

6. Radok, Streton, and Weller, "Atmosphere and Ice," *Oceanus,* **18,** No. 4, p. 16 (Summer 1975).

7. Ed Sobey, "The Ocean-Climate Connection," *Sea Frontiers,* **26,** No. 1 (1980), p. 25.

8. Ed Sobey, "Ocean Ice," *Sea Frontiers,* **25,** No. 2 (1979), p. 104.

9. R. W. Stewart, "The Atmosphere and the Ocean," *Scientific American,* (Sept. 1969).

10. Williams, Hugginson, and Rohrbough, *Sea and Air,* Naval Institute Press, 1973.

13
THE EARTH'S ROTATION AND ATMOSPHERIC CIRCULATION

Hurricane Gladys, October 18, 1968.

In this and the following chapter, we examine large-scale motions of the atmosphere and oceans. They are studied together because both fluids have similar behaviors, and because the circulation of the ocean is strongly coupled to that of the atmosphere.

A description of the atmospheric and oceanic currents would be quite a bit simpler if the earth were not spinning. However, it is, and this causes a rather curious thing to happen, as perceived by earth-bound observers. Fluid masses traveling large distances over the earth's surface seem to follow curved trajectories rather than following straight lines, as they do over short distances or in non-rotating reference frames.

The atmospheric circulation is driven by solar heating, but once under way, these currents follow curved paths caused by the earth's rotation. The circulation of the ocean surface currents is, to a large extent, simply a reflection of the circulation of the atmosphere which drives them.

13
THE EARTH'S ROTATION AND ATMOSPHERIC CIRCULATION

A. INERTIAL MOTION AND THE ROTATING EARTH

When riding in an automobile, we notice that as the car turns a corner, our bodies tend to tip over, or get "shoved" toward the outside of the curve. We sometimes refer to this outward shove as a "centrifugal force," but a closer examination would reveal that this is a fictitious force, being a figment of our false perceptions. We are concerned with our immediate environment—the interior of the car. Relative to that, we are tending to slide outward. But a more global view of our environment would reveal that our body is simply trying to maintain its inertial straight-line motion relative to the trees, houses, and so on. It is our car that is being forced out of its inertial straight-line motion, not our bodies. Consequently, we see that what we call a "centrifugal force" in this respect is really a fictitious force, which we think we perceive only because our immediate environment is turning.

Similarly, children on a spinning merry-go-round think they feel a "centrifugal force" trying to shove them off. A more global view would reveal that their bodies are only trying to pursue the inertial straight-line motion. The merry-go-round keeps spinning and pulling them in; nothing is really pushing them out.

If these children on the merry-go-round were to engage in a snow-ball fight, they would have a further demonstration of curious behavior in a rotating reference frame. One child throwing across the merry-go-round would find that by the time the snowball got to the other side, the target would no longer be there. In fact, if the snowball is thrown slowly enough, or if the merry-go-round is spinning fast enough, it is quite possible that the thrower will be there to receive it. That is, it is **349** quite possible to throw a snowball across a merry-go-round and hit yourself! (The author has done it, when younger and more naïve.) Notice that the child's immediate environment is the merry-go-round. The child will see himself or herself in a stationary position relative to it, and will always be directly facing the target. From the child's point of view, the snowball will have curved to miss the target. Of course, an outside observer elsewhere on the playground will see that the snowball goes straight, and that it is the merry-go-round observers that are turning.

We earth-bound observers are on a large merry-go-round that rotates once every twenty-four hours. Consequently, as we observe the trajectory of something that is under way for several hours, we notice that it will curve rather than go straight. Our first clue that our immediate environment would fool our perceptions is that far away things—sun, moon, stars, and planets—whirl around us once a day. Similarly, the children on the merry-go-round could have been tipped-off regarding the curious behavior of their snowballs if they had noticed the behavior of far-away trees, buildings, and so on, whirling about the perimeter of their vision. Nonetheless, it is understandable that we fix our everyday perceptions, as well as our scientific measurements, in our immediate environment rather than in the stars. Consequently, we must pause to explain the seemingly curious behavior of large-scale currents on the spinning earth.

We divide this discussion into two parts. In the first part, we apply the above description of apparent deflection to the earth in particular. We will see that this "Coriolis deflection" of moving objects will be to the right everywhere in the Northern Hemisphere, and to the left everywhere in the Southern Hemisphere. In the second part, we will be concerned with the particular geometry of the earth. Because of its oblate spheroidal shape, the earth's equator is 25 km "uphill" from the poles. Things on the surface of the earth are subject to the balance of gravity, which tends to pull them "downhill" toward the pole, and the "centrifugal force," which seems to fling them outward toward the equator.[1] Since atmospheric and ocean currents are constrained by the earth's surface geometry, they are subject to the balance of these two forces.

A.1 The Coriolis Effect

As a starting point, we observe that the force of gravity is directed toward the center of the earth, and so things not tied down to the earth's rigid rotating surface tend to orbit around the earth's center. This means that the moon, or an artificial satellite, or a moving air mass, or a batted baseball, for example, all are attracted toward the center of the earth and try to orbit about that center.[2] The elliptical orbit of a batted baseball, neglect-

---

[1]*Most scientists would cringe at our incorrect usage of the term "centrifugal force," although it corresponds to a popular usage (or misusage). As explained earlier, it is really an inertial effect; the object just tries to keep its inertial straight-line motion, but observers on the spinning earth don't see it that way. In any case, we'll put the term "centrifugal force" in quotation marks in this section as a reminder that there really is no "force"—it just looks that way to observers in a spinning reference frame.*

[2]*There are very minor corrections needed in precise calculations that account for the effect of the slight equatorial bulge. But these effects are very small, and we ignore them here.*

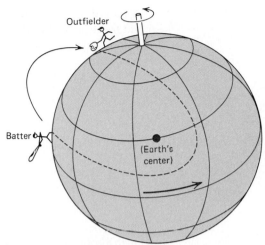

FIGURE 13.1 *A batted baseball goes into an elliptical orbit around the earth's center (until the earth's surface gets in the way).*

FIGURE 13.2 *Pitcher throws ball due east at catcher. Ball goes into orbit around the earth's center. Catcher gets carried to the left by the spinning earth. Ball misses to the right.*

ing air resistance, is shown in Figure 13.1. Of course, the earth's surface gets in the way before the orbit is completed, perhaps with an outfielder standing there to catch it where it comes down.

Although things in motion that are not fastened to the earth's surface tend to orbit the earth's center, things that are tied down do not, and this is the cause of the Coriolis effect. Consider a baseball, a pitcher (with a very strong arm), and a catcher, with the pitcher and catcher being separated by several hundred miles. Suppose the pitcher throws the ball to the catcher to his east. It is shown in Figure 13.2 that when the ball is released, it begins its orbit around the earth's center. The catcher, however, is carried to the left of this by the spinning earth he is on; the pitcher sees that the ball started out in the correct line, but then missed the catcher to the right.

FIGURE 13.3 *Pitcher plans orbit of ball along surface of earth (dashed line). While ball is in the air, the catcher, pitcher, and dotted line all spin eastward with the earth, and the ball still misses to the right.*

Now suppose that the pitcher and catcher change places. The pitcher knows the ball will follow along a "great circle"—a trajectory around the earth's center. So the pitcher marks off this great circle on the surface along which the ball must be thrown to reach the catcher to the west. (See Figure 13.3). Having done this, the pitcher throws the ball along this path, but in spite of the planning the catcher is missed to the right again. The staked-off great circle rotated eastward with the earth, the pitcher, and the catcher, but the baseball ignored the spinning earth below, and found its own, non-spinning, great circle around the earth's center.

Being foiled in his or her pitching accuracy in the east-west directions, the pitcher now tries throwing the ball north. Being at a lower latitude than the catcher, the pitcher is farther from the earth's spin axis than is the catcher. Hence, the pitcher is carried farther and faster than the catcher in the daily eastward motion with the rest of the earth around the axis (Figure 13.4). Now when the pitcher thinks the ball is thrown straight north at the catcher, the ball is actually being thrown north and east, because the eastward motion with the earth gives an eastward motion to the baseball. This is similar to what would happen if you were riding a flatcar on a speeding train, and were trying to throw an apple core into a trash can when you're right beside it, by the time the apple core gets out to where the trash can was, the trash can will be far to the rear, and your apple core will have missed to the front. In the previous example, the catcher moves eastward more slowly than the pitcher, and so the baseball misses the catcher to the right. (See Figure 13.5.) Similarly, if they exchange places, the catcher will be moving eastward faster than the pitcher, and the thrown ball will pass behind the catcher (Figure 13.6), again apparently veering to the right.

It must be reemphasized that the pitcher and catcher will find themselves at rest in their immediate environment. Relative to trees, houses,

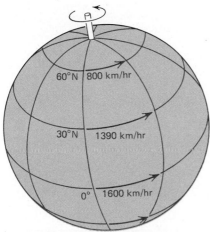

FIGURE 13.4 *The earth spins once a day. Near the equator, the surface has far-*
ther to go to make it around, so it must go faster.

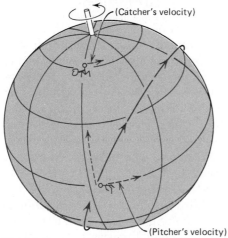

FIGURE 13.5 *Because the pitcher is moving east rapidly along the earth's surface,*
when the pitcher thinks the ball is thrown straight north, it is
actually going north and east. The ball passes to the right of the
more slowly moving catcher.

and so on, they will be motionless. They will be facing each other when
the ball is thrown, and remain there in that position throughout the flight
of the ball. Their conclusion will be that the ball curved! They do not have
the benefit of an astronomical viewpoint, such as ours, as we look down
at the figures. This apparent deflection of a trajectory will always be to the
right in the Northern Hemisphere, and always to the left in the Southern
Hemisphere.

The amount of Coriolis deflection does not depend on the nature of
the object, be it a baseball, bowling ball, automobile, or ocean current.
However, it does depend on the speed. In the pitcher-catcher problem,
the slower the ball is thrown, the longer it takes to get where the catcher

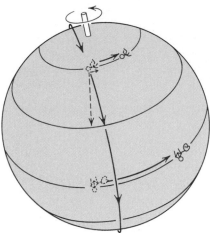

FIGURE 13.6 *The catcher on the equator is moving east much more rapidly than the pitcher. Hence, when the ball gets to where the catcher was, the catcher is gone far to the east and the ball passes behind him. To the pitcher, it looks like the ball curved to the right in flight.*

was, and the more the catcher will have moved by the time the ball gets there. The slower the ball, the bigger the miss. As a more practical example, imagine a car driving at 100 km per hour down a perfectly flat, perfectly straight, highway in the central United States. The Coriolis acceleration would amount to only about 0.0003 times the acceleration due to gravity. Although small, it is measurable. If the driver did not correct for it, the car would be deflected about 2 m to the right after traveling 1 km. Another car going half that speed would experience only half as much Coriolis acceleration. However, it will also take him twice as long to traverse the kilometer. It turns out that half the acceleration acting for twice as long will give the same sideways velocity, and double the total deflection, so the driver of the slower car will find the total deflection to be 4 m for each kilometer driven. The slower the velocity, the greater the deflection for equal distances traveled.

The Coriolis deflection also depends on the latitude, being larger at higher latitudes. We have seen that things not fixed to the earth's surface will tend to orbit around the earth's center. The observer, however, is fixed to the earth's spinning surface, and will follow a different orbit (unless he or she is on the equator). The higher the latitude, the greater will be the difference between these two orbits, and consequently, the greater the apparent deflection. In this book we are not dealing with the orbits of baseballs or cars, but the same phenomenon happens to anything that is moving relatively freely of the earth's surface; such as air and ocean currents.

A.2 Inertial Currents (or a Bowling Ball in a Parking Lot)

Recall that the oblate spheroidal shape of the earth is due to a balance of gravitational and "centrifugal forces," the gravitational force trying to pull

FIGURE 13.7 *Bowling ball on the earth's surface. (a) Moving eastward with the earth, the force of gravity trying to pull it "downhill" toward the north Pole is just balanced by the "centrifugal force" flinging it outward toward the equator. (b) If the ball is going eastward faster than the earth, the "centrifugal force" will be larger, and the ball will deflect toward the equator. (c) If the ball goes eastward slower than the earth, the "centrifugal force" is smaller, and so gravity wins, pulling it "downhill" toward the pole. (d) The trajectory of the bowling ball in the parking lot, as seen by an observer in outer space (i.e., who is not spinning with the earth). (e) The trajectory of the bowling ball, as seen by observers in the parking lot (moving eastward with the earth).*

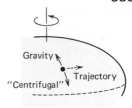

(a) Same speed as earth

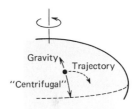

(b) Faster than earth

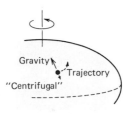

(c) Slower than earth

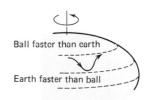

(d) Seen by "Martian"

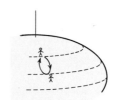

(e) Seen by earth-bound observer

it into a perfectly spherical shape, and the "centrifugal force" trying to stretch it out at the equator. A body that is relatively free to move on the earth's surface (such as a mass of water in the ocean, or a bowling ball in a parking lot) will feel the effects of these two opposing forces (Figure 13.7a). On the one hand, gravity will tend to pull the body poleward, the poles being 25 km "downhill" from the equator. On the other hand, the "centrifugal force" due to the daily rotation of the earth will tend to fling the body outward toward the equator. The shape of the earth's surface has evolved over the ages, so that for bodies rotating eastward with the earth, these two forces are exactly equal and opposite. Consequently, the water in the ocean and the bowling ball placed on the level parking lot will both stay put.

Now we examine what happens to bodies traveling eastward slightly faster or slightly slower than the earth beneath. The key to understanding this is to realize that although the force of gravity does not depend on a body's speed, the "centrifugal force" does. For things going in circles, the greater the speed, the larger will be the "centrifugal force." (Test this by spinning a weight on the end of a string.) Consequently, for things going eastward faster than the earth, the gravity and "centrifugal forces" will no longer be equal and opposite. The "centrifugal force" will be larger, and so the moving bodies will be deflected toward the equator (Figure 13.7b). Conversely, for things moving eastward more slowly than the earth, the "centrifugal force" is smaller, and so gravity will dominate (Figure 13.7c), deflecting them toward the pole. Things going eastward slower than the earth would appear to be moving west as seen by observers on the earth.

So far we've learned that things moving eastward faster than the Earth will be deflected toward the equator, and those going slower than the earth will be deflected poleward. If we add to this our knowledge that the earth's eastward speed varies with latitude (Figure 13.4), being greatest at the equator and decreasing to zero at the poles, we are ready to understand "inertial currents." We will demonstrate that as seen by observers moving with the earth, objects moving along the earth's surface will go around in circles.

We examine the trajectory of a bowling ball in a very large, smooth parking lot somewhere in the Northern Hemisphere. Suppose we bowl the ball eastward. Following the above reasoning, since it is going eastward faster than the earth, it will be deflected southward, toward the equator. But nearer the equator, the earth's velocity is larger, and soon the ball is far enough south that the earth is moving faster than it is. Although its eastward velocity has not changed, that of the earth beneath has. Observers in the lot, moving with the earth, will claim that the ball, which started out going east, was deflected to the south and is now moving westward across the lot.

Similar reasoning shows that the ball will complete the loop. With its eastward velocity less than the earth's, it will be deflected toward the north. As it goes north, the speed of the earth beneath it lessens, and soon the earth's eastward velocity is less than that of the ball. According to the earthbound observers, the ball is rolling east again (Figure 13.7d and 13.7e), having completed a loop. (The student should apply this line of reasoning to a ball originally bowled in some other direction, or in the Southern Hemisphere.)

The looped trajectories, being in the clockwise sense in the Northern Hemisphere and counterclockwise in the Southern Hemisphere, lead one to suspect that this effect may not be entirely independent of the Coriolis effect discussed previously. Indeed, the effect the earth's equator being "uphill" adds to the Coriolis deflection for objects moving west, reduces the Coriolis deflection for, those moving east, and doesn't affect the Coriolis deflection for objects moving north or south. The Coriolis effect alone, however, is not sufficient, because if it were in orbit about the earth's center, the bowling ball, would spend half its time in each hemisphere. Our particular bowling ball, however, never crossed the equator. For things confined to the earth's surface, the fact that the equator is 25 km "uphill" from the poles inhibits their crossing it. As soon as they move far enough equatorward that their eastward velocity is less than the earth's, the "centrifugal force" will be too small, and they will start to accelerate back "downhill" toward a pole.

It is reemphasized that both the "Coriolis force" and the "centrifugal force," as we have used it here, are fictitious forces. The Coriolis force seems to deflect a projectile to the right (Northern Hemisphere), but it is really the observer, not the projectile, who is being deflected as he or she rotates with the solid earth. Similarly, the "centrifugal force" (as we have used it here) is illusory. When sitting on a spinning merry-go-round, nothing is really pushing you out; rather the merry-go-round keeps pulling you in. Your body simply tries to go in a straight line. These fictitious forces that, from the point of view of an observer in a rotating reference frame appear to deflect a projectile that is simply trying to maintain its inertial straight-line motion, are called "inertial forces." Consequently, ocean currents following the characteristic looped paths are called "inertial currents" (Figure 13.8). Many ocean currents receive continual impetus or are confined by some boundary, which tend to reduce these inertial effects. Nonetheless, all currents are influenced by them to some extent.

In this and the following chapters we will examine the motions of two fluids—the atmosphere and the ocean—that, like the bowling ball, have their motions confined to the earth's surface. Both the effect presented in

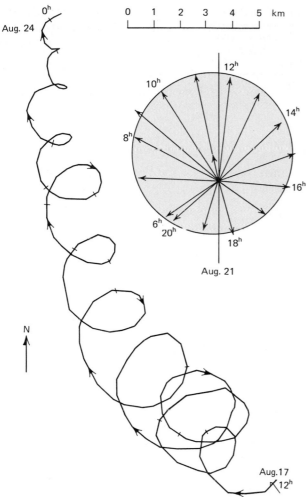

FIGURE 13.8 *Chart of the progress of an inertial current in the Baltic Sea made in 1933, showing the characteristic looped trajectory. (From Sverdrup. Johnson, and Fleming,* The Oceans, *Prentice-Hall, © 1942, renewed 1970.)*

this section and the Coriolis effect of the previous section will influence their motion. However, for convenience we will often refer simply to the "Coriolis effect," but it should be understood that really both these effects are present.

B. THE ATMOSPHERIC CIRCULATION

B.1 Physical Characteristics of the Atmosphere

Many of the ocean currents rely on the circulation of our atmosphere for their propulsion. In particular, the dominant surface currents are wind

TABLE 13.1   The Composition of Dry Air

Gas	Percent of Total Molecules in Air
Nitrogen	78
Oxygen	21
Argon	1
Carbon dioxide	0.03
Neon	0.002
Helium	0.0005
Krypton	0.0001
Hydrogen	0.00001
Xenon	0.00001
Others	(still less)

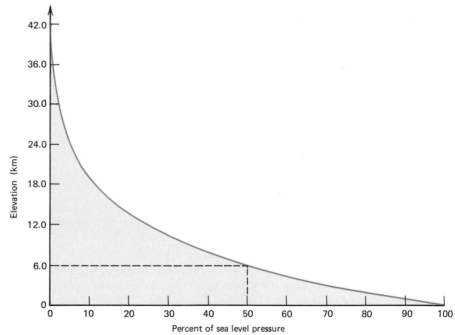

FIGURE 13.9   *Plot of air pressure as a function of elevation. Notice that the pressure reduces by roughly a factor of ½ for every 6 km increase in elevation. (Atmospheric pressure at sea level is about 10^6 dyne/cm², or about 14.7 lb/in².)*

driven, and to some degree the deeper currents are wind generated as well. Hence, a quick study of our atmosphere is necessary for us to understand these currents.

The composition of dry air is given in Table 13.1. It is seen to be mostly nitrogen and oxygen, with only very small amounts of the other gases. The amount of water held by the atmosphere varies from time to time and place to place. As we saw in Chapter 12, warmer air can hold

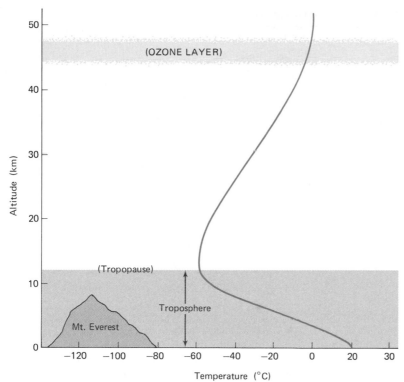

FIGURE 13.10 *Variation of temperature with altitude in the various regions of the atmosphere. The troposphere contains most of the air (thinning out with altitude), and virtually all of the activities of concern to the oceans and ourselves.*

more water than colder air. However, even in the most warm and humid climates, the water vapor never amounts to much more than 3%.

Unlike liquids, gases are easily compressed, and so where the atmospheric pressure is greater, the air is also denser. For example, as one goes to higher altitudes the atmospheric pressure decreases at a rate of about a factor of ½ for every 6 km in elevation, and the air gets thinner at the same rate. (See Figure 13.9). This is in contrast to an incompressible fluid such as the ocean water, where when rising from depths one feels the pressure decreasing, but the density does not change noticeably.

We saw in Chapter 12 that our atmosphere is fairly warm due to its opacity to the infrared radiation given off by the earth. But at higher altitudes the thinner air means less opacity, and so it becomes increasingly easy for radiated heat to pass on out into outer space. Consequently, the atmosphere generally gets cooler with altitude up to a height of about 12 km above sea level, where it is roughly $-60°C$. Above this altitude other effects vary the temperature pattern (Figure 13.10), but since virtually all the activity that affects the oceans occurs in the lower layer, and since most of the mass of our atmosphere is found in this lower layer (the "troposphere"), we won't worry about the upper layers here.

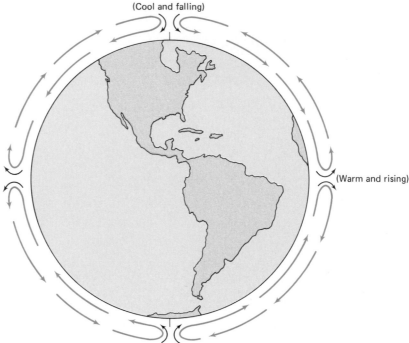

(Cool and falling)

(Warm and rising)

FIGURE 13.11   *If the earth were motionless, and the sun circled its equator, then the warmer, lighter air of the equator would rise, and the cooler polar air would fall. The simple cellular pattern would be that of air rising at the equator, moving poleward aloft, falling at the poles, and returning equatorward along the surface.*

B.2   Cellular Circulation

As is the case with most of the processes on earth, the ultimate source of energy providing the power for the atmospheric circulation is the sun. However, the earth's spin, via the Coriolis effect and atmospheric tides, modifies the circulation patterns established by the sun.

First, imagine the earth were not spinning, but rather that the sun were encircling the earth's equator. The unequal heating would cause the air to rise in the warmer equatorial regions. There are actually two reasons for this. The obvious one is that warmer air is less dense, and therefore lighter. The other reason is more subtle. The water molecule is actually lighter than the nitrogen or oxygen molecule. Since there is a good deal of evaporation from the oceans in equatorial regions, the air there contains relatively more water molecules and less of the heavier molecules than elsewhere (but only by a fraction of a percent). Therefore, it is lighter both because it is warmer and also because it has more water vapor, and so it rises. Similarly, the cooling air in the polar regions will become more dense and fall.

The resulting ideal, non-spinning earth is shown in Figure 13.11. The air rises near the equator, moves poleward aloft, falls in the cooler polar regions, and then returns to the equator along the ground. The surface

(SURFACE WINDS)

FIGURE 13.12 *For the real spinning earth, the atmospheric circulation pattern has three separate cells in each hemisphere. The reason is that air initially moving equatorward is soon going westward, due to the Coriolis deflection, and air initially moving poleward is soon deflected eastward. The direction of the prevailing surface winds in each zone are indicated.*

winds would be prevailing northerlies (going from north to south) in the Northern Hemisphere, and prevailing southerlies in the Southern Hemisphere.

The real earth, however, spins, and the simple picture of Figure 13.11 becomes modified by the Coriolis effect on the moving air masses. In the Northern Hemisphere, for example, the air rising at the equator and heading poleward gets deflected toward the right, and soon is heading east instead. It then falls around a latitude of 30°N, and returns south along the surface toward the equator. But as these surface winds move southward they are again deflected to the right, and so observers on the earth's surface between 0° and 30°N will notice the prevailing winds coming from the east. (See Figure 13.12.)

Near the North Pole, where the cool air falls and begins its southward treck, it soon is deflected to the right, and so people in the Arctic region will also notice the prevailing surface winds coming from the east. In the intermediate latitudes, a third circulation cell is established, with the air sinking near 30°N along with that from the tropical cell, and rising near 60°N along with the air of the north polar cell. The surface winds in this intermediate cell, then, try going north, but become deflected toward the right, and are known to us who live in this region as the "prevailing westerlies" (meaning coming from the west). A similar thing happens in the Southern Hemisphere, except the Coriolis deflection is to the left. The composite picture for the ideal rotating earth is shown in Figure 13.12.

In reality, the circulation cells seem not as simple or well defined as the picture presented here, particularly the middle cell of the three. Nonetheless, the general patterns are observed. The trade winds in the tropics blow surface currents westward across the oceans, and this surface water is returned to the oceans' eastern margins at higher latitudes by the prevailing westerlies.

The equatorial areas where the air is rising and the regions around 30°N and 30°S where the air is sinking, have rather light surface winds and bad reputations among sailors for their rather poor sailing conditions. In the equatorial "doldrums," the warm, moist air rises. As it rises, the atmospheric pressure is less, so it expands and cools. As it cools, the water condenses and it rains, further adding to unpleasant sailing conditions at sea, while making the continents lush in these regions. The regions around 30°N and 30°S latitudes are called the "horse latitudes." In these regions, air is sinking. Because of the higher atmospheric pressure at lower elevation, it gets compressed as it sinks, which makes its temperature rise. The falling, warming air does not lose moisture, so the skies are clear, and the continents are arid in these regions. At sea these regions are sometimes referred to as "maritime deserts," and to a sailor who is stranded by the light winds under sunny skies with no chance of fresh water from the skies, they might be just as deadly as being stranded on a continental desert.

Where warm light air masses are rising, their removal from the region results in low pressures. Similarly, where cooler, heavy air masses are falling, the addition of these air masses in the region creates higher atmospheric pressures. Meteorologically speaking, one expects low barometric pressures and storms in the doldrums, and high barometric pressures and clear skies in the horse latitudes. Figure 13.13 shows the barometric pressures and predominant surface winds averaged over a year. It is seen that the general patterns discussed above are indeed the rule.

B.3 Modifications

The idealized atmospheric patterns presented so far would work best for a world with a uniform oceanic surface and whose spin axis is not tilted. However, the real earth has a tilted spin axis, causing seasonal changes in climate. Furthermore, here and there we find continents on it, which cause some additional modifications of the above idealized patterns. The continental surface reflects a little more of the incident sunlight than does the oceanic surface. Heat absorbed by the oceans is transported in surface currents, whereas that absorbed by continents is immobile. Furthermore, continents influence precipitation from the atmosphere (hence, cloud cover). All these considerations, plus the uneven distribution of the continents between the two hemispheres, should lead us to expect some deviations from the idealized behavior.

Fortunately, most of the important modifications of the general atmo-

FIGURE 13.13   *Yearly averaged sea level pressures, indicating the regions of high and low pressures and the general pattern of the surface winds. (From Alyn C. Duxbury, The Earth and Its Oceans, © Addison-Wesley, Reading, Massachusetts, 1971.)*

spheric patterns are simple to understand. An example is the seasonal changes in high and low pressure centers. In the winter season, the land masses will generally be colder than the oceans, and in the summer the reverse is true (excluding polar regions). Where it is colder, the air will be denser and falling, creating high pressure centers, and where it is warmer, the air will be lighter and rising, creating lows. Hence, in winter there will be a tendency for the highs to be over the land and the lows over the ocean, and in the summer the reverse will be true. This can be seen, for example, at the temperate regions (say roughly 45°N) in Figure 13.14, which shows the mean sea level pressures for July and January. Because continental surface in the Southern Hemisphere is meager, the effect is stunted there.

Among other things, the barometric pattern described above is responsible for the Asian monsoons. In the winter, the high pressures are over the cooler land, so the winds blow toward the lower pressures out to sea, causing the "winter" or "dry" monsoons. In summer, the low pressure centers are over the warmer land, and the high pressure centers are out to sea. Hence, the wind comes in landward (Figure 13.15) from sea, laden with water for the Asian "wet" or "summer" monsoons.

The reason we are interested in the barometric pressure patterns is that the winds flow from highs toward lows, yielding modifications of the circulation described earlier in this section. This flow will not be in a

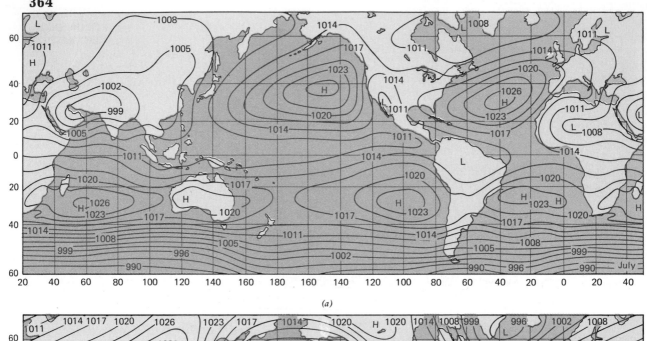

(a)

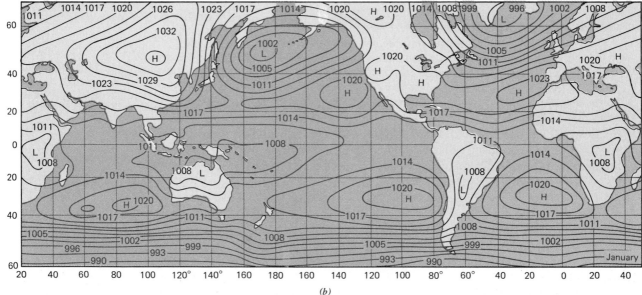

(b)

FIGURE 13.14 *Mean sea level pressures for the months of July (a), and January (b), in units of millibars. (After Y. Mintz and G. Dean, Geophys. Res. paper 17, Geophys. Res. Directorate, Air Force Cambridge Research Center, 1952.)*

straight line, however. Winds leaving high pressure centers are subject to the Coriolis effect, and will deflect to the right (Northern Hemisphere) as they go. Hence, in northern latitudes, winds spiral out clockwise from high pressure centers, as illustrated in Figure 13.16a.

Winds flow in toward low pressure centers, as toward a large vacuum

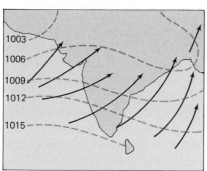
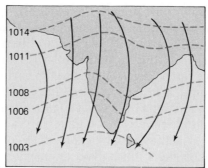

FIGURE 13.15   The monsoons of southern Asia. (left) With higher pressures over
the cooler oceans, the moisture-laden air blows landward, causing
the summer (wet) monsoons. (right) The winter (dry) monsoons are
when the higher pressures are over the cooler land, and the winds
blow out to sea. (Pressure indicated in millibars.)

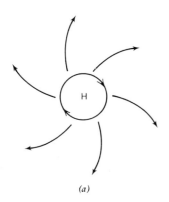

(a)

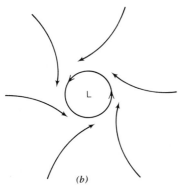

(b)

FIGURE 13.16   (a) In the Northern Hemisphere, winds leaving high pressure centers get Coriolis deflected to the right, causing clockwise motion
around a high. (b) Winds coming toward a low are deflected, and
miss the center to the right, causing counterclockwise motion of
the air around a low.

cleaner, but something happens to them on their way. In the Northern Hemisphere they get deflected to their right, and miss their mark on that side. Thic causes them to spiral in a counterclockwise sense, as shown in Figure 13.16b. (In the Southern Hemisphere this is, of course, reversed.) The terminology applied to the various kinds of winds is "cyclonic winds" around a low pressure center (such as the center of a cyclone), "anticyclonic winds" around a high pressure center, and "gradient winds" that are in transit from a high toward a low.

Another modification of the wind patterns discussed so far is the atmospheric tides. The atmosphere is fluid and just as mobile as the oceans, if not more so, and so the sun and moon create tidal bulges in the atmosphere under which the earth spins, just as they do in the oceans. This will cause daily fluctuations in the above-described patterns, but being oscillatory, it will have little net effect when averaged over a day (just as with tidal oscillations on ocean currents).

C. SUMMARY

Because everything in our immediate environment moves with us, we tend to forget that we are actually riding a kind of large merry-go-round that is rotating at a very high speed. Because of our rotating reference frame, it appears to us that things moving relatively free of the earth's surface follow curved trajectories, curving to the right when viewed from the Northern Hemisphere, and to the left when viewed from the Southern Hemisphere.

For things whose motion is confined to the surface of the earth, there is another effect as well. Because the equator is 25 km uphill from the poles, they experience the two opposing forces of gravity, which tends to pull them downhill toward the pole, and the "centrifugal force," which seems to fling them outward toward the equator. Because the "centrifugal force" depends on the speed, things moving eastward faster than the earth will be deflected toward the equator, and things moving eastward slower than the earth (appearing to go west to earthbound observers) will be deflected toward the poles. The result is that atmospheric and oceanic currents tend to follow looped trajectories in the absence of other constraints.

Due to unequal solar heating, air tends to rise near the equator and fall near the poles. Because of the Coriolis effect, the atmosphere flows in curved trajectories, and circulation patterns are broken up into three circulation cells in each hemisphere, with air tending to rise near the equator and 60°N and S, and tending to fall at 30°N and S and at the poles. The surface winds tend to blow from the east in the tropical and polar cells, and from the west in the temperate cells.

The Coriolis effect also determines the direction of air flow around high and low pressure centers. High pressures are associated with falling air masses, and low pressures with rising air. The unequal heating of ocean and land areas causes predictable seasonal changes in pressures and wind patterns.

1. Why do you suppose a knowledge of the earth's atmospheric circulation is an important preliminary to studying the oceanic circulation? Why do you suppose the earth's rotation is important?

2. When rounding a corner in a car, we think we feel a force trying to tip us over toward the outside. Why is this a fictitious force? What is really happening?

3. When children ride a spinning merry-go-round, is there really any force trying to tip them over or shove them off? Does it feel like there is? Explain.

4. Explain the difficulties involved in trying to play catch on a spinning merry-go-round. Does the ball really curve? Why does it seem to curve to those riding the merry-go-round?

5. For things that are moving with respect to an earthbound observer, the longer they're underway, the more they seem to curve. Why?

6. It is said that moving objects seem to incur a Coriolis deflection that is to the right in the Northern Hemisphere. Now consider a non-moving object, such as the sun or the stars. Do they travel in straight lines across our sky, or is their path curved too? Which way is it curved as seen by people in the Northern Hemisphere?

7. Why do we say that the equator is "uphill" from the poles?

8. If the poles are downhill from the equator, why doesn't the water go there, flooding them and leaving equatorial regions high and dry?

9. If one neglects air friction, then the paths of projectiles are what kind of geometrical figure? Why can't the projectile normally complete the orbit?

10. Suppose a pitcher throws a baseball directly eastward along a great circle route toward a catcher. Explain the difference between the motion of the ball and the motion of the catcher. Why does the pitcher think the ball curved? If the earth were not spinning, what would happen?

11. Repeat the previous question for a ball thrown westward.

12. Most points on the earth's surface are moving at fairly high speeds due to the earth's daily rotation alone. What is this speed at the equator? Why is it less at higher latitudes?

13. When the pitcher throws the ball northward toward the catcher, why does it miss to the right?

14. In Figure 13.5 the sketched orbit of the baseball appears to be straight, but just not in the right direction. Is this what the pitcher sees, or will the pitcher see it starting in the right direction but not going straight? Explain this difference. (*Hint:* Maybe when the ball reaches the head of the first solid arrow, the pitcher will be over by the head of the dashed arrow on the equator.)

15. While the ball is in flight, the pitcher and catcher have remained at

rest with regard to the trees, houses, streets, clouds, mountains, lakes, and so on, in their immediate environment, and they conclude that they haven't moved. Are they right? What in their environment might they look at to find this out?

16. Explain how the following two statements can be in agreement with each other: (1) The slower the speed, the smaller the Coriolis acceleration. (2) The slower the speed, the greater the Coriolis deflection for each kilometer traveled.

17. How does the amount of Coriolis deflection depend on the latitude and why?

18. A mass of water in the ocean feels the effect of two opposing forces, one tending to pull it northward and one southward (Northern Hemisphere). What are these two forces?

19. Of the two opposing forces referred to in the previous question, under what condition are they equal and opposite? Under what conditions is the northward force larger and why? Under what conditions is the southward force larger?

20. If the earth's solid surface were spherical, what would happen to the ocean waters and why? What would happen to the atmosphere?

21. Consider a frictionless bowling ball in a perfectly flat parking lot in the central United States. In the language of the opposing gravitational and centrifugal forces, why does it deflect northward when bowled west? Why does it deflect southward when bowled east?

22. Why does the bowling ball appear to deflect westward when moving south? Does it really deflect westward, or is this just the Coriolis effect?

23. Things confined to the surface of the earth experience a poleward "downhill" gravitational pull. Was this relevant to our previous discussion of the Coriolis effect? (*Hint:* Are projectiles confined to the earth's surface?)

24. Would the Coriolis effect be about the same if the spinning earth were spherical or slightly egg shaped rather than being an oblate spheroid? Would the course of inertial currents be altered much in this case?

25. Why is the exact shape of the earth crucial in understanding the course of inertial currents, but not important in understanding the Coriolis effect?

26. Sketch the path followed by a bowling ball in a parking lot as seen by a "Martian" and as seen by an earthbound observer. Explain the difference between them.

27. Why is the "Coriolis force" a fictitious force?

28. The "centrifugal force," as we have used it in this chapter, is also a fictitious force. Explain why. (*Hint:* Is there really a "force" trying to push you off the spinning merry-go-round, or is your body just trying to go in a straight line?) There really is such a thing as "cen-

trifugal force," but it is something different than what we've used here. Look it up in a physics text.

29. What is inertia? Why are the "Coriolis force" and the "centrifugal force" (as we used it here) called "inertial" forces?

30. What are the three most abundant components of dry air, and what percentage of the total is each?

31. Compare the compressibility of gases and liquids.

32. How high would you have to be above sea level for the air to be ¼ as dense as it is at sea level? Are any mountains that high? Why do mountain climbers sometimes carry tanks of compressed air?

33. Why is it that infrared radiation starting at higher elevations has an easier time getting out of our atmosphere than infrared radiation starting from lower elevations?

34. Why does the temperature of the atmosphere decrease with increased altitude? (This is strictly only true for the lowest 13 km, but that is most of the atmosphere, so don't worry about what happens above that.)

35. What is the source of energy that drives the circulation of the atmosphere?

36. Why is moist air lighter than dry air of the same temperature?

37. Give two reasons why air rises in equatorial regions.

38. Why does air fall in polar regions?

39. Describe what the atmospheric circulation pattern would be if the earth weren't spinning (but somehow the equator were heated all the way around).

40. Sketch the circulation pattern for the real spinning earth. Why does the surface air heading southward toward the equator never make it that far? What happens to the air heading poleward after rising at the equator?

41. What directions are the prevailing surface winds at different latitudes? Explain why they don't go either north or south.

42. What are the trade winds? Why do they flow the same direction on both sides of the equator, even though the direction of the Coriolis deflection is opposite?

43. Why does it rain a lot in the doldrums?

44. Why is there so little precipitation in the horse latitudes?

45. How are high and low pressure centers related to rising and falling air?

46. What are some of the earth's irregularities that might cause deviations from the idealized weather patterns?

47. What seasonal variations in the location of high and low pressure centers might you expect, and why? Why is this effect not as pronounced in the Southern Hemisphere?

48. Explain the Asian monsoons.

49. What will be the direction of air flow in the neighborhood of high and low pressure centers in the Northern Hemisphere? Why?

50. What are "gradient winds"?

*51. From your experiences on merry-go-rounds, cars going around corners, and so on, what are some of the things you've noticed to seem different from your rotating frame of reference?

*52. How do you suppose deeper, subsurface currents might be generated by the wind?

*53. In the middle of the three circulation cells in either hemisphere, the air rises near 60° latitude and falls near 30° latitude. Does this correspond to rising where warmer and falling where cooler? If not, then what does cause this circulation in this cell? Why do you suppose this cell is the least predictable of the three?

*54. At what altitude do you suppose you might find the high altitude return flow in a circulation cell?

*55. Since there are tides in the oceans, would you expect tides in the atmosphere too? How about tides in the solid earth? Why or why not?

*56. Describe the motion of a ball bowled initially westward in a perfectly flat parking lot somewhere in the Southern Hemisphere. Describe the motion as observed by earthbound observers and by extraterrestrial observers. Explain all deflections.

*57. Imagine a planet similar to the earth, but spinning twice as fast. In comparison to the earth, what differences would you expect to find in its atmospheric circulation patterns?

*58. In coastal areas, what daily fluctuations in wind patterns would you expect to find and why?

SUGGESTIONS FOR FURTHER READING

1. J. G. Harvey, *Atmosphere and Ocean: Our Fluid Environments,* Artemis Press, Great Britain, 1976.

2. Keith C. Heidorn, "Land and Sea Breezes," *Sea Frontiers,* vol. **21,** No. 6 (1975), p. 340.

3. Albert Miller and Jack C. Thompson, *Elements of Meteorology,* 2nd edition, Merrill, Columbus, Ohio, 1975.

4. Michael J. Mooney, "Waterspout vs. Marina," *Sea Frontiers,* **24,** No. 3 (1978), p. 159.

5. A. H. Perry and J. M. Walker, *The Ocean-Atmosphere System*, Longman, New York, 1977.

6. F. G. Walton Smith, "Planet's Powerhouse," *Sea Frontiers*, **20,** No. 4 (1974), p. 204.

14
OCEAN SURFACE CURRENTS

An obvious difference in direction between surface and subsurface currents.

The circulation of the atmosphere and of the ocean are intimately connected. Having studied atmospheric circulation and the Coriolis effect in the preceding chapter, we are now prepared to investigate the major circulation patterns of the oceans. We divide this study into two parts: surface currents and the movement of deeper water masses.

In this chapter we study surface currents. The dominant surface currents are driven by the wind, and are heavily influenced by the Coriolis effect. They tend to be intensified along the western margins of the oceans, and the distribution of water masses within them may differ considerably from the usual horizontal stratification pattern with denser waters lying deeper.

The storm surges that ravage some coasts from time to time are more local and temporary types of wind driven currents. Langmuir circulation is a peculiar water motion also induced by winds. Finally in this chapter, we study seiches, which are currents created when water sloshes back and forth in a harbor or other confined region.

14
OCEAN SURFACE CURRENTS

A. WIND DRIVEN CURRENTS

A.1 Comparison with Wind Patterns

The major ocean surface currents simply reflect the wind patterns in the atmosphere. However, the ocean currents have a great deal more inertia, resulting in smaller seasonal fluctuations than those of the capricious winds. Figure 14.1 shows the major surface currents, and should be compared with the yearly average atmospheric patterns of Figure 13.13. Clearly, the westward equatorial currents reflect the trade winds, and the prevailing westerlies return the surface water eastward at higher latitudes. Complementary to this are the anticyclonic current gyres (clockwise in Northern Hemisphere and counterclockwise in Southern Hemisphere) around the prevailing high pressure centers. With a little imagination you can even detect cyclonic tendencies around some low pressure centers, such as the cyclonic surface current gyre east of Greenland or west of Alaska. These cyclonic tendencies are suppressed, however, because the Coriolis deflection of the moving water masses is in the anticyclonic direction (i.e., to the right in the Northern Hemisphere and to the left in the Southern Hemisphere).

The trade winds, blowing westward across the oceans, stack the water up against the western margins of the oceans. As a result, the water surface slopes upward at a rate of about 4 cm per 1000 km toward the western shore. This amounts to a total of 15 cm across the Atlantic, for example. Some of this stacked-up water then returns eastward, flowing "downhill" in what are called "equatorial countercurrents." These are often found just north or south of the equatorial currents, and sometimes the return flow is below the surface, underneath the equa-

373

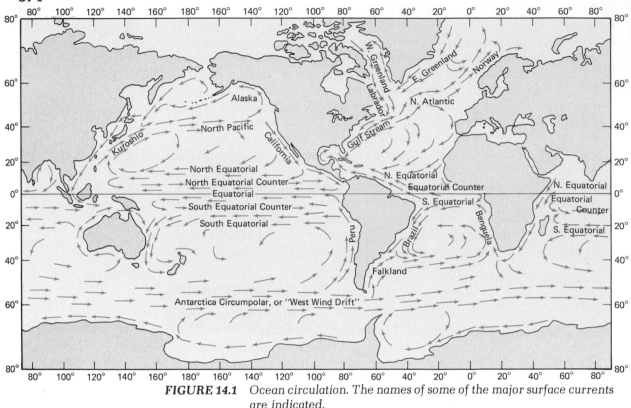

FIGURE 14.1 *Ocean circulation. The names of some of the major surface currents are indicated.*

torial current. One of the more famous of these is the Cromwell Current, which is an eastward flowing, subsurface countercurrent in the Pacific. It has one of the biggest volume transports of all currents, and reaches a maximum velocity only 100 m below the surface. It is rather surprising that these two large currents—the equatorial current and the countercurrent—flow in opposite directions all the way across the Pacific, and are within 100 m of each other.

Off the west coast of the Sahara Desert, the trade winds blow the surface water away toward Brazil. These blown-away surface waters are replaced by upwelled deeper waters from up to 300 m depth. With these deeper waters come nutrients, and with exposure to the sunlight, these nutrients foster abundant biological activity there off the African coast. The upwelled water is colder, but it warms up as it is blown westward along the equator. This temperature pattern is found in the Pacific, too; as one goes westward along the equator, the surface temperature increases.

A.2 The Eckman Spiral (for Northern Hemisphere)

As might be expected, when currents are wind driven, the water on the very surface moves the fastest. At depth, the water must be dragged along

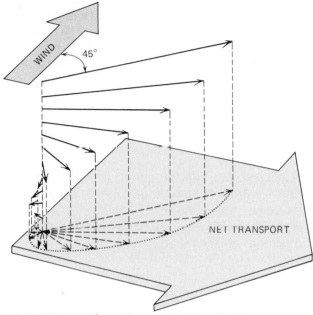

FIGURE 14.2 *The Eckman spiral for the Northern Hemisphere. Due to the Coriolis deflection, the water flows to the right of the wind direction. The surface current will be 45 deg. to the right of the wind, and successively deeper layers will go slower and further to the right. The net water transport will be 90 deg. to the right of the wind direction.*

via its friction with the water above. Water is fairly friction-free (more accurately, it supports very little "shear stress"), and one finds the water at greater depths moving at lesser velocity.

In the Northern Hemisphere, the surface water tends to flow to the right of the wind direction, with the Coriolis effect tending to deflect it further to the right and the wind trying to straighten it back up in line with itself. An equilibrium is reached, which for deep water under perfect conditions would have the surface water flowing at 45° to the right of the wind. In shallower water, the angle would be less than 45°.

Below the surface, each layer of water is being dragged along by the layer above it. The Coriolis effect ensures that each layer tends to flow to the right of the driving force. Hence, the surface layer flows to the right of the wind, and each subsurface layer flows to the right of the layer above it. Furthermore, since the Coriolis deflection increases for slower speeds, this tendency will be even more pronounced in deeper, slower-moving layers. The total picture is illustrated in Figure 14.2. At each depth the water is moving slightly slower and slightly more to the right of the water just above it. This is called the "Eckman spiral."

At some certain depth the water will be moving exactly in the opposite direction of the surface current, or 225° to the direction of the wind. How deep this is depends on the strength of the Coriolis deflection, which

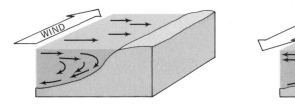

(a) *(b)*

FIGURE 14.3 (a) *The general wind direction (relative to the coastline) causing
downwelling, and (b) the general wind direction causing upwelling
along the coast for the Northern Hemisphere.*

depends on the latitude. Off California this would occur at about 100 m
depth, and on the equator it wouldn't happen at any depth, because the
Coriolis deflection is zero there. According to the theory, at whatever
depth this 180° reversal of the surface current direction is found, the current velocity there should be ⅟₂₃ as great as that at the surface.

In the previous paragraph we saw that for purely wind driven surface
currents, the water at different depths moves in different directions. The
average of all these current components shows that the net water transport
is 90° to the right of the wind direction (and, of course, 45° to the right of
the surface current direction). In shallower water, the bottom of the Eckman spiral is missing, and the net water transport is not so far to the right.

A.3   Upwelling and Downwelling

The Eckman transport contributes to the upwelling and downwelling of
coastal waters. Consider the coast of some continent in the Northern
Hemisphere, for example, with the offshore wind blowing in such a way
that the coast is to the right of the wind direction, as in Figure 14.3a. The
water transport will be to the right of the wind, and the surface water will
stack up against the coast, causing downwelling there. In the case where
the offshore wind blows predominantly in the other direction in such a
way that they coast is to the left of the wind direction as in Figure 14.3b,
the net surface water transport is seaward, being replaced by the upwelling of cooler, deeper waters along the coast. This is the case along the
central and northern California coast. Due to the upwelling of cooler
water, San Francisco can brag about being an "air conditioned city."

The Peruvian fishing industry is particularly keyed to these processes.
Normally the winds off the Peruvian coast are blowing from the southeast.
The Eckman transport (Southern Hemisphere) is to the left, or out to sea.
The nutrients brought up with the upwelled water to the sunlit surface
zone foster increased biological activity and a hearty fishing industry. The
same winds that cause this upwelling continue on across the oceans as
"trade winds," which blow the equatorial surface current with them.

Occasionally, there are prolonged periods during which these winds
slacken. Water masses that were piled up on the western margins begin
flowing eastward back across the ocean. Not only do the reduced winds
cause less upwelling along the Peruvian coast, but the subsequent return
of the nutrient-depleted surface water southeast along the Peruvian coast

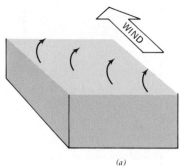 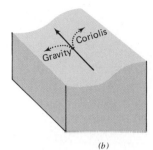

(a) *(b)*

FIGURE 14.4 (a) *As a current begins to flow, it experiences Coriolis deflection to the right (Northern Hemisphere). (b) This continues until the slope in the water surface is sufficient that the force of gravity, pulling the water downhill off the mound, balances the Coriolis deflection into the mound. At this point, the two cancel, so the current flows sideways along the mound in the direction of the wind.*

brings the fishing industry to a screeching halt. This unwelcomed coastal current is called "El Niño," after the Christ child, because it is most likely to occur near Christmas time. We now feel that by studying variations in the trade winds, we can predict occurrences of El Niño and warn the fishing industry many months in advance.

A.4 Slopes in the Water Surface

We have seen that trade winds blowing westward across equatorial oceans tend to stack the water on the ocean's western margins. This creates a slight slope to the water surface, amounting to a rise of about 4 cm for every 1000 km distance. But there are other factors that can cause slopes to the water surface, and some of these are much more dramatic. For example, violent storm winds can drive surface waters into a coast, causing sea level to rise two meters or more over a stretch of a few hundred kilometers.

Among the more interesting and dramatic slopes to the ocean surface are those caused by the Coriolis deflection of water masses sideways to the direction of current motion. The Gulf Stream, for example, has nearly a 2 m rise in surface elevation across its 75 km width. This is because the water is deflected to the right as it goes and tends to stack up on that side. As a current flows, this deflection continues until the surface slope is sufficient that the continued Coriolis deflection into the mound can no longer overcome the force of gravity pulling water back down off the mound (Figure 14.4). At this point, the two sideways forces exactly cancel, so the current continues on in the direction of the driving force, with no further deflection. When facing the direction of the current in the Northern Hemisphere, there will be a slope to the water surface, with it being elevated (a "mound") on the right and depressed (a "trough") on the left.

As we have seen, the major anticyclonic current gyres in the oceans are driven by the trade winds and prevailing westerlies. The slope to the water surface caused by the currents encircling these gyres is such that

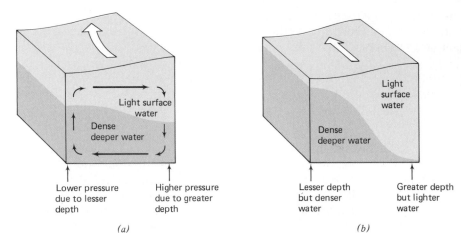

(a) *(b)*

FIGURE 14.5 (a) *As the current begins to flow, Coriolis deflection of surface waters causes them to stack up on the right, forming a mound there. Beneath the mound, the pressure is increased due to the increased depth of the water. Consequently, at depth, the water is forced out from under the mound toward the lower pressure region beneath the trough. (b) As the lighter waters accumulate on the right, the pressure beneath the mound lessens, due to the smaller weight of the overlying waters. Likewise, the pressure beneath the trough increases, due to the increased weight of the denser overlying waters. Eventually, the pressure at depth equalizes, and there is no more sideways flow.*

the central portions of these gyres are elevated one or two meters above the outer regions. In the central portions of cyclonic gyres, where the currents flow in the opposite direction, the ocean surface is correspondingly depressed.

A.5 Slopes in Density Contours

The water at any depth below the surface must support the weight of the overlying layers. The deeper the water and the denser the water, the greater will be the weight of these overlying layers.

In the preceding section, we saw that Coriolis deflection of surface currents to the right (Northern Hemisphere) causes an elevation of the sea surface, or "mound" to form on the right side, and a depression, or "trough" to form on the left side of the moving current. Beneath the mound, the pressure initially increases because of the increased depth of the water there. Likewise, beneath the trough, the reduced depth of the water causes a reduction in the pressure. Since water tends to flow from high pressures towards lower ones, the water at some depth beneath the surface current tends to flow out from under the mound and toward the trough.

Initially, then, there will tend to be a "corkscrew" motion to the current as it flows, with lighter surface water flowing into the mound, and denser deeper waters accumulating under the trough. (See Figure 14.5) Because denser water weighs more, the accumulation of denser water under the trough causes the pressure there to increase. Likewise, the accumulation of lighter water in the mound causes the pressure beneath the mound to decrease. Eventually, the two have adjusted to the point where the pressure beneath the mound is no longer greater than that beneath the trough, and the deep water no longer flows from one to the other.

In equilibrium, then, we would expect no sideways motion of the water either at the surface or at depth. At the surface, the mound has developed to the point where the gravitational pull of water down off the mound exactly counterbalances the Coriolis deflection into the mound. At depth, the lighter surface water has accumulated in the mound, and the denser deeper water has accumulated beneath the trough, so that the pressure beneath the two regions is the same. When this stage is reached the current will flow in the direction of the applied force with no sideways deflection at any depth.

As illustrated in Figure 14.5, the accumulation of light surface waters in the mound and of dense deep waters beneath the trough results in a distinct slope to the contours of constant density taken in a cross section across the current. When facing the current direction in the Northern Hemisphere, the surface slopes upward to the right, and the contours of constant density slope downward to the right. In the Southern Hemisphere, where the Coriolis deflection is to the left, the slopes are reversed.

The difference in densities between neighboring water masses is only very slight. This means that if pressures at depth are to be the same, then several hundred meters of a different water mass are required to compensate for one meter difference in surface elevation. Across the Gulf Stream, for example, the contours of constant density drop about 800 m, while the surface elevation rises only about 2 m. (See Figure 14.6) Because they are much more pronounced, slopes in the contours of constant density are much more easily detected than slopes in the water surface.

In fact, by measuring the slopes of contours of constant density, oceanographers can find the locations and velocities of ocean currents. The strength of the Coriolis deflection depends on the latitude and current velocity in a known way.[1] Higher latitudes and larger current velocities result in greater Coriolis deflection, and therefore in greater slopes in the contours of constant density. Although we have discussed the sloping density contours for surface currents only, the Coriolis deflection occurs at all depths, of course, so the locations and velocities of subsurface currents can also be determined in a similar way.

A.6 Geostrophic Flow

We have seen that the Coriolis deflection of wind-generated surface currents tends to cause the water surface to slope upward to the right and the contours of constant density to slope downward to the right (Northern

---

[1]*The strength of the Coriolis acceleration on a mass moving with velocity* **v** *across the earth's surface is given by* $2\Omega v \sin \psi$, *where ψ is the latitude, and Ω is the earth's rotational speed in radians per second ($\Omega = 7.27 \times 10^{-5}/s$).*

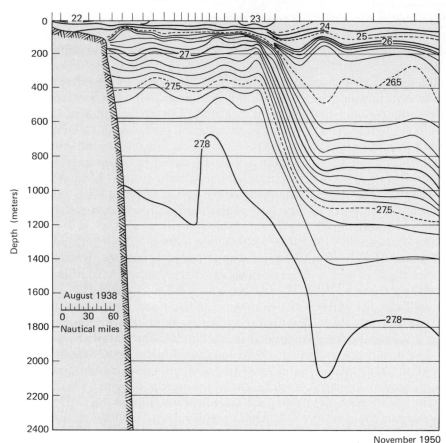

FIGURE 14.6   *Cross section of the Gulf Stream between Cape Cod and Bermuda,
showing contours of constant density. (The numbers should be read
as follows: 27 means 1.027 g/cm³, 27.5 means 1.0275 g/cm³, etc .)
Denser water is on the north (left) side, and lighter water is on the
south (right) side. (F. C. Fuglister, Woods Hole Oceanographic Institution, 1950.)*

Hemisphere). Now consider what would happen if the wind suddenly
stopped.

If the current velocity should diminish, then so would the strength of
the Coriolis deflection, and the water would begin flowing downhill off
the mound (Figure 14.7). But as it flows downhill, it would pick up speed
and be deflected back to the right. Soon its velocity would be sufficient
that again the strength of the Coriolis deflection back into the mound
would exactly counterbalance the downhill pull of gravity. Consequently,
even if the wind should stop blowing, the current would continue to flow
sideways along the mound.

This type of current—which results from the balancing of gravity,
which tends to pull the water downhill—and the Coriolis deflection—
which tends to deflect the water back into the mound—is called "geostrophic flow." Most major ocean surface currents are a combination of

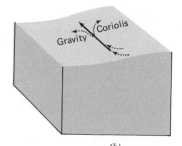

(a)   (b)

FIGURE 14.7   (a) *If the wind and surface currents were to stop, water would begin flowing downhill off the mound. But then it would be Coriolis deflected to the right (Northern Hemisphere). (b) Soon it would be flowing sideways along the mound again, with Coriolis deflection back into the mound exactly offsetting the gravitational pull downhill off the mound. This flow sideways along a mound, which results from the influences of gravity and the Coriolis deflection alone (although some other driving force must have created the mound in the first place) is called "geostrophic flow."*

wind driven and geostrophic. The winds are the basic generator and driving force behind these currents, but geostrophic effects insure that the currents continue to flow at a fairly constant rate even during periods when the wind is stopped.

Once established, geostrophic currents can continue for remarkably long periods in the absence of wind or other driving force. If it weren't for friction, they could continue indefinitely. However, due to frictional losses, the current must flow slightly downhill in order to regain the energy that is being lost. You would think that this downhill flow would reduce the size of the mound and thereby stunt the geostrophic flow.

Remember, however, that light surface waters have accumulated in the mound, and dense deeper waters have accumulated beneath the neighboring trough, until the pressures at depth beneath them are the same. As water is removed, the mound becomes lighter, so it is buoyed back up (Figure 14.8). Similarly, as water flows into the trough this region becomes heavier and sinks. Therefore, even as water is slowly transferred from the mound to the trough, the mound does not disappear, and neither does the trough. The currents persist much longer than might be expected from measurements of surface topography alone.

A.7   Storm Surges

To this point, we have been discussing the large scale, stable, wind driven surface circulation of the oceans. A much more local, temporary, and unpredictable type of wind driven surface current occasionally causes a great deal of concern in some coastal areas. These are called "storm surges," or sometimes "storm tides," and they are caused by a combination of low atmospheric pressures and strong winds associated with storms.

Just as high atmospheric pressures push the ocean surface down slightly, the ocean beneath low pressure centers rises in response to the

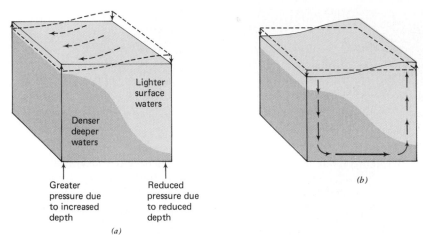

FIGURE 14.8 (a) *As water flows from the mound to the trough, the pressure beneath the mound lessens and that beneath the trough increases. (b) Consequently, at depth, the water will flow out from under the trough and toward the reduced pressure region beneath the former mound, forcing the mound back up again.*

lowered pressure. This causes typically about ½ meter rise in sea level, which by itself wouldn't be too alarming. More frightening are the strong storm winds associated with these low pressure zones. When they occur over coastal regions, strong winds blowing in towards these low pressure zones drive surface waters with them. This landward current stacks water up on the coast, sometimes raising water level several more meters, inundating some coastal areas that are normally dry. In addition, the winds create large storm waves, which added to the high water level, can be quite devastating (Figure 14.9).

Some coastal areas are accustomed to large tidal variations. In these regions, storm surges wouldn't be particularly damaging, unless they happened to be in progress during a period of high tide. But many coastal areas are low and accustomed to small variations in sea level, and can be devastated by storm surges.

The city of Galveston, Texas was built on a low barrier beach in the Gulf of Mexico, where the normal tidal range is very small. In 1900 a storm surge destroyed this city, with a loss of 5000 lives. In 1953 the Dutch dikes gave way to a storm surge, which flooded an area of 3200 km² and drowned 1700 people. The South Asian Coast of the Bay of Bengal has many heavily populated, low-lying regions, that are not accustomed to large variations in sea level. Because of these conditions, storm surges there can be very hazardous. A century ago, a storm surge there took 200,000 lives. Today, although world population is growing, fatalities are diminishing as early warning and communication systems are improving. The only effective way to reduce damage from storm surges is to get out of their way.

FIGURE 14.9 *Strong storm winds blow surface water shoreward, causing a rise in sea level along the coast and driving large storm waves into shore.*

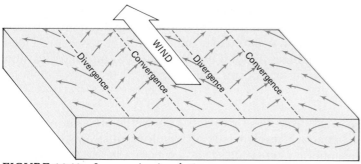

FIGURE 14.10 *Langmuir circulation.*

A.8 Langmuir Circulation

When winds blow across the water at speeds of more than a few kilometers per second, the surface water develops an interesting circulation pattern. Rather than simply flowing directly in line with the wind, the water flows in helical or "corkscrew" motions through long parallel "Langmuir cells." (See Figure 14.10.)

Such cells are each typically a few meters wide and a few meters deep, oriented lengthwise parallel to the wind direction. Neighboring cells have alternate clockwise and counterclockwise corkscrew motions, which means that the boundaries between these cells alternate between zones of converging and diverging surface waters. Debris floating on the surface are carried with the surface water into the zones of convergence, forming long visible slicks, parallel to the wind direction.

We do not yet completely understand why this peculiar circulation pattern develops. It is probably a result of instabilities caused by the transfer of energy from the wind to the water. Of course, since we don't yet completely understand the wind-to-water energy transfer mechanism, it should not be surprising that we don't completely understand the cause of the Langmuir circulation patterns either.

B.   WESTERN INTENSIFICATION OF CURRENTS

After studying the circulation of our atmosphere, with the "trade winds" blowing westward across the oceans in equatorial latitudes, and "prevailing westerlies" blowing eastward across the oceans at higher latitudes, we can understand why the anticyclonic current gyres should form the dominant surface current pattern in the low and mid latitudes of the oceans. What isn't so obvious, however, is the reason why these currents should be so swift and narrow along the western boundaries of these oceans.

The Gulf Stream is just one of these peculiar western boundary currents. A look at Figure 14.1 shows that there are similar intensified boundary currents along the western edges of all oceans, in both hemispheres. From considerations of the winds alone, you would think that in each case the current could be half an ocean wide. But they're not, so we are faced with the problems of explaining why they are so swift and narrow, and why they occur on the western ocean margins only.

B.1   The Causes

There are four related processes that contribute to the creation of strong narrow western boundary currents, all of which result from the earth's rotation and atmospheric circulation. The first is due to the trade winds, which blow westwardly across the equatorial oceans, and which result from the Coriolis deflection of the surface winds in the tropical atmospheric circulation cells. (See Figure 13.12.) These winds drive the westward-flowing equatorial surface currents, which gain momentum as they cross the oceans. When they run into the continents on the western margins, they must squirt out the sides, thereby providing the water masses and generating impetus for these western boundary currents. (Figure 14.11.)

The South American equatorial coast is oriented in a northwest-southeast direction. This geometry deflects most of the Atlantic Equatorial Current northward, making the Gulf Stream in the North Atlantic more

FIGURE 14.11   *Upon hitting the rock, the water squirts out to the sides in the same way that water of an equatorial current squirts out to the sides when it strikes a continent on the western ocean boundary.*

impressive than its South Atlantic counterpart. Studies show that the surface waters of the Atlantic undergo more mixing across the equator than do those of the Pacific, even though the equatorial Pacific has more than twice the width over which this mixing could take place. It is thought that this northward deflection of the Atlantic Equatorial Current from the Brazilian coast may be largely responsible for this, because it deflects to the north some waters that otherwise would have remained in the Southern Hemisphere.

The second factor is related to the prevailing westerlies, which drive the return current eastward across the oceans at higher latitudes. At these higher latitudes, the Coriolis deflection is stronger, so these surface currents are strongly deflected towards the equator. This added water intensifies the westward-flowing equatorial currents. Furthermore, it means that across most of the oceans, the surface currents are forced toward the equator rather than away from it (Figure 14.12). The only place they are allowed to return poleward is along the western margins, where they have not yet begun their eastward return flow, so there is an absence of this Coriolis-induced opposition.

A third related cause of the western intensification is also a result of the stronger Coriolis deflection at higher latitudes. In each major current gyre in each ocean the eastward return flow is at a higher latitude than the westward flow along the equator. Consequently, when the water is returning east, it gets quickly deflected toward the equator, whereas when it is moving westward near the equator, it gets deflected only very weakly. This means that in any one complete cycle, the water tends to flow farther to the west than toward the east (Figure 14.13), so the gyre tends to move

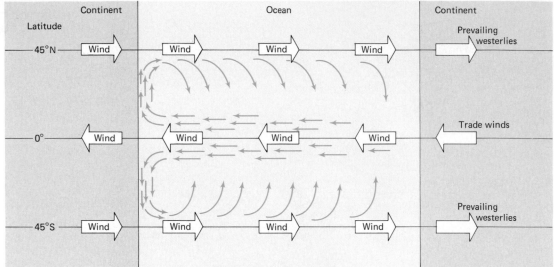

FIGURE 14.12 *Wind and ocean circulation patterns in an idealized ocean basin. Because the Coriolis effect is stronger at higher latitudes, the east-ward flowing waters are strongly deflected back towards the equator. Consequently, surface waters tend to flow towards the equator everywhere except along the very western edge of the ocean.*

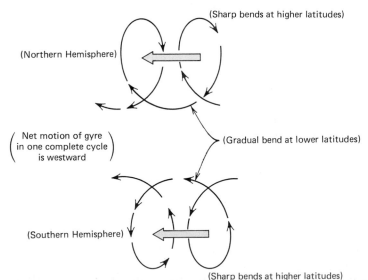

FIGURE 14.13 *Because the Coriolis deflection is greater at higher latitudes than at lower latitudes, the gyres migrate westward, squashing up against the continents on western margins of the oceans.*

westward across the ocean. It gets forced up against the western margins where the currents are correspondingly compressed and intensified.

The fourth cause of the intensified western boundary currents is also related to the earth's rotation through the apparent change in rotational

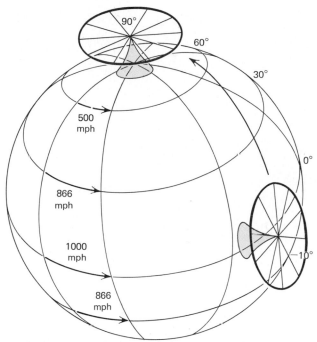

FIGURE 14.14 *A non-spinning bicycle wheel on the equator, when transported very carefully by its base to the North Pole, still won't spin. But the earth spins beneath it, making it seem to an earthbound observer that the wheel has acquired a clockwise spin, making one revolution every 24 hours.*

state of objects transported north or south along the earth's surface. To illustrate this effect, imagine we have a large bicycle wheel mounted horizontally on frictionless bearings, which is sitting motionless on the equator, as illustrated in Figure 14.14. Suppose that we very carefully transport this wheel to the North Pole, touching only the stand and being very careful not to touch the wheel itself, so we don't inadvertently provide some impulse that might start it spinning.

Under these conditions, the wheel will still have no spin at all when we set it down at the North Pole. However, now the earth will be spinning beneath it once every 24 hours. Since we are earth-bound observers, it will appear from our point of view that the earth is still and the wheel is spinning, rather than vice versa. From our point of view, the wheel has somehow miraculously acquired a spin in the clockwise direction that gives it one complete revolution every 24 hours.

The same thing appears to happen to air or water masses as they move north or south along the earth's surface (Figure 14.15). As they go, their rotational state appears to change. A mass of water starting on the equator with no spin (no "vorticity") at all, appears to acquire a spin as it goes. The farther poleward it goes, the faster it appears to spin. At intermediate latitudes, this amount of spin will be less than one complete revolution per

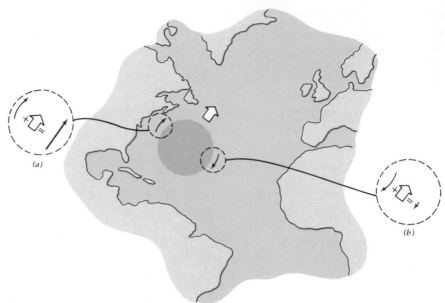

FIGURE 14.15 *As a mass of water moves northward, it appears to pick up a clock-*
wise spin relative to the earth. On the western margin of the mass,
both the velocity due to the spin and that due to the motion of the
mass are in the same direction, so they add up to make a very large
total velocity (a). On the eastern margin, the two velocities are in
opposite directions, so they subtract, leaving little net velocity, if
any (b).

day. But if the water mass is wide enough, it doesn't have to spin very
rapidly for the outer regions to have high speeds.

Because of the dominant wind-induced current gyres, surface waters
in the western portions of all oceans are traveling away from the equator.
As these water masses go, they acquire the appropriate spin (clockwise in
the Northern Hemisphere and counterclockwise in the Southern Hemisphere) to make the current on the very western edge extremely swift, as
illustrated in Figure 14.15.

B.2 The Gulf Stream

One of the best known of these swift, narrow western boundary currents
is the Gulf Stream, which is located just off the eastern continental margin
of North America. Having a width of about 50 to 75 km and a depth of 1½
to 2 km, it flows northeastward with a velocity of 3 to 10 km per hour. Daily
tidal oscillations move the entire Gulf Stream back and forth some 6 to 8
km per day. Sometimes there are counter-currents down its eastern side,
and sometimes spinning eddies move down its western side, bringing cold
northern water down the eastern seaboard of North America.

Before starting up the east coast, the Gulf Stream wanders through
the Gulf of Mexico and then through the Straits of Florida, whereupon its
volume transport increases by some 80%. Just where this additional water

comes from is not certain, but it is known that some water from the Antilles Current—a large, slow-moving, northward current of Antarctic intermediate water—does somehow get into the Gulf Stream. There are seasonal fluctuations in its volume transport, reaching a low in the fall. Attempts to correlate this to seasonal fluctuations in the trade winds, which push the equatorial current, have failed.

The eastern edge of the Gulf Stream is poorly defined, with the current velocity gradually diminishing to zero. However, since the current velocity is largest on the western side (for reasons described in the previous section), there is a sharp distinction between the western edge of the Gulf Stream and the neighboring "slopewater." It is usually even possible to distinguish this edge visually. Much of the water in the Gulf Stream has come from the South and Equatorial Atlantic, and most of has spent a long time near the surface. Consequently, many of its nutrients have been consumed. Being deficient in nutrients and suspended organic matter, there is little biological activity, and so it is a deep, rich blue color. By contrast, abundant plankton make the neighboring slopewater green.

There are other ways of visibly detecting the Gulf Stream as well. Because of its motion, its velocity relative to the wind will be different than that of the slopewater. This often makes a noticeable difference in the sea state (i.e., wave activity) when going from one to the other. For example, if the wind is blowing northward along with the Gulf Stream, then the Gulf Stream will be calm, whereas this same wind will be making waves on the slopewater next door. Since it is warmer, the air above the Gulf Stream will generally be rising and be carrying more evaporated water. This makes clouds, so sailors can sometimes see the Gulf Stream ahead by looking at the sky. Being warmer, it also radiates more infrared radiation, so it can be seen on aerial infrared photographs. (See the photo on the title page of Chapter 12.)

At higher latitudes the Gulf Stream meanders. Off the New England and Nova Scotian coasts these snakelike windings generally have amplitudes of 15 to 45 km and wavelengths of 150 to 400 km, which have the ability to move up to 25 km sideways in one day. There may be several "flops" per month in any one of these loops, and it seems at times that they are somehow bouncing off the continental slope like a basketball on a gym floor. Sometimes a loop gets so pronounced that the Gulf Stream doubles back on itself, and the loop breaks off in an eddy and drifts away, as in Figure 14.16. It isn't yet clear what causes this meandering. Could it be inertial effects as described in Section A of the previous chapter?

C. SEICHES

An interesting current is caused by the water sloshing back and forth in a confined region, such as a harbor or estuary. However, to observe a "seiche," one really need go no farther than the kitchen. The resonant sloshing (i.e., seiching) frequency of an average cup of coffee is slightly

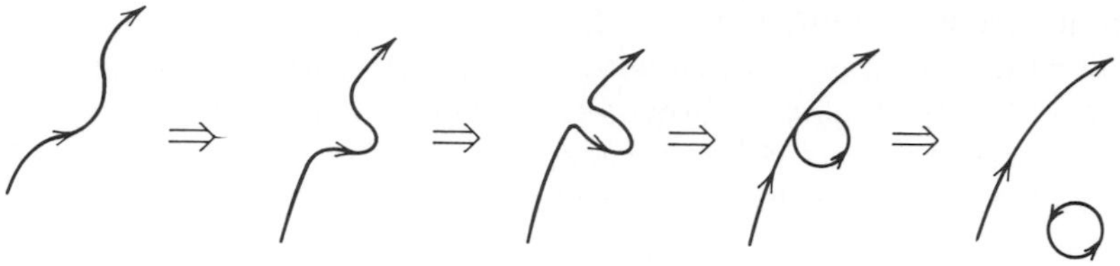

FIGURE 14.16   *The Gulf Stream meanders. Occasionally a loop becomes so pronounced that the Gulf Stream doubles back on itself, and the loop breaks off in a separate eddy.*

faster than one cycle per second. If you would stir your coffee or tea back and forth at this frequency, it would quickly begin sloshing out of the cup and onto the saucer and tablecloth. Next time you have a chance to observe coffee or tea drinkers, notice that they will either stir with a circular motion or with a much quicker back-and-forth motion, so as not to hit the cup's resonant frequency and avoid embarrassment.

Liquids in larger containers have resonant periods for sloshing too— larger containers have longer periods. In many harbors the resonant periods are several minutes long. For example, in the Los Angeles harbor, driving forces with periods of 3, 6, and 12 min will set up resonant oscillations. The driving force often comes from appropriate surf beat or long period swell. Because the period of these seiches is so long, it is difficult to notice, but sped-up motion pictures will show boats harmoniously dancing back and forth on their moorings as the water sloshes back and forth. Because people at the harbor may have difficulty in telling when a seiche is in progress, they may be somewhat bewildered by the high currents, and the ships snapping mooring lines and damaging pier pilings for no apparent reason.

When stirring tea at the fundamental seiche frequency of the cup, you make a wave (technically, a "standing wave") that will at one time have a crest on one side of the cup and a trough on the other side. Then the wave will bounce back across the cup, and the crest and trough will have changed places. If you stir more quickly, you will find another, higher resonant frequency. The waves now having half the wavelength as before, there will alternately be a crest on both sides with a trough in the middle, and a trough at both sides with a crest in the middle (Figure 14.17). Since in this second case the waves are twice as close together, you might expect that the frequency would be twice as great. But you'd be wrong. Since the wave speed decreases with the wavelength, waves twice as close together pass less than twice as frequently. Similarly, you will find resonances at higher frequencies for waves 3, 4, 5, and so on times closer together than the waves of the fundamental sloshing mode, but the corresponding resonant frequencies will be less than 3, 4, 5, and so on times the fundamental frequency. In wave jargon, the higher harmonics will not be simple multiples of the fundamental seiche frequency.

FIGURE 14.17 (a) *For a seiche in a teacup, the wavelength of the fundamental mode will be twice the length of the cup.* (b) *The next resonance will occur for a wavelength half as long as that of the fundamental mode.* (c) *The third resonance will occur for a wavelength ⅓ as long as that of the fundamental mode, and so on.*

(a)

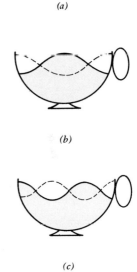

(b)

(c)

The same is the case in estuaries. Because the depth of the water in the estuary varies with location and with the tides, and because estuaries are irregular, having varying lateral dimensions in different directions, it becomes almost impossible to predict what driving frequencies will produce a seiche. To find the various frequencies, one must first wait until there is a seiche, record the frequency, and then next time can predict it coming, providing conditions are similar.

A further complication is caused by the Coriolis deflection of the moving water. During a seiche of the fundamental resonant frequency in an estuary in the Northern Hemisphere, for example, the moving water will pile up everywhere on the right-hand margin as it goes. This is called "amphidromic motion"; the water sloshes counterclockwise around the perimeter of the confinement (Figure 14.18), much as it does in a teacup when you swirl it in circles. Actually, in real embayments, imperfect reflections, varying depths, and irregular shapes often create serious alterations in this simplified picture of amphidromic motion.

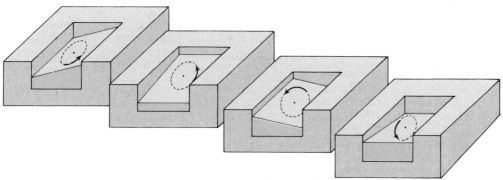

FIGURE 14.18 *Amphidromic motion in a idealized embayment in the Northern Hemisphere. In the first figure, the water entering stacks up against the right side due to the Coriolis effect. In the third figure, the water stacks up against the other side when leaving due to the Cortiolis effect. Similar arguments can be applied to the second and fourth figures. The point at which the water level remains stationary is called the "amphidromic point." In this idealized basin, the amphidromic point is in the very center. The net result is that the water will slosh around the perimeter of the basin, as indicated.*

D. SUMMARY

The major ocean surface currents are direct reflections of the predominant wind patterns, although the ocean currents have much greater inertia and therefore less variation. The surface waters are driven by the wind, and each successive layer of water is dragged along by the layer above. Due to Coriolis deflection, the surface water flows to the right of the wind direction (Northern Hemisphere), and each successive layer of water flows to the right of the layer above, in a pattern called the "Eckman spiral." Divergence of surface waters causes upwelling of deeper waters, and convergence of surface waters causes downwelling.

Due to the Coriolis deflection, surface waters tend to deflect to the right of the wind direction (Northern Hemisphere) causing a mound and an accumulation of surface waters on the right and a trough and corresponding accumulation of deeper waters on the left. Facing the current direction, then, the surface tends to slope slightly upward to the right, and the contours of constant density have a more pronounced slope downward to the right. If the driving force is removed, then the Coriolis deflection of waters flowing downhill off the mound forces them to continue flowing sideways along the mound in what is called "geostrophic flow."

Storm surges are created when strong storm winds drive water and waves into a coast. Winds blowing across the water surface often cause a peculiar corkscrew flow in surface waters in "Langmuir cells" of alternate clockwise and counterclockwise sense.

Surface currents are greatly intensified along the western boundaries of all oceans. This is caused by the deflection of equatorial currents as they strike the continental margins: the strong Coriolis deflection of eastward-flowing waters at higher latitudes that forces water back towards the equator and causes the gyres to tend to migrate westward, and by the increasing rotation of water masses as they flow toward higher latitudes. The Gulf Stream is a familiar and well-studied example of an intensified western boundary current.

Seiches are local surface currents experienced when water sloshes back and forth rhythmically in a harbor or other confined area.

QUESTIONS FOR CHAPTER 14

1. What drives the dominant ocean surface currents?

2. How do the surface currents in the oceans of Figure 14.1 compare to the surface wind patterns of Figure 13.3?

3. Why should surface water motion in the neighborhood of atmospheric low pressure centers correspond less well to the wind patterns than surface currents near high pressure centers?

4. Why do you suppose daily changes in wind patterns have little effect on the surface currents?

5. If one sails westward with the trade winds, there is a slope to the water surface and one is sailing "uphill." How big is the slope, and why is there this slope?

6. Why are there equatorial countercurrents?

7. Why is there an abundance of biological activity off the west coast of Africa?

8. How does the temperature of the equatorial surface water change across an ocean, and why?

9. How does the water velocity vary with depth for wind driven surface currents? Do you suppose the same would be true for equatorial currents? Why or why not?

10. In the Northern Hemisphere, why does the surface current tend to go to the right of the wind direction? What direction does the surface current end up going? This is the compromise between what two opposing effects?

11. Explain the Eckman spiral, both what it is and why.

12. Why do you suppose it is that the depth at which there is a 180° reversal of the surface current direction depends on the latitude?

13. What do we mean by the direction of the "net water transport" in an Eckman spiral? How far to the right of the wind direction is this?

14. Explain how a wind that blows southeast along the California coast can cause upwelling of deep water there.

15. Explain why a reversal of the wind direction along the Peruvian coast could stifle the Peruvian fishing industry.

16. What causes the mound in the surface of the North Atlantic, South Atlantic, North Pacific, and so on? Roughly how high is this mound?

17. What two opposing effects determine the height of the mound?

18. Why is it that the less dense water concentrates under the mound, and the more dense water tends to accumulate under the trough?

19. Suppose you were in a submarine facing the direction of current flow in the Northern Hemisphere. How would the contours of constant density slope (e.g., flat, vertical, from upper left to lower right, etc.), and why?

20. Why does the water pressure beneath the current eventually equalize, even though the water is deeper toward one side?

21. How does the slope of the density contours depend on the speed of the current? How might this be used to measure current speed when there is no land or other fixed reference point in sight?

22. Suppose we were able to take a super-large piece of machinery and instantly stop a current and level off the ocean surface. Why would the mound reappear? Why would the current start flowing sideways along the mound again?

23. Why is there a dip in the surface near the center of cyclonic gyres?

24. What is geostrophic flow?

25. What causes storm surges, and why are they damaging to some coastal areas? What kinds of coastal areas would be the most vulnerable to storm surges?

26. Describe Langmuir circulation. What do you look for on the water surface to reveal Langmuir circulation?

27. Explain why a horizontal bicycle wheel that is not spinning on the equator will begin spinning as it is transported northward? How fast will it be spinning (apparently) when it reaches the North Pole? What would appear to happen to it as we transport it back southward toward the equator?

28. Consider a large mass of water that is being driven northward, as illustrated in Figure 14.15. Why is it that the net velocity of the water on the western side of this mass will be large and the velocity on the eastern side will be small?

29. What is it that drives water northward along the western side of the Atlantic Ocean? What happens to its spin as it goes?

30. How do the trade winds strengthen the narrow western boundary currents?

31. Why does the Gulf Stream carry more water than its counterpart in the South Atlantic (the Brazil Current)?

32. Does the current in a gyre bend more sharply at higher latitudes than at lower ones? Why? How does this cause the gyre as a whole to move?

33. Consider the top part of Figure 14.13, which shows a current gyre in the Northern Hemisphere. It shows that the turn of the current toward the south at high latitudes is sharper than the turn of the current toward the north at lower latitudes. Now consider the fact that the equator is "uphill." How might this affect the turn to the south and the turn to the north (i.e., does it make one slower and one sharper than otherwise)?

34. Summarize the four contributing factors that help intensify the currents along the western margin of each ocean.

35. What is the Gulf Stream?

36. Roughly what are the dimensions and current velocities in the Gulf Stream?

37. Why do you suppose it was thought that seasonal changes in the trade winds might cause seasonal changes in the Gulf Stream?

38. Which margin of the Gulf Stream is most clearly defined? Can you think of any reasons for this?

39. As one sails across the Atlantic Ocean, what are some of the things one might look for to tell when the Gulf Stream has been reached?

40. Why would the waves on the Gulf Stream look different from those on the neighboring slopewater?

41. Describe the meandering of the Gulf Stream?

42. What is a seiche? Give examples. Can you pronounce it?

43. A teaspoon could provide the periodic driving force for a seiche in a teacup. But what could provide the periodic driving force for a seiche in a harbor or estuary?

44. Sketch an idealized estuary cross section, and make a sketch of how the periodic displacement of the water surface (a "standing wave") would look for a seiche with the fundamental frequency. Make a similar sketch for some of the higher harmonics.

45. In an estuary of a certain length, how are the wavelengths of seiches of the second, third, or fourth harmonics related to the wavelength of the fundamental frequency? Why are the frequencies of the harmonics not so simply related to the frequency of the fundamental mode?

46. Explain amphidromic motion.

*47. Would you expect that surface currents or deeper currents would be faster? Why?

*48. We have seen that the surface currents are most intense on the western margins of the oceans. Are the deeper currents more intense on one margin? Which margin? Why? Is it the same margin in both hemispheres?

*49. Can you think of any reasons why the peculiar Langmuir circulation pattern develops as winds pick up?

*50. Why might the fundamental seiche frequency in a certain harbor be different during high tide than low tide? (*Hint:* The frequency of sloshing may be related to how fast the wave travels, which depends on the water depth.)

SUGGESTIONS FOR FURTHER READING

1. Charles H. V. Ebert, "El Niño: An Unwanted Visitor," *Sea Frontiers,* **24,** No. 6 (1978), p. 347.

2. T. F. Gaskell, *The Gulf Stream,* Cassell, London, 1972.

3. C. P. Idyll, "The Anchovy Crisis," *Scientific American* (June 1973).

4. Phillip Mason, "The Changeable Ocean River," *Sea Frontiers,* **21,** No. 3 (1975), p. 171.

5. "Oceanography from Space," *Oceanus,* **211,** No. 3 (Fall 1981).

6. "Ocean Eddies," *Oceanus,* **19,** No. 3 (Spring 1976).

7. Stephen Pond and George L. Pickard, *Introductory Dynamic Oceanography,* Pergammon, New York, 1978.

8. John P. Robinson Jr., "Galveston's Killer Hurricane of 1900," *Sea Frontiers,* **27,** No. 3 (1981), p. 166.

15
DEEP OCEAN CURRENTS AND WATER MASSES

The densest deepest water masses are first formed at the surface beneath Antarctic ice.

Deep water flows are much slower and much more voluminous than those at the surface. The deep water masses generally acquire their particular characteristics when near the surface, and these characteristics determine the flow of these masses over the next several hundred years. In this chapter we study the properties and movements of these deeper waters. At the end of the chapter we also investigate tidal currents and current measurement techniques, which are pertinent to waters at all depths.

A.   THERMOHALINE CIRCULATION

In addition to wind, another driving force for oceanic currents is gravity. Denser waters sink. Those that are densest sink to the bottom, and then flow downhill along the bottom to the deepest accessible points. If the ocean were perfectly still, eventually it would be stratified with densest waters on the bottom and increasingly less dense layers toward the surface. But we have seen many processes that disturb the oceans, and so perfect stratification is never attained. The three factors that determine the density of seawater are the pressure, salinity, and temperature. The density increases with pressure and salinity, but decreases with increased temperature.

A.1   Pressure and Density

The pressure increases by one atmosphere for every 10 m depth below the surface. Consequently, depths are sometimes recorded in "decibars"[1] instead of meters, the two being equivalent. Water is almost incompressible, requiring 200 atmospheres to reduce the volume by only 1%. One must go to a depth of 2 km before this pressure is reached, so the density of water would increase by 1% for every 2 km down due to pressure alone.

Actually, what is of interest to oceanographers is what the densities of various waters would be if they were all at the same depth, because that is what will determine which would sink and which would rise. Consequently, for purposes of studying thermohaline circulation, the densities of the various waters must be compared when under the same pressure. The most convenient pressure to choose for this comparison is atmospheric pressure. The oceanographer then need only send a bottle down to whatever depth wanted for sampling, bring the sample up, and measure the density right there aboard ship.

The density of most seawater lies in the range between 1.024 and 1.028 g/cm³. Since the first three figures will almost always be the same,

[1] 1 decibar = 0.1 bar = 10% of atmospheric pressure

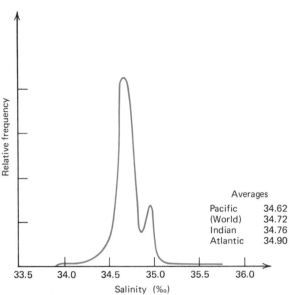

FIGURE 15.1 *Plot of the relative abundances of ocean waters of various salinities. (After R. B. Montgomery, Deep Sea Res.,* **5,** *(1958).)*

oceanographers use a more convenient parameter denoted by "$\sigma_{s,t,p}$," which is defined by

$$\sigma_{s,t,p} = (\text{density} - 1) \times 1000$$

Consequently, an equivalent way of saying the density of a certain sample of water is 1.027, for example, is to say its $\sigma_{s,t,p}$ is 27. The symbols "s, t, p" are intended to signify that the density depends on the salinity, temperature, and pressure. We have seen, however, that oceanographers prefer to compare densities at the same pressure (i.e., atmospheric pressure, for convenience). Hence, the most commonly used parameter to give a water's density is "σ_t" ("sigma tee"), which is defined as $\sigma_{s,t,p}$ at $p = 1$ atmosphere. If, for example, a certain sample has a σ_t of 26.4, then the density of that sample, when measured aboard ship, was 1.0264 g/cm^3.

A.2 Salinity

The salinity is a measure of total dissolved salts in the water, with certain technical and very minor qualifications discussed in Chapter 11. It is measured in grams of dissolved salts per kilogram of water—hence, parts per thousand, denoted by "‰," as contrasted with the familiar "%" for parts per hundred. Generally, seawater has a salinity in the range between 34‰ and 36‰. A better idea of the range and frequencies of various salinities is shown in Figure 15.1. It is seen that the Atlantic is slightly more saline on the average than the other oceans, due primarily to waters received from the Mediterranean sea through the Straits of Gibraltar. These more saline waters are also responsible for the high-salinity bump near 35.0‰ on the curve of Figure 15.1.

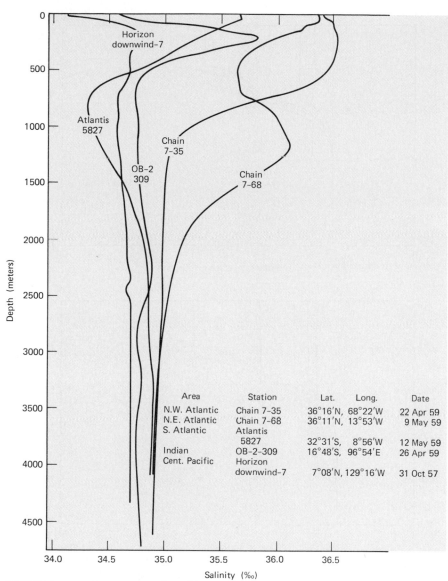

Area	Station	Lat.	Long.	Date
N.W. Atlantic	Chain 7-35	36°16'N,	68°22'W	22 Apr 59
N.E. Atlantic	Chain 7-68	36°11'N,	13°53'W	9 May 59
S. Atlantic	Atlantis 5827	32°31'S,	8°56'W	12 May 59
Indian	OB-2-309	16°48'S,	96°54'E	26 Apr 59
Cent. Pacific	Horizon downwind-7	7°08'N,	129°16'W	31 Oct 57

FIGURE 15.2 *Some examples of how salinity varies with depth for five different oceanic stations. Notice that the two stations with low surface salinity are near the equator where precipitation is high. The others are in the horse latitudes, where evaporation is high. (From Hugh J. McLellan, Elements of Oceanography, Pergamon Press, 1965.)*

A typical plot of salinity vs. depth (Figure 15.2) would show the most rapid variations in salinity are generally within a hundred meters of the surface, where exposure to evaporation or precipitation and subsequent mixing would have the biggest effect on the water's salinity. Regions of rapid change in salinity are called "haloclines." Similarly, a region of rapid temperature change is called a "thermocline," and one of rapid change in density is a "pycnocline."

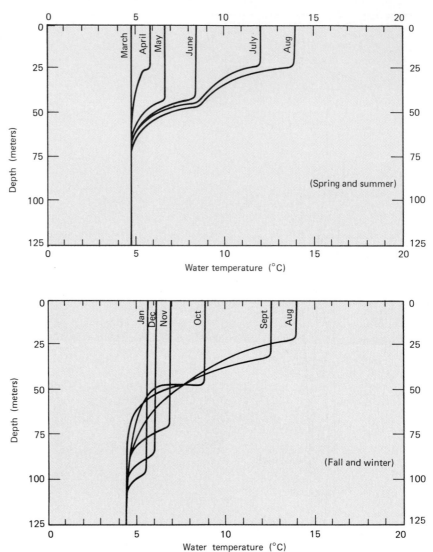

FIGURE 15.3 *Typical seasonal changes in the temperature of near-surface water for temperate latitudes. (top) Notice that the smaller waves of July and August do not mix the surface heat as deeply as did the larger waves of May and June. (bottom) The larger waves of the stormy winter months mix the surface water deeper, reaching about 100 m depth in January. (After A. J. Dodimead, F. Favorite, and T. Hirano, "Review of the Oceanography of the Subarctic Pacific Region," International North Pacific Fisheries Commission, Bulletin 13, 1963.)*

A.3   Temperature

The warm water of the coastal surf tends to give us a false impression of the dominant conditions at sea. Actually, most of the water of the ocean is very cold—93% is below 10°C, and more than 75% is colder than 4°C.

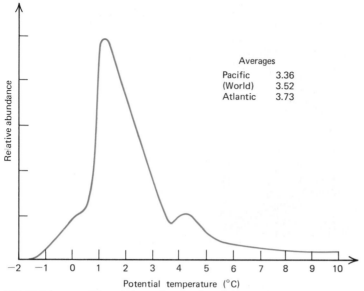

Averages

Pacific	3.36
(World)	3.52
Atlantic	3.73

FIGURE 15.4   *Plot of the relative abundances of ocean waters of various potential temperatures. (After R. B. Montgomery,* Deep Sea Res., **5** *(1958).)*

(For comparison, room temperature is about 23°C.) The exposure of the surface waters to the sun makes them warmer, and the heat is mixed down to 100 m or so. But below this, the temperature decreases rapidly. Above this thermocline, the water temperature changes considerably with the seasons (Figure 15.3), but below it the temperature remains constant. As a result, when we take all waters at all depths into consideration, the world average ocean temperature is found to be about 3½°C (Figure 15.4).

We have seen that the density of water depends on the pressure. So does the temperature. As we bring a water sample up from depths, the reduced pressure causes it to expand slightly, and upon expanding it cools. This "adiabatic" temperature change isn't much, amounting to about 0.1°C per kilometer depth (ranging between 0.04°C and 0.16°C, depending on the initial temperature of the water), but nonetheless it is measurable. In fact, where water is homogenous and well mixed, such as in some trenches, oceanographers find that the temperature increases slightly with depth due to this effect. Faced with a dilemma similar to that faced in measuring densities, oceanographers can either record the temperature of the sample when it gets to the surface, called its "potential temperature," or they can record its *in situ* temperature. Usually the former is chosen, in line with the convention chosen to record densities.

A.4   The T-S Diagram
If the densities of various waters are to be compared at the same pressure, then they depend only on the salinities and the temperatures, and we need only measure the salinity and temperature of a sample to determine its density (Figure 15.5). This is often done with the help of a "T-S diagram" such as that of Figure 15.6. On this diagram, we have located points

FIGURE 15.5 *A salinity-temperature-depth sensor being lowered into the Weddell Sea.*

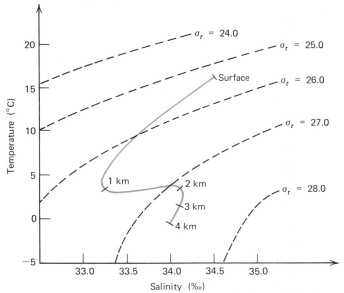

FIGURE 15.6 *An example of how a T-S diagram might appear at some oceanographic station. Deeper water is neither necessarily always colder, nor necessarily always more saline, but it is necessarily always denser.*

of equal density (i.e., equal σ_t) by dashed lines. It is seen that as the salinity of the water increases, so must its temperature, if the density is to remain the same.

At any one oceanographic station, we retrieve water samples from many depths. The temperature and salinity of each sample is measured and then plotted on a T-S diagram to determine its density. In Figure 15.6, the solid line indicates these points on a T-S diagram for samples

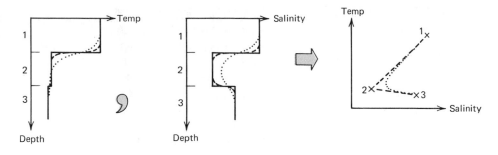

(a) *(b)* *(c)*

FIGURE 15.7 *Beginning with three distinct batches of water (nos. 1, 2, and 3) layered vertically with temperatures and salinities given by the solid
lines in figures (a) and (b), the position of these three batches on the
T-S diagram is shown by the "Xs" in figure (c). As they mix across
the boundaries, the resulting temperature and salinity pattern
would be given by the dashed and then the dotted lines in these
figures.*

retrieved from several different depths at one station. It is seen that the
water in this particular diagram is stable against vertical thermohaline
flow, not because the deeper water is always more saline, nor because the
deeper water is always colder, but rather because the deeper water is
always more dense. By making many such plots of columns of water at
many points at sea, oceanographers can get an idea of where and in what
directions the thermohaline currents are flowing without measuring their
flow directly.

It is interesting to study what happens when various distinct waters
of different temperatures and salinities are placed on top of each other.
In Figure 15.7, for instance, we have three distinct waters arranged vertically as they should be, with the densest on bottom and the least dense
on top. Originally, each would be represented by the appropriate "x" on
the T-S diagram, but as they mix across the interface between layers, they
go to the dashed and then the dotted lines of that figure. Hence, if the T-
S plot from any one station looked like the dotted line in that figure, we
could conclude that below that station were these three distinct waters
with some mixing across the boundaries. Indeed, we do find regions in the
oceans that resemble this (Figure 15.8). Similarly, at stations with different
T-S plots, we can analyze other vertical combinations of different waters
that would make the appropriate curve.

A.5   Formation of Deep and Intermediate Waters

It is seen in the T-S diagram that the curves of constant density are not
straight lines, but rather are concave downward. This is because the thermal expansion of salt water is not linear in temperature. The density of
saltwater is plotted in Figure 15.9 as a function of the temperature. It is
seen that near zero, heating up the water has very little effect on its density

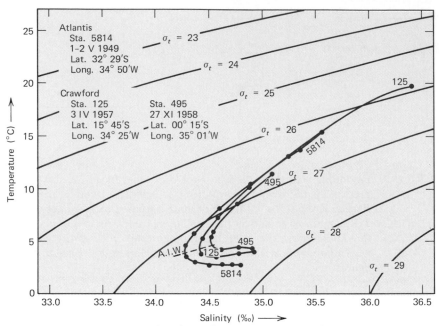

FIGURE 15.8 *T-S curves for three different stations in the South Atlantic, show-ing the layer of Antarctic intermediate water sandwiched between layers with higher salinities. (From Hugh J. McLellan,* Elements of Oceanography, *Pergamon Press, 1965.)*

whereas at higher temperatures the thermal expansion is more pro-nounced. At lower salinities this non-linearity is even more pronounced, and for fresh water the density actually increases with temperature from 0°C to 4.3°C.

When two waters of equal densities, but of different temperatures and salinities, are mixed, the resulting water will be denser than the original waters. This curious effect is illustrated in Figure 15.10. The two dots "A" and "B" represent the temperatures and salinities of the original two waters. The temperature and salinity of the mixture will lie somewhere on the straight line between them; just where depends on the relative amounts of the two waters in the mixture. (For example, if the mixture was ¼ of *A* and ¾ of *B*, then the mixture would lie ¼ of the way toward *A* from *B*.) But notice that every point on this line is denser than the original two batches. Thus, where two or more wind driven surface currents con-verge and mix, there will be a tendency for the mixture to be denser and undergo downward thermohaline flow. This process is referred to as "caballing."

Even if they did not become denser, surface waters would still sink in zones of convergence due to the thrust of the incoming water masses. Similarly, where surface waters diverge, deeper waters must come up as replacement. Consequently, winds not only drive the surface currents, but they may generate deeper currents as well. Major regions of converging and diverging surface waters are shown in Figure 15.11.

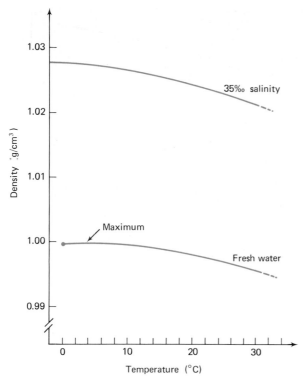

FIGURE 15.9   *Plot of density vs. temperature for fresh and salt water.*

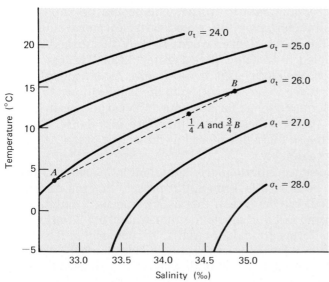

FIGURE 15.10   *When mixing two different waters of the same density (e.g., waters
A and B), the resulting mixture is denser than the original waters.*

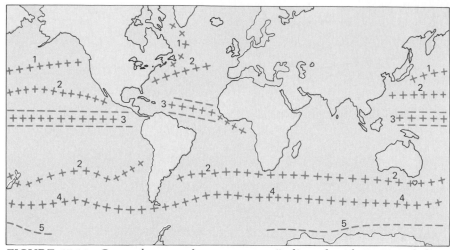

FIGURE 15.11 *General areas of convergence (indicated with "+") and divergence (indicated with "−") in the oceans. They are named: 1, Arctic Convergence; 2, Subtropical Convergence; 3, Equatorial Convergence and Divergence; 4, Antarctic Convergence; and 5, Antarctic Divergence.*

A.6 Formation of Sea Ice and Antarctic Bottom Water

The presence of a shallow halocline is necessary for the formation of sea ice (Figure 15.12). If the water were of all the same salinity, then as the surface water cooled, it would sink to the bottom and be replaced at the surface by the next layer down. Through this mixing mechanism, then, the ocean would have to be brought down to the freezing point throughout its entire depth before it would start to freeze. This would be impossible. With a shallow halocline, however, the cooled surface water would only mix down as far as the saltier denser water, so only the surface water would need to be cooled to freezing. (See Figure 15.13.) The less water that needs to be cooled, the more easily it will freeze. As an example, the Arctic sea ice extends twice as far southward in the Pacific (down to 60°N latitude) as in the Atlantic (only as far as about 75°N). The reason is the surface water of the Pacific is less saline, therefore less dense. Consequently, as it cools it doesn't mix downward as far as that of the Atlantic, enabling it to cool more quickly.

The formation of sea ice is particularly important in the study of thermohaline circulation, because it is through this process that the Antarctic bottom water is formed. We have seen that as ice freezes, it only incorporates about 30% of the salt and ejects the remaining 70%.[2] During the southern winter as ice forms around Antarctica, particularly in the Weddell Sea, the water below it becomes more saline in addition to being cold. This combination makes it the densest of all waters, so it sinks to the bot-

---

[2]There are large variations in this figure, depending on the conditions under which the ice freezes.

FIGURE 15.12 *An ice breaker and its ice reconnaissance helicopter leading supply ships near the Palmer Penninsula, Antarctica.*

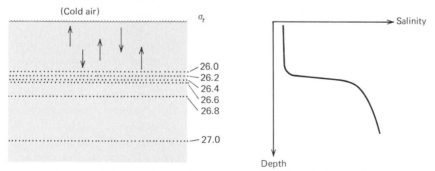

FIGURE 15.13 *The cooling surface water will sink relative to the warmer water
of the same salinity below. If there is a shallow halocline, then the
cooling surface water only mixes down to the denser water below.
Thus, only the surface water need cool (not the entire ocean) in
order to freeze over.*

tom of the ocean, and stays at the bottom for a long period of time.

The same process makes dense bottom water in the Arctic regions
during the northern winter, but as the Arctic Ocean is mediterranean,
being surrounded by land and shallow sills, this water doesn't make it out
of the Arctic Ocean basin, and doesn't show up in our oceans.

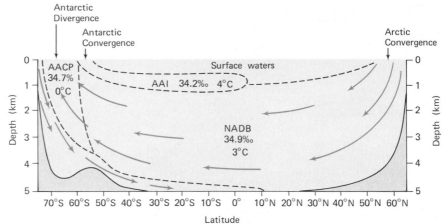

FIGURE 15.14   *Vertical cross section of the Atlantic Ocean from Antarctica to Greenland, showing the locations of the various water masses. Labels are as follows: AAB = Antarctic Bottom Water, AACP = Antarctic Circumpolar Water, AAI = Antarctic Intermediate Water, NADB = North Atlantic Deep and Bottom Water.*

A.7   Survey of Major Water Masses

In temperate and tropical latitudes, surface waters tend to be quite warm and fairly salty. Typical temperatures for these water masses range from 8°C to 18°C, and salinities are typically in the range between 34.5‰ and 36‰. The high temperatures make these water masses considerably less dense than other oceanic water masses, so they remain at the surface, seldom reaching below 1 km in depth. Because of their contact with atmosphere and sunshine, their properties change considerably from place to place and from season to season.

The vertical mixing of the surface waters, due to their interaction with the sun and atmosphere, is seasonal at higher latitudes but not at lower latitudes. Near the equator, in seasons when evaporation is large, the surface becomes more saline, but also warmer when compared to seasons where precipitation is large. Consequently, the density of the surface water doesn't change much. At higher latitudes, however, precipitation is largest when it is warmest. Thus, the water becomes both less saline and warmer at the same time. There is considerable change in the surface density between summer and winter at higher latitudes then, and so the amount of vertical mixing of surface waters is strongly dependent on the season.

By contrast, the characteristics of each of the deeper water masses (Figures 15.14 and 15.15) remain quite constant over large distances and long time periods. These masses form at or near the surface, and then sink, carrying with them the characteristics acquired while they were forming. At depth, they are insulated from the capricious surface environment, so they remain pretty much the same as when they were formed. There may be some slow slight changes due to slow mixing with neighboring water masses, respiration of organisms, hydrothermal activity, or heat received

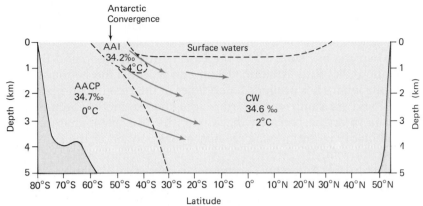

FIGURE 15.15 *Vertical cross section of the eastern Pacific Ocean from Antarctica to the Aleutian Islands, showing the locations of the various water masses. Labels are as follows: AACP = Antarctic Circumpolar Water, AAI = Antarctic Intermediate Water, CW = Common Water.*

directly from the ocean bottom. However, the deep water masses are so voluminous that these processes have very little effect overall.

All oceans are interconnected at their southern ends. In the region north of Antarctica, from about 60° to 45° South Latitude, the prevailing westerlies drive the "Antarctic Circumpolar Current" eastward completely around the globe (see Figure 14.1), uninterrupted by any land mass.[3] The water in this current is called "Antarctic Circumpolar Water," and it is cold, dense, and quite deep. Its temperature is around 0°C, its salinity is around 34.7‰ and because of its high density it extends all the way down to the ocean bottom in the Indian and Pacific oceans, but not in the Atlantic, where the denser "Antarctic Bottom Water" stays beneath it.

The Antarctic Bottom Water is the coldest and densest of all water masses in our oceans. It forms beneath the freezing ice of the Weddell Sea in the Southern winter, and has a temperature of about −1°C and salinity of about 34.6‰. It flows downward and northward along the Atlantic Ocean bottom, reaching beyond the equator. It is fairly well confined to the Atlantic Ocean by the ocean bottom topography. Similar water masses form beneath the northern ice cap in the Northern Hemisphere's winter, but these waters do not reach the major oceans due to obstruction by continents and sills.

On Figure 14.1 you can see that there are converging surface currents in the North Atlantic near Greenland. Winter cooling and evaporation make the waters produced by convergence in this region quite cool, salty, and dense. This region is the most prolific producer of deep water in all

---

[3]*It does get "squeezed" a bit however in passing between Cape Horn and the Antarctic Peninsula.*

oceans. The water produced here, called the "North Atlantic Deep and Bottom Water," has a temperature and salinity of about 3°C and 35‰, respectively. This makes it less dense than the Antarctic Bottom water, but roughly the same density as the Antarctic Circumpolar Water.

As you can see in Figure 15.14, the North Atlantic Deep and Bottom Water fills the bulk of the Atlantic Ocean Basin. It fills the bottom of the North Atlantic where it is produced, and flows at intermediate depths past the equator and on south until it runs into the Antarctic Circumpolar Water. At the interface between these two, the two water masses mix, aided undoubtedly by turbulence created by the motion of the Antarctic Circumpolar Current.

This mixture of Antarctic Circumpolar Water and North Atlantic Deep and Bottom Water is called "Common Water." It is carried into the Indian and Pacific Oceans by the Antarctic Circumpolar Current, where it fills the major portion of these two oceans. You see, the winter cooling and evaporation of converging water masses near Greenland are such prolific producers of deep waters that these waters not only fill most of the Atlantic Ocean, but through mixing and transport by the Antarctic Circumpolar Current, they fill most of the other two oceans as well.

One further important water mass that appears in all three oceans is the "Antarctic Intermediate Water." As the Antarctic Circumpolar Current flows eastward, it experiences Coriolis deflection toward the north. This provides a major impetus for the convergence of surface waters north of this circumpolar current. (You may wish to look at Figure 14.1 and see if you can identify any other factors contributing to this convergence.) The Antarctic Intermediate Water formed at these convergences has temperature and salinity around 4°C and 34.2‰, respectively, which means that it is not as dense as the North Atlantic Deep and Bottom Water or the Common Water. Consequently, it flows above these, but beneath the surface waters.

The net flow speed of the deep water masses is usually very slow. Typical waters take several hundred or even a thousand years to flow 10,000 km (¼ of the way around the earth). This means that the Coriolis deflection is very pronounced. In the Southern Atlantic, the northward-flowing Antarctic Bottom Water tends to bank up on the western (left-hand) ocean margin, flowing beneath the swift surface current, but in the opposite direction. Likewise in the Northern Atlantic, the southward flowing North Atlantic Deep and Bottom water tends to be intensified on the western (right-hand) ocean margin, flowing southward beneath the northward-flowing Gulf Stream at the surface.

B. TIDAL CURRENTS

We have seen so far that the net drift of most currents, especially deeper currents, is relatively slow. However, if one were to measure their speeds at any one instant, they would be found surprisingly high. How can this be?

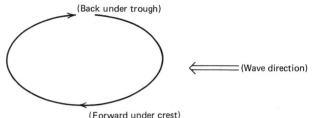

FIGURE 15.16 *Motion of a water molecule under a tidal wave as seen from the top. Width and breadth of the orbit would be several kilometers, but the height (in or out of this page) would be only about 70 cm (Northern Hemisphere).*

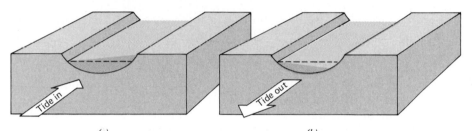

(a)   *(b)*

FIGURE 15.17 *Cross section of an estuary, showing the flow of the tide in and out of it (Northern Hemisphere). (a) As the tide flows in, it banks up on the right side of the estuary, due to the Coriolis deflection. (b) As it flows out, it banks up on the other side, and the overall water level at the mouth must be lower than when it flows in. This makes the tidal variations larger on the right side than on the left.*

The solution to this apparent paradox is that the tides cause significant motion of water molecules. However, this motion is cyclical, and the water molecules end up where they started from. The tides are long, and everywhere they are very shallow water waves. The presence of the bottom squashes the would-be circular orbitals into very long ellipses, being typically 70 cm high and as much as 8 km long.

As with any other wave, when the tide crest passes overhead, the water molecules are moving forward in their orbitals, and when a trough is overhead, they are moving back. But since the tidal motion is large, the Coriolis deflection is significant. So when following the water motion from above, one would notice continual deflection to the right (Northern Hemisphere), traversing an ellipse and ending up where it starts from (Figure 15.16). Consequently, if an oceanographer wishes to measure the net drift velocity of a certain current by placing something in it and following its motion, it must be followed for several days so the effect of the tidal current can be sifted out.

When the tides enter an estuary along a coast in the Northern Hemisphere, the Coriolis deflection banks them up on the right side (facing into the estuary), and it banks them up against the other side as it leaves. This causes both the high tide to be higher and the low tide to be lower on the right-hand side of the estuary (facing in) than on the left. (See Figure 15.17.) In the Southern Hemisphere, of course, the opposite is the case.

TABLE 15.1 Current Measurement Techniques

1. Direct	(1*a*) Eulerian (fixed to solid earth, measuring water go by)
	(1*b*) Lagrangian (fixed in water, measuring solid earth go by)
2. Indirect	(2*a*) Geostrophic (measuring slope of density contours due to current flow)
	(2*b*) Electromagnetic (measuring deflection of ions flowing through the earth's magnetic field)

C. CURRENT MEASURING TECHNIQUES

In addition to helping satisfy our curiosities, current measurement also has some commercial value. Helping us understand the weather, protecting ships and piers from seiches, tracking ice bergs and fish larvae, and finding good shipping lanes are among the things of immediate economic interest. Of course, as is the case with most scientific research, how society can best use it will best be determined in retrospect rather than prospect.

Current measuring techniques can be put into two general classes: direct and indirect. Each of these, in turn, can be split into two subclasses (see Table 15.1). The direct techniques would either have an apparatus fixed to the solid earth, measuring the water go by ("Eulerian"), or would have an apparatus fixed in the water, measuring the solid earth go by ("Lagrangian"). An example of the former might be some floating device, tied to a weight at the ocean bottom by a wire of appropriate length, depending on the depth of the current to be measured (Figure 15.18). The water speed might be measured by the speed with which a propeller is turned, and the current direction might be detected by comparing the direction of a weathervane-type pointer to a compass needle. An example of a Lagrangian device would be a float of just the right buoyancy to float at the desired depth. It would be deposited one day and then retrieved perhaps several weeks or months later. Of course, it would have to have some signal-generating device so it could be found. The ocean current would then be known by how far and in what direction the float had gone.

Of the two subdivisions of the class of "indirect" current measurement techniques, one is already known. We have already studied (in the preceding chapter) how we can deduce a current's velocity and direction by measuring the slope of the density gradients and knowing the latitude. The other subclass involves electricity and magnetism. When charged particles move through a magnetic field they get deflected—opposite charges in opposite directions. Ocean currents carry many charged particles—the salt ions—that are moving through the earth's magnetic field. When two metal plates are put into this current (e.g., behind a boat), the positive ions will go toward one plate and the negative ions toward the other. By measuring the electrical voltage thus generated between these two plates, and the orientation of the plates for maximum voltage, oceanographers can tell the speed and direction of the current.

FIGURE 15.18   *Current measuring devices ready for launching.*

D.   SUMMARY

The deeper ocean currents are driven primarily by gravity. Denser waters tend to sink and flow downhill, displacing lighter waters upward. The density of seawater depends on the pressure, salinity, and temperature, increasing with increased pressure and salinity, and decreasing with increased temperature. For comparative purposes, densities are all measured at the same pressure, and are recorded in terms of a quantity called "σ_t."

Once the salinity and temperature of a sample have been measured, it can be represented by a point on a T-S diagram. By plotting on a T-S diagram the characteristics of the waters in a vertical column beneath an oceanographic station, we can determine the characteristics of the various water masses, and the amount of mixing that has occurred.

When surface waters of similar densities, but different temperatures and salinities mix, the resulting water tends to be denser than either of the parent masses. Deep and intermediate waters tend to form at the convergence of surface waters, aided by winter cooling and evaporation. Bottom water is formed beneath freezing polar ice in wintertime. Antarctic

Bottom water is the densest water of all, but is fairly confined to the Atlantic Ocean by bottom topography.

The Antarctic Circumpolar Current communicates with all oceans. The North Atlantic Deep and Bottom Water, prolifically produced at a convergence near Greenland, mixes with Antarctic Circumpolar Water to form Common Water, which fills the bulk of the Pacific and Indian Ocean Basins. Antarctic Intermediate Water is formed in all oceans at the Antarctic Convergence.

The gravitational attraction of the sun and moon cause tidal currents. Due to Coriolis deflection, these oscillatory currents flow in elliptical patterns when viewed from above.

Currents may be measured directly either by observing the motion of water relative to the solid earth, or by observing the motion of the solid earth relative to the water. Indirect techniques include measuring the slope of the density contours, and measuring the deflection of salt ions moving through the earth's magnetic field.

QUESTIONS FOR CHAPTER 15

1. What drives the thermohaline currents? Why are they called "thermohaline"?

2. How rapidly does pressure increase with depth in the ocean?

3. How deep must one go before the water is compressed enough that its density is increased by 1%? What is the pressure at this depth in terms of atmospheric pressure (1 atm = 14.7 lb/sq. in.)?

4. In studying thermohaline flow, why does one need only to study the density of various samples at *one* pressure? What pressure is chosen, and why?

5. How does the density of seawater compare to that of fresh water?

6. What is $\sigma_{s,t,p}$? What is σ_t?

7. What is the typical range of salinities for seawater?

8. Why are the largest variations in salinity found near the surface?

9. What name is given to a region showing rapid change of salinity with depth? How about a rapid change in temperature? In density?

10. Explain the general features of the plot of water temperature vs. depth in temperate latitudes for different months, as in Figure 15.3. As the surface water temperature cools in November, December, and January, why do you think the thermocline gets deeper? (*Hint*: The constant temperature of the near-surface waters may be due to mixing. What mixes the surface waters, and why might this mechanism mix the waters deeper in the winter months?)

11. About what is the average temperature of the ocean water? Why is it so much colder than surface temperatures?

12. All ocean waters form their characteristics (such as temperature) when they are at the surface. Since the deep and intermediate waters are so cold, what conclusion can be drawn regarding the latitudes in which they are formed?

13. Explain why the "potential" temperature of a sample would be different from its *in situ* temperature. Which would be cooler, and why?

14. On the plot of Figure 15.6, explain why the water at each of the depths indicated is always more dense than the water above it. Is it always due to cooler temperature? Is it always due to higher salinity?

15. Sketch some T-S plot for a vertical water column. Then indicate how many different water masses were involved in this result, and what their individual characteristics were before mixing at the interfaces.

16. When one says that thermal expansion is "not linear in temperature," what does that mean?

17. How does the thermal expansion for salt water differ qualitatively from the thermal expansion for fresh water? (For example, do they both start out as liquids at the same temperature? Do they both always increase in volume with increased temperature?)

18. Explain why it is that when various surface waters converge and mix, the resulting water often is denser than any of the surface waters that went into the original mixture. (*Hint:* Assume that surface waters are all of roughly the same density, though differing in salinity and temperature, and then show on a T-S plot what the result is when two such waters are mixed.)

19. Besides the mixing of surface waters, how else can the surface winds cause deeper flow in the oceans?

20. Why is North Atlantic deep water (formed at the intersection of the Gulf Stream, Labrador Current, and East Greenland Current) rich in oxygen and poor in nutrients?

21. Why is a shallow halocline necessary for the formation of sea ice?

22. How does Antarctic bottom water form, and why is it so dense?

23. Why doesn't Arctic bottom water make its presence felt in our oceans like Antarctic bottom water does?

24. We have seen that the surface currents are most intense on the western margins of the oceans. Are the deeper currents more intense on one margin? Which margin? Why? Is it the same margin in both hemispheres?

25. Are the deeper currents intensified on the same margin as the surface currents? Do they flow in the same direction?

26. Why is there some vertical mixing of the surface waters in lower

latitudes, that is, mixing somewhat deeper than what waves alone could accomplish?

27. Explain why the amount of vertical mixing of the surface waters in temperate latitudes is so strongly dependent on the seasons.

28. Why is it that if one measures the speed of a deep water current at any one instant, one will find it to be considerably faster than the net speed of the current averaged over a long time?

29. What is a typical height and length of the orbit of a water molecule under a tidal wave (i.e., a tide)? Why is it so squashed?

30. With the help of the Coriolis effect, explain why the orbit of a water molecule in a tidal wave (i.e., a tide) is an ellipse oriented almost horizontally rather than vertically. Why isn't it the same for smaller waves?

31. Explain why one side of an estuary has larger tidal fluctuations than the other side.

32. Can you think of some immediate practical benefits that measuring and understanding ocean currents has?

33. What is the difference between "direct" and "indirect" techniques for measuring ocean currents? Give an example of each.

34. What are the two subdivisions of direct current measuring techniques? Give an example of each.

35. What are the two subdivisions of indirect current measuring techniques? Explain how each works.

*36. Why do you suppose the average salinity of the Atlantic is slightly higher than that of the other two oceans?

*37. A water sample is 1.8°C at its position on the ocean bottom, 4 km down. It is quickly brought to the surface, and aboard ship its temperature is measured again. About what would it be this time?

*38. Sketch a T-S diagram for any vertical column of water that you wish and that you think might represent a real column somewhere in the oceans. Dividing the column into three parts—surface, intermediate, and deep waters—discuss where each of these three waters must have formed to display the characteristics they have on your plot. Is your water column stable against vertical thermohaline flow?

*39. Repeat what was done in Figure 15.7, except for three different batches of water with different characteristics. (Yours needn't be stable against vertical thermohaline flow, so you might have denser waters on top, if you wish.) Indicate with "x" where each of your three water masses would be on a T-S plot initially, and then sketch what the T-S plot would look like after the water masses had mixed slightly across their interfaces.

*40. From what you know of the trade winds, and of the direction of the Coriolis effect in the Northern and Southern hemispheres, explain why you might expect divergences to appear just north and south of the equator, as in Figure 15.11. Why wouldn't there be a divergence right on the equator? (*Hint:* Is there any Coriolis deflection on the equator?)

*41. Given that there are divergences just north and south of the equator, why would you then expect to find a convergence on the equator? (*Hint:* You know that at a divergence, deeper water must flow upward to replace the exiting surface waters. What must be happening between two regions where water is flowing upward? Sketch the circulation cells in a cross section, if helpful.)

*42. If common water is formed from a mixture of Antarctic Circumpolar Water and North Atlantic Deep and Bottom Water, shouldn't its characteristics be intermediate between those of its two parents? Are they? If not, could you offer an explanation?

*43. When deep water is well mixed, its temperature rises slightly with depth. Why do you suppose this is?

*44. Why does bottom topography restrict the flow of Antarctic Bottom Water more than other deep water masses?

SUGGESTIONS FOR FURTHER READING

1. Laurence Armi, "Mixing in the Deep Ocean—The Importance of Boundaries," *Oceanus*, **21,** No. 1 (Winter 1978), p. 14.

2. D. James Baker, Jr., "Currents, Fronts, and Bottom Water," *Oceanus*, **18,** No. 4 (1975), p. 8.

3. Henry Stommel, "The Circulation of the Abyss," *Scientific American* (July 1958).

16
SEAS AND ESTUARIES

New York Harbor. (Skyviews)

The coastal interaction between continent and ocean is frequently moderated by intermediary water masses of varying size, which have somewhat restricted communication with the open ocean. The larger intermediary bodies of water are referred to as "seas," and the smaller ones as "estuaries." The greater portion of the United States coastline, for example, incorporates such intermediary water masses, including nearly 20% of the West Coast, 80% of the East Coast, and all of the Gulf Coast, of course.

Due to their smaller size these marginal waters tend to be more heavily influenced by interactions with the neighboring continent than do oceanic waters. Daily and seasonal changes in temperature, salinity, circulation, and biological activities are much more pronounced.

Seas and estuaries come in a wide variety of shapes and sizes, each with its own peculiarities. In this chapter we will present some general patterns, basic underlying processes, and idealized models for these marginal bodies of water, with the hope that these generalities will help the student better understand any particular sea or estuary of interest.

A. SEAS

Compared to the open ocean, most seas are somewhat shallower and have more protected waters. Most are large enough that they have their own characteristic pattern of wind-induced surface currents, reflecting the local dominant wind patterns. Some, such as the Gulf of Mexico and the Caribbean Sea, communicate well enough with the open ocean that the dominant surface currents are driven by ocean processes. Others, like the Black Sea, are so isolated that the ocean exerts no direct influence at all.

Because of the variety of influences exerted by winds and neighboring continents and oceans, there are as many different surface water properties and motions as there are different seas. Superimposed on these particular local patterns, however, are some general patterns influencing all seas, and involving the deeper water masses as well. The general pattern in each sea depends on whether or not the fresh water removed through evaporation exceeds the fresh water received through precipitation. We intend "precipitation" to include precipitation received both directly and as runoff from the nearby land mass.

A.1 Where Evaporation Exceeds Precipitation

Where evaporation is greater than precipitation, there is considerable vertical mixing of the waters. The surface water becoming more saline will sink, carrying oxygen that it acquired at the surface with it. (See Figure 16.1.) The oxygen fosters more animal life at depth than would be present otherwise. This water will then flow along the bottom of the

419

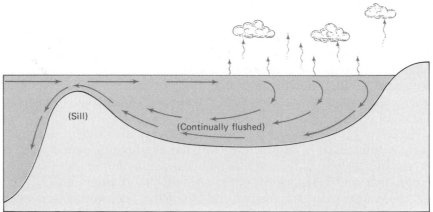

FIGURE 16.1   *Where there is large evaporation, surface water becomes saline and sinks, continually flushing out the bottom of the sea.*

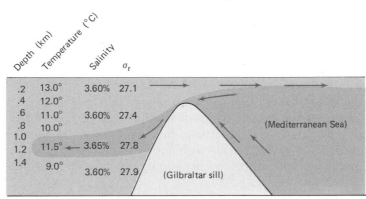

FIGURE 16.2   *Spilling over the Gibraltar sill, Mediterranean water finds equilibrium about 1.2 km down in the Atlantic. At this depth it is both warmer and saltier than the surrounding waters of the Atlantic. (After P. H. Kuenen, Realms of Water, Wiley, 1963.)*

sea and out into the neighboring ocean. Often, a sea is separated from the ocean by a shallow sill. In this case, the sunken saline water fills up the basin behind the sill to sill depth, and then it spills over the sill and out into the neighboring ocean, being replenished on the surface by incoming, lighter ocean water, which is then subjected to evaporation, increased salinity, sinking, and so on, and so goes the cycle. It is seen that such seas are continually flushing out their bottom water. The average length of time the water stays in the sea, between entering at the surface and being flushed back out from the bottom, is called the "residence time," a term we encountered in a slightly different context in a previous chapter.

The Mediterranean is an example of this kind of sea, having a residence time of roughly 80 years. Although the flushed out bottom water is more saline than the waters of the Atlantic, it is also warmer, so its density is not that much different, going down to an equilibrium depth of roughly 1.2 km after spilling out over the Gibraltar sill (Figure 16.2). It is mostly

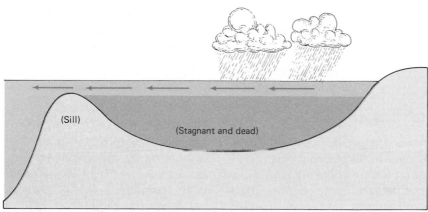

FIGURE 16.3 *Where precipitation is large, the surface water remains fresher and lighter than deeper water, and there is little vertical mixing. Once the oxygen is depleted from the deeper water, it is not replenished from the surface, and so it is lifeless.*

due to this contribution from the Mediterranean that both the mean temperature and the mean salinity of the Atlantic are slightly higher than those of the other two oceans. During the Second World War, German submarines made use of the two-way currents to slip in and out of the Mediterranean Sea, undetected by the allied forces on Gibraltar. They could shut off their engines and float just under the surface, entering the Mediterranean with the surface current, or when leaving, they could float a little deeper with the outward flowing current.

A.2 Where Precipitation Exceeds Evaporation

A different situation exists, however, in seas where precipitation exceeds evaporation. The lighter fresher water from precipitation remains on the surface. If the sea is separated from the main ocean by a sill, then the denser bottom water is stagnant and dead. It is dead because once animal life has used up the oxygen, it does not get replenished by more from the surface. Furthermore, beyond a depth of a hundred meters or so, it does not get the sunlight needed for the survival of plant life, which could otherwise manufacture oxygen. Hence, life is stopped. The remains of surface life, falling into this water, will only decay incompletely. The situation is depicted in Figure 16.3.

Examples of this type of sea include the Black Sea, and to some extent, the Arctic Ocean. Because of a small amount of mixing with the deeper salt water, the surface water of the Black Sea is not completely fresh when it leaves through the Bosporus and Dardanelles, but rather its salinity is about 20‰. Replacing this small outflow at the surface is a subsurface salt water inflow over the sill at a depth of a few tens of meters. This inflow is small compared to the volume of the Black Sea, so the residence time of the deep water is nearly 2000 years. Being funneled through these narrow straits, the current velocity is fairly swift. In fact, a few centuries ago fishermen in the area would drift toward the Mediterranean with the surface water, and when wishing to return home, they would simply lower their nets into the deeper currents, which dragged them back.

B.   ESTUARIES

Estuaries are smaller and shallower than seas. They are generally less influenced by large-scale oceanic processes and more influenced by processes on the neighboring land mass, such as seasonal variations in fresh water runoff and temperatures. Because of their smaller size they are also more influenced by the products of local human populations, such as excavation, construction, and pollutants.

In estuaries, fresh water runoff from the land mixes with salt water from the oceans. The "head" of an estuary is where the fresh water river or stream enters, and the "mouth" is where it communicates with the ocean. An estuary may have more than one head and more than one mouth.

Estuaries are influenced both by oceanic processes, such as tides and salt water intrusion, and by the fresh water flowing in from the streams. But there are wide variations in the relative influence of these two factors. On one hand, clean stream-cut channels, such as the mouths of the Mississippi, Congo, and Columbia Rivers are dominated by the forceful river current. Oceanic influence is overshadowed by the seaward flow of fresh water.

On the other hand, some estuaries may have no inflowing streams at all, or the stream may dry up during some seasons. In this case, the fresh waters enter through the slow flow of ground water only. Also in some estuaries, bay mouth bars may cut off direct communication with the ocean during some seasons, so that the interaction is limited to waters flowing through the sand (Figure 16.4).

B.1   Flooded Coastal Plains

The East Coast of the United States is dominated by estuaries characteristic of flooded coastal plains (see Figure 6.12). During the last ice age, sea level was about 135 m below its present level, and the present East Coast was then inland on a long flat coastal plain extending hundreds of miles farther east. Rivers cut valleys across these plains and carried their sediment load out to the edge of the continental shelf where they entered the ocean.

As the ice melted back, sea level rose and the coastal plains began to be flooded. As the coastline moved inland, the wave motion brought some of the sediment with it through processes described in Chapter 9. The rising waters extended farther up the low-lying river valleys, creating the broad drowned river valleys that we find there today (Figure 16.5).

The combination of relict sediments brought up with these rising

FIGURE 16.4 *Sometimes baymouth bars may close off the mouths of estuaries completely, as is the case for these estuaries along the south shore of Martha's Vinyard.*

waters and new sediments created by ongoing processes was sufficient for extensive bay mouth bars and barrier beaches to form in these relatively shallow coastal regions. The longshore current flows southward along the East Coast, driven by counterclockwise eddies from the Gulf Stream, and the flow patterns are reflected in the cuspate shapes of many of these barrier beaches (Figure 16.6).

The waters behind barrier beaches are called "lagoons." Because barrier beaches form in shallow water, lagoons are shallow, which means that they have a large surface area in comparison to their volume. As a result, they heat up quickly, cool down quickly, and their salinities vary greatly with changes in the weather. For example, evaporation is so heavy on the Gulf Coast of Texas, that fresh water evaporates from Laguna Madre (Figure 16.7) faster than it can be replaced by inflowing streams. The result is that it is actually *more* saline than seawater, with a typical range of 40 to 50‰. Salinities as high as 110‰ have been recorded there, and after rainstorms they have been as low as 2‰. Laguna Madre is exceptionally shallow, even for a lagoon, so these variations are more extreme than would be possible for most. Nonetheless, it does demonstrate the general feature of lagoons that their large surface area and small depth make their water properties quite variable.

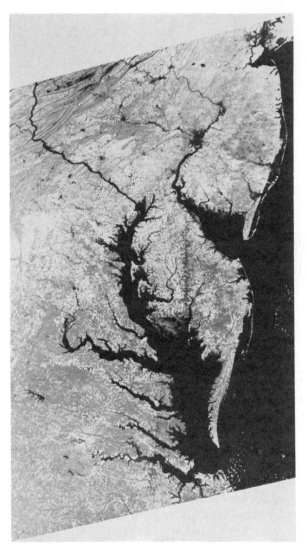

FIGURE 16.5 *High altitude photo showing some drowned river valleys along the East Coast of the U.S., including Deleward Bay (above) and several included in the Chesapeake Bay system.*

The barrier beaches are occasionally interrupted by tidal inlets that may come and go periodically as changes in tides or current patterns change the need for communication with the open ocean. On the East Coast, the tidal ranges are relatively small, so tidal inlets are rather few and far between (see Figure 16.6). This further protects lagoons from oceanic influences. Wind-induced currents and seiches frequently cause greater changes in sea level along lagoon shores than do the tides. Extensive salt marshes, alternately exposed and submerged during the tidal

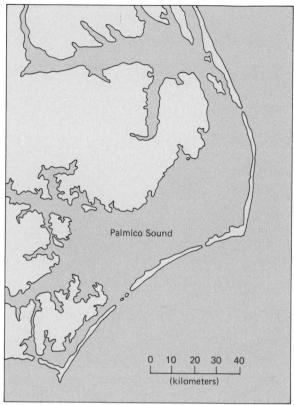

FIGURE 16.6   *Palmico Sound on the North Carolina coast. The large fresh water input from rivers maintains the salinity of the water in the sound roughly in the range of 15‰ to 20‰.*

cycle, support heavy biological activity based on high plant productivities in these nutrient-rich areas.

Often, the presence of barrier beaches means that drowned river valleys may empty into lagoons rather than directly into the ocean. This is quite common along flooded coastal plains. One such system is illustrated in Figure 16.6.

B.2   Mountainous Coasts

Mountainous coasts, such as the North American West Coast, offer different kinds of estuaries. There are few large rivers discharging into these coastal waters because the mountain ranges would block their seaward flow. Furthermore, because of air flow patterns described in Chapter 13, the eastern sides of these ranges are quite arid, so there wouldn't be much river discharge even if the mountains didn't block their flow.

Unlike flooded coastal plains, mountainous coasts are rough, irregular, have coastal waters that often get very deep very quickly, and lack large amounts of sediment influx from large rivers. These conditions pre-

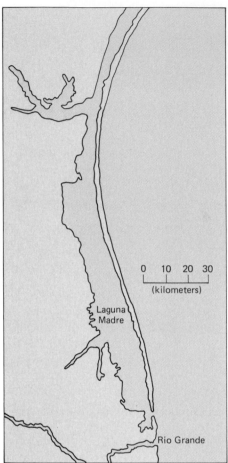

FIGURE 16.7   *Laguna Madre in southern Texas. Because there is virtually no fresh water input, and because the rate of evaporation is high, most of the water has a salinity in the range of 40‰ to 50‰.*

vent the formation of extensive barrier beaches as are common on the East Coast. Bays are relatively narrow, corresponding to the irregular topography. Although bay mouth bars frequently form (Figure 16.8), the relatively small size of these bays and the larger tidal range means that they are much more heavily influenced by the tidal cycle than are the large East Coast lagoons. The tidal range in many West Coast estuaries is sufficient that large marshy regions, called "sloughs" (Figure 6.6), are alternately inundated and exposed as the tide rises and falls.

The mountain building processes sometimes cause blocks of crustal material to fall below sea level. San Francisco Bay (Figure 16.9) and the Strait of Juan de Fuca in Northern Washington are examples of such tectonic-formed estuaries. These are generally deeper than lagoons, and can accommodate ocean-going ships without extensive dredging.

Glacially formed embayments, called "fjords," are common along mountainous coasts at higher latitudes (Figure 16.10). The slow but forceful and incessant flow of glaciers over the ages have carved clean and

FIGURE 16.8 Aerial photo of a bay on the mountainous West Coast of the U.S. Their rather small size, small fresh water inflow, and large tidal range means that these bays are more heavily influenced by the tidal cycle than are those on the East Coast.

FIGURE 16.9 San Francisco Bay is an example of a bay formed by the sinking of crustal material caused by tectonic processes.

FIGURE 16.10 *The steep walls of this fjord are typical of glacially-cut valleys.*

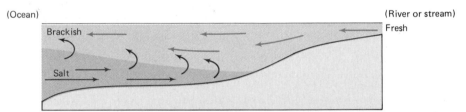

FIGURE 16.11 *Cross section of a typical estuary, showing the salt water intruding in a wedge along the bottom, mixing with the fresh water above, and then the mixture exiting at the surface.*

impressive U-shaped valleys down through the mountains leading to the coast. These have melted back somewhat since the last ice age, and the rising ocean has moved in leaving these magnificent steep-walled estuaries.

B.3   General Circulation Patterns

If there were no tides, we would expect a stable current pattern in estuaries similar to that illustrated in Figure 16.11. The fresh water from the stream would flow out along the surface. Denser salt water would flow in along the bottom, mix with the fresh water above, and then flow out with the fresh water at the surface in a somewhat diluted form.

In most estuaries, however, this simple salt wedge picture is considerably modified by a number of factors. One of these is the tides. Rising and falling twice a day, flood tides entering the estuary and ebb tides leaving cause the interface between salt and fresh waters to move rhythmically back and forth along the length of the estuary. Actually, mixing across the interface between the two waters causes the transition between the two to be gradual rather than abrupt.

This mixing is caused by a variety of factors. The two water masses moving in opposite directions create turbulence across their interface, which speeds up mixing. The mixing is greatly enhanced in estuaries with large tidal ranges, where large tidal currents induce large-scale turbulence and mixing. In broad, shallow estuaries, such as lagoons, wind can create currents and waves that help mix the waters, and heavy surface evaporation can cause further mixing.

In most estuaries, the mixture of waters exiting the estuary's mouth at the surface is nearly as salty as the ocean waters. The large amount of salt water in this mixture indicates the rate of flow of salt water entering the estuary in the salt wedge is considerable larger than the rate of flow of fresh water from the river or stream.

Because there is a net flow of fresh water into the estuary at the head, there must be a net flow out of the estuary at the mouth. Of course, currents flow both ways through the mouth as the tides come and go, but on the average, slightly more must flow out with the ebb tide than flows in with the flood tide. Thus, the two parts of the tidal cycle are not quite symmetrical. Ebb tide usually lasts slightly longer than flood tide.

Because large tidal currents and wind- or weather-induced surface effects tend to create turbulence and mixing, the ideal salt-wedge model for currents in estuaries is often severely modified in reality. The salt-wedge model is usually most accurate in deep stream-cut channels where tidal influences are overshadowed by the fresh water flow, and surface effects are minimized by the great depth. Nonetheless, the basic salt-wedge idea should underly the analysis of patterns in most estuaries, even when large modifications must be made to accommodate tidal and surface effects.

Some estuaries have submarine sills that isolate their bottom waters from communication with the ocean. Many fjords are like this because the glacier that carved the fjord dumped its sediment load at the mouth of the valley where the glacier met the ocean. In these estuaries, the sill prevents the salt water wedge intrusion. Dense water fills the basin behind the sill, but does not get flushed with the tidal cycle as it does in most estuaries. Of course, there is some mixing with the fresh water flowing out along the surface, which slowly removes the denser deeper water. This gets replaced by a corresponding small flow of salt water in over the sill. But usually this is a very slow process, and the turnover time for the water trapped behind the sill is quite long. For the most part it can be considered to be stagnant and dead. It is usually anoxic, because it is too deep for plantlife to replace the oxygen that has been depleted through respiration and decomposition of detritus from the surface. The situation is a smaller-scale analogue of that illustrated for some seas in Figure 16.3.

B.4   Classification of Estuaries

The "tidal volume" or "tidal prism" of an estuary is the volume of water between the high and low tide surfaces (Figure 16.12). It corresponds to

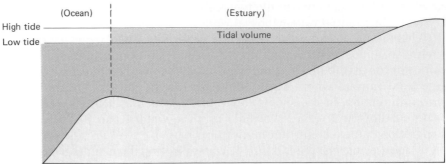

FIGURE 16.12 *The tidal volume, or "tidal prism," of an estuary is the volume of water contained between high and low tide levels.*

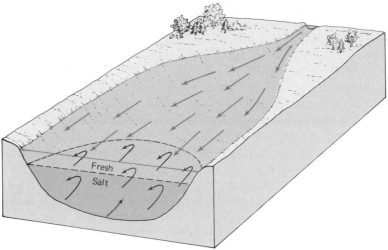

FIGURE 16.13 *Type I estuary. Large fresh water flow resists the intrusion of the salt water wedge, causing a rather steep interface.*

the volume of water entering and leaving the estuary during any one tidal cycle. Since the size of the tides vary during the month, so does the tidal volume.

Estuaries are sometimes classified according to how the tidal volume compares with the rate of fresh water influx. This is essentially a measure of how strongly they are influenced by tidal effects. The scheme uses idealized conditions that are seldom met in reality, as discussed previously. Nonetheless, it does provide basic guidelines for understanding similarities and differences among various estuaries. When the fresh water input is much greater than the tidal volume, the estuary is classified as "type I." Type I estuaries are generally relatively deep, and the fresh water flow is sufficient to maintain a steep interface with the intruding salt water wedge (Figure 16.13). The mouths of large rivers, such as the Columbia, Mississippi, and Congo, generally fall into this class.

When the tidal volume is comparable to or somewhat larger than the fresh water inflow, the estuary is "type II". The salt water wedge won't

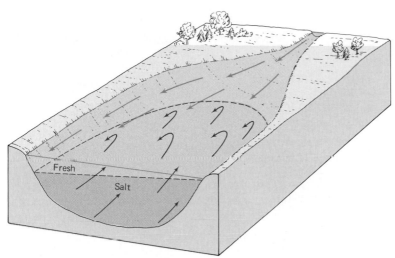

FIGURE 16.14 *Type II estuary. The fresh water flow is more moderate. The salt water wedge intrudes farther, and so its Coriolis deflection is more pronounced, causing the slanted salt water-fresh water interface.*

meet so much resistance, and the interface will not be so steep; hence, the wedge will extend relatively farther upstream. The banking of the inflowing and outflowing currents due to the Coriolis effect is more noticeable (Figure 16.14), since the salt water is not so stunted in its upstream flow as was the case with type I estuaries. Because the salt water in the wedge mixes with the fresh water above and flows out with it, the outflow of brackish surface water will be much greater than the fresh water input from rivers and streams. Examples of type II estuaries are Chesapeake Bay, Puget Sound, and most fjords.

When the fresh water inflow is extremely small in comparison to the tidal volume, the salt water will intrude far upstream into the estuary. Often the Coriolis deflection of the incoming salt water is sufficient that the tilted interface of the two waters intersect the surface (Figure 16.15). The right side of the estuary (looking upstream, Northern Hemisphere) has ocean water at the surface, and the fresh water exits along the left side. These are called "type III" estuaries. Delaware Bay is an example of one of these.

Mixing between the two water masses makes the transition between salt and fresh water masses more gradual than implied in the idealized figures 16.13 through 16.15. In spite of mixing, the effect can still be seen when salinities are measured and plotted as in Figure 16.16. Where tidal influence is greater, the mixing is more thorough. The tilted salt-fresh interface in type III estuaries, for example, is demonstrated through gradual change in surface salinity across the width of the estuaries.

Most bar-built estuaries are shallow and so heavily influenced by surface processes that they don't at all fit into any of these three categories. Bar-built estuaries are therefore lumped into a fourth class all by themselves, and display a wide variety of characteristics. For example, in some

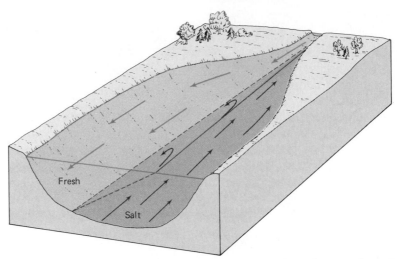

FIGURE 16.15 *Type III estuary. With a very small fresh water flow, the salt water intrudes far up into the estuary. The Coriolis deflection is so pronounced that the salt water-fresh water interface intersects the surface.*

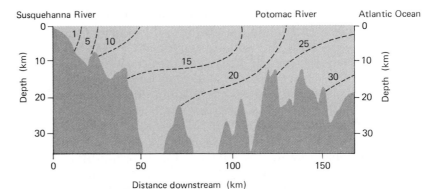

FIGURE 16.16 *Cross section along the length of the Chesapeake Bay, starting from the Susquehanna River and ending at the Atlantic Ocean. The salinity profile is indicated in parts per thousand (‰). (Data from the Chesapeake Bay Institute, The Johns Hopkins University.)*

the rate of evaporation exceeds the fresh water input, so that the estuarine water is actually saltier and denser than seawater. The circulation in these is similar to that illustrated for seas in Figure 16.1. Seawater flows in through the mouth along the surface, and the outward flow is along the bottom, just the opposite of the flow patterns in most estuaries. Such lagoons are called "negative" or "inverse" estuaries. Laguna Madre is an example of one of these.

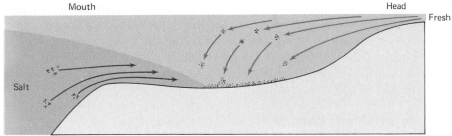

FIGURE 16.17 *Nutrients tend to collect in estuaries, being brought in through the mouth by the salt water wedge, and being brought in through the head with the fresh water runoff from land. Nutrients tend to collect on the bottom with sinking organic detritus, and the intruding salt wedge tends to keep them from leaving.*

C. ESTUARINE ECOLOGY

C.1 Biological Importance

Most estuaries are exceptionally rich in biological activity, and are crucial to the lives of many organisms. More than half the fish and shellfish harvested in the United States spend a major portion of their lives in estuaries. Seven of the ten commercially most valuable species are dependent on estuaries. These include salmon, clams, oysters, shrimp, flounders, and menhaden.

The reason for the large biological productivities of estuaries involves their relatively shallow, protected waters, and the availability of nutrients. We will see in later chapters that the sinking of organic detritus tends to take nutrients out of the sunlit surface waters and into darker depths where they cannot be reused in photosynthesis. In shallow coastal waters, however, sunlight may penetrate down to the bottom where these nutrients accumulate, so they can be recycled immediately by the plants. Even where sunlight is dim, coastal turbulence can bring these nutrients up to the surface where plants can reuse them. This availability of nutrients is the reason why coastal waters are so much more productive than the waters over the deep ocean basins.

Estuaries are particularly rich in nutrients because the salt water wedge entering the mouth along the bottom tends to bring the nutrients and organic detritus with it. Other nutrients enter at the head with the fresh water runoff from land (Figure 16.17). Thus, nutrients tend to accumulate in estuaries, and the intruding salt water wedge tends to keep them there. Furthermore, most parts of most estuaries are sufficiently shallow that sunlight can penetrate clear to the bottom. The abundance of sunlight and nutrients promotes vigorous plant growth, both planktonic and attached.

The abundance of plants provides an abundance of food for the animals. As a result, estuaries provide homes, breeding grounds, and/or nurseries for a large number of animal species. The abundant shallow water vegetation also provides protection for many juveniles, and support for many tiny molluscs, crustaceans, and larvae of many species. Among the larger, more mobile animals, usage of estuaries varies. Some spawn in the open ocean, but their juveniles move into estuaries to feed and mature. For others, the pattern is reversed.

C.2   Importance to Humans

Not only are estuaries particularly important to the lives of many marine organisms, but they also seem to be particularly attractive to human populations as well. Conflicts of interest between the two groups arise that are becoming particularly acute as human populations mushroom.

Most of the world's largest metropolitan areas are centered on estuarine systems, for reasons that are easily understood. The protected waters have provided good ports for ocean-going vessels. Connections with rivers have provided routes for further commercial connections with inland areas. Access to goods has spawned industry and related services, and the biological productivity of these estuaries has supported healthy fishing industries. Consequently, fishing, industry, and commerce have provided the economic base for growing populations.

The conflicts of interest between human and marine biological populations tend to take several forms. Humans tend to crowd the water's edge, filling tidal flats and marshlands for human construction, or dredging them for commerce. These areas tend to be the most productive of all oceanic regions. Some are several times more productive of plant materials than the finest agricultural croplands, and they support much of the marine life in deeper parts of the estuary. Dredged channels change the current patterns and muddy the waters to the point where many native organisms can no longer survive.

C.3   Pollution

One of our major concerns over the influence of humans on estuarine systems involves the addition of "pollutants." Pollutants are things humans add that alter the environment (Figure 16.18). They generally are of two forms: heat or materials.

It must be remembered that the addition of foreign heat or materials to the environment by itself is not necessarily bad. Estuaries are places having relatively large seasonal temperature changes, and the makeup of dissolved and suspended materials also changes seasonally with seasonal changes in foliage and runoff from the neighboring land. But over the ages, the surviving organisms have developed the ability to incorporate these changes into their own life cycles. For some, appropriate changes in life stages are triggered by these seasonal changes in their environment.

But when we add heat or materials to their environment, we interrupt the normal seasonal rhythm. The organisms may not be able to cope with this interruption, or the interruption may prematurely trigger a change in their life cycle, so they are unable to cope with later conditions. Further-

FIGURE 16.18 *The products of our growing population are particularly injurious to the valuable and fragile estuarine environment.*

more, we often pollute in excess, exceeding the thermal or material tolerances of these organisms.

Power plants are a major source of thermal pollution as they use coastal waters as a coolant. In addition, they destroy some coastal life through antifouling agents (usually chlorine) or mechanical abrasion used to keep the plumbing free of fouling organisms. The coolant waters are raised typically by 15°C in temperature, which causes high fatalities, but the net effects are not necessarily all bad from our human viewpoint. For example, in one plant, the rise in water temperature was found to kill off both the copepods, microscopic animals important as a food source for commercial species, and the ctenophores, which are voracious consumers of copepods in competition with the commercial fish. It turned out in this case that the reduction in ctenophore population permitted a growth in the remaining copepod population that exceeded the copepod loss in the coolant waters. This provided an enlarged food base for the commercial fishes, and actually increased the fishery production in the estuary. More frequently, however, the net effect of our thermal pollution is bad.

Of even greater concern are the material pollutants we discharge into our estuaries (Figure 16.19). Most of these fall into the following categories:

FIGURE 16.19 Pollution.

1. Bacterial contamination from human or animal wastes, which often are more hazardous to the human population than to the marine organisms.

2. Decomposing organic matter, which depletes the dissolved oxygen (bacteria use up oxygen in decomposing organic wastes) and thereby suffocates the native animal population.

3. Materials that fertilize some plants at the expense of others, thereby shaking up the entire chain of life ultimately dependent on these plants for their sustenance.

4. Inert things, such as silt or clay, that muddy the waters and smother the bottom-dwelling organisms.

5. Chemical toxins. Natural selection has favored the survival of species that could collect and hoard certain trace nutrients important to their survival. Unfortunately, when human pollution makes these more available, they can overconsume. Many of these are poisonous in excess. Furthermore, the hoarding ability sometimes also applies to trace poisons that were not normally present in their environment and therefore previously were of no concern. Some of these chemical toxins become quite concentrated in organisms, and many become even more concentrated as they are passed along the food web.

It takes time to develop a stable intricate biological community, with various organisms filling various important ecological niches. Altering any component upsets the entire community. Often our pollution is not constant in time, so that stable communities of organisms cannot develop.

Fortunately, there are some natural processes that help cleanse estuaries of our pollutants. Some settle out in the sediments, and some are carried out to sea. Many of our pollutants are surface-active, meaning they tend to adhere to surfaces. The fine sediments suspended in turbid coastal waters provide a large amount of surface area altogether onto which these pollutants can collect. As these sediments settle out of suspension, then, these pollutants tend to go with them. The concentration of pollutants in the bottom sediments is a concern, of course. Although removed from the water, these pollutants are still there in hibernation, waiting for future turbulence, waves, or currents to stir them up again.

Most of our estuarine pollutants end up getting flushed out to sea. In estimating the amount of damage to be done to a particular estuary, one must know not only the rate at which pollutants are being dumped into it, but also the rate at which they are being flushed out.

The "flushing time" is a measure of the rate of turnover of the fresh water in an estuary. It answers the following question: "If one were to suddenly remove the entire fresh water content of an estuary, how long would it take for the inflow from rivers and streams to replace it?" To determine the flushing time, one must measure the entire fresh water content of the estuary, and the rate of fresh water influx.

$$\text{flushing time} = \frac{\text{fresh water content}}{\text{fresh water influx}}$$

This can then be compared to the rate at which pollutants are entering the estuary to estimate the damage being done. Typical values range from days to decades.

Since most estuaries are flushed primarily by the flood and ebb of tides, you may wonder why we use fresh water flow, rather than tidal volume in measuring flushing time. The two are related, of course. Where tidal flushing is large, the estuarine water is quite salty and the fresh water content is low. According to the above formula, the small fresh water content results in short flushing time. Furthermore, the pollutants enter the estuary from land, just like the fresh water. Therefore, their residence in the estuary would be similar to that of the fresh water.

Not all the fresh water or all the pollutants leave in any one flushing time (Figure 16.20). As you pour water into an overflowing glass of milk, not all the milk leaves. The residue left in the glass will still have a lot of milk in it. Even after three or four glasses of water have been poured in, still some milk remains in the mixture. Likewise, many flushing times are required to remove old pollutants from our estuaries. Similarly, many flushing times are required before the entry of new pollutants reaches a steady state (Figure 16.21). To observe the result of any change in our pollution, we must wait many flushing times before a steady state is reached, and then many generations of organisms before their adjusted steady states are reached. Of course, pollutants flushed out to sea do not disap-

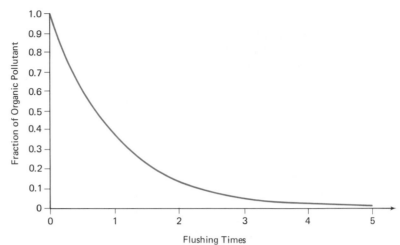

FIGURE 16.20   *Plot of remaining pollutant as a function of time, assuming perfect mixing. Many flushing times are required before the pollutant is completely removed.*

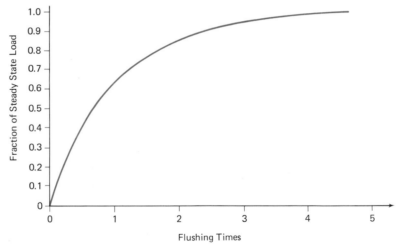

FIGURE 16.21   *Plot of concentration as a function of time for a new pollutant being introduced at a steady rate. Many flushing times are required before it reaches its final steady-state concentration. (Assuming perfect mixing.)*

pear; they just go somewhere else. Because of its size the ocean is able to absorb much more pollution than are the estuaries. Nonetheless, pollutants flushed to sea spread through the ocean to other areas, especially along coastlines where they are aided by strong coastal currents. Such pollution cannot be kept local and is therefore of global concern. Unfortunately, international agreements to try to protect the world ocean from pollution have a few weak links. These are the few coastal countries willing to compromise the quality of their coastal environment for the short term economic gain offered by certain international industries, who seek a haven for their polluting operations.

Along a large portion of the world's coastlines, interaction between land and sea is moderated by intermediary bodies of water. The larger of these are called seas, and the smaller ones estuaries. The smaller the body of water, the more heavily it is influenced by climate and processes on the neighboring land mass.

Although the details vary from one sea to the next, there are two general circulation patterns for the water in seas. Where evaporation exceeds precipitation, surface waters become more saline and sink, causing continual flushing of the bottom water, with replacement water flowing from the ocean into the sea along the surface. Where precipitation exceeds evaporation, the surface waters have reduced salinity, and flow out of the sea along the surface. Mixing between surface and deeper waters causes some deeper water also to flow out along the surface, but for the most part the deeper water is fairly stagnant and gets flushed only very slowly.

Estuaries are smaller and more variable than seas. Flooded coastal plains tend to have drowned river valleys and barrier beaches, which create large lagoons. Mountainous coasts have relatively little river discharge, and the jagged terrain creates many relatively small embayments. Tectonic-formed estuaries and fjords (at higher latitudes) are common along mountainous coasts.

In most estuaries, the basic underlying circulation pattern involves a salt wedge intruding from the ocean through the mouth of the estuary, and fresh water flowing in at the head. After mixing, the brackish water then flows out of the mouth along the surface. A great deal of mixing, especially where the tidal volume is large, or in shallow estuaries where surface processes are important, causes a great deal of variation from this ideal salt wedge picture. Estuaries are sometimes classified as types I, II, or III, according to the relative influences of tidal flushing and fresh water inflow. Bar-built estuaries don't fit well into any of these categories, and so they are usually lumped into a separate category all by themselves.

Abundance of nutrients and shallow water cause most estuaries to be extremely productive. They are crucial to the lives of many organisms, and their biological importance is much greater than most other oceanic regions. Humans also tend to collect around estuaries, and conflicts of interest between the two communities arise.

Human pollutants are thermal or material. The material pollutants are generally of the following five types: bacterial contamination, decomposing organic matter, fertilizers, inert things, and chemical toxins. Estuaries are normally places with rather large seasonal variations in temperature and materials. The organisms have adapted to accommodate these regular natural variations in their life cycles. Human pollution, even when not in excess, tends to interrupt these natural cycles. Disruption of any portion of the biological community affects the rest of the community as well.

Pollutants often are surface-active and collect in the sediments. The flushing time is a measure of the time required to flush the fresh water from an estuary, which would be similar to that for dissolved or suspended pollutants. Many flushings are required before a steady state is reached. The flushing of pollutants into the oceans is an international concern.

QUESTIONS FOR CHAPTER 16

1. What is the difference between a sea and an estuary? (The actual dividing line is a bit arbitrary.)

2. How many different ways can you think of through which "precipitation" (as used in this chapter) enters a sea?

3. What is the general circulation pattern in a sea where evaporation exceeds precipitation?

4. Why might you expect to find a more flourishing community of deep-sea fauna in a sea where evaporation exceeds precipitation than in a sea where precipitation is larger?

5. How is it that German submarines could go either way past Gibraltar without the use of their engines?

6. How do the salinity, temperature, and density of the water leaving the Mediterranean Sea compare with that of the Atlantic Ocean water that it encounters?

7. Describe the general circulation pattern in a sea where precipitation exceeds evaporation.

8. How could fishermen of antiquity drift either way through the Bosporus and Dardanelles? How does this differ in principle from the currents used by the German submarines mentioned in Question 6?

9. Why is it that the surface water leaving the Black Sea is not completely fresh?

10. What are the "mouth" and "head" of an estuary?

11. Would seas or estuaries be more affected by pollution in general? Why?

12. How were drowned river valleys formed?

13. How were the barrier beaches formed along the United States East Coast?

14. Why should lagoons experience greater daily and seasonal changes in temperature and salinity than other types of estuaries?

15. Briefly explain why wind-induced currents and seiches should frequently cause greater changes in water level along East Coast lagoon shores than the tides.

16. Summarize the types of estuaries typical of flooded coastal plains and how they were formed.

17. Why is there relatively little river-water discharge into the United States West Coast? (2 reasons.)

18. Why are there no extensive barrier beaches on the United States West Coast like there are on the East Coast?

19. Why are West Coast estuaries generally more heavily influenced by the tides than East Coast estuaries?

20. What is a tectonic-formed estuary? Give an example of one.

21. How are fjords formed?

22. Summarize the kinds of estuaries common to mountainous coasts, and how they are formed.

23. Briefly describe the basic salt-wedge model of current flow in an estuary.

24. What are some of the mechanisms that cause mixing between salt and fresh water masses in estuaries?

25. Briefly explain how it is that by measuring the salinity of the water leaving the estuary along the surface, we can conclude that more salt water enters the estuary than fresh water.

26. In most estuaries, does the flood tide or ebb tide last longer? Why?

27. How did the sills form at the mouths of many fjords?

28. Why is the deep water in many fjords rather stagnant and dead?

29. What is the tidal volume of an estuary?

30. What is the distinction between type I, type II, and type III estuaries?

31. In which of the above estuarine types is the Coriolis effect most pronounced? Why?

32. Give an example of each of the above types of estuaries.

33. Is the transition between salt and fresh water masses in estuaries abrupt or gradual? Why?

34. What effect has the Coriolis effect on the salt and fresh water flows in an estuary in the Northern Hemisphere?

35. What is an inverse estuary? A lagoon? A slough?

36. How does the shallowness of estuaries enhance their biological productivities?

37. How are nutrients brought into estuaries?

38. What is a pollutant?

39. What are some of the ways that humans alter estuarine environments?

40. If estuaries naturally have large seasonal changes in temperature and dissolved materials, why is human thermal and material pollution such a concern?

41. Are the net effects of thermal pollution by power plants necessarily all bad from a human viewpoint? Explain.

42. List the five basic types of material pollutants that humans contribute to estuaries.

43. Briefly explain how natural selection has made organisms particularly vulnerable to chemical toxins.

44. Why do many dissolved or suspended pollutants tend to collect in the sediments? Why is this of some concern?

45. What does the flushing time measure?

46. How would you go about determining the flushing time for a particular estuary?

47. Briefly explain why estuaries with large tidal flushing would have a short fresh water flushing time according to the formula.

48. Why would the residence of dissolved pollutants be similar to that of the fresh water?

49. Are all the pollutants removed from an estuary in one flushing time? Explain.

50. What are the weak links in present international agreements protecting the world ocean from harmful pollutants?

*51. Briefly explain why seasonal changes in temperature, salinity, circulation, and biological activity should be more pronounced in estuaries than in oceanic waters.

*52. Briefly explain why the basic salt-wedge model should work better in type I estuaries than in types II, III, or bar-built estuaries.

*53. How is it that the salt wedge tends to trap nutrients in estuaries? Why don't these nutrients flow out of the estuaries with the outward flow of brackish waters along the surface?

*54. Why is the conflict between human and marine populations more acute in estuaries than along open coastlines?

*55. Briefly explain how a change in plantlife, such as one species replacing another, can affect animal population in an estuary. Also explain how the removal of a predator might be harmful to other plant or animal populations.

*56. Why must organisms in coastal embayments be more adaptable than their ocean-dwelling relatives?

*57. About what fraction of the age of the earth is the age of the Industrial Revolution?

*58. The effect of pollution is frequently not the destruction of the planktonic community, but rather the alteration of its character (e.g., a change in species present). What further ramifications might this have?

*59. Many people claim that even more destructive than the pollution of the estuaries is the fact that our pollution varies considerably from time to time in both quantity and composition. How might this view be supported from consideration of the welfare of the marine biological community?

SUGGESTIONS FOR FURTHER READING

1. Willard Bascom, "The Disposal of Wastes in the Ocean," *Scientific American* (Aug. 1974).

2. Des Connell, "Australia's Great Barrier Reef National Park," *Sea Frontiers,* **26,** No. 4 (1980), p. 200.

3. Dolan, Hayden, and Lins, "Barrier Islands," *American Scientist,* **68** (1980), p. 16.

4. Sebastian A. Gerlach, *Marine Pollution,* Springer-Verlag, New York, 1981.

5. E. W. Seabrook Hull, "Oil Spills: The Causes and the Cures," *Sea Frontiers,* **24,** No. 6 (1978), p. 360.

6. George H. Lauff (ed.), *Estuaries,* American Association for the Advancement of Science, publication no. 83, 1967.

7. Haynes R. Mahoney, "Imperiled Sea Frontier: Barrier Beaches of the East Coast," *Sea Frontiers,* **25,** No. 6 (1979), p. 328.

8. "Marine Pollution," *Oceanus,* **18,** No. 1 (Fall 1974).

9. "The Ocean as Waste Space?," Oceanus, **24,** No. 1 (Spring 1981).

10. Martin L. Wiley, *Estuarine Interactions,* Academic Press, New York, 1978.

17 BIOLOGICAL PRODUCTIVITY IN THE OCEANS

Marine diatoms

In the following several chapters we will be studying marine life. With the very minor exception of a few bacteria that can use the energy released in inorganic chemical reactions to synthesize organic materials, essentially all the energy required by living organisms to perform their various functions is derived from the sun.

Plants have the ability to convert solar energy into stored chemical energy through a process called "photosynthesis." This energy is contained in the organic matter produced, and it can be released when this matter is oxidized through "respiration," either by the plants themselves or by other organisms farther along the food chain to whom this organic matter is passed (Figure 17.1).

Very few rays of sunlight penetrate beyond the first hundred meters of water, so the marine plants that use sunlight to synthesize organic materials can survive in the surface waters only. These plants are called the "primary producers," because the whole web of life in the oceans depends ultimately on them (Figure 17.2). Herbivores feed on them, carnivores feed on the herbivores and on each other, and omnivores can feed either on other animals or on plants. (See Figure 17.3) But without the plants, none could exist. Because of their crucial role of supporting the entire biological community, we begin our study of marine biology with a look at the plants.

A. THE BASICS OF LIFE

A.1 Ingredients in the Production of Organic Tissue

Carbohydrates are produced during photosynthesis. Basically, the solar energy is used to remove the oxygen from CO_2 molecules, replacing it with water molecules. The end product of photosynthesis, then, is free oxygen (O_2) and hydrated carbon atoms, or "carbohydrates." Some of the carbohydrates produced in photosynthesis are then converted into fats and proteins via subsequent chemical activity.

Molecules of organic materials can be quite complex, due to the flexibility of the carbon atom. Each carbon atom can share four electrons with neighboring atoms, which enables carbon atoms to form complex molecules. These generally include long chains of carbon atoms, which provide the basic structure to which other things are attached. (See Figure 17.4.)

As a general rule, the carbohydrates are the simplest organic molecules, and the proteins are most complex. The simplest carbohydrate would be one single hydrated carbon atom, CH_2O, although larger carbohydrate molecules are more common. For example, sucrose is six hydrated carbon atoms ($C_6H_{12}O_6$) and the molecules of common table sugar are made of twelve carbon atoms hydrated with eleven water molecules ($C_{12}H_{22}O_{11}$). By contrast, most protein molecules are **445**

FIGURE 17.1   *The energy supply for life processes comes ultimately from the sun. It is converted into stored chemical energy by plants in a process called "photosynthesis." This energy can later be released and used either by the plants themselves, or by animals which feed on these plants, or by animals which feed on the animals which feed on the plants, and so on.*

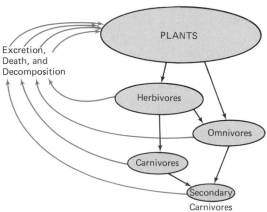

FIGURE 17.2   *The food web. All life is ultimately supported by the plants.*

extremely complicated. We normally describe their structure in terms of the "amino acids," which are the basic building blocks from which they are constructed.

As we have seen, the various carbohydrate molecules are made up of

(a)

(b)

FIGURE 17.3 *Most of our ocean harvest is presently from the top of the food chain—the secondary carnivores. (a) 60-lb yellow fin tuna being caught from a NOAA research vessel off West Africa. (b) English sole catch.*

carbon, hydrogen, and oxygen. Fat and protein molecules are also composed mostly of these three elements, but they also have small amounts of other elements. These include some common and easily accessible salt components, such as potassium (K), calcium (Ca), sodium (Na), sulfur (S), chlorine (Cl), and so on. But they also include some relatively rare or inaccessible elements, particularly nitrogen (N) and phosphorus (P). In most organic matter nitrogen is somewhat more abundant than phosphorus. The nitrogen is particularly important in the amino acids and in the more complex proteins.

Although nitrogen is the major constituent of our atmosphere, and also is amply abundant as a dissolved gas in seawater, it is not very abundant in a form usable by most plants in photosynthesis—as a nitrate (NO_3^-), a nitrite (NO_2^-), or in some cases as ammonia (NH_3). Phosphorus is relatively scarce in the biosphere. In the ocean it is normally encountered in the form of a phosphate, $H_2PO_4^-$ or HPO_4^{--}. Fortunately, plants can use it in either form.

The carbon generally enters the biosphere in the form of carbon dioxide (CO_2). Most of the carbon dioxide has been dissolved in the ocean. It combines with water to form carbonic acid ($H_2O + CO_2 \rightarrow H_2CO_3$) and then loses one or two of its hydrogens to form the bicarbonate (HCO_3^-) or carbonate (CO_3^{--}) radical, respectively.

$$H_2CO_3 \rightarrow H^+ + HCO_3^-, \text{ or } \rightarrow 2H^+ + CO_3^{--}$$

In the oceans, under normal conditions, the bicarbonate ion is the most common, being roughly 10 times more prevalent than the carbonate ion, and roughly 10,000 times more prevalent than the nondissociated carbonic

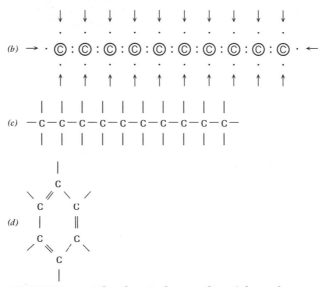

(a)

FIGURE 17.4 *The chemical versatility of the carbon atom makes it ideal for forming the basis of the large, complicated organic molecules. (a) It has
four electrons in its outer shell, which is four short of the eight, and
four more than the zero needed for the preferred ideal gas electronic configurations. Consequently, it eagerly shares its four electrons with those of four neighboring atoms. (b) One thing the carbon
atom can do is to form long chains, each carbon atom sharing electrons with the carbon neighbor on either side, and leaving two electrons to share with other things (indicated by arrows), such as
hydrogen atoms, OH^- radicals, and so on. (c) In carbon chemistry,
the sharing of electrons with a neighbor, called a "chemical bond,"
is indicated by a line. So this chain is the same as the one above. (d)
Another common configuration of carbon atoms is the hexagonal
ring.*

acid form. Therefore, marine plants that assimilate carbon dioxide in photosynthesis generally encounter it in the bicarbonate form.

A.2 Ingredients in Short Supply

Now let's review these materials and ask the question, "If we had unlimited sunlight available to allow photosynthesis to go unchecked, which of
these materials would we run out of first?" Since the animals are dependent on the plants for their livelihood, the question amounts to asking
which of the above materials would limit the biological activity of the
ocean.

We saw that carbon, hydrogen, and oxygen formed the bulk of the
organic materials, with nitrogen and phosphorus present in smaller

TABLE 17.1 Relative Abundances of Various Elements in Organic Tissues and in Seawater. The abundances are measured relative to that of nutrient nitrogen in order to illustrate that carbon, oxygen, and hydrogen are found in seawater in more than ample supply. It is the relatively short supply of nutrient nitrogen that limits the production of organic tissue.

| | Relative Abundances by Weight | | | | | |
	N	P	C	H	O	Si^a
Organic matter	1	0.11	6	1.2	12	5.3
Seawater	1	1.4	56	216,000	1,700,000	6

| | Relative Abundances by Number of Atoms | | | | | |
	N	P	C	H	O	Si^a
Organic Matter	1	0.05	7	17	10	2.6
Seawater	1	0.6	65	3,000,000	1,500,000	3

^{a}The figures for silicon (Si) are relevant for diatoms only.

amounts. The water (H_2O) itself furnishes ample amounts of hydrogen and oxygen for the plants, as well as ample supplies of whatever salts are needed. This leaves carbon, nitrogen, and phosphorus, which are present in organic matter in the ratio of about 50 to 9 to 1 by weight, or 150 to 20 to 1 by number of atoms.

Of these three remaining materials, then, it is carbon that is in greatest demand for the production of organic tissue. However, it is also in greatest supply in the marine environment. In fact, it is much more abundant than necessary to support maximum productivity. (See Table 17.1.) Instead, it is the small abundance of nitrates and phosphates that limit the amount of biological growth possible, and so they are called the "nutrients." Because the nutrients furnish only a small part of the total biological materials, addition of a small amount of nutrients will often stimulate a large amount of biological activity.

The productivity of all plants is contingent on the availability of nitrates and phosphates. But certain plants are constrained by the availability of other materials as well. For these plants, the term "nutrient" can be broadened to include these additional materials. Important examples are the microscopic plants called "diatoms" (Figure 17.5), which are responsible for a large portion of the ocean's total primary productivity. They require silicates for the production of their tiny silica exoskeletons. Silicates are not very abundant in the oceans, and can be as limiting to diatom productivity as are the nitrates and phosphates. For these important microscopic plants, then, the term "nutrient" would have to be broadened to include silicates.

There are some other materials that are also found in organic matter, particularly in the proteins and in organs with specialized functions, whose abundance is well below the 1% level. These are refererred to as "micronutrients," and include some vitamins and "trace elements," such as sulphur, chlorine, iodine, fluorine, potassium, sodium, calcium, magnesium, iron, copper, manganese, zinc, and others. The exact function of each of these trace elements is being heavily researched at present. Many seem to be particularly important to more complicated organisms with specialized systems, such as the iron needed by the hemoglobin of our red blood cells. In the marine environment, these trace elements are usually available in ample quantities, although terrestrial organisms frequently suffer from lack of one or more of these.

A.3 Photosynthesis and Respiration

The basic equation for life may be written as

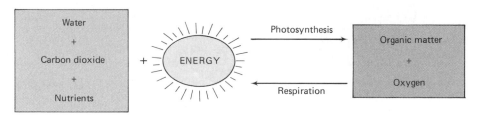

In photosynthesis, the energy used is sunlight. When plants and animals respire, they use up oxygen and produce mostly water and carbon dioxide. The energy produced is used in their own biological processes, such as locomotion, tissue production or repair, digestion, body heat, and so on. (Incidentally, the arrow labeled "respiration" is also the route followed when we produce energy by burning hydrocarbons, such as wood in our fireplace or gasoline in our autos.)

A specific example of this equation involving the production of a simple sugar would be

$$6H_2O + 6CO_2 + energy \xrightarrow{\text{Photosyntheses}} C_6H_{12}O_6 + 6O_2$$

If we found this sugar molecule produced by this plant and ate it, we would get energy for our bodies by driving the above equation the other way.

$$C_6H_{12}O_6 + 6O_2 \xrightarrow{\text{Respiration}} 6H_2O + 6CO_2 + energy$$

B. THE NUTRIENTS

B.1 The Nutrient Cycle

The equation of life of the previous section represents a cycle for the materials involved—carbon, hydrogen, oxygen, and the nutrients. These are synthesized first into organic matter through photosynthesis, and then are returned to the environment during respiration to be used again. In the near-surface marine environment where light is available for photo-

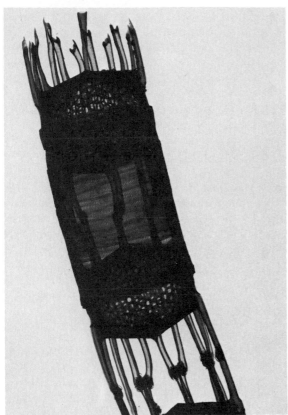

FIGURE 17.5   *Photo of the chain-forming diatom, Skeletonema costatum, one of the most important diatoms in the ecology of temperate coastal marine waters. (Magnification about 3000 ×.)*

synthesis, carbon, hydrogen, and oxygen are plentiful. However, the nutrients are not, and so the productivity of an area depends upon them getting returned to the environment and recycled as quickly as possible.

Perhaps due to the scarcity, organisms seem to have become very proficient at collecting and hoarding these materials. In some instances up to 80 or 90% of the nutrients available in an area have been found to be already incorporated into organisms. Some microscopic plants are so proficient at hoarding phosphates that they can survive and reproduce for several generations after phosphates have been depleted from their environment. Unfortunately, the crucial ability to hoard nutrients and important trace elements also frequently extends to trace materials that are not needed and that may even be deleterious to the organism's health. Consequently, in polluted coastal waters, relatively small concentrations of these harmful materials may cause widespread devastation in biological communities.

The nutrient scarcity is further aggravated by the fact that acquired nutrients are only very reluctantly returned to the environment. When in need of energy, organisms (including humans) generally oxidize the car-

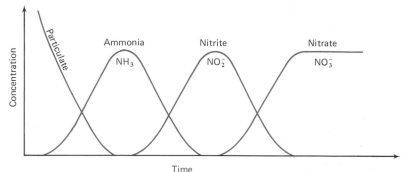

FIGURE 17.6 *Plot of concentration vs. time for the various products of bacterial decomposition, beginning with suspended particulate organic detritus, and ending with the release of nitrates usable by plants in photosynthesis. It is a three-step process, involving three different types of bacteria.*

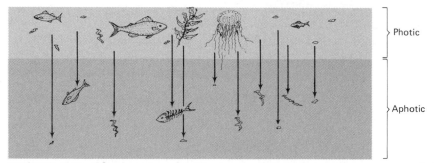

FIGURE 17.7 *The remains of surface organisms sink into deeper waters before decomposing. In this manner, nutrients are removed from the photic zone. The deeper, aphotic waters become enriched in nutrients, but these nutrients must await some process that returns them to the surface before they can again be incorporated into organic matter via photosynthesis.*

bohydrates first, and the fats next. Only when in dire need of energy and when their carbohydrate and fat stores have been exhausted will they begin oxidizing protein, in which the bulk of the nitrogen and phosphorus is contained.

Some nutrients become available through the excreted wastes of organisms, especially when feeding is heavy and digestion incomplete. About half of the nitrogen excreted by animals is in the form of ammonia. This and the phosphates tend to dissolve directly from the fecal material and get back into the food cycle rather quickly. However, much of the nutrients must await death and decomposition of the organism, or appropriate decomposition of the fecal material, before they are converted to a form useable by most plants. The release of nutrient nitrogen generally takes the most time, requiring a three-step process involving three separate types of bacteria (Figure 17.6). This takes so long that much of the detritus has sunk below the productive photic surface waters before this nutrient is released (Figure 17.7). It must await some process to return it to the surface where it can be recycled by the plants, and this may take years.

Because of the difficulties described above, once a nitrate has been incorporated into organic matter, it can be a long time before it becomes available as a nutrient again. The other materials are recycled more quickly. Phosphorus, for example, requires only a single-step bacterial action in the decomposition of organic matter in order to return it to the environment as a nutrient. Because of their slowness in completing the cycle, it is generally nitrates, rather than phosphates, which restrict the biological productivity. Because the other materials are always abundant in surface water, we are usually concerned only with the nutrient cycle in dealing with the productivity of an area, in particular the nitrates.

B.2   Upwelling

As we have seen, an important restriction of biological productivity is caused by removal of nutrients from surface waters as organic detritus sinks into deeper waters, carrying the nutrients with it. In some areas of the ocean, these nutrient-laden deeper waters are able to come to the surface in a process called "upwelling." Of course, these areas are rich in biological activity.

Upwelling occurs frequently along continental margins, where winds drive surface currents out to sea and deeper waters surface to replace them. It also occurs in some places at sea where surface currents diverge, again meaning that deeper waters are rising to replace the departing surface waters. (See Figure 17.8.)

Upwelling is quite pronounced in the equatorial regions, where the trade winds blow the surface water from east to west across the ocean. The surface water vacating the eastern margin must be replaced by the nutrient-laden deeper water from below. This happens in the Pacific off Peru, and in the Atlantic off Equatorial Africa, making these areas particularly good fishing grounds. Similarly, most of the western coast of North America experiences winds coming predominantly from the northwest. This drives the surface currents southeasterly along the coast, but due to the Coriolis effect, they flow out to sea as well, allowing deeper water to flow up as replacement.

Also having exceptional biological activity are the coral reefs. Because of their steep sides, the daily tidal currents must shoal abruptly to make it across the reef, thus bringing deep water to the surface.

The surface currents diverge in several regions of the ocean. Along the equator, the trade winds blow the surface currents from east to west. Due to the Coriolis effect, however, the equatorial current north of the equator is deflected northward, and that south of the equator is deflected southward, causing the pronounced equatorial divergences. Another such area is the Antarctic divergence, where deep water first formed in the Northern Hemisphere finally resurfaces, accounting for flourishing bio-

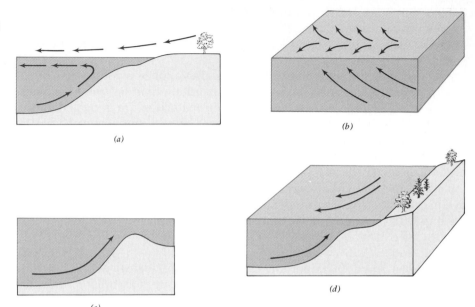

FIGURE 17.8 *Illustration of some mechanisms causing upwelling. (a) Trade winds
blowing out to sea from the western tropical margins of continents
drive the surface waters out to sea; they are replaced by deeper
waters. (b) Where surface waters diverge, such as the equatorial or
Antarctic divergences, deeper waters must come up as replacement. (c) Large bumps in the ocean bottom make deep currents
shoal. (d) The Coriolis effect may divert longshore currents out to
sea, requiring replacement by deeper waters.*

logical productivity in this area, too, especially during the months of summer sunshine.

B.3 Turbulence

Turbulence can also bring nutrients to the surface from deeper waters.
Turbulence usually results from waves or swift surface currents. Some
surface currents reach a depth of 1 km or so, but in general, turbulence is
most pronounced near the surface, and is not very effective in retrieving
nutrients from depths greater than a few hundred meters.

The most common turbulent mechanism is the waves, which can mix
the upper 100 m or so quite well. This is not enough to retrieve materials
below one of the ocean thermoclines, but it is enough to keep the waters
of the continental shelves well mixed. This accounts for the high biological productivity of the shelves, particularly in tropical and temperate latitudes, where sunshine is abundant. (See Figure 17.9.)

Turbulence is really a mixed blessing, because it can reduce the
thickness of the productive photic zone as well as bring nutrients to the
surface. In coastal waters, turbulence brings fine sediments to the surface
along with the nutrients, making the water more turbid and reducing the

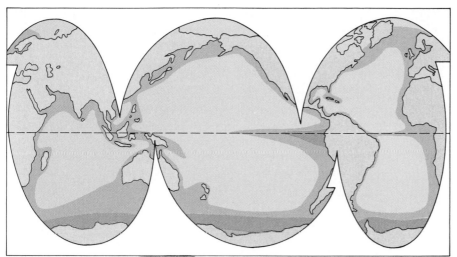

FIGURE 17.9 *Regions of high biological productivity are darkened. The darkest areas indicate the areas of extremely high biological productivity.*

depth to which sunlight penetrates. This negative aspect of turbulence can be especially important in coastal waters of high latitudes where lack of sunlight rather than lack of nutrients is the major factor limiting productivity.

C. GEOGRAPHICAL VARIATIONS IN PRODUCTIVITY

C.1 High Latitudes

In some parts of the world it is the lack of sunlight, rather than lack of nutrients, that inhibits primary productivity in the ocean's surface waters. This is particularly true in polar regions, where the sun is low in the sky at best, spending most of the time either just above the horizon or just beneath it. Furthermore, the ice cover in the north polar area screens out much of the already meager sunlight there.

Nutrients are quite plentiful in polar surface waters, because the water is cold and there is no thermocline separating it from deeper nutrient-rich waters. Therefore, there is better mixing between surface and deep waters, bringing nutrients to the surface. There is especially good circulation in the Antarctic regions, and so there is particularly good mixing of surface and deeper waters there.

C.2 Tropical Latitudes

Where nutrients are plentiful, tropical waters are extremely productive. We have seen that waters overlying continental shelves and areas of upwelling are particularly rich in nutrients. But away from these areas, the situation is quite different.

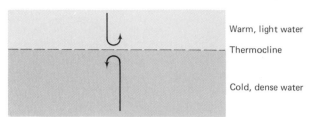

FIGURE 17.10 *Being lighter, the warm surface waters float atop the colder deeper waters, and the two don't mix with each other. As a result, the thermocline presents a barrier to the return of nutrients once they have sunk beneath it.*

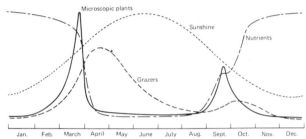

FIGURE 17.11 *Typical seasonal variations in the abundance of microscopic plants nutrients, and grazers in surface waters of temperate latitudes.*

Tropical waters enjoy plenty of sunshine, and so it is the scarcity of nutrients that limits primary productivity. This scarcity is particularly aggravated by the thermocline always present. Due to solar heating, the surface water is considerably warmer and less dense than the water below the thermocline. Therefore, the two water masses don't mix (Figure 17.10). This means that as organic detritus sinks below this thermocline, the nutrients are lost to the surface water and do not get recycled. Tropical waters and waters derived from them, such as the Gulf Stream, tend to be a beautiful deep blue in color, which is a testimony that they are actually a biological desert. Biologically active waters would be filled with tiny microorganisms and organic debris, which would give them a turbid green or yellow-green appearance.

C.3 Temperate Latitudes

In the polar latitudes it is scarcity of sunlight that limits primary productivity, whereas in the tropics it is the lack of nutrients. It should be no surprise that in intermediate latitudes a combination of the two are involved. A typical pattern for primary productivity in temperate latitudes is illustrated in Figure 17.11, and is as follows.

In winter the productivity is low due to the lack of sunlight. As the days begin to lengthen, there is an early spring "bloom," which reaches a peak and starts to decline in the late spring and early summer. The bloom begins at the surface, and then proceeds to greater depths, as the nutrients at the very surface are used up (Figure 17.12).

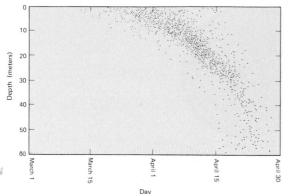

FIGURE 17.12   *Illustration of the vertical distribution of the microscopic plants during the spring bloom. The spring bloom starts at the surface and proceeds to greater depths, apparently as the surface nutrients are used up.*

There are several causes for the late spring decline. One is that the nutrient supply is finite and cannot support unlimited growth. Another is the proliferation of the tiny herbivores or "grazers," which thrive on the new abundance of microscopic plants. Perhaps the most important, however, is the development of the thermocline through the heating of the surface waters. We have seen that this prevents the nutrients that have sunk into deeper waters from reaching the surface again. Hence, biological productivity remains low due to scarcity of nutrients until the late fall when the surface waters have cooled down, the thermocline disappeared, and the nutrients can reach the surface again. However, this fall bloom is short-lived; the days are getting short, sunlight is reduced, and soon the winter low returns.

D.   OTHER FACTORS AFFECTING PRODUCTIVITY

We have seen that the primary factors affecting the productivity of various parts of the ocean are the availability of nutrients and the availability of sunlight. There are other factors that have a lesser, but quite noticeable, effect also. Like sunshine and nutrients, most of these tend to be seasonal, too.

D.1   Grazers

Most of the photosynthesis in the ocean is carried out by microscopic single-celled plants that float in the upper sunlit regions (Figure 17.13). The spring bloom in these plants fosters an increased population of the grazers, which feed upon them. Most of these grazers are also microscopic organisms, the largest group of which are copepods—microscopic arthropods resembling tiny shrimp. (See Figure 17.14.) The increased population

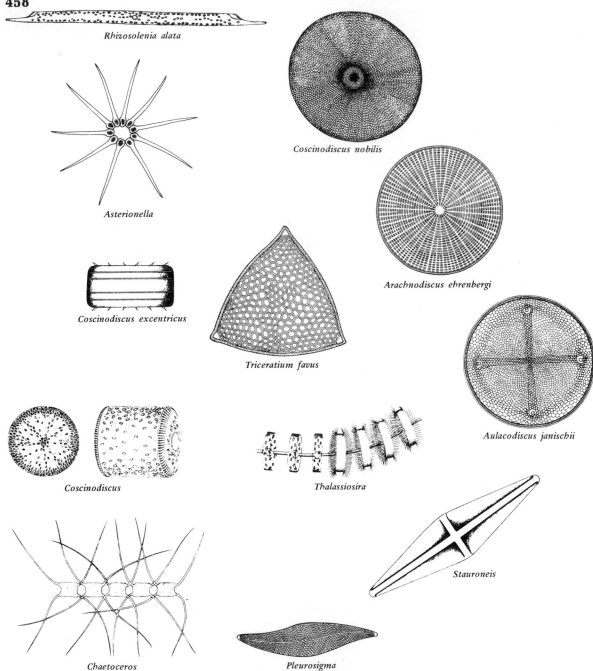

Rhizosolenia alata

Asterionella

Coscinodiscus nobilis

Coscinodiscus excentricus

Triceratium favus

Arachnodiscus ehrenbergi

Aulacodiscus janischii

Coscinodiscus

Thalassiosira

Stauroneis

Chaetoceros

Pleurosigma

FIGURE 17.13 Some representative species of diatoms. (Bayard H. McConnaughey, Introduction to Marine Biology, 3rd ed., The C. V. Mosby Co., St. Louis, 1978. Redrawn from various sources.)

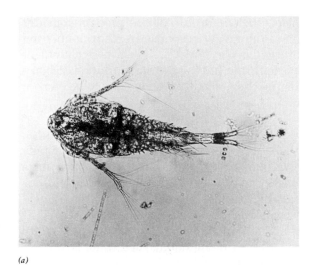

(a)

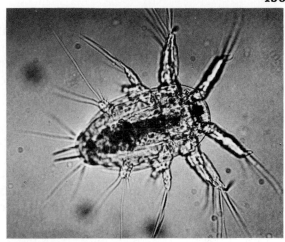

(b)

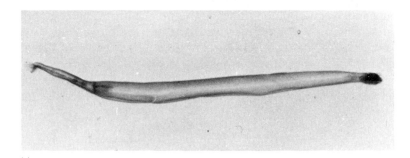

(c)

FIGURE 17.14 *Some microscopic grazers:* (a) *copepod,* (b) *crustacean larva, and* (c) *arrow worm.*

of grazers helps curb the bloom of plants, but how important they are in stopping the plant growth as compared with other factors, such as the removal of nutrients from surface waters, still remains an open question.

It is the "host–parasite" problem, familiar in many fields of science and social sciences. An increase in the host population will result in a subsequent increase in the parasites. Depending on how sensitive the parasite population is to changes in the host population, the situation may be unstable. An increase in the host may cause a much larger increase in the parasite population, which in turn kills off the host to a point where the parasites can no longer survive. This overreaction in the parasite population, then, leads to its own demise, which allows the host to regenerate, and so goes the cycle.

Applying these models to the plant-grazer problem in the marine environment, it is possible to adjust the sensitivity of the grazers to the plant population in such a way that we can duplicate the spring bloom in plant population in temperate latitudes, the summer lull, and the brief fall bloom again, without regard to nutrients at all. These models consider only availability of sunlight, and the grazer response to plant population changes in duplicating the observed cycle. These models prove that grazer response *can* play a very important role in the seasonal variations of productivity, although how influential they actually are remains to be shown.

D.2 Temperature

Although most of the water in the oceans is fairly uniform in temperature, the surface waters, where there is sunlight for photosynthesis, display large geographical and seasonal temperature variations. The temperature determines what kinds of plants can survive and how well the various species can function. But it also influences productivity in ways understandable even without familiarity with the various plant species.

A rule of thumb is that chemical reactions double their rate for every 10°C increase in temperature. All else being equal, then, photosynthesis should proceed more rapidly in warmer waters. However, "all else" is rarely equal. Except in shallow coastal regions, warm surface waters are deficient in nutrients because of the thermocline separating them from the cool deep waters below. Warm waters also tend to be deficient in dissolved gases, such as CO_2, needed in photosynthesis. It is because of these "other" things that high productivity is usually associated with *cold* surface waters rather than warm ones. Productivity is especially high in upwelling regions, where cold and nutrient-laden deep waters come to the surface.

D.3 Coastal Variations

Coastal waters are the most variable of the marine environment (Figure 17.15). Being shallow, they display the largest seasonal temperature changes. The fresh water runoff from the nearby land mass causes variations in the salinity, and also variations in the nutrient supply, since an important source of nutrients in coastal waters comes from the nearby land masses. During storms (Figure 17.16), coastal waters become very turbid, and plants receive much less sunlight. Clearly, coastal waters favor plants that can tolerate such changes in temperature, salinity, nutrients, and sunlight.

Most microscopic planktonic plants, such as diatoms and coccolithophores, are fairly intolerant of changes. They do best in the more uniform surface waters. Plants attached to the bottom, such as kelp, sargassum, eel grass, or red algae, have evolved in waters that are shallow enough that light reaches the bottom. Consequently, they are more toleratnt of these coastal variations than are the microscopic planktonic plants, and so they produce better in these waters. Nonetheless, each plant has a certain set of conditions under which its productivity is highest. Variations of these optimum conditions will always result in reduced productivity.

Much of the substrate in nearshore regions consists of loose and drifting sediment, which is not suitable for attached plants. In these regions, plant productivity is particularly low.

FIGURE 17.15 *Due to the availability of nutrients, coastal waters support a healthy biological community. However, the organisms in these regions must be able to tolerate relatively large variations in the temperature and salinity of the water. They must also withstand variations in the turbidity of the water, which affects the penetration of the sunlight. The organisms in this photo are exposed during a very low tide.*

FIGURE 17.16 *Storms increase the turbidity of coastal waters, which reduces the amount of sunlight reaching marine plants.*

D.4 Currents

We have already seen one important effect that currents have on productivity is when they produce upwelling of deeper waters. Together with the lack of a thermocline, turbulence caused by the Antarctic Circumpolar Current is instrumental in bringing nutrients to the surface, causing the extraordinarily large productivity in this region of the earth. Currents are also important to fixed nearshore plants as they ensure a continual flow of nutrients to the plants and the removal of waste products. Currents can also have a detrimental effect on productivity by removing planktonic forms away from the geographical region where conditions were optimum, or by removing them from the sunlit surface waters, or by making coastal waters turbid and reducing the available sunlight.

D.5 Waves and Weather

Waves and foul weather have an ambivalent effect on plant productivity, hurting it in some ways and helping it in others. Clouds, of course, reduce the available sunlight. Choppy waves tend to reflect sunlight, allowing less to penetrate the marine environment. Increased wave activity usually increases the dissolution of atmospheric gases, and also increases the turbidity of coastal waters, further reducing the supply of sunlight. There is an inverse correlation between the thickness of the mixed surface layer and the thickness of the photic zone. Deeper mixing generally results in more turbid water, which reduces the depth to which sunlight penetrates. In coastal waters, then, increased wave activity generally increases the supply of nutrients and dissolved gases, but decreases the supply of sunshine. Whether this increases or decreases plant productivities depends on whether lack of nutrients or lack of sunshine was previously the limiting factor.

D.6 Plant Efficiencies

From our discussion so far, we have learned that the principal constraints on productivity are the availability of nutrients and the availability of sunlight. You might expect, then, that where nutrients are abundant, the productivity should be directly proportional to the amount of sunlight, being greatest at the surface and decreasing with depth. This turns out to be false, however, because productivity depends both on the availability of sunlight and on the efficiency of the plants in using it. Figure 17.17 shows a typical plot of the productivity of a plant culture as a function of the light available to it. At high intensities, plants become very inefficient at using the light available. In fact, beyond a certain point, productivity actually *decreases* when the light intensity is increased.

In the marine environment, maximum productivity occurs about ⅕ of the way down through the sunlit surface waters. (See Figure 17.18.) In clear water this would be at a depth of roughly 20 m, but in turbid coastal waters this would be as little as 4 or 5 m below the surface.

E. MEASURING PRODUCTIVITY

E.1 Productivity and the Abundance of Life

One way to determine the productivity of the various regions of the ocean would be to sample the organisms living there. Where there is more life, productivity must be higher. Since the entire biological community is

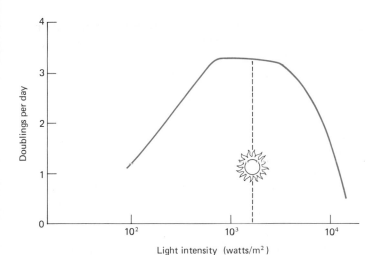

FIGURE 17.17 *Growth rate vs. light intensity for a green algae (Chlorella pyrenoidosa) at room temperature. The intensity of sunlight on a clear day with the sun directly overhead is indicated for reference. (C. Sorokin and R. W. Krauss, Plant Physiol.,* **33** *(1958).)*

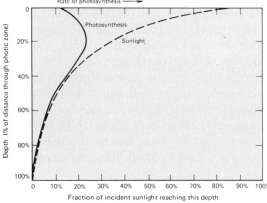

FIGURE 17.18 *Plot of rate of photosynthesis as a function of depth. The photic zone may extend to a depth of 100 m in oceanic waters, or to 20 m or less in turbid coastal waters. There is an inhibition of photosynthesis near the surface due to too much light. With increasing depth, the available sunlight decreases, but the plants become more efficient at using it. Maximum productivity takes place about 20% of the way down through the photic zone.*

dependent on the productivity of the plants, the distribution of any organism would give some indication of plant productivities. For example, the distribution of fishermen or pelicans would be some reflection of primary productivity, since they depend on the fish, which depend on the production of organic material by the plants. Likewise, the distribution of worms or bottom scavengers would reflect the productivity of the surface waters above. Higher productivity means more detritus falling to the bottom, which means more worms and bottom scavengers.

However, the farther along the food chain we go, the more variables enter that might make the distribution of that organism different from that of the primary producers. Porpoises and whales, for example, don't spend all their time catching fish and krill. They have other activities, too, that remove them from the good food areas. Furthermore, other variables besides food sources influence the distribution of a species. In the ice pack, for example, we would find fewer fishermen, but more polar bears, fewer parrot fish, but more krill, and so on.

Even sampling the plants themselves is not a very accurate way of determining productivity, as some plants in some waters may be producing lots of organic material, and others in other waters may be producing very little.

In short, sampling the organisms of a region is not a very accurate way of measuring productivity, although it does give some rough qualitative indication. When life samples are taken, they are usually done with other objectives in mind, such as determining the extent of a species, its relative importance in the local biological community, its migratory habits, and so on.

We have seen that an accurate sampling of any organism is not a very accurate reflection of the rate of primary productivity in the region. To compound the problem, it is difficult to get an accurate sampling of any organism in the first place. Certain species are particularly adept at avoiding capture. Sending down a camera or a diver to observe larger forms is also misleading, as some organisms are curious and some skittish. Some can hide and some can't. Organisms tend to school or cluster, so two samples taken within a few meters of each other may show quite different results. This is even true of planktonic forms (those having limited mobility), because winds or surface currents tend to make them congregate in certain regions.

It is clear that sampling marine organisms is not as simple as picking apples from a tree. The complications force us to use a little ingenuity in obtaining and interpreting our results. Since no collection mechanism is without faults, several are in use, some of which will be described here.

E.2 Taking Life Samples

For collecting plankton, nets are generally used (Figure 17.19). They are conical and are towed behind boats. Many have valves in front that enable them to be closed during their descent to the appropriate depth and also during their retrieval, so that the sample isn't contaminated with plankton from other depths than the one desired. Some are equipped with flow meters, which indicate the volume of water that passes through the net. Comparing this to the size of the sample collected, we can calculate the population density of the various plankton.

Nets of various mesh-widths are available, the finest being of silk with mesh-widths of about 60 microns (1 micron = .001 mm = 10^{-6} m).

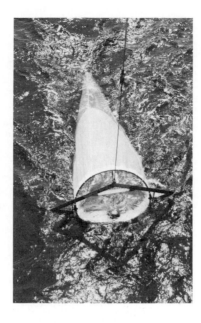

FIGURE 17.19   *Plankton nets.*

Many plankton, however, are smaller than this and will not be caught in the net. To study these very tiny plankton, water samples mut be brought aboard the ship and put in a centrifuge. Being slightly denser than water, they will settle to the bottom in the centrifuge, and can be separated in this way.

Another way of sampling plankton is to pump the water aboard ship and through a filter. This is particularly useful for water near the bottom, where a plankton net would have trouble gathering samples.

Plankton nets must be towed slowly in order to avoid tearing the net or crushing the plankton by the pressure of incoming water. This has the disadvantage of tying up precious time of the research vessel. For this reason, high-speed samplers have been developed. They are towed behind any ship at cruise speed. They have a small opening in the front. Water entering this opening flows into a large chamber at reduced speeds. It passes through a gauze filter before exiting. The gause is on a roll that turns continuously, so a continuous lateral sample of the plankton community is rolled up on the gauze roll for study when the ship ends its journey. (See Figure 17.20.)

Commercial fishermen have developed the best means for catching larger and more mobile members of the marine community. Gill nets, purse seines, trawling gear, line gear, and dredges are a few of the most common tools for catching fish. Both gill nets and purse seines are long nets with floats on the top side and weights on the bottom side to keep them hanging vertically in the water. The gill nets are strung out in regions where the fish of interest are known to be. When attempting to pass through the nets, they get stuck, with the netting staying between their gill cover and pectoral fin, so they can't go either way. The purse seine is

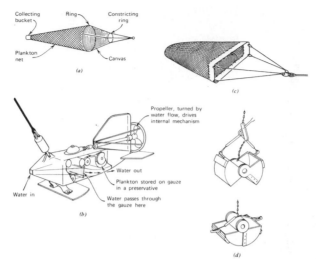

FIGURE 17.20 *Various types of sampling devices. (a) An ordinary plankton net with a restricted opening to reduce the flow rate through the delicate netting. (b) A Hardy continuous plankton sampler. (c) A dredge. (d) A grab sampler.*

draped completely around the school of fish. A string around the bottom is then drawn in, which pinches the bottom of the net together (like a "purse-string"), and prevents any of the fish from escaping (Figure 17.21). In either case the fish are trapped, and the net is brought aboard with the harvest. Trawling gear are nets towed behind moving boats, and line gear is simply hooks on a line. Line gear is used only for large fish, of course.

Dredges are made of rigid nets pulled along the bottom, and are used to collect samples of bottom-dwellers. This is only good for those living on the bottom rather than *in* the bottom, and many of these are mobile enough to avoid capture. For sampling organisms living within the bottom sediment, grab samplers are used to take a bite of the bottom sediment. Only the first 10 cm or so need be taken, as lack of oxygen prohibits life below this.

E.3 Oxygen Production

During photosynthesis of organic materials, oxygen is released. Consequently, measuring oxygen released by plants is a direct measure of their productivity.

In principle, such oxygen-monitoring experiments are simple to perform. We can collect a sample from the area of interest, measure its oxygen content, and then put it into a clear bottle and lower it to the depth of interest (Figure 17.22). After a while, we retrieve the sample and measure the oxygen concentration again to see how much it has changed.

Unfortunately, during the course of our experiment, some oxygen will be consumed through respiration by some of the organisms within our bottle. Consequently, the change in oxygen concentration we measure will be equal to the difference between how much was produced through pho-

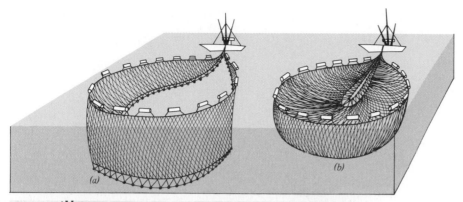

FIGURE 17.21   *Purse seine. (a) The net is laid around a school of fish. (b) Then a
"purse string" through the bottom is drawn to seal off the bottom
and prevent escape. Then the net is hauled in. (c) Photo of a Canadian fleet operating with nets of this sort.*

tosynthesis (P) and how much was consumed through respiration (R).

$$\text{Net Change} = P - R$$

Since we wish to know the total amount of oxygen produced through
photosynthesis (P), we must independently measure the amount of that
which was consumed through respiration (R), so we can make the appropriate correction on our measured net change. We do this by lowering an
identical sample in an identical bottle, which has been painted black to
prevent the admission of sunlight. With no sunlight, there can be no photosynthesis in the black bottle, but respiration will continue. When we
retrieve the black bottle, the measured oxygen depletion gives us the factor "R" in the above equation, with which we can make the correction.

For example, suppose we measure a net gain of 5 ml of oxygen in the

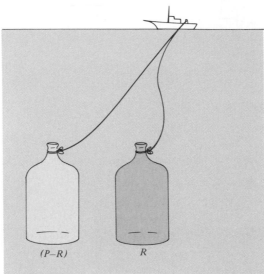

FIGURE 17.22 *Illustration of why two bottles are needed when measuring oxygen production to determine the rate of photosynthesis. In the clear bottle, both photosynthesis and respiration are going on, so the net oxygen production is a result of the difference between the two rates, (P-R). In the dark bottle, only respiration is occurring, so the oxygen consumption in this bottle determines R. Knowing R, we can now determine what P in the first bottle was.*

clear bottle, and a depletion of 3 ml in the black bottle. Then we know that photosynthesis actually produced 8 ml in the clear bottle, 3 ml of which was consumed through respiration, leaving us with the measured net gain of 5 ml. The 8 ml figure would then be the one attributable to photosynthesis alone, and the one we would use to determine the rate of primary productivity.

This method can be used to determine the productivity of individual macroscopic plants, by putting them in plastic sacks or by covering them with bell jars (for those attached to the bottom), and monitoring their oxygen production. It also enables us to study productivities of individual species by making pure cultures of the species of interest, putting them in bottles, and measuring the oxygen production under a variety of temperatures, salinities, and light intensities.

Using this method to estimate the productivities of various regions of the ocean involves several assumptions. One is that the sample used is representative for the area. Because biota show varying concentrations over a distance of just a few meters, the particular sample used may have too many of one species and too few of another, for example. For more accurate results, many different samples must be used, so that these fluctuations will average out. Many microbes show daily vertical migrations, causing cyclical daily changes in the populations at every level. Because of this and because of variation in light intensity during the day, samples should be taken during all hours of the day for more accuracy. There are

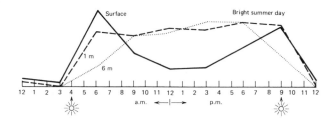

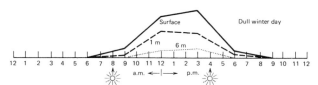

FIGURE 17.23 *Productivity of diatoms at about 60°N latitude. The results were obtained by measurements of oxygen produced by diatoms in bottles. The times of sunrise and sunset are indicated. Measurements were made at the surface and at depths of 1 m and 6 m as indicated. How would you explain the midday drop in surface productivity during a bright summer day? (From A. P. Orr and S. M. Marshall,* The Fertile Sea, *Fishing News Ltd., London, 1969.)*

also seasonal changes in light intensity and water conditions that must be considered before individual measurements of oxygen production may be used to project productivity for any region of the ocean. (See Figure 17.23.)

Among other things, productivity depends on the availability of sunlight, which decreases with depth. There is a depth beyond which there is not enough sunlight for the plants to sustain themselves. This is called the "compensation point." Beneath the compensation point, photosynthesis may continue, but the plants cannot photosynthesize at a rate commensurate with their own respirational needs. Consequently, they cannot really be considered "producers" in this environment. The depth of the compensation point depends on many things, of course. It depends on the plant species, as some are more efficient in their use of meager sunlight than others. Also, it depends on the availability of nutrients. But it is most sensitive to the intensity of sunlight, and is, therefore, related to the season of the year and the turbidity of the water. Typical depths for the compensation point are 100 m in clear ocean water and 20 m in coastal waters.

E.4 Carbon Assimilation

The most accurate method for measuring the productivity of a given sample involves using a radioactive isotope of carbon, C^{14}. A measured amount of carbon dioxide, made with C^{14}, is dissolved in the water. After a certain amount of time, some of this carbon will have been assimilated into new organic tissue, and the rest is driven off (e.g., by heating). By measuring the radioactivity of the remaining sample, we can determine the rate at which the organic material was being produced.

This method has the particular advantage that we don't have to

TABLE 17.2 Comparison of Productivities in Terrestrial and Marine
Environment

	Average Productivity (g Carbon/m²/ yr)	Fraction of Earth's Surface	Total Production, Tons of Carbon Assimilated/yr
Terrestrial	160	28%	25 billion
Marine	50	72%	20 billion

account for the carbon lost to CO_2 through respiration during this time.
The radioactive "tagged" carbon atoms will be a very minute fraction of
the total carbon in the plants, so respiration of tagged carbon atoms will
be only a negligible fraction of the total respiration. Virtually all the C^{14}
assimilated into new materials will still be in those materials as the end
of the experiment. Consequently, we don't have to make a separate measurement of the rate of respiration, as we do in the oxygen experiments.

E.5 Results

Using the above techniques, it is found that the average productivity for
the entire ocean is about 50 g of carbon assimilated per square meter of
the ocean per year. Values range from five or six times this amount in
areas of upwelling to considerably less than this amount in the open
ocean, away from upwelling. Under the Arctic ice cap, the productivity
amounts to only about 1 g of carbon assimilated per square meter of ocean
per year, due to the scarcity of sunlight.

The productivity of plants in the surface waters restricts the amount
of animal life that can exist at all depths. For example, in regions where
surface productivity is 50 g of carbon per square meter per year, the total
respiration by plants and animals at all depths cannot exceed this amount.
Where productivity is greater, correspondingly more respiration can be
supported.

The average productivity on land is more than three times as much
as the average productivity in the oceans, in spite of the fact that a lot of
land is quite barren and non-productive. In fact, even though the land
amounts to only 28% of the area of the earth, the total productivity on land
is slightly greater than the total productivity of the oceans (Table 17.2). On
land, about 25 billion tons of carbon are assimilated per year altogether,
whereas only about 20 billion tons of carbon per year are assimilated by
marine plants. One result of this is that those who plan to look to the ocean
for a solution to the world food problem will probably be disappointed.

The relatively low productivity of the oceans raises the question,
"Why?" Conditions on the continent are much harsher than those in the
ocean. Most land plants must develop extensive root systems in order to
obtain water, and extensive leaf systems in order to get carbon dioxide
and oxygen. Ocean plants, in contrast, are continually bathed in these
materials, and do not need these complicated special systems. Surely the

same sun shines on the ocean that does on the continents. What then, can be wrong with the ocean environment?

The primary reason for the low productivity of the ocean is its great depth. Nutrients released by decaying detritus have difficulty returning to surface waters where they may again be used by plants. On the continents, by contrast, detritus remains on the surface, and released nutrients are immediately available for use again by the plants. Clearly, if we hope to develop aquaculture in response to the world food problem, we're going to have to solve the problem of keeping nutrients near the surface.

F. SUMMARY

The energy for life is provided by the sun. Plants convert this energy into stored chemical energy, which is later released during respiration by either the plants themselves, or by organisms further along the food chain.

The bulk of organic tissue is composed of carbon, hydrogen, and oxygen, all of which are amply abundant in the marine environment. Nutrient nitrogen and phosphorous are needed in smaller amounts, but their scarcity often restricts organic production. In addition to the supply of nutrients, the rate of organic synthesis also depends on the depth and stability of the photic zone, and the supply of micronutrients and toxic substances.

Since the nutrients are in short supply, the organic productivity of a region is usually quite sensitive to how rapidly they are recycled and returned to the photic zone from decaying organic detritus. Nutrient nitrogen is particularly slow at being recycled, and frequently sinks beneath the photic zone before it is released from the detritus. Because of this removal of nutrients from surface waters, deeper waters tend to be rich in nutrients. Wherever these deeper waters are brought to the surface, there tends to be a large amount of biological activity. Upwelling and turbulence are two mechanisms that help accomplish this.

In high latitudes the lack of sunshine restricts biological productivity. In tropical latitudes, the warm surface waters are isolated from the nutrient-laden deeper waters by a thermocline. Consequently, oceanic productivity is low due to lack of nutrients. In temperate latitudes, the thermocline is seasonal, resulting in seasonal variation in nutrient supply. This, coupled with seasonal variation in sunlight, produces a seasonal pattern in primary productivity that typically displays a peak in the early spring and a smaller peak in the fall. Over the continental shelves, detritus cannot sink beneath the surface zone, and so productivity is higher over the shelves in temperate and tropical latitudes.

In addition to nutrients and sunshine, other things affecting productivity are the microscopic grazers, which feed on the primary producers, the water temperature, which influences the rate of chemical reaction, the variability of the coastal environment, currents, and waves. Although available sunlight decreases with depth, plant efficiencies increase, and so the most productive region is about 20% of the way down through the photic zone.

The productivity of an area may be inferred from the amount of biological activity there, although there are many variables in making such estimates. More accurate determinations are made by measuring rates of oxygen production or of carbon assimilation.

QUESTIONS FOR CHAPTER 17

1. Where does the energy for life originally come from?

2. What kinds of organisms can capture this energy and store it in a form that can be released later during respiration?

3. What is the name of the process through which plants convert solar energy into stored chemical energy? What is the name of the process through which this stored chemical energy is released (by oxidation)?

4. Why is it that productive marine plants can only be found in the upper 100 m of seawater?

5. What are "herbivores"? "Carnivores"? "Onmivores"?

6. What is synthesized in photosynthesis?

7. What is it about the carbon atom that makes it able to form the basis for very large and complicated molecules?

8. What are carbohydrates made of? What is the simplest carbohydrate? What are fats and proteins made of?

9. How are amino acids related to proteins?

10. There is lots of dissolved N_2 gas in the ocean. Can this be used by plants? What forms of nitrogen can?

11. In what form is phosphorus normally encountered? Can the plants use it in this form?

12. The carbon outgassed from the earth's interior usually arrives in what form? In what forms does it appear in seawater? Which of these is most common?

13. Of the materials needed for photosynthesis, which element is most abundant in the ocean? (*Hint:* Water is H_2O.) What element is second most abundant? Third?

14. If sufficient sunlight were present that photosynthesis proceeded unchecked, what material(s) would be used up first?

15. In addition to nitrates and phosphates, what else might be considered to be a nutrient for diatoms? Why?

16. What are "trace elements"? List some of them. Can you state what functions in your body some of these trace elements are related to?

17. Why do we terrestrial creatures suffer more frequently from lack of trace elements than marine organisms?

18. Write out the basic equation for life. Where does the energy come from in photosynthesis? What are some of the things you can think of for which this energy might be used when it is released during respiration?

19. Is the release of energy in your car engine or in a coal-burning power plant basically the same as a fundamental biological process? Which one?

20. Write down the specific chemical equation for the production of, or the oxidation of, the simple sugar, $C_6H_{12}O_6$.

21. Why is the productivity of the near-surface waters dependent on the recycling of the nutrients, and not the recycling of carbon, hydrogen, or oxygen?

22. Is it the carbohydrates, the fats, or the proteins that organisms oxidize first when in need of energy? Which do they oxidize last? Why is this bad from the point of view of recycling the nutrients? Why do you think organisms do it in this order?

23. Discuss some of the things that cause depletion of nitrates from surface waters. Is the same true of phosphates? Which is usually more critical, and why?

24. As organisms die and decompose, why is it that the nitrogen takes particularly long to get back into the environment as a nitrate? Why is it that even after it reenters the environment as a nitrate, it frequently must still wait a while before it can be used in photosynthesis?

25. Although nitrogen is far more abundant in the biosphere than is phosphorus, it is the lack of nitrates rather than phosphates that usually limits the biological productivity of an area. Explain why this is so.

26. Why is it that biota are particularly abundant in regions of upwelling waters?

27. What are some of the things that might cause upwelling of deeper waters? Could you locate on a map some of these regions where upwelling is likely?

28. Why are the coastal waters of much of the western coast of the United States cool and rich in biological activity?

29. Why are waters near coral reefs nutrient-rich?

30. What causes turbulence?

31. How is it that turbulence can affect biological productivity?

32. Why is it that turbulence is much more effective in bringing nutrients to the surface over continental shelves than over deep ocean basins?

33. What is the main factor limiting biological productivity in high latitudes? Why are nutrients particularly abundant in the surface waters there?

34. In what latitudes would you expect the greatest contrast between the productivity over the shelves and that over the deep ocean waters? Why?

35. How does the presence of a thermocline beneath the surface waters restrict the productivity of these waters?

36. Do you suppose upwelling in tropical waters or upwelling in polar waters would have a bigger effect on the biological productivity of the area? Explain your answer.

37. What is the difference in color between biologically active waters and those that are biologically dead?

38. Why is the Gulf Stream such a beautiful deep blue color?

39. Describe the seasonal variation of productivity in temperate latitudes, and explain it.

40. In the host–parasite problem, how might a slight increase in the host population eventually lead to the demise of the parasites?

41. Describe how the spring and fall blooms and midsummer lull in temperate latitudes can be explained in terms of the host–parasite problem.

42. How does temperature influence the rate of chemical reactions? How should this bear on productivity if all other factors are equal?

43. If chemical reactions proceed faster in warmer waters, why is it that highest productivity is usually found in *colder* waters?

44. What are some of the things that make environmental conditions in coastal waters so variable?

45. Why are attached plants usually more tolerant of environmental changes than planktonic plants?

46. Why is productivity low in nearshore waters with loose sediment substrate?

47. What are some of the positive and negative effects that currents may have on productivity?

48. What are some of the negative effects on productivity caused by waves and foul weather? What are some of the positive effects?

49. Is it true that productivity is usually highest at the surface where sunlight is most plentiful? Explain.

50. How far down is the depth of maximum primary productivity? Roughly how deep would this be in clear ocean water? In turbid coastal waters?

51. What is a typical depth to which the photic zone extends in the deep ocean waters? In turbid coastal waters?

52. Explain how studying the distribution of any single organism would

shed some light on the productivity of various regions of the ocean. What are some of the things wrong with using this method to determine productivities?

53. Why is sampling the plants themselves still not a very accurate way of measuring productivities?

54. Why might photographs give a poor representation of even the various large forms of life present?

55. Describe some of the gear used for sampling planktonic life.

56. Why must plankton nets be towed slowly? Why is this undesirable?

57. Describe some of the more common commercial gear used to catch fish.

58. How can samples of organisms living on or in the bottom be obtained?

59. Why is life found only within the upper 5 or 10 cm of bottom sediment?

60. Why do you suppose that water trapped in the sediment of the ocean floor usually loses all its oxygen?

61. Suppose we wanted to determine productivity by measuring the rate of oxygen production. How could we find the *net* rate of oxygen produced by a sample? How is this related to the rate of photosynthesis? What else must be known? How is this measured?

62. How can this method be modified to determine the productivity of a single plant attached to the bottom?

63. Why is a study of biological productivity primarily concerned with the plants?

64. Suppose we wanted to determine the productivity in a certain region of the ocean by monitoring the oxygen production in samples of the ocean water. What are some of the variables that make inaccurate the results from any one sample? What can be done about these things to increase the accuracy of our results?

65. What is the "compensation point"?

66. What are some of the things that the depth of the compensation point depends on? Explain why each is relevant.

67. How can radioactive C^{14} be used to measure productivity?

68. In using C^{14} to measure productivity, why is it *not* necessary to make a separate measurement of respiration, as it was in the oxygen-monitoring experiments?

69. On the average throughout the ocean, about how much carbon is assimilated per year from each square meter of area? How much more than this is assimilated in upwelling regions?

70. How does the average productivity per square meter on land compare to the average productivity per square meter in the ocean? How do the *total* productivities of these two regions compare?

71. What is the main reason why the productivity of the ocean is low compared to that on land? Explain.

72. Why is the return of nutrients not such a problem on land as it is in the ocean?

*73. Can you think of any possible reason why plants at the surface use sunlight more inefficiently than those a few meters deeper?

*74. Why do you suppose that the spring bloom begins in the very surface water and then progresses deeper, as illustrated in Figure 17.12?

*75. Exactly how does the thermocline "prevent" the surfacing of nutrients?

*76. According to Table 17.1, hydrogen and nitrogen have nearly equal abundances by weight in organic matter. Does this mean that there are roughly as many nitrogen atoms present as hydrogen? Explain.

SUGGESTION FOR FURTHER READING

John H. Steele, "Patterns in Plankton," *Oceanus*, **23**, no. 2 (Summer 1980).

18
THE SPECTRUM OF MARINE ORGANISMS

This chapter describes some of the more common organisms found in the oceans.

There is a wide variety of organisms inhabiting the oceans, ranging from porpoises and seals, which frolic on the surface, to bacteria and worms working within the deep ocean sediment. This wide range in appearance and behavior of marine biota offers us a variety of criteria according to which we may describe them. Among these criteria are how the organisms feed, where they live, and how well they move.

One means of describing marine organisms is based on their mode of nutrition (Table 18.1). Those that can be independent of outside sources of organic matter, having the ability to use solar energy to produce their own food, are called "autotrophic." Those that cannot are "heterotrophic." Most plants are autotrophic, and all animals are heterotrophic.

The organisms seem always to be more complicated than the categories into which we try to put them. Many plants, for example, can produce most, but not all, of their requirements. These plants depend on absorbing some materials (e.g., certain vitamins) that have been produced by other plants in the community. Strictly speaking, these plants cannot be considered completely autotrophic.

Some organisms can use inorganic chemical reactions to provide the energy for the synthesis of organic tissue rather than using sunlight. Many bacteria do this, but often produce only a fraction of their total requirement in this manner. Many bacteria, then, are in this borderline region, being neither completely dependent on, nor completely independent of, outside sources of organic material.

Another way of characterizing marine life is according to the environment it inhabits. Those organisms that live on the ocean bottom or within the sediment are called "benthic." Those that live in the water are called "pelagic."

As we have seen earlier, the pelagic environment is conveniently subdivided into the photic and aphotic zones, as different types of organisms tend to be found in these two regions. Productive plants are found only within the photic zone, although various forms of animal life can be found at all depths. The lower edge of the photic zone is conveniently defined by the compensation point, which is the point beneath which plants consume more than they are able to produce with the meager sunlight there. Sometimes oceanographers refer to an inter-

TABLE 18.1   Some Classification Schemes for Marine Organisms

Classified According to	Classes	Characteristics
Mode of nutrition	Autotrophic	Can make their own food
	Heterotrophic	Cannot make their own food
Mobility	Nekton	Swimmers
	Plankton	Drifters ⟨ phytoplankton (plants), zooplankton (animals)
Habitat	Pelagic	Live in water ⟨ intertidal, neritic, oceanic
	Benthic	Live in/on bottom ⟨ littoral, sublittoral, bathyal, abyssal, hadal

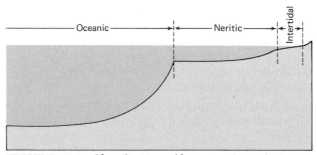

FIGURE 18.1   *Classification of biozones for pelagic organisms.*

mediate dimly lit "disphotic zone," extending a few hundred meters beneath the photic zone, where some lighting still exists, but is insufficient for plant productivity. In clear ocean waters, the photic zone extends to a depth of roughly 100 m. In the more turbid coastal waters, light cannot penetrate as well, and the photic zone is thinner than this, 20 m being typical.

Waters in and near the photic zone show much greater seasonal and geographical variations in temperature and salinity than do deeper waters. This is reflected in much greater geographical variation in biological species found in surface waters than is found in deeper waters.

Another way of subdividing the pelagic zone involves geographical considerations instead of sunlight (Figure 18.1). According to this scheme, the pelagic zone is divided into the "neritic" environment, which encompasses the waters of the continental shelf, and the "oceanic" environment, encompassing the waters of the deep ocean. There is much more water in the oceanic region, but for reasons we have already studied, the neritic environment is quite productive biologically.

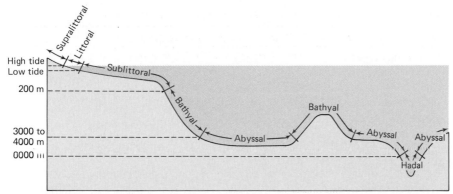

FIGURE 18.2 *Biozones for benthic organisms.*

The benthic community is subdivided geographically into "littoral," "sublittoral," "bathyal," "abyssal," and "hadal" biozones, as illustrated in Figure 18.2. The littoral zone is the intertidal region, and the sublittoral zone is the rest of the continental shelf. Together they accommodate the majority of the benthic life, as they don't have the life-limiting light and food deficiencies of the deeper waters. The sublittoral zone is customarily further subdivided into the "inner shelf" and the "outer shelf." The inner shelf is above the compensation point, and so some plants will be attached to the bottom, except where moving sediment grains prevent this. The outer shelf is below the compensation point, and so it is lacking in attached plants.

The bathyal zone is the intermediate range of depths, encompassing the continental slope, rise, and much of the oceanic ridge system. The abyssal plains and ocean basins comprise the "abyssal" zone, and the relatively small areas of the ocean trenches make up the "hadal" biozone. We have already seen that the benthic creatures in these deeper regions must put up with reduced food supplies. Another problem is that the great pressure increases the viscosity of the water, which reduces the mobility of these organisms.

A.3 Mobility

Pelagic organisms can be categorized according to whether they are swimmers or drifters (Figure 18.3). The swimmers are called "nekton," and the drifters are called "plankton." The distinction is not as clear as it might seem, as a careful study reveals that few organisms are without some weak means of mobility, such as the ability to change their density slightly, or having tiny microscopic hairlike whips for locomotion.

To resolve these borderline cases, a commonly used distinction is that nekton are capable of "fast extended motion" and plankton are not. The distinction if often a matter of size and maturity rather than breed. Many species that are nekton as adults are plankton during their egg and larval stages.

There is much more diversity among plankton than nekton. Most nek-

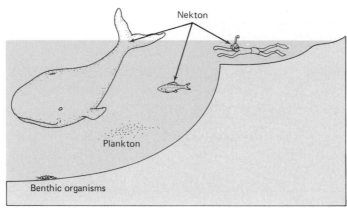

FIGURE 18.3 *Oceanic life can be divided into two classes—the benthic organisms, which live on or in the bottom, and the pelagic organisms, which live in the water relatively free of the bottom. The pelagic organisms are further divided into the swimmers, the nekton, and the nonswimmers (or rather, the not-so-obvious swimmers), the plankton.*

ton belong to the animal phylum Chordata, which are those animals that have some sort of a segmented spinal column. These would be fish, reptiles, and mammals, for example, with perhaps a few members of other phyla present, such as squid or octopus.

The plankton include all plants that are not attached to the bottom, and most animals ranging in size from large jellyfish, which may be 1 m in diameter, down to the microscopic one-celled organisms that make up the bulk of the ocean's animal community.

The distribution of the various species of plankton is controlled by currents and wind, and is limited by the environmental conditions, such as light, temperature, and salinity that each can tolerate. The species of nekton are, in general, more widely distributed than the individual planktonic species, which is a reflection of their greater mobility as well as their control over their body fluids, which moderate the cellular environment.

The Benthos are also subdivided according to mobility. Those that are attached to the bottom are called "sessile," and those that can move are said to be "vagrant." The vagrant benthos can be further subdivided into those that live on the bottom but are not confined to it ("nektobenthos," or "demersal fish,"), such as flounders and rays, those that crawl over the bottom, such as starfish and crabs, and those that burrow within the bottom such as worms and clams. (See Table 18.2.)

C.4 The Plankton

The great diversity among the various types of plankton makes further description useful (Table 18.3). One of these further descriptions is on the basis of size, and is related to how they can be separated from seawater when biological samples are taken. Those that are so small that they cannot be caught even in the finest-meshed silk nets are called "nannoplank-

TABLE 18.2   Ways of Characterizing Benthos

Benthos	sessile (attached)	
	vagrant (mobile)	live on, but not confined to, the bottom
		crawl over the bottom
		burrow in the bottom

TABLE 18.3   Ways of Characterizing Plankton

By size	Nannoplankton—smaller than 60 microns
	Microplankton—larger than 60 microns, but smaller than 1 mm
	Macroplankton—larger than 1 mm
By duration of planktonic life	Meroplankton—part of life as plankton
	Holoplankton—all of life as plankton
By kingdom	Phytoplankton—plants
	Zooplankton—animals

ton." They are less than 60 microns in diameter, and include coccolithophores and smaller diatoms (Figures 18.4 and 18.5), among other things. Plankton that are larger than this, but still are too small to be seen easily with the unaided eye, are called "microplankton." They range from 60 microns in diameter to about a millimeter, and they include the bulk of the plankton in the marine environment. Plankton easily seen with the unaided eye are called "macroplankton" (or sometimes "megaplankton"), and include such things as krill, jellyfish, and floating sargassum weed.

Plankton may also be distinguished according to how much of their life they spend as plankton. Those that are plankton their entire life are "holoplankton," and include most planktonic plants. Those spending only part of their lives as plankton are called "meroplankton." Most meroplankton are juveniles, and the vast majority never reach adulthood, as they provide a favorite food supply for many larger organisms.

Planktonic plants are called "phytoplankton," and the planktonic animals are called "zooplankton." The phytoplankton are responsible for the bulk of the primary productivity, receiving some help from the attached plants in coastal regions. The most important marine producers are the diatoms, followed by coccolithophores and various kinds of dinoflagellate algae. The primary importance of the zooplankton in the ecological community is that they consume and concentrate the materials produced by the phytoplankton, facilitating its passage up the food chain. The most important zooplankton, in this respect, are the various copepods (Figure 18.6), due to their very large numbers. Krill are important in Antarctic waters.

Some species of zooplankton are extremely sensitive to the conditions of their environment. These "indicator species" may be used to trace the origin and extent of various water masses. Their skeletons in the bottom sediment also help us to know environmental conditions of past ages.

C. LACINIOSUS

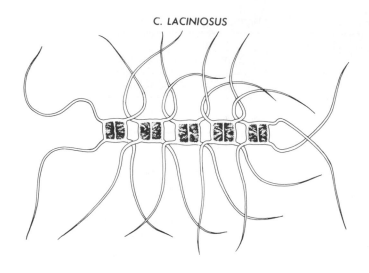

CHAETOCEROS

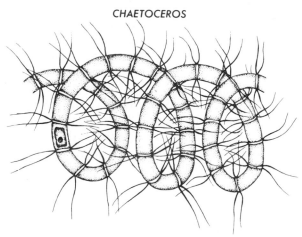

FIGURE 18.4 *Some diatoms form chains, such as those sketched above. (Bayard H. McConnaughey,* Introduction to Marine Biology, *3rd ed., The C. V. Mosby Co., St. Louis, 1978. Redrawn from various sources.)*

B. THEIR CLASSIFICATION

The organisms of the world are extremely varied, and are far too numerous for our minds to comprehend. Order aids our comprehension, and so we have established systems for classifying biological organisms, which also helps us in communication with colleagues and in attempting a deeper understanding of the biosphere.

Of course, there are an infinite number of possible classification schemes. We could classify organisms according to color, length, weight, habitat, and so on. But the idea that forms the basis of the most commonly used classification schemes for all the various species is that the devel-

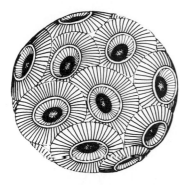

COCCOSPHAERA ATLANTICA

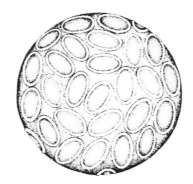

SYRACOSPHAERA BRASILIENSIS

DISCUSPHAERA TUBIFEV

FIGURE 18.5 *Sketch of various kinds of coc-colithophores, enlarged about 2000 times. The individual skeletal plates are called "coc-coliths." (Bayard H. McCon-naughey, Introduction to Marine Biology, 3rd ed., The C. V. Mosby Co., St. Louis, 1978. Redrawn from various sources.)*

opment of the individual organism reflects the history and development of its ancestors. Thus, most species in one class have some developmental similarities, and probably had similar ancestry.

It turns out that biological organisms are so varied that no classification scheme yet devised is sufficient to find a unique spot for each one without ambiguity. But we can devise classification schemes that handle the bulk of nature's organisms fairly well, and so that is what we do.

All biota are divided in a few broad categories called "kingdoms," according to broad general characteristics. The members of any one kingdom are then subdivided according to more specific characteristics into "phyla." The members of any one phylum are then further subdivided into "classes," according to even more specific characteristics, and so on. The order of the various steps of classification, from the largest, broadest groups to the most specific categories is as follows:

1. Kingdom   4. Order   7. Species
2. Phylum   5. Family
3. Class   6. Genus

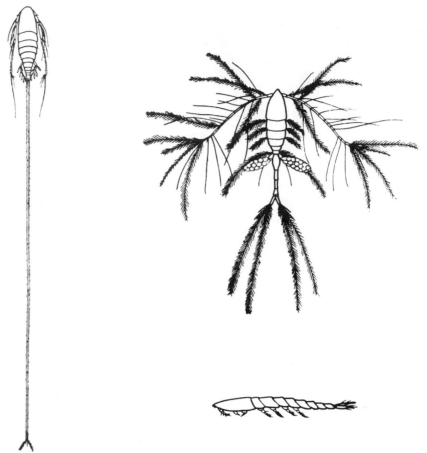

FIGURE 18.6 *Sketches of some copepods with interesting, intricate appendages. Copepods range in size from about 0.2 mm to 2 mm, and are the most abundant herbivores in the ocean. (Sverdrup, Johnson, and Fleming, The Oceans: Their Physics, Chemistry, and General Biology, Prentice-Hall, © 1942, renewed 1970.)*

It is emphasized that this classification scheme is according to similarities in individual characteristics, and not according to numbers of members or relative importance in the biosphere. There are some species (especially of arthropods) whose members outnumber those of an entire phylum. In the pages that follow, where we describe the classification of the more important marine organisms, we may on one hand discuss some members of some classes in detail, and on the other hand ignore entire phyla completely, due to the smaller number of importance of the members.

A common and historically important scheme of classification divides all organisms among just two kingdoms—plants and animals. Plants are characterized by ability to photosynthesize, and animals by their mobility. This causes some troubles, though, as there are some plants, such as yeasts and fungi, which do not photosynthesize. Many animals are planktonic and no more mobile than some plants. Some bacteria are autotrophic and some aren't. Some single-celled dinoflagellates can photosynthesize and some cannot. Should two microorganisms that are clearly structurally and ancestrally related be put into two different kingdoms just because one can photosynthesize and one cannot?

Most of these ambiguous organisms are microscopic, so some classification schemes have been developed to provide separate kingdoms for these microscopic borderline organisms. One of these divides all organisms among four kingdoms—Monera, Protista, Metaphyta, and Metazoa. We describe this scheme briefly here. It should be emphasized, however, that the complexity of organisms ensures that they will not all fit perfectly into any classification scheme. There is not only frequent disagreement as to the classification of a particular organism, but also disagreement as to which classification scheme should be used in the first place.

The kingdom Monera contains primitive, one-celled organisms who do *not* have a nuclear membrane separating the cell nucleus from the rest of the cell material. It is divided into two phyla, one of which includes bacteria, and the other blue-green algae. We saw in Chapter 1 that primitive blue-green algae provide the earliest plant fossil record on earth.

The kingdom Protista includes organisms whose cells do have nuclear membranes and chromosomes, and who may have primitive powers of mobility. This kingdom includes all single-celled animals, single-celled plants aside from those in the kingdom Monera, and larger plants that do not have true leaves, roots, and stems. Important marine plants in this kingdom are the green algae, brown algae, red algae, diatoms, and dinoflagellates. Important animals are the protozoans Foraminifera and Radiolaria.

The kingdom Metaphyta includes green, multicellular plants with true leaves, stems, and roots. These are very common on land, but play a minor role in the ocean, only being found in restricted coastal areas.

The kingdom Metazoa includes all multicellular animals. There are large varieties of these in the oceans, of course, and we'll examine several of them in succeeding pages.

For the remainder of this chapter, we will use the more familiar scheme of division between plants and animals. For those who prefer division among the four kingdoms—Monera, Protista, Metaphyta, and Metazoa—the above description should be sufficient to enable you to see how each of the phyla described below fits in.

C.   Marine Plants

We will begin our discussion with the plants, because they support the entire biological community through their production of organic materials. They are confined to the upper 100 m or so of water by the availability of sunlight, and the production in this thin layer supports the animal populations at all depths. Ocean plants can be considerably simpler than their terrestrial relatives, since they are continually bathed in the medium that brings them their requirements. They don't need roots to bring them water and nutrients, nor do they need leaves to obtain O_2 or CO_2.

The plant kingdom is subdivided into four subkingdoms. Listed in order of increasing sophistication, they are:

1. Thallophyta (algae and fungi)
2. Bryophyta (mosses)
3. Pteridophyta (ferns)
4. Spermatophyta (flowering and seed producing plants)

Of these subkingdoms, only the first and last have members in the marine environment.

C.1   The Thallophytes

The overwhelming majority of marine plants belong to the subkingdom Thallophyta. These are the most primitive plants, and they are commonly referred to as "algae" rather than "thallophytes." Some of these are anchored to the ocean bottom ("sessile"), and others are planktonic. The sessile thallophytes are found in shallow water only, of course, rarely in water deeper than 100 m. They cannot be anchored where there is drifting sand, near sandy beaches, for example. They generally cling to solid surfaces with holdfasts that are not true roots, but rather are only for purposes of anchoring the plants. Coastal areas having rocky bottoms make up less than 2% of the total oceanic area, so the anchored plants contribute considerably less to the total oceanic productivity than do the planktonic forms, which can drift anywhere. The anchored coastal plants are more familiar to us, both because of their proximity to land and because of their size. Most planktonic producers are microscopic, and so we don't notice them, in spite of their large numbers and importance.

The various kinds of algae are by far the most important producers in the marine environment. Below are listed the most important phyla. The first four are named according to the color of their more prominent members. The fifth group includes diatoms and coccolithophores, which are by far the largest contributors to the ocean's total productivity, in spite of their small microscopic sizes. Unfortunately, there is no universal agreement at the moment regarding what phylum they should belong to, or even if both should belong to the same one. Sometimes they are put in a phylum called "Chrysophyta," and are sometimes called "golden-brown algae," or sometimes "yellow-green algae." The sixth phylum includes the dinoflagellates that can photosynthesize.

a. Blue-green algae (Phylum Cyanophyta)
b. Green algae (Phylum Chlorophyta)
c. Brown algae (Phylum Phaceophyta)

d. Red Algae (Phylum Rhodophyta)

e. Diatoms and coccolithophores

f. Dinoflagellate algae (Phylum Pyrrhophyta)

The first four phyla are predominantly sessile plants, and so they are fairly well restricted to shallow coastal waters. This is a very small fraction of the total ocean area, so they contribute only a small fraction to the total productivity of the oceans, in spite of the abundance of nutrients in the nearshore waters.

The two greatest contributors to the total productivity in the oceans are diatoms and coccolithophores. The third most important contribution to total production probably comes from the various dinoflagellate algae. All of these algae types are microscopic, single-celled plankton, and so we do not see them when we visit the beach. Far more familiar to us are the visible, nearshore forms, and so we'll discuss them first.

(a) Blue-Green Algae. The blue-green algae are extremely primitive, nonnucleated, single-celled plants. They are especially common in fresh water, where they tend to form slimy mats and surface scum. They are nowhere particularly important in the oceans. Some species are found in the brackish waters near river mouths, and some species are found in tropical regions. Compared with other more advanced algae types, they show remarkable tolerance to changes in the salinity and temperature of their environment.

Although the more prominent members of this phylum are blue-green in color, there is a rather famous marine planktonic member (Trichodesmium) whose color is red, due to the presence of additional pigments in the cell. Large concentrations of these are found in the tropical waters of the Red Sea, and they are responsible for the name of this body of water.

(b) Green Algae. Green algae reproduce by spores, and are found mostly at shallow depths of less than 10 m. They are found both in fresh water and in salt water, although more commonly in fresh water, where they are the dominant form of algae. They often appear as small, "leafy" attached plants, and are familiar as the small leafy slime attached to old boats, pilings, and so on, which have been sitting still in fresh water for a while. A common form in the marine environment is "sea lettuce," (Figure 18.7) which derives its name from its appearance.

(c) Brown Algae. Brown algae are almost entirely marine, and are considered the most advanced of the marine algae (Figure 18.8). They generally grow in deeper water than do the green algae, and as a rule prefer cooler waters and rocky bottoms. One important member of this group is kelp (Figure 18.9), which is commercially harvested for iodine, potassium salts, food, animal food, and fertilizer. The leaflike "fronds" are held near the sunlit surface by a bulb full of gas, which the plant produces. The bulb is anchored to the bottom by a long hollow tube called a "stipe." These plants can attain lengths greater than 80 m, if needed. They also have an interesting "two-step" reproduction mechanism. The offspring of the mature kelp is a small bottom-anchored plant that otherwise barely

FIGURE 18.7 *Foreground: sea lettuce (a green algae). Background: marsh grass.*

resembles the parent. The offspring of this small plant is again a long kelp of the familiar appearance. Thus, each generation resembles its grand-parents, but not its parents.

Another important member of this phylum is the sargassum, or "gulf weed," which grows in the Caribbean area. It is much smaller than the kelp, but is anchored, and produces buoyant bulbs for the fronds, like the kelp. The bulbs tend to come in grapelike clusters, which accounts for the name; "sargassum" is derived from the Portuguese word for "grapes." It is not always benthic, as it may break off during storms and float at the surface via the gas filled floats, in this way becoming planktonic. These small planktonic mats of tangled sargassum can grow and reproduce as they travel, producing a unique environment for small marine organisms, until they lose their floats and sink. They tend to be carried by currents to a region of the North Atlantic that is understandably called the "Sargasso Sea."

(d) Red Algae. Red algae (Figure 18.10) are small compared to the more prominent brown algae, and can be found from the intertidal zone out to depths of greater than 100 m in some cases. Thus, red algae extend out to greater depths than brown algae, and brown algae extend farther

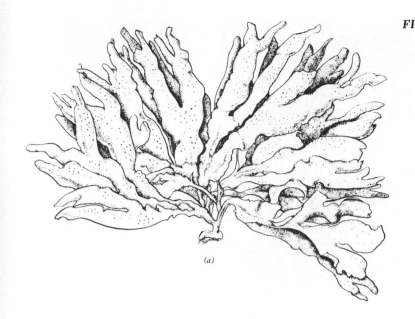

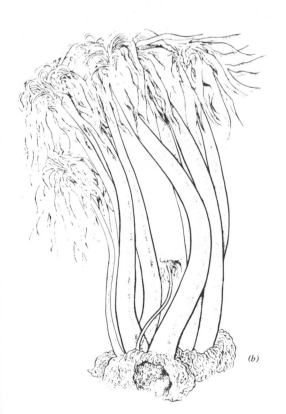

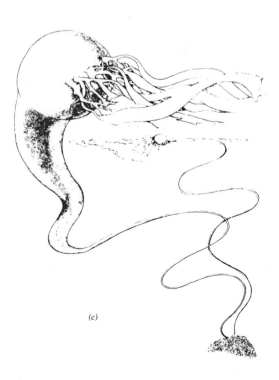

FIGURE 18.8 *Sketch of some brown algae. (a) Fucus. (b) Sea palm. (c) Kelp. (Bayard H. McConnaughey, Introduction to Marine Biology, 3rd ed., The C. V. Mosby Co., St. Louis, 1978.)*

FIGURE 18.9   *Kelp (a brown algae).*

(a)

(b)

FIGURE 18.10   (a) *Sketch of the red algae, Gigartine.* (b) *Photo of the red algae* Chondrus crispus *(Irish moss).* ((a), *Bayard H. McConnaughey,* Introduction to Marine Biology, *3rd ed., The C. V. Mosby Co., St. Louis, 1978.)*

out than do green algae. Their red color means that they are more efficient at absorbing sunlight in the blue and green portion of the spectrum, which is the only part of the spectrum that will penetrate to the lower regions of the photic zone.

Red algae are very pretty due to their coloring, which is usually variations on red or purple, depending on the other pigments present. They

  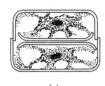 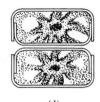

(a)　　　　　　　　(b)　　　　　　　　(c)　　　　　　　　(d)

FIGURE 18.11   *Illustration of successive stages of cell division for a pillbox type of diatom.*

are most abundant in warm waters, although some are found in cold areas, too. Some are harvested for minerals or fertilizer. Some are harvested to extract a thickening agent used in such things as ice cream, salad dressings, soups, cosmetics, and paints.

(e) *Diatoms and Coccolithophores.* Diatoms belong to the class Bacillariophyceae, and are the most abundant phytoplankton in the ocean, both in terms of total mass and in terms of numbers of individual members. Each diatom is a single cell, measuring between 20 and 80 microns in length. Diatoms display a wide variety of perforated geometrical shapes, such as tiny hockey pucks, pincushions, butter dishes, or tin cans, which are frequently stuck together in short strings. Each diatom secretes a perforated silica shell, which allows it access to nutrients and other needed materials in the environment without exposing it to the small predators. Also, the rough shell increases friction with the water, which is important to the denser-than-water diatoms, which continually face the problem of sinking too rapidly out of the photic zone.

The diatoms reproduce by cell division. First the nucleus splits, then the cell itself. This creates a problem, of course, since the two offspring are still bounded by the original hard silica shell. It turns out that the silica shells come in two pieces. One end slips inside the other end, like the two halves of a pillbox (Figure 18.11). When the cell divides, each new cell takes half of the pillbox, and must secrete the other half to make its new skeleton complete. The secreted half always is the "inside" half of the "pillbox." This means that the secreted half will always be smaller than the original. Consequently, as this process continues from generation to generation, the silica "pillboxes" become smaller and smaller. Since the cell needs a certain minimum volume to function properly, a point eventually comes when the "pillbox" is too small for the diatom. At this point, the cell discards its skeleton entirely, and secretes a new larger one, starting the cycle over again.

Diatoms have the ability to produce "resting spores," which enable them to wait out extended periods in an inhospitable environment. When conditions again permit, the resting spore will develop into a diatom, and normal activity is resumed.

Coccolithophores are extremely tiny phytoplankton, measuring 10 to 20 microns in diameter. They escaped detection for years, as their small size enabled them to slip through the filters used to obtain samples of microscopic plankton for study. Their relative importance is still unde-

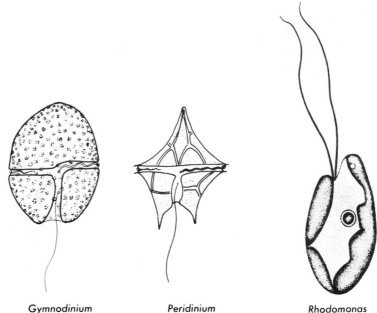

Gymnodinium   Peridinium   Rhodomonas

FIGURE 18.12   *Some dinoflagellates. (Bayard H. McConnaughey,* Introduction to Marine Biology, *3rd ed., The C. V. Mosby Co., St. Louis, 1978, Redrawn from various sources.)*

cided, although the abundance of their skeletons among the sediments makes it clear that they are important members of the marine community. They have two flagella for minor locomotion, and their bodies are covered with very tiny calcite plates. They prefer warmer climates. This is in contrast to diatoms, which have siliceous skeletons and prefer cooler waters. This correlation between water temperature and skeletal material is understandable in light of what we know about solubility and water temperature. Most solid materials, including silica, are less soluble in cold water. Calcium carbonate, however, is less soluble in warm water, due to the reduced carbon dioxide content and lower acidity of the water.

(f) Dinoflagellate Algae. Dinoflagellates are one-celled organisms similar to diatoms in size. They have cellulose outer shells, and have one or more whiplike tails called "flagella" that provide locomotion (Figures 18.12 and 18.13). Because they don't have a mineral skeleton, they don't contribute to the bottom sediment. Some can photosynthesize and others cannot. Many are luminescent, and can make the water seem to glow at night when agitated, as in the wake of a ship. Dinoflagellates seem to prefer warmer waters, in contrast to the diatoms, which are most abundant in cooler waters.

C.2   The Spermatophytes

The flowering and seed producing plants are the most sophisticated, having true roots, stems, and leaves. They apparently evolved on land first, and then some members returned to the sea, probably through fresh or

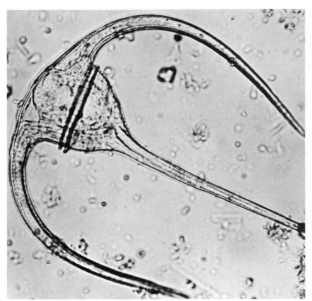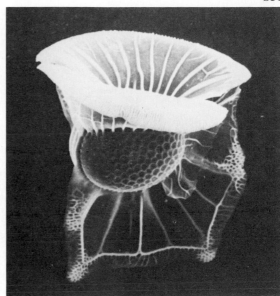

FIGURE 18.13 *The dinoflagellate Ceratium (left) and Ornithocerus (right). The magnification is about 300 ×. ((left) Walter Dawn.)*

brackish water inlets. They constitute a very minor portion of the total marine plant community, and are found only in the upper subtidal region, from the low tide down to a depth of less than 7 m. Mangrove trees, such as those which populate the Florida coast, are partially submerged marine members of this subkingdom. They can trap sediments in their roots, forming small mangrove reefs. They also have the interesting ability to move slowly along the ocean bottom by discarding old roots on one side and growing new roots on the other side.

Another important marine member of this subkingdom is eelgrass (Figure 18.14). This can reproduce either by producing seeds or by extending a special stem (called a "rhizome") along the ocean bottom, which can produce new leafy shoots and root systems at various points along it. The student may be familiar with terrestrial plants, such as strawberries, spider plants, or quack grass, which do similar things. Where eelgrass is found, it forms a nice thick matting that furnished protection for many marine animals that would otherwise be too vulnerable to predators.

D. Marine Animals

Unlike plants, animals can inhabit all regions of the ocean, being limited only by food supply and oxygen. These two things are produced by plant life in the photic zone, but mix downward a few hundred meters by surface mixing mechanisms. So animal activity is most vigorous in these upper few hundred meters, and water at depths greater than a few hundred meters is relatively barren of life. Some detritus from the surface

FIGURE 18.14   *Eelgrass at low tide (a spermatophyte).*

accumulates on the ocean bottom, and so the bottom supports a collection of benthic organisms. Consequently, we expect to find most animal life within a few hundred meters of the surface, and the second most active area will be the ocean bottom. Over the continental shelf, of course, these two regions merge, and so we find vigorous animal activity at all depths there.

All animal phyla have marine representatives. In fact, there are some phyla that are entirely marine. We will describe below not all of the animal phyla, but rather those that are most heavily represented in the marine community. These phyla are listed below, roughly in order of increasing complexity, with brief descriptions or examples where appropriate.

1. Protozoa (Phylum Protozoa)
2. Sponges (Phylum Porifera)

3. Organisms consisting of a gut with tentacles at one end, such as jellyfish, polyps, and sea anemones (Phylum Coelenterata)

4. Small luminous jellyfish-like animals with some self propulsion (Phylum Ctenophora)

5. Various kinds of worms, including flatworms (Phylum Plathyhelminthes), segmented worms (Phylum Annelida), and arrow worms (Phylum Chaetognatha)

6. Molluscs, such as chitons, snails, clams, mussels, octopus, and squid (Phylum Mollusca)

7. Animals with jointed legs and external skeletons, such as copepods, krill, shrimp, crabs, and lobsters (Phylum Arthropoda)

8. Spiny-skinned radially symmetric bottom dwellers with an internal skeleton, such as starfish, sea urchins and sand dollars (Phylum Echinodermata)

9. Animals with gill slits and a cartilaginous skeletal rod at some stage in their development, such as sharks, fish, reptiles, birds, and mammals (Phylum Chordata).

D.1 Protozoans

Protozoans are single-celled animals that can be either benthic or planktonic. They are most common in warmer waters and in environments containing decomposing organic matter, because they rely heavily on the direct absorption of organic molecules for their nutrition. Many of the more plentiful kinds of protozoans secrete tests for protection, but can extend their protoplasm in "pseudopodia" through perforations in these tests in order to capture food.

The most numerous marine protozoans belong to the order "Foraminifera" (Figures 18.15 and 18.16), which have amorphous protoplasmic jelly within the cell walls, and are related to the fresh water amoeba. They have calcareous outer shells that they can dissolve or break through in order to ingest captured particulate matter or to excrete particulate waste. These are often large for single-celled organisms, sometimes being quite visible to the unaided eye. One genus of Foraminifera, called "Globergina," secretes spherical tests, deposits of which make up the famous White Cliffs of Dover on the southern coast of England.

The second most abundant marine protozoans belong to the order "Radiolaria" (Figure 18.17). These secrete very beautiful and intricate silica tests. Also sometimes placed in this phylum are some bacteria, and dinoflagellates, which do not photosynthesize.

The particular species of Foraminifera or Radiolaria that survive and proliferate are quite sensitive to the environmental conditions. Consequently, by studying their tests in the sediment, oceanographers can learn something about the climate conditions during earlier times when these layers were deposited.

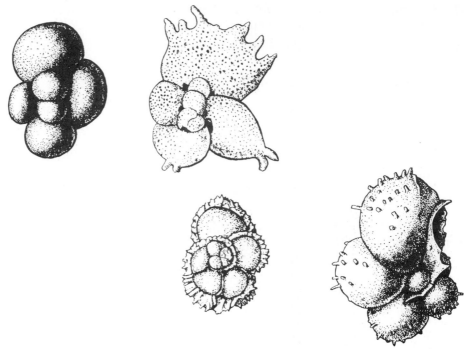

FIGURE 18.15 *Some varieties of Foraminifera, enlarged approximately 100 times. (From J. A. Cushman, Foraminifera: their Classification and Economic Use, Harvard University Press, © 1959.)*

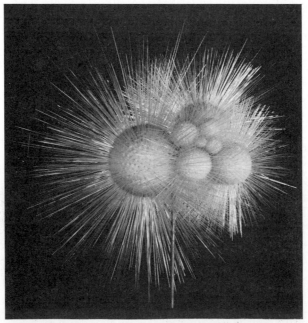

FIGURE 18.16 *A model of the foraminifer, Globergina bulloides (Phylum Protozoa).*

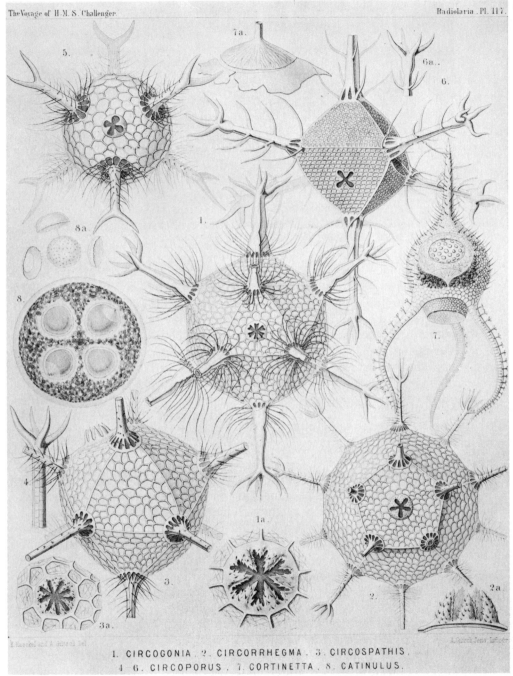

1. CIRCOGONIA. 2. CIRCORRHEGMA. 3. CIRCOSPATHIS.
4–6. CIRCOPORUS. 7. CORTINETTA. 8. CATINULUS.

FIGURE 18.17   *Sketches of various radiolarians from the Challenger expedition (Phylum Protozoa).*

FIGURE 18.18   *A sponge (Phylum Porifera).*

D.2   Sponges

The sponges (Figure 18.18) are multicellular benthic animals that have few natural enemies and can be found attached to the sea floor at all depths. They show very little organization among the cells, and their structural support is furnished by small spicules distributed fairly randomly throughout their volume. These skeletal spicules can be made of calcium carbonate, silica, or a cellulose-like material called "spongin," and can take various geometrical shapes, typically measuring 1 or 2 mm in length. Most commercial sponges have the spongin spicules. Sponges are filter-feeders, passing water through their bodies and filtering out the microorganisms and organic debris.

D.3   Coelenterates

All members of this phylum are carnivores (Figure 18.19). They all are fairly primitive organisms, consisting of a gut lined with protoplasm, and having tentacles at one end. They all have stinging cells in their tentacles, each of which consists of a bladder containing a poison and a hollow needle, which distends when the cell is disturbed. The sting is intended to paralyze small prey, and the tentacles help bring the prey into the gut.

There is a good variety of members in this phylum. Some are fixed to the bottom and others are freefloaters. In fact, this phylum exhibits a high degree of polymorphism; that is, a single species may appear in a variety of forms. For example, different generations of a species may alternate between sessile ("polyps") and free floating ("medusa") forms.

Examples of coelenterates attached to the bottom are corals and anemones, which form the largest class within this phylum. The anemones resemble flowers of 1 to 10 cm in diameter, and are popular and colorful inhabitants of tide pools. Children often enjoy agitating the center of the anemones with their fingers and feeling the tentacles close in. This is how they capture small prey, but as yet, there has not been an anemone victorious over a finger, in spite of countless contests.

(a)   (b)

FIGURE 18.19   Colenterates: (a) A Portuguese Man of War jellyfish. (b) An anemone eating a fish.

Another class includes the jellyfish, which are planktonic members
of this phylum. In contrast to the anemones, the tentacles of some species
of jellyfish have registered victories over humans, sometimes inflicting
pain to unsuspecting swimmers, and causing fatality in a few rare cases.

D.4   Ctenophores

The members of this phylum are not very well known to the average person, primarily due to their small size. The largest ones are only a few centimeters in diameter. They do have an important impact on the marine
biological community due to their heavy feeding. They are something like
small jellyfish, except they have some mobility created by eight rows of
flagella long their sides (Figure 18.20). They have two long tentacles for
capturing prey. All members of this phylum are bioluminescent, and so
they can be seen at night if near the surface. They are sometimes given
the common names "comb jellies," "sea walnuts," and "sea gooseberries."

D.5   Worms

(a) Flatworms   The flatworms belong to the phylum Plathyhelminthes.
They are usually a centimeter of so in length, but can reach lengths of

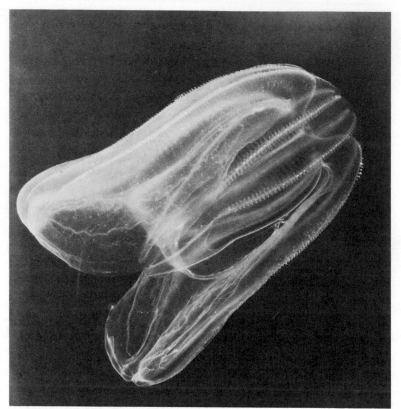

FIGURE 18.20   *A ctenophore. (Ctenophores are sometimes called "comb jellies.")*

several meters, such as the tapeworm parasite found in humans. Marine species usually inhabit the rocks of the ocean bottom or burrow in the mud, their mobility being created by tiny hairs covering the body. Some species have body coloring to mimic their environment, providing camouflage, which can also make the flatworm quite colorful.

(b) Segmented Worms   The segmented worms belong to the phylum Annelida. Earthworms and fresh water leaches are familiar nonmarine members of this phylum. Most have planktonic larval stages, and most are benthic deposit feeders or filter feeders as adults. A few species have developed special appendages to facilitate swimming. Some build tubes attached to the bottom. They inhabit these tubes, spreading tentacles resembling a flower from the top, and with which they capture small particles of food. Benthic burrowing species play an important role in mixing the upper few centimeters of bottom sediments. Scarcity of oxygen prevents them from going deeper into the sediments.

(c) Arrow Worms   Arrow worms (Figure 18.21*b*) belong to the phylum Chaetognatha. They are pelagic and exclusively marine, unlike the flatworms and segmented worms that are primarily benthic, and have nonmarine members in their phyla. The adults are 2 cm to 8 cm long, transparent, have fins for locomotion, and are fast swimmers. They are sometimes called "bristle-jawed worms," because they have movable

(a)

(b)

(c)

(d)

FIGURE 18.21 (a) *Flatworms (Phylum Plathyhelminthes).* (b) *An arrow worm (Phylum Chaetognatha).* (c) *A segmented worm (Phylum Annelida).* (d) *Feather duster tube worms (Phylum Annelida).* ((b) *Walter Dawn.***)**

bristles around the mouth, which they fan out and use to seize their tiny prey. Unlike most other worms, they exhibit no larval stage, going directly from egg to small adult.

D.6 Molluscs

Molluscs all have soft bodies, a soft "foot" used for locomotion, and many secrete calcareous protective shells from a tissue called the "mantle." There are a wide variety of molluscs separated into several classes, the most important of which we mention here. Some are shown in Figure 18.22.

Chitons are small, flat molluscs that have eight calcareous plates protecting their back. They live in shallow water or intertidal regions, and

(a)

(b)

(c)

FIGURE 18.22 *Molluscs: (a) A chiton, a snail, and several mussels. (b) The cockle attempts to dislodge the starfish via a series of maneuvers. (c) An octopus.*

feed on algae scraped from rocks as they slowly creep along.

Another class of molluscs are the "gastropods," which are snails or slugs that also move about on one "foot." In a few species, the foot has evolved into a means of propulsion through the water, but most gastropods are confined to the ocean bottom, consuming whatever appropriate organic material may be found there. Those students familiar with aquariums may know that when gastropods are put inside, they creep over the glass, eating the algae and keeping the aquarium glass clean.

The bivalves, such as mussels, clams or oysters, make up another class of molluscs. The adults of this class live on or in the ocean bottom, or attached to solid surfaces. Like the gastropods, they can use their foot for propulsion along the bottom, but unlike the gastropods, they are filterfeeders.

Another class of molluscs, the "cephalopods," includes the octopus, squid, and chambered nautilus. In these, the "foot" seems to have evolved into tentacles with suction cups for holding onto things. The members of this class have a chewing beak, and have a means of propulsion by squirting a water jet. The squid normally swims with its tentacles toward the front, but when frightened, it uses this water jet to propel itself in reverse. Unlike most other molluscs, the cephalopods have no separate larval form. The juveniles are tiny plankton, but they have the same form as the adults, just much smaller. At the other extreme, the adult giant squid can attain lengths greater than 15 m, and is the largest invertebrate animal on earth.

D.7 Arthropods

The arthropods are characterized by having jointed legs and an external skeleton, like an ant or a crab. They are the largest group of animals, both in terms of number of different species and number of individuals. The majority of all land animals are arthropods belonging to the class "Insecta," and the majority of all marine animals are arthropods belonging to the class "Crustacea." Over 75% of all animal species belong to this phylum, and the fraction of individuals is even greater.

The crustaceans are further divided into various subclasses. One subclass contains the copepods, which resemble tiny shrimp of about 0.5 to 5 mm length. The copepods are extremely numerous, and make up the bulk of the animal mass in the ocean in spite of their small size. They feed primarily on diatoms and particulate organic matter. They are the most important single link between the microscopic phytoplankton and larger animals in the ocean food chain.

Slightly larger than the copepods are the euphasids, or "krill," which belong to another subclass of crustaceans, and range from 1 to 5 cm in length as adults. They also feed on microscopic plankton, and play a very important role in the link between the microscopic producers and the larger animals. Krill are very abundant in cooler waters of higher latitudes, and are a favorite food for some whales.

Krill seem like a good thing to harvest for human consumption, especially since the whale population has been so drastically reduced. Copepods would also be a target for human consumption due to their large abundance, if an efficient means could be found for harvesting these tiny creatures.

The barnacles are in another subclass, and seem to bear absolutely no resemblance to the other crustaceans, all of whom seem obviously related to shrimp or crabs of various sizes. The link is that barnacles have a larval stage, and these tiny larvae do resemble the other crustaceans. These larvae are quite mobile, in contrast to the adult barnacle, which accounts for the reason why barnacles can appear on such unlikely places as boats, whales, crabs, sharks, and so on.

FIGURE 18.23   *A spiny lobster (An arthropod).*

Lobsters (Figure 18.23), crabs, shrimp, and prawns belong to the order of crustaceans called "decapods," whose name has obvious significance. These creatures are mostly benthic, and are popular food for humans.

D.8   Echinoderms

The members of this phylum are exclusively marine, and most are bottom dwellers. They all have internal skeletons, and have a noticeable radial symmetry, usually five-sided. Their larvae are usually planktonic and have bilateral symmetry.

One member of this phylum is the sea cucumber, whose five-sided symmetry is only visible in the skeleton, and then as seen from the front or back rather than from the top or bottom. Starfish (Figure 18.24), sea urchins, and sand dollars are other common members of this phylum. Most are filter-feeders, but starfish feed on oysters and clams. Suction cups on their arms enable them to grasp and pull open the bivalves. Then they extend their own stomachs inside the shell to digest the soft tissue there. One species of starfish, called the "crown of thorns," feeds on coral, and has caused considerable destruction on some coral reefs in the Pacific.

Some echinoderms have peculiar ways of dealing with hardship or injury. When attacked, the sea cucumber can discharge some of its entrails, leaving a meal that hopefully satisfies the predator. Later it can regenerate replacements. The starfish can regenerate missing parts as well. At one time, the people who fish for oysters and clams would cut up

FIGURE 18.24   *Several different species of starfish (an echinoderm).*

the starfish they captured with the hope of killing this predator of the clam and oyster beds. Unfortunately, this did not accomplish what was intended.

D.9   Chordates

The members of this phylum have all had gills or gill slits and a cartilaginous segmented skeletal rod, called a "notochord," at some time in their development. Some examples are illustrated in Figure 18.25. The most common and well-known members of this phylum belong to the subphylum "Vertebrata," which is characterized by having an internal skeleton with a spinal column of vertebrae, a brain, red blood, and two pairs of appendages. People, of course, belong to this subphylum. (Did you know you had gills once during your development?)

The most primitive vertebrates belong to a class that includes the sea lamprey (Figure 18.26), which has no scales or jaws, and lives by attaching itself to larger creatures and sucking its nutrition from the body of the host. Hagfish are similar parastic members of this class, which feed on dead animals.

Primitive fish, such as sharks and skates, belong to another class. They have cartilaginous skeletons and no gill covers. True fish belong to another class. They have bony skeletons, scales, gill covers, and a swim bladder, in contrast to the primitive fish.

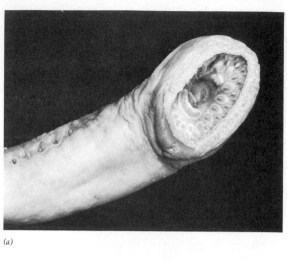

(a)

(b)

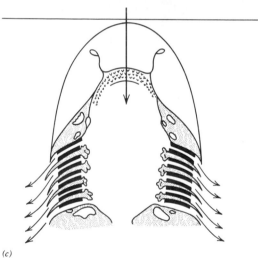

(c)

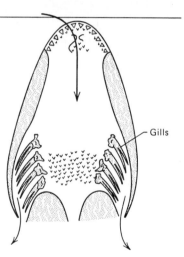

Gills

(d)

FIGURE 18.25 *Chordates: (a) The rasping, suction cup-like mouth of a sea lamprey. (b) The underside of a young sting ray. Rays are close relatives of sharks and skates. (c) The gills of bony fish (right) are protected by hard gill covers. Water passing through the gills of primitive fishes (left) leaves through several individual gill slits. (d) Dolphins are air-breathing mammals, and therefore belong to the same biological class as do people.*

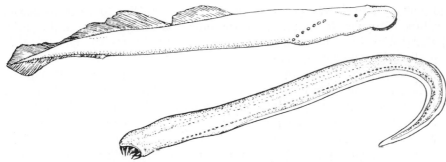

FIGURE 18.26 *Sketch of a sea lamprey (above) and a hagfish (below). Actual lengths are about 60 cm. (Bayard H. McConnaughey,* Introduction to Marine Biology, *3rd ed., The C. V. Mosby Co., St. Louis, 1978.)*

Another class are the reptiles. Common marine representatives include snakes and sea turtles, which breathe air and consequently must live near the surface. Birds constitute another class, but there is some disagreement regarding whether any of these should be considered "marine organisms." Some, such as penguins, pelicans, gulls, albatrosses, and auks are extremely dependent on the ocean in any case.

Another class includes the mammals. Seals, walruses, sea lions, sea cows, sea otters, whales, and porpoises are some marine representatives of this class. Some whales, of course, are the largest of all mammals. For example, blue whales can attain lengths greater than 35 m and weights greater than 150 tons. Such large mammals would find it impossible to live on land, as bony skeletons could not support such heavy weight.

E. SUMMARY

We describe marine organisms according to several characteristics. They may be autotrophic or heterotropic, according to whether they can manufacture their own food. We also describe them by their habitat. Benthic organisms inhabit the bottom, whereas pelagic organisms inhabit the waters. Both benthic and pelagic domains have further geographical subdivisions. Nekton are swimmers and plankton are drifters. The plankton are extremely populous, and so further subdivision is helpful. The subdivisions of plankton can be according to size, amount of life spent as plankton, or according to their photosynthetic ability.

The most widely used classification schemes are based on the idea that the development of an organism reflects the evolution of its ancestors. The broadest categories are called "kingdoms," which are further subdivided into "phyla." Each phylum is further subdivided into "classes," and so on. One common classification scheme divides all organisms into two kingdoms: plant and animal. No classification scheme is without ambiguity.

Of the four plant phyla, only two have marine members. The thallophytes, or algae, are the most numerous marine plants. Blue-green, green, brown, and red algae are more visible and familiar to us from coastal waters. However, the microscopic diatoms, coccolithophores, and dinoflagellates contribute more to the total productivity in the oceans.

We also describe several of the animal phyla more prominently represented in the oceans. Protozoans are single-celled animals that thrive on decomposing organic matter. Sponges are primitive multicellular filter feeders. Coelenterates have tentacles for capturing prey and bringing it into a primitive gut. Ctenophores are like small bioluminiscent jellyfish with rows of cilia for propulsion and having voracious appetites. Flatworms and segmented worms are benthic scavengers, burrowers, and filter feeders. Arrow worms are small, transparent, fast swimmers with bristles on their mouths for capturing prey. Molluscs include chitons, snails, clams, mussels, octopuses, and squid. The majority of all marine animals are crustaceans, which is a class belonging to the phylum "Arthropoda." They have jointed legs and external skeletons. They include copepods, krill, crabs, shrimp, and lobsters. Echinoderms are radially symmetric benthic organisms with internal skeletons, such as starfish, sea urchins, sand dollars, and sea cucumbers. The Chordates are animals having gill slits and cartellaginous skeletal rods at some time in their development, and include fish, reptiles, birds, and mammals.

QUESTIONS FOR CHAPTER 18

1. Give some examples of heterotrophic and autotrophic organisms. Why is the distinction between the two not always very clear?

2. What are benthic organisms? What are pelagic organisms?

3. What are the two subdivisions of the pelagic environment based on the depth to which sunlight penetrates? Could the benthic environment also be subdivided in this way? In what region of the benthic environment would you find plants attached to the bottom? (This region is sometimes called the "inner shelf.")

4. Why is the environment of the photic zone more variable than that of the aphotic zone? Is this reflected in the varieties of organisms found?

5. What is the pelagic environment over the continental shelf called? What is the pelagic environment farther out to sea called?

6. What are the names of the various geographical subdivisions of the benthic community? To what regions does each correspond?

7. How does the water's viscosity change with depth?

8. What is the difference between nekton and plankton? What is a commonly used distinction for resolving borderline cases?

9. Give some examples of nekton. Of plankton. Which group has greater diversity among its members?

10. What are the three subgroups of plankton, based on their size? Give the range of sizes for each.

11. What are "holoplankton" and "meroplankton"?

12. What are planktonic plants called? Planktonic animals? What is the primary role played by each, as viewed by interested parties farther up the food ladder, such as ourselves? What are the names of the most plentiful members of each of the two types of plankton?

13. What is it about some species of zooplankton that makes them good "indicator species"? Explain.

14. What idea forms the basis of the schemes used to classify organisms?

15. Give the names of the various steps in the classification schemes, listed in order from the largest, broadest groups down to the most specific.

16. What is the distinction between the Monera, Protista, Metaphyta, and Metazoa kingdoms? Give an example of a member of each kingdom.

17. Why are marine plants generally much simpler than their terrestrial relatives? Why don't they need roots, stems, or leaves, usually?

18. What two subkingdoms of plants have marine members?

19. The members of what subkingdom go by the name "algae"?

20. Why are sessile algae only found in water less than about 100 m deep?

21. What are the six most important phyla of algae in the ocean? Which are predominantly sessile? Which phyla are predominantly planktonic?

22. What are the two most important contributors to the total productivity in the oceans? To which group of algae do they belong?

23. What makes the Red Sea red? To which group of algae do these plants belong? Is the name of this phylum a misnomer? Explain.

24. Which is usually the dominant form of algae in fresh water?

25. What is kelp? How does the sargassum weed get its name?

26. On most world maps you will find the words "Sargasso Sea" written across the central western portion of the North Atlantic. How does it get its name?

27. Compare the four phyla of sessile nearshore algae. Where might they be found and what color might they be? Give at least one example of each kind.

28. Describe diatoms.

29. Why do succeeding generations of diatoms get smaller and smaller? What do they do when their skeletons are too small?

30. How do coccolithophores differ from diatoms?

31. Explain why there should be some relationship between water temperature and the skeletal material used by microscopic organisms.

32. What are dinoflagellates? How do they compare to diatoms in size? In locomotion?

33. Give an example of a marine member of the subkingdom Spermatophyta. A terrestrial member.

34. What is a "rhizome" and what can it do?

35. Why can animals live in places where plants cannot?

36. Where in the oceans do you expect to find most animals? Why? Where would you find the second largest abundance of animals? Why?

37. What two orders of the phylum Protozoa are most plentifully represented in the oceans? What are their skeletons made of?

38. How is it that foraminifers and radiolarians are particularly helpful in helping oceanographers learn of the earth's climatic conditions in times past?

39. What provides the structural support for sponges? How do they get their food?

40. What are the general characteristics shared by all coelenterates? Give some examples of members of this phylum. If you don't know what an anemone or a jellyfish is, look it up somewhere.

41. What is "polymorphism?"

42. How do ctenophores differ from jellyfish?

43. Compare and contrast the members of the three worm phyla. Which phylum is exclusively marine? Which is exclusively pelagic? Which plays an important role in mixing bottom sediments?

44. Why don't worms burrow more than a few centimeters into the bottom sediments?

45. What are molluscs? What common features do most molluscs have?

46. What are chitons? Gastropods? Bivalves? Give examples of some gastropods and bivalves. Does the octopus or squid have a soft "foot" like the other molluscs? Explain.

47. What is the phylum of animals having jointed legs and an external skeleton?

48. What's the difference between "arthropods" and "crustaceans?"

49. How can barnacles be crustaceans when they don't even vaguely resemble the others?

50. What are some crustaceans that are frequently eaten by humans? Are they more or less expensive than finfish? What are some of the smaller but more numerous crustaceans called?

51. What can the starfish do with its stomach that most other animals cannot do?

52. What remarkable thing can a sea cucumber do when being harassed by a predator?

53. What are common characteristics of all chordates?

54. What phylum do vertebrates belong to?

55. List various important classes of vertebrates and give examples from each.

56. Can you pronounce the 11 animal phyla described in this chapter? Have you memorized their names? (You'll be excused if you miss a worm or two.)

57. List the 11 animal phyla described in this chapter. Briefly describe the important general features of each. Give an example or two of animals in each.

*58. Why is it false reasoning to think that the severe pressures would "crush" animals in the deep ocean? The atmospheric pressure on us is about 14.7 lb per /sq. in., or roughly 100,000 lb (50 tons) over the total area of our bodies. Why doesn't this crush us? (*Hint:* Do you suppose inside pressure pushing out equals outside pressure pushing in? Can you "crush" our body fluids?)

*59. How would you distinguish between plants and animals? To which kingdom would you say a toadstool, which doesn't photosynthesize, belongs? A coral which doesn't move?

*60. Describe briefly how you'd expect to find the plant life and animal life distributed in a vertical column of seawater. Explain why you'd expect this.

*61. Because arthropods are so numerous, the jointed legs and external skeletons must be giving them some considerable advantages compared to other not-so-populous animals. Do you have any ideas what these advantages could be?

*62. If mobility is advantageous for animals, why are plankton so much more abundant than nekton?

SUGGESTIONS FOR FURTHER READING

1. Willard Bascom, "Ocean Waves," *Scientific American* (Aug. 1959).

2. Leonard Engle, *The Sea,* Life Nature Library, Time-Life Books, New York, 1969.

3. Martha Howbert, "Some Midwater Fishes—A Picture Essay," *Oceanus,* **24,** No. 2 (Summer 1981), p. 39.

4. John D. Isaacs and Richard A. Schwartzlose, "Active Animals of the Deep-Sea Floor," *Scientific American* (Oct. 1975).

5. John D. Isaacs, "The Nature of Oceanic Life," *Scientific American* (Sept. 1969).

6. "Sharks," *Oceanus,* **24,** No. 4 (Winter 1982).

7. Bob Wallace, "A Galerie of Sponges," *Sea Frontiers,* **25,** No. 2 (1979), p. 66.

19
MARINE ECOLOGY

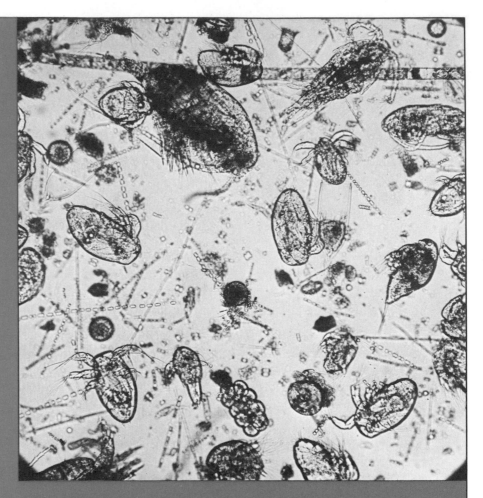

In the last chapter we briefly described the various kinds of organisms common in the oceans. Our interest extends further than merely a knowledge of what is there. We also wish to know how they function, how they interact with their physical and biological environments, and what role each organism plays in the welfare of the larger community. The study of these areas is called "ecology," and it is the subject of this chapter and the next. In this chapter we focus on fundamental processes and general ideas underlying the interaction of all marine organisms with their environment. In the following chapter we become more specific, and examine the ecological roles and distributions of each particular kind of marine organism.

19
MARINE
ECOLOGY

A. THE CELL

To understand the interaction of an organism with its environment, it is helpful to have a basic understanding of how an organism works. What is it that makes it different from all the inanimate things around it (Figure 19.1)? How are plants and animals able to carry out all those special functions, such as photosynthesis, metabolism, and movement? Needless to say, complete answers to these questions are still being researched, and their proper explanation would require much more space than exists between the covers of this book. Consequently, the following paragraphs can provide only brief, sketchy, and very incomplete explanations. Hopefully, they will stimulate more questions than they answer.

The basic unit of any organism, large or small, is the biological "cell" (Figure 19.2). Most cells are quite small, for reasons we will soon learn. Although there is great variation in size among the different kinds of cells, dimensions on the order of 20 microns (.02 mm), or about a hairwidth, are typical. Each cell is surrounded by a membrane that separates it from its environment, and most cells each contain a central "nucleus" within it that regulates the cell's activities.

Different cells do different things. Whether it is a single-celled organism all by itself, or a small component of a larger multicellular organism, each cell has a set of duties it must perform if the organism is to survive. These duties are carried out through physical and chemical processes involving the molecules within the cell. We have already studied (Chapter 17) the chemical interactions involved when a cell photosynthesizes or respires. Even motion, such as muscular contraction, is caused by electrostatic forces between molecules whose electrical charge distributions have changed.

A.1 The Transport of Materials

Out of the cell's immediate environment and through the cell membrane must come the materials needed by the cell to carry out its func-

515

FIGURE 19.1 *How do these creatures differ from the rock they live on? What distinguishes the living from the nonliving?*

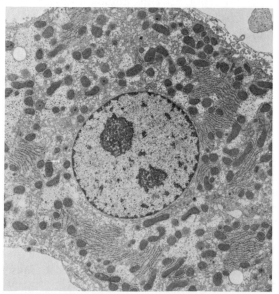

FIGURE 19.2 *Stained cross section of a liver cell as seen through a microscope. The large circular central region is the cell nucleus. Many other units of the cell's substructure are also clearly visible in this photo.*

tions. If it is involved in photosynthesis, then it must have water, carbon dioxide, and nutrients coming in. If it is involved in respiration, then it must be appropriately supplied with organic materials and oxygen. In addition to the appropriate raw materials being transported in, the waste products must also be carried back out.

Because of its excellent ability to dissolve materials, water generally serves as the medium through which these raw materials and waste products are transported. The fluid within a cell is water-based, and cells generally require a water-based external environment as well. Most single-celled organisms go dormant or die when removed from water, and most multicellular organisms ensure that their cells are immersed as much as possible in water-based body fluids. The reason your body fluids are made of water, rather than air or some other fluid, is due to the excellent solvent properties of water.

The transport of materials in and out of the cell is accomplished by "diffusion." Diffusion is the tendency for molecules to go from regions of higher concentration toward regions of lower concentration, and is a result of their random thermal motions. Examples of diffusion are familiar to anyone who has smelled breakfast cooking or has smelled ammonia fumes.

If you uncap an ammonia bottle, soon you can smell it from anywhere in the room. This is because the ammonia fumes that were once trapped within the bottle are now free to go anywhere, and so they do. There is nothing to prevent an ammonia molecule in the room from going back into the bottle, and some do. But since their concentration in the room is smaller than in the bottle, there are fewer in the room ready to go into the bottle than there are in the bottle ready to go into the room. Consequently, more molecules diffuse out of the bottle than into it. (See Figure 19.3.) If left alone, the diffusion out of the bottle will continue until the concentration of ammonia molecules in the room is the same as the concentration in the bottle. Then there will be as many molecules going back in as out, and the net change will be zero.

The same thing happens within a cell. As the raw materials are used up, their concentration within the cell becomes smaller. Consequently, new raw materials diffuse through the cell membrane, restocking the depleted supply inside. Similarly, as waste products build up within the cell, they become more concentrated than in the outside medium, so they diffuse out.

If diffusion alone accounted for the transportation, then the concentration of the water soluble substances within the cell would always tend to be the same as that in the outer environment. Often this is not the case. Frequently, cell membranes are quite sophisticated, allowing some materials to pass in one direction more easily than the other. Although diffusion is the basic transporting mechanism, these discriminating cell membranes can cause the concentrations of some materials to be considerably different within the cell than outside.

Many large multicellular organisms isolate their internal body fluids from the external environment. Communication between the two is

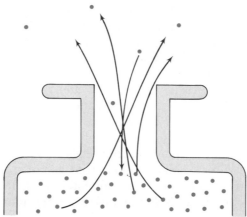

FIGURE 19.3 *Illustration of the diffusion of ammonia molecules out of an open ammonia bottle. Any molecule may move in any direction. Any single molecule just outside the bottle may diffuse in, and any single molecule just inside the bottle may diffuse out. But since there are more molecules just inside the bottle than there are just outside it, the total number diffusing out will be larger than the total number moving back in. This same argument applies to all materials, and so all materials tend to diffuse from regions of higher concentration toward regions of lower concentration.*

allowed only through certain special multicellular membranes, such as lungs, gills, and guts. These membranes discriminately allow the passage of certain materials ensuring a composition of their internal body fluids, which provides an optimal environment for the functions of their cells. Specialized systems, such as hearts, blood vessels, red corpuscles, and so on, also help ensure a good flow of materials to and from the cells in some organisms.

A.2 How a Simple Cell Can Do Complicated Things

Once it has the raw materials, how does a cell know what to do with them? For example, how does it know how to generate food stores, energy, add to its own growth, or reproduce itself? This question is generating heavy research efforts, and interesting results.

Within a cell are some protein molecules, called "enzymes," which serve as catalysts for certain chemical reactions. Each enzyme has a chemical structure that makes it attractive to certain molecules. Its own thermal motion, as well as the thermal motions of the other molecules in the protoplasm, causes it to roam about within the cell, undergoing countless collisions each second. Occasionally, a collision will be with the kind of molecule that is attracted to that particular region on the enzyme, and so it will stick there. This molecule's chemical nature may be changed through interaction with the enzyme directly, or it may wait until the enzyme has encountered more molecules of particular kinds; different molecules stuck in different places. These molecules may then interact with each other,

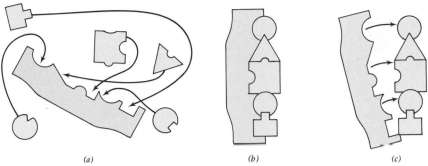

(a)   *(b)*   *(c)*

FIGURE 19.4   *Schematic illustration of how an enzyme or a molecule of RNA can serve as the scaffolding on which materials are assimilated. (a) Random thermal motions eventually bring each of the needed raw materials to the appropriate position on the scaffolding. (b) These materials then interact with each other to form larger molecules. (c) The newly formed material then peels off the scaffolding, leaving it free to start the process again.*

often with the help of some interactions with the enzyme itself. This new substance is then released, leaving the enzyme to start the process all over again. Of course, the substances produced through these interactions may or not be of any use to the cell. But by natural selection, only those kinds of organisms whose enzymes have produced the needed materials have survived.

The molecules produced in these enzyme-induced reactions may be used by the cell directly, or may be used to form larger more complicated molecules. For complex functions, such as the manufacture of complicated proteins, there are large "RNA" molecules within the cell that serve as the scaffoldings on which the necessary raw materials are assimilated and the particular proteins are fabricated. Various materials are attracted to various portions of a RNA molecule, so the ordering of the components on the RNA molecule determines the order of the ingredients in the new molecule being produced (Figure 19.4). Some of these ingredients are produced by enzymes, some are collected and transported by enzymes, and some come directly from the outer environment, through the cell wall, and run into the RNA molecule all by themselves. When the ingredients are finally assimilated, they begin interacting with each other, often aided by certain enzymes present, and then peel off the parent RNA molecule.

You can see that for a cell to function properly, it not only needs the correct raw materials, but it needs to be able to produce the appropriate enzymes and RNA molecules as well. The information needed to do this is contained in long, thin protein molecules, each twisted in the form of a double helix (Figure 19.5). These are the molecules of which genes are made. They are found in cell nuclei and are called "DNA." By the ordering of the various chemical groups along its length, a DNA molecule can produce the needed enzymes, RNA molecules, and it can even reproduce itself (Figure 19.6) through processes similar to those described for enzymes and RNA above. Consequently, through the ordering of the

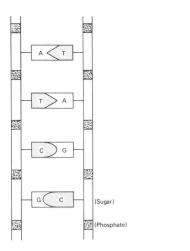

A = adenine
T = thymine
C = cytosine
G = guanine

FIGURE 19.5   *DNA is like a microscopic ladder that has been twisted into the form of a double helix. The edge of each element of the ladder is formed of a sugar plus a phosphate, which helps it couple to the next sugar of the next element. The coded information that regulates the cell is carried in the "rungs" of the ladder. There are four different kinds of "rungs," consisting of pairs of the chemical structures called adenine and thymine (in either order), or guanine and cytosine (in either order). One of each kind of rung is illustrated in the drawing above. The ordering of the rungs on this ladder carries the instructions for the cell, much like the ordering of the holes in a paper tape can carry the instructions for a computer.*

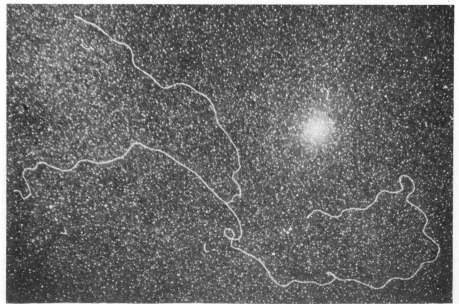

FIGURE 19.6   *Electron microscope photo of DNA undergoing replication.*

TABLE 19.1   Characteristics of an Efficient Cell

An Efficient Cell Would	Implication
Require very little incoming nourishment	Small volume
Produce very little outgoing wastes	Small volume
Get its raw materials easily	Large surface area
Get rid of its wastes easily	Large surface area

chemical groups along DNA molecules, they contain all the information necessary to run cell activities.

To summarize, the behavior of a cell is influenced by its outer environment, which determines what materials are available to diffuse into the cell. It is also influenced by the cell membrane, which may have the ability to selectively concentrate some of these materials within the cell. Furthermore, there are enzymes and RNA molecules within the cell that facilitate the chemical reactions needed for the cell to perform its functions. The information necessary to produce these enzymes and RNA molecules is carried in DNA molecules found within the cell nucleus, which also is able to reproduce itself.

A.3   Cell Efficiency

A cell is in contact with its environment only along its outer surface. Consequently, both the rate at which raw materials can get into a cell and the rate at which the waste products get out depend on the surface area of the cell. The larger the area, the more exchange can take place. However, the amount of raw materials *required* by a cell depends on its *volume*; bigger cells need more materials. The most efficient cell would receive much but require little. Hence, it would have a large surface area and a small volume.

The same is true of the waste products. The smaller the volume, the fewer the waste products would be produced inside. The larger the surface area, the more quickly these would diffuse out. So in getting rid of wastes, large surface areas and low volumes would also be best. (See Table 19.1.) If we are interested in cell efficiency, we must ask what characteristics would maximize a cell's surface-to-volume ratio.

To answer this question, notice that whenever you break something, you expose new freshly made surfaces without changing the total volume (Figure 19.7). The more you break it up, the smaller the pieces become, and the more surfaces are exposed. Smaller pieces mean larger surface-to-volume ratio for the material. Consequently, the most efficient cell would be one that is as small as possible, providing it is still large enough to hold all the materials needed to perform its functions.

There is also a second important reason for the efficiency of a cell to depend on small size. This is that the diffusive transport of materials in water (or protoplasm) proceeds very slowly over large distances. In contrast to the rather rapid diffusive transport of molecules in air, which enables the smell of frying bacon or burned toast to quickly penetrate every room in your house, diffusive transport in water proceeds much

FIGURE 19.7 *Chalk: whole, broken once, and broken many times. Each time it is broken, new surfaces are exposed, with no change in the total volume. Therefore, the smaller the pieces, the greater will be the surface area in comparison to the volume.*

more slowly because the molecules of liquid water are much more densely packed than those of air, and provide a great deal more obstruction. A blindfolded elephant would have a great deal more difficulty getting out of a forest if the trees were closely spaced, than if they were few and far between. This is basically the reason why diffusion over macroscopic distances proceeds much more slowly in water than air.

The frequent collision with water molecules makes it increasingly difficult for the random thermal motion of diffusive processes to distribute molecules over greater distances (Figure 11.5). For example, the characteristic time required for the diffusion of a typical salt in water is about 1 second for a distance of 0.1 mm (100 microns, or the size of a large cell), 1 day for a distance of 3 cm, and one year for a distance of only 60 cm. Clearly, if the cell relies on diffusion for the transport of materials within it, then small size is a must.

B. THE MARINE ENVIRONMENT

With this basic understanding of how a cell works, we are better able to understand how some of the features of the marine environment affect the organisms that live there. Before going into detail on some of the problems posed by the marine environment, it is perhaps worthwhile to give some overall perspective by comparing it to the terrestrial environment, with which we humans are more familar.

B.1 Comparison of Marine and Terrestrial Environments

It is no surprise that life first developed in the oceans, and only recently was it able to move onto the land. By comparison, the terrestrial environ-

FIGURE 19.8 *Some intertidal organisms exposed while the tide is out. Above, a
colony of mussels and some algae attached to a rock await the
water's return. Below, some closed sea anemones (closed because
the author's child stuck his fingers in them); looking like doughnuts,
sit in a tide pool. The white flecks on the rocks are barnacles. Some
algae make the dark splotches on the rocks.*

ment is extremely harsh and cruel. For example, terrestrial creatures must
conquer gravity, dessication, temperature extremes, and many other hardships not normally encountered by marine life. Life on land has had to
develop very specialized systems and appendages in order to cope with
the environment. For example, the ability to grow thick winter coats of
fur, strong legs, claws, complicated root systems, or fruits are a few of the
many terrestrial innovations that would be unnecessary in the marine
environment. Plants, in particular, have reached a high degree of development on land. Marine plants are relatively simple by comparison, not
having to endure such an abrasive environment as the terrestrial plants
do.

When walking along the seashore, we frequently marvel at how
"hardy" the intertidal organisms must be (Figure 19.8). They endure the

FIGURE 19.9 *Humans can adapt to a wide variety of environments.*

pounding of the waves, and can go for hours at a time without water when the tide is out. It is true that the intertidal environment is harsh when compared with other marine environments, but for us to marvel at them rather than at ourselves shows that we have lost our perspective. We think it is amazing that they can develop gelatinous coatings or crusty shells to fight dessication for a few hours, but we forget that we ourselves have developed the means to go for days without water, and even then we need only a trickle by comparison. We can survive weeks without food and months without sunlight. We can survive continual 45°C below zero or 45°C above zero climates (Figure 19.9). We can live in cold snow or in warm mud. By comparison, all marine organisms are spoiled.

Nonetheless, having put them in perspective as a group, we can now compare them to each other. The nearshore marine organisms, as a group, are generally more adaptable and specialized than those found farther out to sea. Estuarine inhabitants, including some more advanced plants, oysters, and many fish, must endure large, daily fluctuations in salinity in their environments as the tides enter and leave. Intertidal creatures must endure hours without water and relatively large temperature fluctuations compared to those farther to sea. Farther away from shore, the daily and seasonal temperature and salinity fluctuations are smaller. The deep sea creatures are generally most intolerant of change.

B.2 Density Problems

The life-giving sunlight is found in the near-surface waters only. Even under optimum conditions, there is not enough sunlight below about 100 m depth for plants to survive and carry on their photosynthesis of organic materials. Naturally, it is advantageous for animals to be near the food source, so most animals must stay near the surface waters as well.

However, the biological cell contains many special materials necessary to help it carry out its particular functions. Many of these special materials are large organic molecules or special salts that tend to make the

FIGURE 19.10 *Sketch of a shark, indicating how its body shape, pectoral fins, and tail all provide an upward thrust when swimming.*

organism denser than water. Furthermore, many organisms have dense mineral skeletons for protection or support, which further compound the organism's density problem. How can these organisms remain in the near-surface waters where the life-giving sunlight is found?

The various organisms display a wide variety of ways of dealing with this problem. Fish have gas-filled "swim bladders" that help neutralize their buoyancy. Mammals and reptiles have lungs, whose air content helps hold them near the surface as well. The primitive fishes, such as sharks, skates, and rays, don't have any sophisticated internal mechanism to neutralize their buoyancy, and must either swim or sink (Figure 19.10). This is also true of cephalopods, such as the octopus and squid. Most of these do spend a large portion of their lives resting on the bottom. Some animals have no means of overcoming their density problems, neither through neutralizing their buoyancy nor through ability to swim. These are confined to the bottom, remaining permanent members of the benthic community.

The microscopic plants and animals also have developed a variety of techniques for dealing with their density problems. Some have tiny pouches of secreted wax, oil, or air that help neutralize their buoyancy. Others have long thin protrusions or flagella (Figure 17.14), which may serve two purposes. First, they retard the rate at which these tiny organisms sink, much like the fluff on milkweed or cottonwood seeds retard the rate at which they sink through air. Also movement of these appendages may give the organism sufficient mobility to remain near the surface, and even to change its position in response to daily changes in sunlight.

Of course, it is especially crucial that plants be able to remain near the surface. The large attached algae often have gas-filled bladders or hollow portions to help keep their photosynthetic parts near the surface (Figure 18.9). But this is not the case for the microscopic single-celled algae which accounts for the bulk of the ocean's productivity. Most of these also secrete protective mineral exoskeletons, or "tests," that compound their density problems. How do these microscopic plants combat their tendency to sink out of the sunlit surface waters?

They conquer this problem in two ways. First, they maximize their friction with the water in order to retard the rate at which they sink, and second, they have high rates of reproduction so that those organisms remaining in the surface waters can reproduce fast enough to replace those lost to the dark depths below.

Friction is proportional to surface area. This is why a man with a parachute falls more slowly than one holding a handkerchief over his head,

FIGURE 19.11   *Electron microscope photo of the reticulated exoskeleton of the diatom* Cyclotella.

or why duck down falls more slowly than grains of sand. As we saw in the previous section, one way of increasing surface area is through small size. The more pieces you break something into, the more surface areas are exposed. Small size means large surface-to-volume ratio, or, equivalently, large friction in comparison to weight. This is why fine dust or chalk dust settles out of the air more slowly than stones or pieces of chalk. It is also why gravel settles out of water more quickly than fine clays. In addition to small size, another way these microscopic plants increase their friction with water is by making their tiny exoskeletons reticulated with cavities and spicules (Figure 19.11). In warmer, less viscous waters, the microscopic plants tend to be smaller and their skeletons tend to have longer spicules, which demonstrates that friction is an important ingredient in their survival.

Although friction retards the rate of sinking, it cannot prevent it. By itself, then, friction cannot keep the microscopic plants in the surface waters. Fortunately, waves and surface currents ensure sufficient turbulence to continually mix the surface waters, keeping some microscopic plants in the photic zone. Of course, turbulence is a two-way process, keeping some in the photic zone, but removing others. This is why high rates of productivity (i.e., reproduction) are an important ingredient in the survival of the microscopic producers. Those remaining in the surface waters must reproduce at a sufficient rate to be able to replace those lost to the dark depths. Under optimum conditions, some of these microscopic plants can quadruple their numbers in a single day.

B.3   Salinity Problems

With the exception of certain coastal regions, the marine environment is of fairly uniform salinity. Most marine organisms spend their entire life

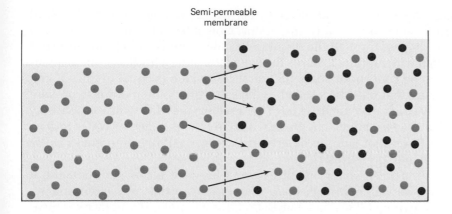

Semi-permeable
membrane

● = Salt molecule

● = Water molecule

FIGURE 19.12   *When salt water and fresh water are separated by a semi-permeable membrane, the force of attraction between the salt ions and
the polarized water molecules causes water to flow through the
membrane from the fresh side toward the salt side. The amount
of pressure that must be applied on the salt side to stop this flow is
called the "osmotic pressure."*

cycle in waters of the same salinity, gaining from these waters all the
materials they require. Since they have evolved and survived in this environment, it is no surprise that the body fluids of most marine organisms
are isotonic (i.e., of the same salinity) with their environment.

However, many of the more advanced organisms, particularly the
vertebrates, have body fluids of different salinity than the fluids around
them. The body fluids of fresh-water and terrestrial vertebrates, such as
yourself, tend to be saltier (hypertonic) than the water of their environment, and marine vertebrates tend to have body fluids less saline (hypotonic) than their environment. The reason the more advanced animals
exert this control over their body fluids is probably related to their greater
mobility. Perhaps their mobility brings them into a wider variety of environments, some of which would otherwise have been detrimental to their
health. Perhaps it has been important to their survival that their cells be
maintained in an environment that maximizes their efficiency.

In any case, the fact that the body fluids are maintained at a salinity
different from that of the environment poses special problems. Biological
membranes, including those surrounding individual cells and those separating the organism's body fluids from the external environment (e.g., the
gut), tend to allow water molecules to pass through them more easily than
the dissolved salts. Such membranes are referred to as being "semi-
permeable."

Because of the electrical charge distribution on the water molecules
and the dissolved salts, the two are strongly attracted towards each other.
When fresh water and salt water are separated by a semi-permeable
membrane (Figure 19.12), the water molecules will be strongly attracted

towards the salt side, and the salt ions will be strongly attracted toward the fresh side. Since only the water molecules and not the salts are allowed to pass through the membrane, there will be a net flow of water through the membrane towards the salt water side. This flow can be quantified by the concept of "osmotic pressure," which is equal to the amount of pressure that must be applied to the salt water side in order to prevent this flow from the fresh side. The greater the difference in salinity between the two fluids, the greater will be the tendency of water to flow towards the saltier side. In other words, larger salinity differences cause larger osmotic pressures.

When a fresh water organism is put in a salt water environment, it tends to lose water through these membranes to the salty environment, and it dehydrates. This happens to you, for example, if you drink seawater. Similarly, if a salt water organism is put in a fresh water environment, it tends to absorb water from that environment, resulting in bloating and eventual rupture of the membranes surrounding the swelling cells. You have special mechanisms to keep your body fluids more salty than your environment, and a tuna fish has special mechanisms to keep its body fluids less salty than its environment. But you are both in trouble when you switch environments, because neither's mechanisms work in reverse.

Intertidal organisms are particularly tolerant to changes in the salinity of their environment (Figure 19.13). Some find themselves in tide pools on hot days, where heavy evaporation increases the salinity markedly before the tide returns. On other days it may be raining, and the runoff flushes out their pool leaving it fresh.

The number of species that can tolerate large changes in salinity is relatively small. There are far more species of fresh water animals and of salt water animals than animals that can tolerate intermediate or varying salinities. Brackish estuaries may sometimes be teeming with aquatic life, but relatively few species will be represented. In estuaries that have large changes in salinity, fresh water species don't venture as far downstream, nor do salt water species venture as far upstream, as in estuaries where the salinity gradients are stable. Salinity is clearly a decisive factor in determining the distributions of organisms.

B.4 Temperature

Since temperatures in marine environments remain fairly constant, the ability of a marine organism to regulate its temperature would not provide such an advantage as it would to its terrestrial counterparts, who must endure large daily and seasonal temperature changes. Consequently, most marine organisms have temperatures controlled by their environment, perhaps slightly elevated due to their own metabolism, absorption of sunlight, and so on. The exceptions to this are some of the more advanced organisms, such as seals, sea otters, walruses, polar bears, and so on, who can spend prolonged periods out of water.

We know that temperature must have some influence on the behavior of the organisms and their cells, both from theoretical and from empirical

FIGURE 19.13 *Tidepool residents such as these must be able to withstand large changes in the temperature and salinity of their environment.*

considerations. We know that increased temperatures increase the rate of chemical reactions, and this must have some bearing on the activities of the cells. Indeed, we do notice some differences when comparing warm water species with their cold water counterparts. The warm water organisms tend to metabolize and photosynthesize at a higher rate. They also tend to grow, reproduce, and age more quickly. For some reason, they also tend to be smaller and more colorful than their cold water counterparts (Figure 19.14).

Warm water biological communities tend to hold a greater diversity of species, each occupying a narrower ecological "niche," or functional role in the community. The reason for this is not yet quite clear. One theory is that because the higher temperatures foster higher rates of chemical reactions and of growth and development of individuals, they also foster greater rates of mutation and biological evolution. The result would be greater species diversification in warm waters, as is observed. Another theory involves the observation that tropical climates have been quite uniform over most of earth's history, whereas temperate and polar climates have been quite varied. The relatively few cold water species presently observed may be the result of large kills resulting from the climate variations of recent ice ages. According to this theory, only the hardiest sur-

FIGURE 19.14 *A clownfish and a porcellanid crab on a giant anemone. Tropical animals such as these tend to be smaller and more colorful than their cooler-water counterparts.*

vived, and there has not yet been enough time for much species diversification in the temperate and polar regions.

In spite of fewer species, cold waters generally support larger numbers of individuals due to higher rates of primary productivity. The schools of large numbers of relatively few species makes for better fishing in cooler waters (Figure 17.21).

As a result of its evolutionary history, each organism has a preference for a certain temperature that is optimum for its operation. Sometimes the optimum temperature is different for different stages of life. In fact, changes in water temperature can trigger changes in the life stages of some organisms, such as inducing moulting, migration, reproduction, or feeding habits. One of the dangers of our thermal pollution of coastal waters is that we may disrupt the life cycles of some of these organisms to the point that they can no longer survive (Figure 19.15). Although each organism can survive a range of temperatures to either side of optimum, there is generally greater tolerance to temperature variations on the low side. Low temperatures tend to foster a state of quiescence and lower metabolism.

Polar and sub-polar waters tend to be well-mixed, with both surface and deeper waters having similar temperatures and salinities. As a result, surface and deep animals are quite similar in polar areas (Figure 19.16). On the other hand, there is a great deal of difference between surface and deep water species in tropical waters, due to the very different temperatures of these water masses. The species that inhabit temperate waters change greatly with the seasons due to migration and spawn of fishes from other areas, and due to changes in life stages triggered by seasonal changes in these waters.

FIGURE 19.15 *Because of its effects on the lives and life cycles of coastal organisms, we are particularly concerned with the thermal discharge from industrial operations and power plants. On the left is a photo of a model used to study the flow of coolant waters from the Diablo Canyon nuclear power plant on the West Coast. On the right is an aerial photo of the actual discharge during a trial run.*

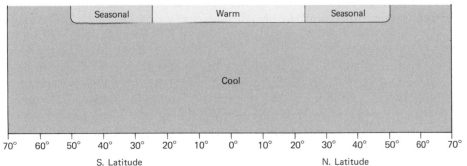

FIGURE 19.16 *Schematic illustration of ocean temperatures as a function of depth and latitude. These differences are also reflected in differences in the animal populations found. In surface waters at high latitudes, the same species are found as in deeper waters. In tropical surface waters, different species are found than in deeper waters, and in temperate surface waters, the animal population varies seasonally, as does the temperature.*

B.5 Light

In addition to salinity and temperature, light intensity also influences the distribution of marine organisms. Daily variations of light intensity in surface waters are reflected in the productivity of plants (Figure 17.22), indicating that each species has an optimum range of light intensity for its operation. Its productivity declines when the light is either more or less intense than this. Daily vertical migrations of tiny animals are also obviously in response to variations in light intensity. For some, seasonal changes in light intensity trigger different stages of life.

Experiments indicate that good vision may not be as important to many marine animals as is sensitivity to water motion. Many can school, navigate, catch their prey, and so on, even in complete darkness. On the other hand, evidence that vision must be of some importance is reflected in body coloration and in the fact that most large motile marine animals have eyes. Small surface animals are usually transparent, and larger ones are usually dark on top and light on the bottom, so that they blend in with the background both as seen from above and from below. Deeper animals tend to have reddish or black coloration. Since the red portion of the solar spectrum doesn't penetrate seawater very far, the red colors look black at dimly lit depth.

Bioluminescence is fairly common among marine organisms. Many flagellates, jellyfish, ctenophores, bacteria, and animals hosting these bacteria are luminescent. About 80% of all animals at mid-depths have some light-producing capability, and some larger animals seem to be able to turn it on and off at will. The exact reason for bioluminescence is not known. It may be for recognizing each other, mating, luring prey, or schooling, for example, and the reason may differ from one species to the next. But these organisms probably wouldn't do it if there wasn't some reason, so bioluminescence is another indication that light and vision are of some importance to many creatures, even in deep waters.

B.6 Interdependence

The welfare of marine organisms is influenced not only by their physical environment, but also by their biological environment. There is a complex web of interdependencies, with each organism having its own little niche that supports the welfare of others in the community. One important set of such interdependencies involves the supply of food. This will be explored in the next section.

But there are other functional interrelationships between marine organisms (Figure 19.17). In addition to their role in the release of the main nutrients during decaying of organic detritus, bacteria are also important producers of vitamins and other trace nutrients, which are needed by many marine organisms. Some algae, sponges, corals, coelenterates, and others, produce antibiotics needed by other organisms. Some rely on reefs built by others for protection, support, or for an environment needed for it to produce or catch sustenance. Some animals clean parasites off of others in return for protection. Some small fish swarm around jellyfish tentacles for protection from larger predators. Some algae, snails, and other small creatures in estuaries depend on eelgrass for support. In mature stable marine communities, these interdependencies become quite extensive and complex. In the next section we focus on one important set of interdependencies—the food web.

FIGURE 19.17   *An intricate set of interdependencies among marine organisms has evolved. Here, a small wrasse cleans a large sea bass. The bass gets cleaned of tiny parasites, and the wrasse gets a meal.*

C.   THE FOOD WEB

The sustenance of the entire marine community comes ultimately from the primary producers, most of which inhabit the surface waters and use sunlight to photosynthesize organic materials. A very minor amount of organic synthesis is carried out by bacteria using energy released in inorganic chemical reactions. The energy stored in organic materials can be released later through oxidation (equivalently, "respiration," or "metabolism") either by the plants themselves or by other organisms in the food web to whom these materials have passed. This released energy can be used to carry out whatever activities the organism requires, including the fabrication of other needed organic materials.

Less than 2% of the ocean is sufficiently shallow and has sufficiently firm substrate to accommodate large attached plants. Consequently, seaweeds (Figure 19.18) such as those we find washed up on our beaches, do not represent a very large component of the plant population. Most of the ocean's primary productivity is carried out by microscopic single-celled phytoplankton.

Most of the organic matter produced by these plants is stored and subsequently oxidized by the plants themselves. But some is passed onto animals feeding on these plants, some to animals feeding on the animals that feed on these plants, and so forth. Each step along the way is called a "trophic" level. For example, the plants constitute the "zeroth trophic level," the animals that feed directly on the plants are the "first trophic level," and so on.

In general, the individuals at each trophic level tend to be larger and more complex than those at lower trophic levels on which they feed. But

FIGURE 19.18 *Seaweeds such as these require shallow water for sunlight pene-tration, and firm substrate for anchorage by their holdfasts. For these reasons, seaweeds are found mainly in rocky coastal areas, which are only a tiny fraction of the total ocean area. Therefore, they make up only a small (although quite visible) fraction of the total marine plant population.*

this isn't always true. For instance, simple microscopic parasites may feed on large fish or mammals. Also, it isn't always clear what level a certain organism belongs to. For example, some may be able to feed either directly on the plants or on other animals. Should they belong to the first trophic level or higher ones? Often the trophic level is a matter of maturity rather than breed. The larvae or juveniles of large carnivores may be microscopic zooplankton, feeding directly on the plants.

A general rule of thumb is that each trophic level uses up about 90% of the food it consumes, and stores only about 10% for passage on up to higher trophic levels. That is, most of the food materials consumed by any organism stop there and do not get passed on further. After these materials have been metabolized, the organism returns the waste products back into the marine environment, where they can be reused by plants to start the cycle all over again. Some of these, such as carbon dioxide, are ready for reuse immediately, but others, such as the nutrients, tend to still be bound up in chemical complexes, which must await the work of bacterial decom-posers before being released in a useable form.

The bacterial decomposers, then, are instrumental in the recycling of materials through the food web (Figure 19.19). They take the organic det-

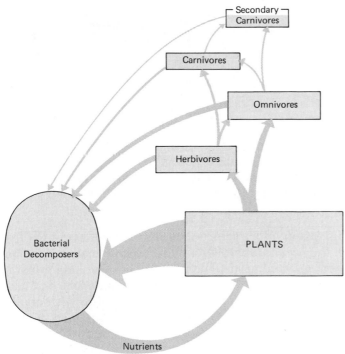

FIGURE 19.19 *Passage of food through the food web. On the average, only about 10% of the food consumed by an organism gets passed on to the next higher trophic level. The remaining 90% is eventually returned to the bacterial decomposers.*

ritus and waste products from each organism and transform them back into forms needed by the plants in photosynthesis. At each trophic level, the amount of material meeting this fate is considerably larger than the amount of material passed on to the next trophic level.

The bacteria gain the energy needed to carry out these functions primarily through the further oxidation of organic materials. Where the water is anoxic (i.e., no dissolved oxygen), some special bacteria may acquire oxygen through chemical manipulation of nitrate (NO_3^-) or sulfate (SO_4^{--}) ions dissolved in the water, giving off ammonia, nitrogen, and hydrogen sulfide (NH_4, N_2, H_2S) as by-products. But this is not a very efficient way of getting oxygen, so in anoxic areas, the decomposition of organic materials proceeds slowly and often incompletely. In such areas, considerable organic matter usually remains within the sediments.

Although bacteria are vigorously active, their total biomass is small relative to that of other organisms. Consequently, they are not an important food source in the oceans, aside from the diets of a few small bottom-dwellers.

Because the primary producers support the entire marine biological community, you might think that the bulk of the organic materials in the ocean is contained in phytoplankton. However, this is wrong because the vigorous microscopic biological activity quickly converts organic materials

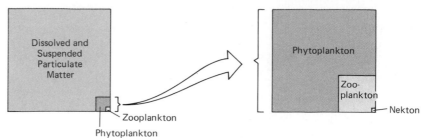

FIGURE 19.20 *Schematic illustration of the relative amounts of organic matter in the oceans in various forms. 98% of it is in the form of dissolved and suspended particulate organic organic matter, and most of the rest is phytoplankton. About 0.2% is zooplankton, and only 0.002% is nekton.*

produced by phytoplankton into organic materials of other forms. On the average throughout the oceans, about 98% of all organic matter is in the form of dissolved or suspended particulate matter. These include protein, lipid and carbohydrate molecules, tiny fragments of plants and animals, wastes, dead or dormant bacteria and phytoplankton, and other organic detritus. Only about 2% of all organic matter is actually contained in phytoplankton, and even smaller amounts in other forms (Figure 19.20). For example, only about 0.2% is stored in zooplankton, and 0.002% in nekton. There are, of course, large regional variations in these figures.

The zooplankton, which feed on the phytoplankton and suspended particulate matter, are mostly microscopic multicellular organisms, such as copepods, arrow worms, or the larvae or juveniles of larger organisms. They have special adaptations for filtering these tiny food particles out of the water. They serve the larger community by concentrating the disbursed particulate food for easier passage on up the food web. For example, we could not hope to gain our sustenance by filtering these fine dispersed particles from the water. Nor could a pelican nor a tuna fish. We could, however, feed on the anchovies, which feed on the zooplankton, which feed on these tiny particles. You can see that one ecological function of each trophic level is the collection and concentration of food for passage on to higher levels.

The abundance of organic materials in the ocean's surface waters varies greatly from one region to the next, but concentrations of a few parts per million by weight are typical. They tend to be "surface active," which means they tend to collect on any foreign surfaces introduced into the marine environment. You have probably noticed that rocks, sticks, pilings, boats, and so on, soon acquire a slimy coating after being put in the water. If left, the slimy coating will soon attract bacteria and algae, followed by barnacles, mussels, and a host of other organisms intent on making the new food-rich reef their new home (Figure 19.21). In addition to its accumulation on solid surfaces, some organic materials accumulate in "slicks" on the water surface. Wind-induced Langmuir circulation patterns may cause these slicks to form in long parallel rows, accompanied

FIGURE 19.21   *Suspended organic matter tends to be surface active, forming a slimy coating on pilings and other structures and serving as the basis for the development of a larger community, such as this one photographed during a very low tide.*

by driftwood and other floating debris. Like the zooplankton, surfaces also serve the function of concentrating the surface-active dissolved and suspended particulate organic matter for passage on up the food web.

Suspended sediments also serve as nuclei for the accumulation of surface-active organic materials. As we have seen, the smaller the pieces that something is broken into, the more surface area is exposed. Consequently, the finest sediments are the most efficient at accumulating these organic coatings. Also, finer sediments stay in suspension longer, giving them more time to acquire thicker organic coatings. For these two reasons, deposits of finer sediments tend to be richer in organic matter, which makes them preferred regions for deposit feeders ("mud-eaters"), and more likely sources for future petroleum deposits, among other things.

Most organic detritus is consumed in surface waters before settling deeper. This is the cause of the oxygen minimum layer below the photic zone; below the photic zone, photosynthesis cannot occur, so oxygen used up in the metabolism of detritus is not replaced. Relatively small amounts of organic detritus reach deeper waters, so deeper waters are still oxygen-rich. The decay and digestion of organic material within the bottom sediments depletes the oxygen supply there, and the oxygen-depleted water trapped within the sediments does not get replaced. Consequently, only in the upper few centimeters of sediment is there sufficient oxygen for animal or bacterial activity.

D. SUMMARY

The basic functional unit of any living organism, large or small, is the cell. Most cells are microscopic. They are separated from their environment by a cell membrane, and most have a cell nucleus that regulates the cell's activities. The transport of materials is accomplished through diffusion, and concentrations of materials can be influenced by discriminating cell membranes. Many larger organisms are able to control their body fluids in order that their cells are continually bathed in a favorable environment.

Within a cell are enzymes, which facilitate certain chemical reactions, and RNA molecules, which serve as the scaffolding on which large complicated protein molecules are assimilated. Through the ordering of molecular groups along it, DNA contains all the information necessary to control the cell's activities.

The most efficient cells have a large surface-to-volume ratio. For this reason, and because diffusion works rapidly only over small distances, it is advantageous for cells to be small.

Compared to terrestrial environments, the marine environment is quite mild. Special problems faced by marine organisms include those posed by their density and by the salinity of their environment. Organisms tend to be denser than water, and display a variety of mechanisms to help themselves stay near the productive surface waters. These include mobility and air bladders for larger organisms, and buoyant secretions, flagella, and friction-enhancing protrusions for microscopic organisms. High rates of reproduction and water turbulence are also important to the survival of microorganisms.

Organisms are particularly sensitive to the salinity of their environment. Marine invertebrates are generally isotonic and vertebrates are generally hypotonic.

Chemical reactions proceed faster at higher temperatures. This seems to be reflected in faster development and greater diversification among warm water species. In addition to its obvious importance in photosynthesis, light also seems important to marine animals, as reflected in their coloration, the fact that many have eyes, and the phenomenon of bioluminescence.

There are many different kinds of interdependencies among marine organisms, including supply of protection, support, vitamins, antibodies, hygienic care, and favorable environments. One important and intricate set of interdependencies is the food web.

The bulk of the ocean's primary productivity is accomplished by microscopic phytoplankton. On the average, only about 10% of the food material consumed by any animal gets passed on to the next trophic level. The bacterial decomposers are instrumental in transforming organic detritus produced at various trophic levels back into materials that can be used by the plants and cycled through the food web again. In spite of their vigorous and important activity, the total bacterial biomass is relatively small, and they are not an important marine food source, in general.

Most of the organic materials in the ocean are in the form of dissolved and suspended dead organic matter. Only about 2% is phytoplankton, on the average, and the zooplankton and nekton are even much less than this. The dissolved and suspended organic matter is surface-active, which serves to help concentrate it onto surfaces for easier consumption and passage on through the food web. This property also increases its concentration in sediments, especially the fine-grained muds.

QUESTIONS FOR CHAPTER 19

1. What is "ecology?"
2. What is the basic fundamental unit of any organism called, and roughly what is its size?
3. Review what chemical processes occur during photosynthesis and respiration. What raw materials are needed for each, and what waste products are produced during each?
4. Why do most multicellular organisms ensure that their cells are immersed as much as possible in water-based body fluids?
5. Briefly explain how it is that random thermal motions of molecules tend to make them move towards regions of lower concentrations.
6. If molecular diffusion is a random process, and if a molecule just outside an ammonia bottle is just as likely to go in as one inside is likely to go out, then why do more go out than in?
7. Consider two horse corrals separated by a gate that swings open one way only. That is, horses pushing on it from one side may push it open and go through, and horses pushing on it from the other side cannot. If the horses all tend to move about randomly, will they tend to end up evenly distributed between the two corrals, or will they tend to become concentrated inside one of them? Explain. How is this related to the ability of discriminating cell walls to concentrate some materials within the cell?
8. What do enzymes do, and how do they do it?
9. How are the appropriate chemical ingredients delivered to an enzyme within a cell?
10. What is RNA? Where is it found? What does it do? How do some enzymes help it do this?

11. Briefly describe how rather long and intricate molecules can be put together precisely in the right order, if the motion of the raw materials in the cell is random.

12. What is DNA? Where is it found? How does it carry "instructions?"

13. Explain why large surface area and small volumes are important for a cell to be efficient.

14. Explain how surface-to-volume ratios are increased when things are broken into smaller pieces.

15. What are the characteristic times required for diffusion to carry materials dissolved in water over distances of 0.1 mm, 3 cm, and 60 cm, respectively?

16. What are some of the factors that make terrestrial environments rather harsh and cruel in comparison to marine environments?

17. What are some specialized systems, developed by particular terrestrial plants, that would be unnecessary in the marine environment? Do the same for certain specialized systems in particular land animals of your choice.

18. How does the nearshore marine environment differ from that of the deep sea waters?

19. Why is it advantageous for plants and most animals to be able to remain near the ocean surface?

20. Why do organisms tend to be denser than water?

21. What are some of the ways in which large plants and animals combat their tendency to sink?

22. What are some of the ways in which microscopic plants and animals combat their tendency to sink?

23. How does the geometry of some microscopic exoskeletons increase their water friction?

24. Why should the effect of water friction be more pronounced for smaller creatures? (*Hint:* Weight is proportional to volume, and friction to surface area. How should friction per unit weight depend on the size?)

25. Why are the spicules on skeletons of diatoms inhabiting tropical waters longer than the spicules on those living in colder waters?

26. Consider the following three things that keep a thriving diatom community in the photic surface waters: (a) small size, (b) turbulence of the water, and (c) rapid reproduction. For each of these three things, explain how it helps and why it is not sufficient by itself to solve the problem of attrition of the community members into aphotic waters.

27. Why is it that most marine organisms need not regulate their body fluids? Why do the fast swimmers do it?

28. What do "isotonic," "hypotonic," and "hypertonic" mean?

29. What is a "semi-permeable" membrane?

30. From a molecular point of view, why do water molecules tend to flow through semi-permeable membranes toward the salt water side?

31. What is "osmotic pressure?" How would you measure it?

32. What usually happens when salt water organisms are put in fresh water environments? When fresh water organisms are put in salt water environments?

33. Why must tide pool organisms be particularly tolerant of changes in temperature and salinity of their environment?

34. How are fish (both marine and fresh water) able to maintain body fluids having salinities different from that of their environment? How is this related to the horse corrals with the one-way gate of question 7?

35. Why is it that regulation of body temperature is generally not as advantageous for marine animals as it is for terrestrial animals?

36. How does temperature influence chemical reactions?

37. In comparing warm water animals to their cold water counterparts, how do they tend to be different?

38. Briefly describe two theories for why we observe greater species diversification in warm waters than in cooler waters.

39. Why is fishing generally more efficient where there is less species diversification?

40. Comparing polar, temperate, and tropical regions, where would you find the greatest difference between surface and deep animal species? Where the least? Where would the differences be most seasonal? In each case, explain why.

41. What are some of the responses shown by marine organisms to daily changes in light intensity?

42. Explain how the coloration of marine animals demonstrates that vision is of some importance in the ocean.

43. About what fraction of the animals at mid-depth have some light-producing capabilities? Do many animals live there, relative to the number of animals near the surface?

44. List some of the types of interdependencies that exist among members of the marine biological community.

45. What ecological roles do bacteria play in the marine community?

46. Your body synthesizes some of the proteins it needs from some of the materials you eat. Where does it get the energy required to perform this synthesis?

47. How much of the ocean bottom is able to accommodate attached plants? Why so little?

48. What is a "trophic level?" Plants make up which trophic level? How about herbivores?

49. Roughly what fraction of the food material consumed by one organism is passed on up to the next trophic level, on the average?

50. In anoxic regions, where do bacterial decomposers get their oxygen from?

51. Most of the organic materials in the ocean are found in what form?

52. About what fraction of the total organic material in the ocean are the phytoplankton? The zooplankton? The neckton?

53. What service do zooplankton perform for anchovies and pelicans?

54. What does it mean to be "surface-active?"

55. Why do finer sediments tend to be richer in organic matter?

56. Briefly describe the expected variations in dissolved oxygen concentration starting from the surface of the ocean and going down into the bottom sediment. Explain the pattern.

*57. What are some of the specialized systems in your body that help ensure your cells have a favorable environment? Can you say what some of these systems do?

*58. If the raw materials used by cells are water soluble, and if the cell uses these materials to produce a new cell (i.e., in reproduction), then why won't this new cell dissolve away in the water?

*59. Why do you suppose it is that a diatom community could double its size in one day, and an oak forest (of equal total mass) couldn't? (*Hint:* Perhaps the ability to synthesize depends both on the availability of the nutrients and the surface-to-volume ratio of the producers.)

*60. Why should coffee cool faster if poured onto the saucer?

*61. Why might mild thermal pollution of coastal waters be fatal to some species, even if they normally live in waters of that temperature during part of the year?

SUGGESTIONS FOR FURTHER READING

1. Drake, Imbrie, Knauss, and Turekian, *Oceanography*, Chapter 11, Holt, Rinehart and Winston, New York, 1978.

2. M. Grant Gross, *Oceanography*, 2nd ed., Chapter 17, Prentice-Hall, Englewood Cliffs, New Jersey.

3. Scott Johnson, "Crustacean Symbiosis," *Sea Frontiers*, **27**, No. 6 (1981), p. 351.

4. John R. Moring, "Pacific Coast Intertidal Fishes," *Sea Frontiers,* **25,** No. 1 (1979), p. 22.

5. "Marine Pollution," Oceanus, **18,** No. 1 (Fall 1974).

6. "Senses of the Sea," *Oceanus,* **23,** No. 3 (Fall 1980).

20
DISTRIBUTIONS AND LIFESTYLES OF MARINE ORGANISMS

We continue our study of marine ecology with a slightly more detailed treatment of the ecological niches of the more important marine organisms. First we shall investigate the organisms of the pelagic environment, and then we shall look at the benthos.

In deep water, these two communities are fairly distinct. Most pelagic organisms live near the surface, as the sunlight is needed by the plants, and the animals naturally tend to be found near the food source. Consequently, throughout most of the ocean, pelagic and benthic communities are quite distinct, being separated by two or more kilometers of relatively barren waters. In shallow coastal waters, however, the two communities merge, and there is considerable direct interplay between them. In some coastal areas, the presence of large attached plants adds a further dimension to both environments that is not present in most oceanic waters (Figure 20.1).

A.   THE PELAGIC ORGANISMS

The majority of the ocean's living organisms are pelagic. Although the benthos shows greater species diversification, the total mass of the animals of the ocean bottom is only a few percent of that in the waters above, on the average.

Except for a few coastal areas that accommodate attached plants, the great majority of marine plantlife is necessarily pelagic, because they must remain within the photic waters to survive. It is no surprise that the greatest animal activity is near the food source, so most marine animals are pelagic also.

For convenience, we will divide our study of pelagic organisms into three parts according to their positions in the food web: the phytoplankton, the zooplankton, and the nekton.

A.1   Phytoplankton

With a little thought of the differences between marine and terrestrial environments, we can understand why the plants of these two environments should be so different (Figure 20.2). Marine plants are continually bathed in a fluid that brings them all their nutritional requirements. They have no need for complicated systems, such as roots or leaves, which probe neighboring regions to locate the various needed raw materials. Furthermore, the ocean's buoyant force supports their weight, so they have no need of stems or other structures to lift their photosynthetic parts against gravity, as do land plants. It is no wonder that marine plants are quite simple by comparison.

Furthermore, we have seen there are two reasons why it is advantageous for marine plants to be tiny. Smaller size implies larger surface-to-volume ratio. This means that smaller plants have greater efficiency in gaining raw materials and excreting wastes, and that they have

545

FIGURE 20.1 *Throughout most of the ocean, benthic and pelagic communities are separated by two or more kilometers of relatively barren water. In shallow coastal waters the two communities merge, however, and are sometimes joined by large attached plants as well. In this photo, the jellyfish, comb jelly, and fish are pelagic, whereas the crabs, clams, and attached plants are benthic.*

greater friction with the water, which reduces the danger of extinction through rapid sinking out of the photic zone. Consequently, the bulk of the phytoplankton is microscopic and single-celled.

The most important microscopic marine producers are the diatoms and the coccolithophores (Figures 18.4 and 18.5). The diatoms tend to be larger, sometimes banding together in chains, and have exoskeletons of silica. The smaller coccolithophores have calcareous exoskeletons.

The distribution of phytoplankton in the oceans is variable, depending on light intensity, availability of nutrients, intensity of grazing, currents, and many other factors. Their smaller size tends to make coccolithophores more efficient producers and slower sinkers. Therefore, in nutrient-deficient, calm waters, such as those of the open tropical oceans, coccolithophores have a definite advantage. Although relatively little primary productivity is accomplished in these waters, that which is done is done primarily by coccolithophores. They are not found in the colder waters at higher latitudes, however, probably because the greater solubility of calcium carbonate in these waters would prevent them from acquiring and keeping their skeletons.

FIGURE 20.2 *Compared to marine plants, land plants must endure a much harsher environment and therefore tend to be more highly developed and complex.*

Where there is greater turbulence and mixing, there is less danger of extinction through sinking, because the water motion ensures that some members will remain in the photic zone, able to reproduce and replace those lost to depth. Furthermore, larger abundance of nutrients in these waters means that efficiency is not at such a premium. Consequently, the larger diatoms thrive in waters of the continental shelves, regions of upwelling, and higher latitudes where there is better mixing and larger nutrient concentrations.

The tiny siliceous exoskeletons of diatoms are quite intricate and porous (Figure 20.3). After the organisms die, these tiny "tests" sink to the bottom and may accumulate in large deposits called "diatomaceous earth." Diatomaceous earth has many commercial uses including fine filters, deodorizing and decoloring agents, cleaners, polishers, and paint removers. When absorbed into diatomaceous earth, the very "touchy" explosive, nitroglycerine, becomes the "safe" explosive, dynamite. The discovery of this was the basis of the Alfred Nobel fortune, from which the annual Nobel prizes are paid.

Various kinds of dinoflagellate algae (Figure 20.4) are found in most oceanic regions, but we believe that their total contribution to the ocean's primary productivity is less than that of the diatoms and coccolithophores.

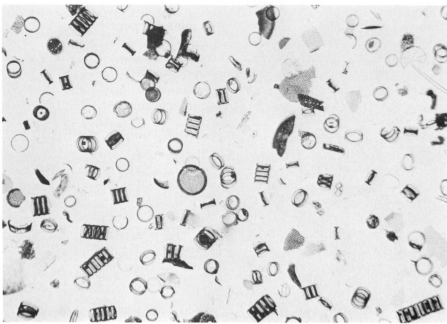

FIGURE 20.3 *Some fragments of the exoskeletons of various types of diatoms, which collect as sediment on the ocean bottom in some regions, forming "diatomaceous earth."*

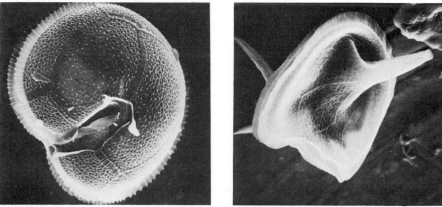

FIGURE 20.4 *Electron microscope photo of two types of dinoflagellate algae. (Peridinium oratum is on the left and Peridinium depressum on the right.)*

Their cell walls are made of cellulose-like materials, which are more buoyant than the mineral exoskeletons of the diatoms and coccolithophores. This, plus their minor locomotive abilities gained by whip-like flagella, give them an advantage in very calm waters or regions where there is gentle downwelling, where diatoms and coccolithophores are particularly threatened by extinction through sinking. Many species of dinoflagellates are consumers rather than producers, some are bioluminescent, some give the water a reddish color, and some are poisonous.

The feared "paralytic shellfish poisoning" is associated with red tides in some coastal areas. When there is a bloom in dinoflagellate algae, some of these give the water its reddish color and some are poisonous to humans. Filter-feeding shellfish may feed on these algae, which are not toxic to the shellfish, because they use non-acid digestion. But when we eat these shellfish, the acid environment of the human stomach breaks down some of the long molecules produced by the dinoflagellates into shorter toxic ones. Depending on the amount consumed, the effects can be simple food poisoning, with symptoms similar to flu, or in more severe cases can cause paralysis starting with the lips, followed by the stomach, total paralysis, and then death.

Because the plants support the entire biological community, you may think that the total biomass of the plants must be greater than that of the higher trophic levels that rely on the plant-produced materials. In general, this is true, although it is not necessarily true. In fact, there are some local regions where the total plant biomass is actually less than that of the higher trophic levels.

It is the rate at which organic materials are produced, which determines the size of the animal community that the plants can support. Of course, more plants can produce more materials, so there is a correlation between the biomass of the plants and the size of the community which they support. But other factors are important too. In some very productive environments, rich in nutrients and sunlight, relatively few plants can produce large amounts of organic materials and support an animal community of biomass much larger than their own.

As an example, suppose you decided to eat microscopic algae. Suppose you kept this algae in a tank with plenty of nutrients and sunlight so they could produce enough to double their mass each day. Since you would consume about 1 kg of food each day, you would need a "standing crop" of only 1 kg of algae to support your much larger body. That is, each day, the 1 kg of algae would become 2 kg, so you could harvest half of it and still be left over with enough to produce food for the days that follow. In this hypothetical example, a relatively small amount of algae supports a much larger biomass at your higher trophic level.

There are some very highly productive regions of the oceans where the single-celled phytoplankton can double or even quadruple their mass each day. In these very productive regions the biomass of the higher trophic levels can actually exceed that of the phytoplankton. However, most of the ocean is not supportive of such rapid phytoplankton growth, so on the average in the oceans, the biomass of the plant producers is actually about 10 times *greater* than that of the higher trophic levels which they support.

A.2 Zooplankton

The zooplankton are animals that are not capable of fast extended motion. These vary in size from large jellyfish that may be a meter in diameter, to

FIGURE 20.5 *Miscellaneous zooplankton, including arrow worms, copepods, and the larvae of other crustaceans. These microscopic animals play an essential role in the marine community by collecting and concentrating plant materials for passage on up the food chain.*

small single-celled protozoans. However, the bulk of the zooplankton in the oceans are small, microscopic, but multicellular organism (Figure 20.5) that feed on phytoplankton and other suspended organic particulate matter. They generally have adaptations for causing water movement and filtering the water in order to make their feeding more efficient.

Whether an organism is a member of the zooplankton population depends on more than just its breed. Some can be either benthic or planktonic, depending on environmental conditions. Some ("meroplankton") spend only a portion of their life cycle as plankton. Some of these undergo abrupt changes, called "metamorphoses," such as changes from egg to larva, or larva to adult, that bring them abruptly into or out of the planktonic community. Others undergo more gradual changes in passing through the planktonic stage, such as juveniles transforming towards adulthood.

In terms of total biomass, copepods of various types are undoubtedly the single most important kind of zooplankton (Figure 20.6). At higher latitudes, especially near Antarctica, their larger cousins, the krill, dominate. Other types of zooplankton of considerable importance include arrow worms, comb jellies, small jellyfish, and the larvae or juveniles of many larger organisms, including segmented worms, gastropods, cephalopods, echinoderms, and fish.

Compared to those of the oceanic environment, there is considerably

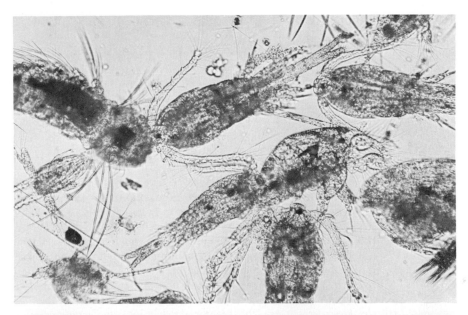

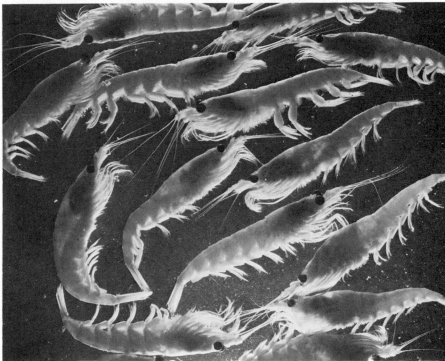

FIGURE 20.6   *Copepods (top) and krill (bottom).*

larger variation of species among coastal ("neritic") zooplankton, reflecting their more varied environment. Coastal zooplankton also tend to be larger than those in oceanic waters. We have seen that small size means

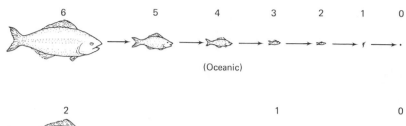

FIGURE 20.7 *In the less productive oceanic waters, emphasis is on efficiency, so the members of each trophic level are of minimum size necessary to carry out their functions. Predator is just slightly larger than prey, and so many extra steps are involved in the food chain before the food reaches large predators, such as tuna, harvested by humans. The heavy losses at each step mean that there is very little left by the time it reaches the large predators. In the more productive coastal and upwelling waters, by contrast, there is less emphasis on efficiency, and fewer steps are involved before the food reaches the large predators. The greater productivity and smaller losses along the food chain combine to make coastal and upwelling waters much richer in large predators harvested by us.*

greater efficiency. Phytoplankton of the oceanic environment tend to be smaller than their coastal relatives, because scarcity of nutrients places a greater premium on efficiency in the oceanic environment. The same is reflected in zooplankton. Scarcity of food places a greater premium on efficiency for oceanic zooplankton as well, which is responsible for their being smaller than coastal species.

The emphasis on small size and efficiency in oceanic regions leads to a long food chain there. Since food is scarce, efficiency demands that each organism be as small as possible consistent with its particular functions. Predator need be only slightly larger than its prey, so a large number of trophic levels are required before food is passed up to large fish, such as tuna, for example (Figure 20.7). Since 90% of the remaining food value is lost at each trophic level, not much is left by the time it reaches large predators. Consequently, large fish are extremely scarce in most oceanic waters for two reasons. First, there is not much plant productivity there, and second, the small amount of food that is produced must pass through many more trophic levels along the way, with heavy losses at each level. Our fishing industry is geared toward the larger predators for human consumption, and because of these two reasons, very little of our catch comes from oceanic waters.

In polar and subpolar regions, there are large seasonal variations in zooplankton populations, in response to the short one or two month bloom of phytoplankton. We think that there is a deep dormant "seed crop" of zooplankton during the summer, fall, and winter months. These are probably brought to the surface during winter mixing of deep and surface

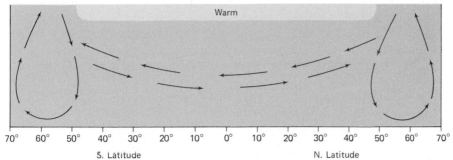

FIGURE 20.8 *The seasonal cycle of zooplankton at high latitudes. They spend the summer, fall, and winter as a small seed crop in deep water, returning to the surface during winter mixing to feast and flourish on the spring phytoplankton bloom. Similarities among species in Arctic and Antarctic waters indicates some zooplankton must pass (or have passed) between the two beneath the warmer surface waters at tropical and temperate latitudes.*

waters (Figure 20.8), where they are ready to respond to the spring bloom of phytoplankton. Many of the same species are found in Arctic and Antarctic waters, and we think that they must have crossed *under* the warm surface waters of the low and intermediate latitudes.

In temperate waters there are large seasonal changes in the size and makeup of the zooplankton population. The size varies in response to the spring and fall blooms in phytoplankton productivity. The makeup changes in response to the breeding patterns of various kinds of nekton. Some migrate down from high latitudes to breed during some seasons, others migrate up from lower latitudes to breed during other seasons, and the large native animals tend to breed still at other times. The juveniles of these various species then constitute an important and seasonally changing component of the zooplankton.

Some species of zooplankton require very special conditions under which they thrive and multiply. Looking for these "indicator species" in the water, we can quickly identify the characteristics of a water mass. We can use them to trace the motions of water masses of interest to us, such as those associated with good fishing.

Most zooplankton migrate vertically in response to light conditions, showing a desire to stay just beneath the sunlit surface waters. They can climb or sink at rates of typically 10 to 40 m per hour, coming to the surface to feed at night, and then sinking to a depth of several hundred meters by midday (Figure 20.9). This daily vertical migration was first observed during World War II when sonar echo sounders recorded daily changes in the position of this then unexplained "deep scattering layer."

Their vertical migration pattern allows the zooplankton to remain as close to the food source of the photic surface waters as possible, without exposing themselves in the sunlight to predators. They can come up to feed during nighttime and then rest during the daytime hours in deeper cooler waters where their metabolism is lower and they are not carried

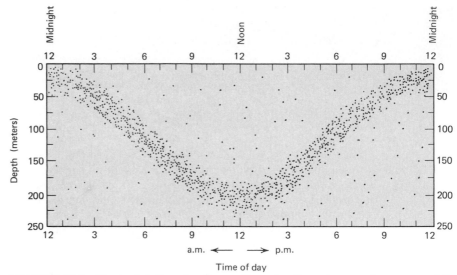

FIGURE 20.9 *Typical daily migration pattern for the microscopic grazers of the "deep scattering layer."*

away by quick wind-induced surface currents. Among the ecological effects of this vertical migration is the transfer of food materials from the surface where it is consumed, to depths where it is metabolized. This one-way removal of nutrients from surface waters, of course, reduces plant productivity. The respiration at depth also reduces the oxygen there, creating the oxygen minimum layer, which makes their own survival more difficult (Figure 20.10).

Some zooplankton remain at the surface, not partaking in the vertical migrations of others. As you might imagine, their greater visibility there means these species are consumed more rapidly by larger carnivores, so they must reproduce more prolifically in order to survive. Also because of the warmer waters they inhabit, they tend to metabolize, develop, and reproduce more quickly than the vertically migrating species. They tend to be carried around more by surface currents. In some regions, these currents are cyclical, and the zooplankton have evolved to time changes in their life stages to coincide with the current cycles.

As most schoolchildren know, fish tend to form large groups, or "schools" (Figure 20.11). This schooling behavior is not restricted to nekton, however. Plankton also tend to come in clumps, which is referred to as "plankton patchiness." Although some of this patchiness can be attributed to local water motions, bringing suspended materials together in some regions, not all of it can. There is good evidence that even the microscopic plankton are social creatures and form groups by design rather than by accident.

The reasons for these schooling behaviors are not completely understood. It is known that group behavior tends to sharpen individual feeding behavior. You may have noticed that when people around you are eating,

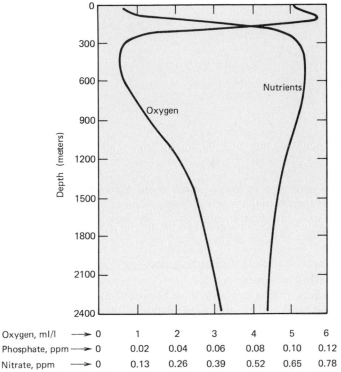

FIGURE 20.10 *Plots of nutrient concentration and oxygen concentration as a function of depth for the North Pacific off Southern California. The effect of the plants depleting the nutrients and enriching the oxygen in the surface waters is clearly displayed. Also, the oxygen minimum region testifies to the animal activity there.*

FIGURE 20.11 *There are probably several reasons for fish to school, including advantages in feeding, reproduction, rearing of young, and reduction in losses to predators.*

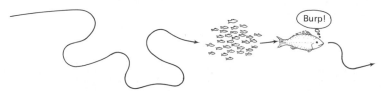

FIGURE 20.12 *A predator can consume much more through frequent small snacks than through a few infrequent large feasts. (So can we.) Therefore, a species can reduce its losses to predators by schooling.*

you start getting hungry yourself. Schooling also has some benefits for reproduction and rearing of young.

One of the most interesting benefits of schooling has to do with protection from predators, and the reason behind this is interestingly subtle. You have probably noticed that the way to gain weight is to eat constantly. If you eat a little "snack" every few minutes, you consume a lot of food altogether and gain weight quickly. On the other hand, if you eat only one meal every day or two, you will lose weight. Because of the finite size of your stomach (and some shrinkage between meals) there is a limit to how much you can eat at one sitting, no matter how hard you try. Through frequent snacks you can consume much more food altogether than you can through infrequent banquets. Applying this to the marine environments, if organisms school, their predators will have a few infrequent banquets, as they encounter these schools. But in this fashion, the predators consume less altogether than if the prey were dispersed and the predators could enjoy frequent small snacks (Figure 20.12).

A.3   Nekton

Pelagic animals capable of fast extended motion are called "nekton." These are predominantly fish, although some mammals, reptiles, and a few fast-swimming invertebrates such as squid (Figure 20.13) are included among marine nekton. Of the various components of the marine biological community, the nekton are the most familiar to us because of their size and visibility. This is misleading, however, because on the whole they make up only about 0.1% of the total ocean live biomass.

Their distribution in the oceans is similar to that of the zooplankton. Clearly it is advantageous for the nekton to stay near their food source, which is either the zooplankton, or other smaller nekton that feed on the zooplankton. For example, much of the nekton are found in the deep scattering layer, migrating vertically each day along with the zooplankton.

Included in the diets of most nekton are their own juveniles. This may

FIGURE 20.13   *Squid are among the relatively few invertebrates that could be considered to be nekton, or "fast swimmers."*

seem cruel or cannibalistic, but it is necessary. Most of the primary food materials are microscopic and well dispersed. Most large nekton would be incapable of harvesting this. This is the role that their juveniles and other zooplankton must play if this food is to be passed on up to higher trophic levels.

As a result, juvenile mortality is extremely high. Some nekton hide or bury eggs in order to protect them. Those that don't must have extremely large spawns; some produce more than a million fertilized eggs in a single spawn. When you consider that only two individuals need to survive to replace the parents, you can see that the juvenile mortality rate must be very high.

Although most nekton are found in the surface waters near the primary food source, some species are found in mid-water and deep water environments. The meager food supply makes it especially advantageous to be efficient at collecting and utilizing food. Most species of fish in these waters are small and have mouths that are huge in comparison to their bodies (Figure 20.14). The large mouths allow them to take advantage of infrequent feasts falling down from surface waters above or resulting from encounters with less fortunate neighbors.

Because the food supply is meager, infrequent, and mostly detritus from the surface waters above, deep sea creatures tend to be small efficient scavengers rather than large predators. Large "deep sea monsters" are quite impossible in this environment.

B.   THE BENTHOS

Less than 2% of the ocean bottom is shallow and stable enough to support attached plants. Consequently, throughout most of the oceans, the benthic

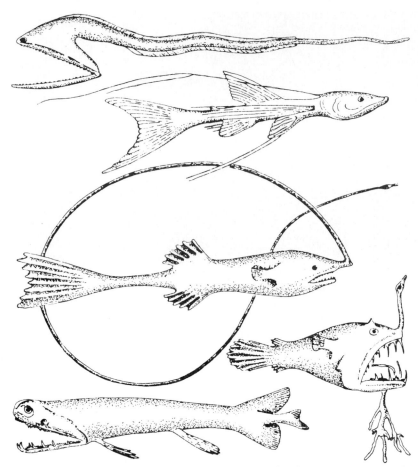

FIGURE 20.14 *Some varieties of deep sea fish, drawn approximately to scale, with the length of the top one being about 15 cm. (Redrawn from Sverdrup, Johnson, and Fleming, The Oceans: Their Physics, Chemistry, and General Biology. Prentice-Hall, © 1942, renewed 1970.)*

community is made up entirely of animals. As we have seen, most of the organic materials produced in the surface waters are also consumed there. Only a very small fraction of these materials ever make it down to the ocean bottom, so the biomass of the benthic community which this detritus supports is usually only a tiny fraction of that inhabiting the surface waters above. In some coastal waters, however, the bottom is shallow enough that sunlight reaches it, and the benthos find themselves within the productive surface waters, rather than beneath them. In these regions, the benthos are much more populous. In some shallow coastal areas, such as within the intertidal zone or just below it, the benthos can be the majority of all marine life.

Although the benthos overall represent only a very small fraction of all marine life, they do come in a wide variety of forms. Of all marine

FIGURE 20.15 *Benthic animals display three different feeding behaviors. Some, such as starfish, crabs, or snails, actively search for prey or other sustenance. Others, such as clams, barnacles and tube worms, filter their sustenance from the water. Still others, such as some worms, ingest bottom sediments to extract residual nutriments.*

animal species, 98% are benthic. This great amount of species diversification displayed by the benthos is a reflection of the wide variety in benthic environments.

B.1 Overview

The ocean bottom inhabitants display three very different kinds of feeding behaviors (Figure 20.15). Some actively search for prey or scavenge, some filter-feed, and some ingest sediments rich in organic matter.

The ocean bottom also offers three very different kinds of habitats, or lifestyles. Some organisms crawl over the bottom, some are attached to the bottom, and some live within the bottom. Those that actively crawl over the bottom are generally scavengers or predators, such as crabs or starfish. Those that are attached to the bottom may be predators, such as anemones, or may be filter feeders, such as tube worms, barnacles, oysters, or mussels. Those that live within the bottom may be deposit feeders, such as many segmented worms, or they may be filter feeders, such as clams.

The benthos also includes some marginal members, called "nektobenthos," which spend large amounts of time on the ocean floor, but are also good swimmers (Figure 20.16). Examples of these include flounders, rays, and some shrimp. As we learned before, many bacteria also thrive on the organic matter contained in the bottom sediments.

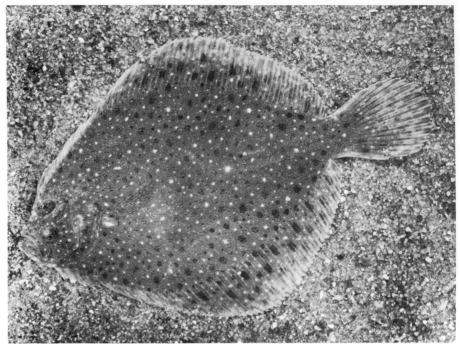

FIGURE 20.16 *This flounder would be considered "nektobenthos," because it is capable of fast extended motion, even though it spends most of the time on the ocean bottom.*

Deposit feeders and bacteria tend to favor fine-grained muds because they tend to be richer in organic matter. Not only do the finer grains present more surface area for surface-active organic adherents to collect on, but also they tend to collect in lower energy environments where organic detritus can also settle out directly. Too much organic matter in the sediments causes too much oxygen depletion, so much of the organic matter goes unused. It is found that optimum conditions are when about 3% of the sediments are organic matter.

The deposit feeders help aerate the sediments, which benefits bacteria and microscopic animals. Still, oxygen depletion begins taking its toll at depths of about 2 or 3 cm into the sediments, and there is seldom any animal activity below 10 cm depth. The bacteria are an important source of lipids and proteins for deposit feeders, although their carbohydrate content is low.

It appears that deposit feeders tend to force out filter feeders, perhaps because their activities stir up the fine sediments, some of which then clog filter feeders' filtering mechanisms. Consequently, filter feeders are more common in sandy or rocky environments than in organic muds.

B.2 Coastal Waters

Coastal environments are quite varied, and support a corresponding variety of benthic biological communities. Except in protected estuaries and embayments, coastal sediments tend to be devoid of attached plants. The

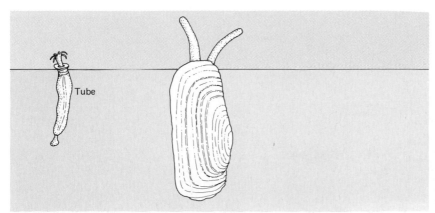

FIGURE 20.17   *The sandy nearshore benthic environment is excellent for bur-*
rowers, such as some clams and segmented worms, who get their
food from the water. Longshore currents ensure continual supply
of food and the rapid removal of wastes.

animals found in these regions are almost exclusively burrowers, with
deposit feeders dominating in the muds, and filter feeders predominant
in the coarser sands (Figure 20.17).

By contrast, in the sediments of protected estuaries and embayments,
where there is relatively little water motion, we tend to find extremely
high plant productivities. This supports a very active benthic community,
some of which may live directly on the plants themselves. The grasses in
some estuaries are covered with algae, bacteria, protozoans, snails, small
crustaceans, and other small creatures, which together may have a total
mass even greater than that of their host.

High-energy rocky coastal areas support an entirely different kind of
community. Above low tide level there is marked visible zonation in this
community, corresponding to varying abilities of the organisms to withstand prolonged periods out of the water (Figure 20.18). In the splash zone
above high tide level are found lichens, and some snails and crabs that
feed on them. In the intertidal region we find barnacles, snails, limpets,
tube worms, sea anemones, mussels, and some kinds of algae, growing in
zones varying noticeably with altitude above the low tide line. Anemones,
snails, crabs, barnacles, and mussels are frequent benthic tide pool inhabitants. Just below the low tide line is the most populous benthic area,
where all of the above are common, plus starfish, sea urchins, nudi-
branchs, and many kinds of crustaceans.

The tropical and temperate coastal areas tend to have more populous
benthic communities. This is probably a result of the shorter growing season and reduced overall productivities of the coastal plants at higher latitudes, which reduces the size of the community they can support. In seasonally changing environments, we tend to find benthos that grow,
mature, and reproduce more rapidly. In more stable environments, these
are forced out by hardier species, because in these environments, the ability to grow and reproduce rapidly is not as advantageous as other
attributes.

FIGURE 20.18   *In and above the intertidal region is a noticeable vertical zonation of organisms, reflecting the varying abilities of the different species to tolerate prolonged periods out of the water.*

B.3   Reefs

Reefs provide unique local environments that support various kinds of benthic communities. Some reefs are man-made (Figure 20.19), such as breakwaters, oil platforms, and sunken ships. Some are created by the roots of mangrove trees, and some by the cemented-together tubes of generations of sabellarian worms.

Of special commercial interest are reefs created by oysters that grow as new oysters fix themselves to the remains of previous generations. Oysters prefer brackish waters of estuaries, with salinity in the range of 7 to 18%, where they can survive but most of their natural enemies cannot. They prefer flowing water that brings them oxygen and large supplies of nutriments, such as plankton and organic detritus. These filter feeders have cilliary action for causing water motion and helping them catch food in a mucous net. This creates an environment enjoyed by other filter feeders, such as barnacles, worms, mussels, sponges, and crabs, which join the reef community. This biological activity is also attractive to some predators, including some kinds of starfish and snails, which become ungrateful guests by feeding on the oysters.

The largest reefs are coral reefs (Figure 20.20). Many are large enough to cause substantial alteration of the surrounding ocean environment, such as causing deflection or surfacing of ocean currents. As we have seen, these prefer warm tropical waters, where there are only small seasonal changes. As a result of the constancy of the environment, inhabi-

FIGURE 20.19 *Underwater photo of an oil drilling rig near the Philippines. Whether they are natural or man made, reefs provide a favored environment for many organisms, both surface dwellers (benthic) and free swimming (pelagic).*

tants of coral reefs have evolved to fill in many varied narrow ecological niches. This means there is large species diversification in coral reefs, but relatively few members of each species.

B.4 The Deep Ocean

Unlike their coastal relatives, the deep ocean benthos (Figure 20.21) are far removed from the productive surface waters and must rely on detritus that somehow has escaped consumption there. Furthermore, the surface waters over the deep ocean are generally less productive in the first place than are coastal waters. For these two reasons, the deep ocean bottom is very sparsely populated, compared to the benthic population in coastal waters.

The deep ocean benthos also has considerably less diversity of species than that of coastal waters. There are presently two popular theories to explain this. One is that the more uniform deep ocean environment provides fewer unique ecological niches, thereby not fostering such a diversity of species. The second theory is based on the observation that in the uniform deep ocean environment live species become rather intolerant of change. Ice ages may have caused 3°C to 8°C changes in deep water temperatures, killing off many of the deep water inhabitants. According to this theory, there has not been sufficient time since the last ice age for evolution to have regenerated much diversification among these deep water species.

The deep ocean sediments are mostly mushy muds, so deposit feeders are more abundant than filter feeders. Also found here are scavengers,

FIGURE 20.20   *Aerial photo of a portion of Australia's Great Barrier Reef.*

which must be very proficient at sensing the presence of rather infrequent
and widely scattered food caches coming from the surface.

The vigorous benthic communities surrounding hydrothermal vents
(Figure 20.22) of the oceanic ridge are particularly interesting because
they are completely independent of the organic matter produced in surface waters. The primary producers are bacteria that seem to use the
water's thermal energy, or the oxidation of reduced elements or H_2S in
the vented water, as the energy source for the synthesis of organic materials. The bacteria seem to be exceptionally efficient and productive,
which may be a result of the higher rates of chemical reactions at these
higher temperatures.

The animal community near a vent seems to be primarily filter feeders, such as some clams, mussels, crabs, tube worms, and various other
worms, including one that looks like spaghetti. A few species are found in
the mouth of the vent, but most are found at the outer fringes at rather

FIGURE 20.21 *Photo of the ocean floor at 1.5 km depth, showing two brittle stars, a deep sea fish, and numerous animal tracks.*

normal ocean bottom temperatures (around 2°C), relying on turbulence to bring them their bacterial sustenance. Although many of them perform similar functions, the species of animals found in the vent communities tend to be different from the normal sea floor species. The vent benthos tend to have higher metabolic rates, mature more rapidly, and reproduce more quickly. These qualities are clearly advantageous in a temporal vent environment, which may require relocation of vent communities as some vents fizzle and others appear.

C. SUMMARY

Although there is greater species diversification among the benthos, the largest number of individuals are pelagic. Throughout most of the ocean, the pelagic and benthic communities are fairly well separated, but in coastal waters they merge into one.

FIGURE 20.22   *A vent community on the Galapagos Ridge 2.8 km deep. Seen in
the photo are very large tube worms (whitish tubes with bright red
plumes), small limpets, crabs, seaworms (left foreground), and a
fish.*

Phytoplankton are quite simple compared to land plants, and most
are microscopic. The most important producers are diatoms, which are
especially prominent in the most productive and nutrient-rich waters,
such as those at high latitudes, in upwelling regions, and over continental
shelves. Coccolithophores are probably the second most important marine
producers, being prominent in warmer surface waters where their small
sizes and calcareous skeletons give them the advantage. Although the total
phytoplankton mass is larger than the total mass of higher trophic levels
that it supports, in some very productive regions the reverse is true.

Most of the zooplankton is microscopic. There are large numbers of
different species. Some spend their entire lives as plankton, and others
don't. Copepods are extremely important zooplankton, as are the krill at
higher latitudes. Coastal zooplankton are generally larger and more varied than are those of oceanic waters. Efficiency is at a premium in the less
productive oceanic waters, which means organisms there tend to be small,
and the food chain tends to be long compared to coastal waters. In polar
and subpolar waters, there are large seasonal fluctuations in zooplankton
populations. In temperate waters there are large seasonal fluctuations in
both numbers and species of zooplankton. Most zooplankton show daily
vertical migration, causing the "deep scattering layer," the oxygen mini-

mum layer, and accelerating the depletion of nutrients from surface waters.

Most marine organisms, large and small, tend to school. This has advantages in stimulating feeding and reproduction, and in the rearing of young. It also helps reduce the loss to predators because predators can consume less altogether in infrequent feasts than in frequent snacks.

The nekton comprise a very minor (although familiar) component of the ocean's total biomass. Their distribution is similar to that of the zooplankton, their main food source. Mid-ocean and deep ocean fishes tend to be small and to have large mouths.

The benthos display three general kinds of feeding behaviors, including actively searching for food, filter-feeding, and deposit feeding. They also have three general kinds of lifestyles, including crawling over the bottom, being attached to the bottom, and living within the bottom sediments. Deposit feeders favor fine-grained muds and filter-feeders favor coarser sediment environments. Oxygen depletion is a life-limiting factor within the bottom sediments.

Except for protected embayments, most coastal sediments are devoid of attached plants, and are inhabited primarily by burrowers. There is noticeable horizontal zonation among benthic species in the intertidal region on rocky coasts. There is also considerable variation with latitude in the coastal benthic population.

Oysters tend to form reefs in flowing brackish waters. Many other animals also join the reef community. Coral reefs are frequently very large and cause significant alteration of the surrounding marine environment. They grow in warm tropical waters, which have remained pretty much the same over the ages, permitting considerable species diversification and the development of an intricate set of interdependencies.

By contrast, the deep ocean benthos shows minimal species diversification. They are also rather sparse due to the small productivity in the surface waters and large distance from the food source. Deep ocean sediments are mostly muds inhabited primarily by deposit feeders. Benthic communities around hydrothermal vents depend on food synthesized by bacteria using energy derived from inorganic chemical reactions, or the water's thermal energy, in place of sunlight.

QUESTIONS FOR CHAPTER 20

1. Why do most pelagic animals live near the ocean surface?
2. In what oceanic regions is there significant interplay between benthic and pelagic organisms? Why not elsewhere?
3. Are there more species of pelagic or benthic organisms? Which has the greater number of individuals?
4. Why are most marine plants pelagic rather than benthic?

5. Why do terrestrial plants tend to be more complicated than marine plants?

6. What are two reasons that make it advantageous for marine plants to be tiny?

7. What are some of the factors that influence the distribution of phytoplankton in the oceans?

8. Why do coccolithophores do better than diatoms in calm, warm surface waters?

9. Where do diatoms outperform coccolithophores, and why?

10. What is diatomaceous earth and what are some of its commercial uses?

11. How was the Nobel fortune made?

12. Under what conditions might dinoflagellate algae outperform diatoms and coccolithophores? Why?

13. Why does paralytic shellfish poisoning affect humans but not shellfish?

14. Does the total biomass of the plants necessarily outweigh the total biomass of the higher trophic levels they support? Explain.

15. On the average in the oceans, the phytoplankton biomass is how many times larger than the mass of the higher trophic levels it supports?

16. What are zooplankton? Are they all microscopic?

17. What is a metamorphosis?

18. What kind of animal forms the largest component of the zooplankton?

19. Is there more variation among oceanic or neritic zooplankton? Why?

20. Do oceanic or neritic zooplankton tend to be smaller? Why?

21. Why does very little of the annual world fish harvest come from oceanic (as opposed to coastal) waters? Give two reasons.

22. When is the zooplankton bloom in polar and subpolar waters? What happens to them during the other seasons?

23. What is the evidence that cold water zooplankton have crossed under tropical surface waters?

24. Why is there seasonal variation in the size of the zooplankton population in temperate latitudes? Why is there seasonal variation in its composition?

25. What are "indicator species," and what do we use them for?

26. What is the "deep scattering layer?" How and why does its position vary during the day?

27. What advantage is there for zooplankton to spend the daytime hours

deeper below the photic zone? How does this injure the larger biological community?

28. What is "plankton patchiness"?

29. What advantages are there for organisms to school?

30. List some nekton that you can think of.

31. The nekton altogether make up about what fraction of the ocean's total biomass?

32. Why is the distribution of nekton similar to that of zooplankton?

33. Why are the spawns of nekton so large?

34. Why are mid-water and deep water fish small? Why do they generally have large mouths?

35. Why are deep sea monsters unlikely?

36. What fraction of the ocean bottom can support attached plants? Why not more?

37. Why is the benthic community generally small (in total biomass) compared to the pelagic community above? In what regions is the benthic community larger?

38. What fraction of marine animal species are benthic?

39. What are the three kinds of feeding behaviors displayed by benthos? Give an example of each.

40. What are the three general kinds of lifestyles exhibited by the benthos? Given an example of each.

41. What are nektobenthos?

42. Why do deposit feeders and bacteria generally do better in finer sediments than coarser ones?

43. Why does some organic matter go unused if the sediments have too much organic matter in them? What is the optimum amount for supporting benthic organisms without wasting any?

44. How deep into the sediments does benthic life extend, roughly? Why not further?

45. Are filter feeders more common in fine sediments or coarse ones?

46. In coastal sediments do we find mostly animals that crawl over the bottom, burrow, in the bottom, or that are attached to the bottom?

47. Why is there marked horizontal zonation of marine benthic species in the intertidal region?

48. Why do oysters grow best in brackish waters?

49. What are some kinds of animals that join oyster reefs? Why do you suppose they like the reef environment?

50. What conditions do corals prefer?

51. Are the ecological niches filled by the organisms on a coral reef broad or narrow? Why?

52. Why are deep ocean benthos sparse compared to those of coastal waters?

53. Describe the two popular theories explaining the lack of species diversification among deep ocean benthos.

54. Are deposit feeders or filter feeders more common on the deep ocean bottom? Why?

55. How is the food produced for benthic communities around hydrothermal vents?

56. How do vent benthos tend to differ from other deep ocean benthos?

*57. If smaller size means greater efficiency, why do you suppose the smaller coccolithophores aren't more productive than the diatoms altogether?

*58. Would you expect the phytoplankton to be a larger fraction of the total biomass in regions of upwelling, or in warm tropical oceanic surface waters? Why?

*59. Why do you suppose we do not find dinoflagellate skeletons in sedimentary deposits, like we do skeletons of diatoms and of coccolithophores?

*60. Why are there more steps in the food ladder in oceanic environments than in neritic environments?

*61. Although we are discovering new species of animals all the time, none of these new discoveries is very large. It seems that all species of large animals (e.g., whales, porpoises, walruses, etc.) have already been discovered. Explain why it is unlikely for any undiscovered large creatures, such as the "Loch Ness monster," or "bigfoot," to exist.

*62. Why do you suppose productivity by attached plants is extremely high in the sediments of protected estuaries? Why is it small in unprotected coastal sediments?

*63. We find we can attract a variety of animals to an area simply by putting something in the water. In fact, such objects, called "fish aggregation devices," or "fads," are now being considered for commercial production. What do you think it is about such artificial reefs that is so attractive to marine animals?

SUGGESTIONS FOR FURTHER READING

1. J. Wesley Burgess and Evelyn Shaw, "Development and Ecology of Fish Schooling," *Oceanus*, **22**, No. 2 (Summer 1979), p. 11.

2. Bruce E. Chalker and Ronald E. Thresher, "The Multitudinous Reef," *Sea Frontiers*, **26**, No. 2 (1980), p. 66.

3. Neville Coleman, "Nature's Architects," *Sea Frontiers*, **26**, No. 6 (1980), p. 330.

4. Barrie Dale and Clarice M. Yentsch , "Red Tide and Paralytic Shellfish Poisoning," *Oceanus*, **21**, No. 3 (Summer 1978), p. 41.

5. Donald P. deSylva, "Injuries from Corals," *Sea Frontiers*, **27**, No. 4 (1981), p. 237.

6. J. Frederick Grassle, "Diversity and Population Dynamics of Benthic Organisms," *Oceanus*, **21**, No. 1 (Winter 1978), p. 42.

7. M. Grant Gross, *Oceanography*, 2nd ed., Chapter 14, Prentice-Hall, Englewood Cliffs, New Jersey, 1977.

8. "Galapagos '79: Initial Findings of a Deep-Sea Biological Quest," *Oceanus*, **22**, No. 2 (Summer 1979), p. 2.

9. Geoffrey A. J. Scott, "Mangroves and Man in the Malay Archipelago," *Sea Frontiers*, **27**, No. 5 (1981), p. 258.

21
OCEAN FOOD RESOURCES

The oceans are both large and largely unexplored. Indeed, they must hold a wealth of secrets that we are just beginning to discover. Because of this, there is the hope among many that the discovery of ocean treasures will somehow be able to save us from impending disaster, as our mushrooming population with seemingly insatiable material appetites rapidly depletes valuable terrestrial resources.

Undoubtedly, exploitation of ocean resources will be able to help bail us out of many impending shortages, at least temporarily. In other cases it will be of little help. We have no crystal ball, and cannot make detailed predictions regarding exactly what surprises and disappointments lie ahead for us in the realm of ocean discoveries and technologies. Nonetheless, from very general considerations we can at least get a rough idea of what might be possible and what would not. This chapter and the next deal with basic processes and concepts involved in our ocean's resources. Hopefully, study of this material will help the student differentiate between what are realistic expectations, and what are only pipe dreams.

The basic problem, of course, is that our population is growing exponentially (Figure 21.1 and Table 21.1) and our rate of consumption of resources is growing even faster. The earth we live on, however, is of finite fixed size, and not growing with us. The combination of heavy and increasing consumption with fixed finite resources, spells disaster. If we wish to survive as a race, we must find a way to make due with the finite existing resources; we will not be getting any more. We must think of ourselves as voyagers on "spaceship earth." We are earthlings confined to our own earthly "spaceship," and are unable to survive elsewhere in space. Wherever this journey takes us, mankind must finish the trip with the same resources that were on board at the beginning. We will be getting no more.

Writing two chapters on ocean resources is challenging because the situation is changing rapidly, and surely the details presented in these chapters will be outdated even before the book leaves the press. Our ever increasing population, coupled with new directions in our expanding technology, forces increased demand for most exploited resources and a constant search for new ones. In addition to increased demand, the supply of each resource varies considerably from time to time as old sources are exhausted and new sources are found.

As an example of the variation in value of a resource, the most valuable single resource extracted from the ocean in 1978 was petroleum, whose wellhead value was around 30 billion dollars per year. Just eight years earlier (1970), its value was only 4 billion dollars per year, being only half the annual income of the world's fishing people and only ¼ the world's annual ocean shipping bill.

Even if the demand and supply of a resource didn't change it, it would still be difficult to give an accurate appraisal of its dollar value.

573

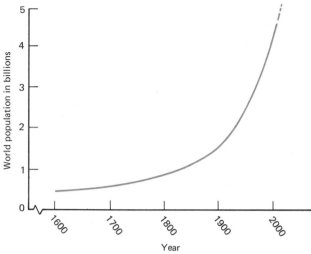

FIGURE 21.1 *World's population since 1600.*

TABLE 21.1 World's Population Since 1
A.D.

Year	Population (billions)
1	0.25
1600	0.45
1700	0.59
1800	0.90
1900	1.55
(2000)	(6.00)

Different sellers and different markets mean different prices. It is considerably more expensive, for example, for an oil company to purchase a barrel of oil from Saudi Arabia than to extract it from a borehole in the Santa Barbara Channel. Nonetheless, our energy-hungry society needs both. Which price should be used in appraising the value of our petroleum resources? Similarly, a kilogram of fish intended for the American fresh fish market is considerably more expensive than a kilogram of Peruvian anchovies intended for conversion into fish meal. Which price should we use in judging the value of the world's fish harvest? Clearly, some kind of compromising and averaging must be used in discussing the values of various resources.

There is also ambiguity in reporting the *quantity* of a resource used. For example, of fish caught for human consumption, less than half of the original weight actually reaches our tables, the rest being removed in processing. Should we report the weight that reaches our tables? Either way may be misleading. For other processes, the numbers are different. When processed for fish meal, some weight is lost to dehydration. In making fish protein concentrate (FPC), the weight of the product is only about ⅛ the original fish weight.

Because of these uncertainties, it is difficult for a scientist, being trained in accuracy, to write about marine resources, knowing the information is approximate and will be quickly outdated. This is in contrast to the subject matter of the first 20 chapters of the book, which covered the *science* of oceanography. The *science* doesn't change with time (hopefully!), although our understanding of it broadens.

Nonetheless, our society will be placing increased reliance on ocean resources, and so it will be increasingly important for the world citizenry to be informed on the subject. For this reason I've decided to give a survey of the ocean resources, after cautioning the reader that the numbers are approximate and will be changing with time.

A.   THE ECONOMICS OF FOOD

The world population is nearly 5 billion. We produce nearly 4 billion tons of food per year, which amount to ⅘ ton per person, or about 2 kg per person per day. For comparison, this is about half the rate at which we consume fossil fuels, and about four times the rate of our steel production.

Two kilograms per person per day may seem like ample food for us, and it is. Of course, a considerable fraction of this food goes into feeding domestic animals, but still there would be ample food to nourish our society, *if it were distributed properly.* Unfortunately, it is not distributed evenly, and nearly half the world population suffers a protein deficient diet.

Over the long run, our stomachs are not nearly as elastic as we are led to believe from our day-to-day experiences. Comparing the diets of the best fed societies to the diets of the most poorly nourished peoples shows a variation of less than a factor of two in food weight or calories consumed per person. (See Table 21.2.) For example, the world average is about 2570 kilocalories[1] per person per day. The United States averages 3580 kilocalories per person per day, 40% over the world average. Of course, there are big variations in the kinds of foods eaten. For example, the wealthier nations eat much more expensive animal protein, whereas poorer nations rely more heavily on cheaper vegetable protein. But on the whole, the demand for food supplies does not vary nearly so much from one person or one society to the next as does the demand for other resources.

In the language of economics, the "demand curve" (Figure 21.2) for food supplies is quite "inelastic." A large change in price has little effect on the amount of food we consume. That is, we'll consume a certain amount regardless of the price. This means that the economics of food is quite different from that of our other resources, and it is one contributing

---

[1]*What is normally called a "Calorie" in food energy is really a "kilocalorie," according to the scientific definition, and so there is a movement in the world of nutrition to switch over to the more proper "kilocalorie" designation.*

TABLE 21.2 Per Capita Daily Consumption of Food Energy and Protein
(Data from Terry N. Barr, Science, vol. 214, no. 4525, pp. 1087–1095.)

	Energy (kcal/day)	Protein (g/day)
Developed countries (average)	(3350)	(97)
United States	3580	106
Canada	3370	101
Western Europe	3380	95
South Africa	2920	77
Japan	2950	88
Oceania	3400	107
Centrally planned (average)	(2680)	(74)
Eastern Europe and Soviet Union	3480	103
China	2390	63
Developing countries (average)	(2200)	(55)
Latin America	2560	66
Far East (excluding Japan and China)	2030	49
Near East	2620	74
Africa (excluding South Africa)	2210	55
World (average)	(2570)	(69)

factor in the inequitable distribution of the present world food supply. A small reduction in the food supply (for example, from a drought in some country) creates a large increase in price (since the demand doesn't change), as illustrated in Figure 21.3. The wealthier nations can pay more for food, and can pay more for the fertilizers and agricultural technology necessary to increase their own production. Consequently, in times of anticipated food shortages, the available supply tends to flow toward the wealthier nations, and the poor nations suffer first. At the present time the average American spends 13% of his or her disposable income on food, whereas the average Indian spends somewhere between 60% and 90% of his or her income on food. If the price of food should double tomorrow, it's clear who will suffer most.

As the above discussion indicates, the present food supply is adequate to feed the world, but the distribution of this food is quite inequitable. What is the prediction for the future? Although details of predictions vary, all experts agree that the situation will get worse. The supply will not be able to keep up with demand, and the distribution will become even more inequitable than it is now.

In the three decades between 1950 and 1980, increased food supplies kept well ahead of the population (Figure 21.4). Food production increased 102% in these three decades, whereas the population increase was only 64%. These overall figures are a bit misleading, as the wealthier countries tended to have the lowest population growth and the biggest increase in food supplies. In the poorer countries, the reverse happened, and food supplies barely kept ahead of the population as is shown in Figure 21.5 and Table 21.3.

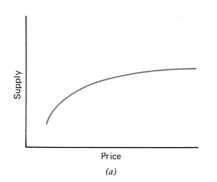

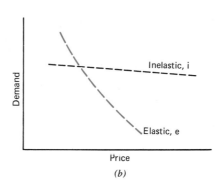

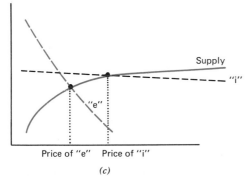

FIGURE 21.2 *Supply and demand curves. (a) Supply curve for a typical resource.*
Higher prices will encourage producers to increase production.
The curvature is caused by the fact that it becomes increasingly dif-
ficult to produce more as the resource is exhausted. (b) Demand
curves showing that higher prices will cause consumer demand to
drop. The "inelastic" curve is for a resource such as food, which
we'll consume at about the same rate regardless of the price. Con-
sequently, higher prices will have small effect on our consumption.
(c) In a free market, the price of the product is determined by
where the supply and demand curves intersect.

In the last decade, the situation has deteriorated, being aggravated by
widespread droughts in India, Africa, and the U.S.S.R. and by big
increases in the price of fertilizers. The wealthier nations tend to hold
more tillable land in reserve, and they can pay higher prices for agricultural technology and fertilizers. However, the increased production in bad
times doesn't benefit the poorer peoples. The wealthier nations tend to sell
only to each other, as they can pay the highest price.

The long range future looks very bleak. Although details vary, most
experts agree that even with expanding present technology, and utilizing
all tillable land, the world food production from both land and sea cannot
increase more than twofold over its present output. Yet the world population will double in less than 35 years. Therefore, even in the unlikely

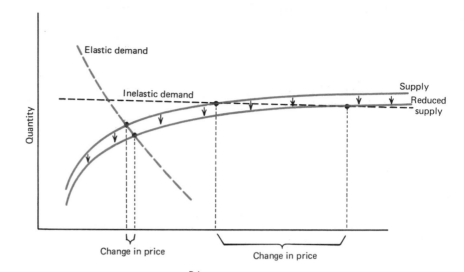

FIGURE 21.3 *Illustration of the effect of a slight reduction of the supply of a resource. If the demand curve is elastic, there will be only a small price increase. However, if the demand curve is inelastic, there will be a large price increase. Small reductions in the supply of food and oil have resulted in large price increases, demonstrating that our demand for these things is quite inelastic.*

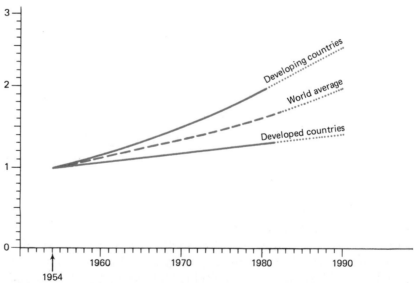

FIGURE 21.4 *World population relative to the 1954 population. Growth has been especially large in the developing countries.*

event that the food could be distributed equitably, there will still be widespread starvation within a few decades.

Starvation can be deterred somewhat through education. Most people are not efficient eaters, overeating certain nutritive areas in order that their bodies get the minimum requirements in others. Through balanced

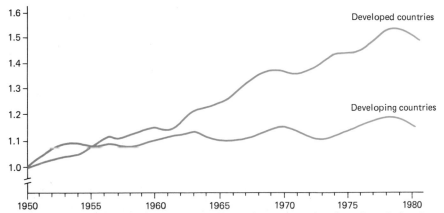

FIGURE 21.5   *Plot of per capita food production in developed and developing*
countries since 1950. (Data from Terry N. Barr, op. cit.)

TABLE 21.3   Average Annual Percentage Increases in Population, Food
Production, and Per Capita Food Production for the Years 1950 to
1980. (From T. N. Berr, op. cit.)

	Developed Countries	Developing Countries
Population	1.1%	2.4%
Food production	2.5%	2.9%
Per capita food production	1.4%	0.5%

diets these wastes could be eliminated, and a limited food supply could
be stretched further. Nonetheless, it is just a matter of time before population catches up with maximum possible food production. Because of
economics, poorer countries will suffer first, and some already are suffering. Children are much more sensitive to malnutrition than are adults, so
the most sickening scenes will appear first among children in poor
countries.

This doomsday forecast, unfortunately, represents the opinions of
most experts. It is not something that may happen, it is already happening.
We are victims of our own numbers. There is no way we can avoid the
problem. The best we can hope for is to reduce the suffering as much as
possible. Along this line, the oceans do offer some possibilities. They
would require new technology, and that we change from "hunter" to
"farmer and herder" in the oceans as we did on land. This will, of course,
take time.

B.   SEAFOODS IN OUR DIETS

Presently, the life in the ocean is used almost entirely as a source of animal protein. Some kelp is harvested for human consumption, and some
research is being undertaken to try to develop grains that can be grown in

salt water. But for the ocean to ever become a major source of vegetable matter in our diet must be considered only a dream at the moment.

The vast majority of the ocean plant life is microscopic and well dispersed. These two things make its harvest unfeasible with present technology. As a result, our best hope for the ocean is that we might be able to increase our marine animal harvest to a point that it could replace the animal protein presently in our diet from land sources. This would release for human consumption the grains presently fed to domestic animals. That is, the most efficient use of our resources would be to use the land almost entirely as a source of vegetable matter in our diet, and the ocean primarily for the production of animal protein.

B.1 Nutritional Value
In addition to the above consideration of maximizing production from limited agricultural resources, there are also nutritional reasons for relying more on seafood as a source of animal protein. Compared to land animals, finfish have the following advantages:

1. They have the amino acids in the correct ratios for our use.
2. They are a better source of Vitamin B-12. (Vitamin B-12 is needed in the manufacture of DNA and in the growth of all tissues.)
3. They are low in cholesterol and saturated fats.
4. They are high in polyunsaturated fats and the essential fatty acids.

The importance of these last two items is that cholesterol and saturated fats have been correlated to heart disease, whereas consumption of unsaturated fats tends to reduce blood cholesterol.

Although purely vegetarian diets are possible, most nutritionists agree that a small amount of animal protein in our diet is advisable, primarily because of its contribution to balanced amino acids and as a source of Vitamin B-12. Although not much is needed for this purpose, a minimum of 7 to 9 kg of animal protein per person per year (or about 20 to 25 g per day) is recommended.

B.2 Products of World Fisheries
About 70 million tons of seafood are being harvested per year, which amounts to about 15 kg per person. About 90% of this is finfish. The rest is mostly molluscs, crustaceans, and whales. (See Tables 21.4 and 21.5.) Nearly half of the identified catch consists of the small clupeoid fish, such as herring, anchovies, and pilchard.

The 70 million tons of seafood are about 2% of the total world annual food production (nearly 4 billion tons). More than half of this is processed to produce fish meal, which is used to enrich the feed of poultry and other livestock. Before World War II, less than 10% of the catch was processed for fish meal, but since then the fraction has been steadily increasing. This tendency concerns many people, as it would be a much more efficient use

TABLE 21.4 Marine and Fresh Water Harvest in 1967. The Numbers Indicate the Percent by Weight of the Total Harvest, Excluding Whales.

Fresh water fish		13.6
Marine fish		77.6
Herring, sardines, anchovies, and so on	32.5	
Cod, hake, haddock, and so on	13.5	
Redfish, bass, congers, and so on	5.2	
Mackerel, billfish, cutlassfish, et cetera	4.4	
Jack, mullet, and so on	3.4	
Tuna, bonito, skipjack	2.4	
Flounder, halibut, sole, and so on	2.2	
Sharks, rays, chimaeras	0.7	
Unsorted and unidentified fishes	13.7	
Crustaceans, Molluscs, and other marine invertebrates		7.4
Molluscs	5.1	
Crustaceans	2.2	
Sea cucumbers, sea urchins, and so on	.1	
Aquatic mammals (porpoises, dolphins, seals, et cetera)		0
Turtles and frogs		0.1
Plants		1.3

TABLE 21.5 1967 Whale Harvest by Number of Animals

Blue whales, fin whales, sperm whales, and so on	52,000
Minke whales, pilot whales, and so on	8000
Total whales harvested	60,000

of limited food resources to eat the fish directly than to eat the poultry which eats the fish.

Of the fish intended for human consumption, only about ½ the original weight actually reaches our kitchens, the rest being lost in processing. As a result fish constitute about 1% of our total diet on the average, or about 17% of the animal protein. There are big geographical variations in this. For example, in the United States fish are only about 5% of the animal protein eaten, whereas in Japan the figure is approximately 50%. (See Figure 21.6.)

A wide variety of gear is used to harvest marine organisms. Line gear and trawls are illustrated in Figure 17.3, purse seines are illustrated in Figure 17.11, and some other types of gear commonly used are illustrated in Figure 21.7. Because the bulk of the commercial harvest is small schooling finfish, nets of various types account for the largest portion of the world catch (Figure 21.8).

In recent years, the Soviets and Japanese have deployed large impressive fishing fleets. Some of these take with them their own factory ships for processing their catches right there on the open seas. This gives them the range to fish anywhere in the world and the ability to stay at sea for extended periods of time. As is illustrated in Figure 21.9, this has

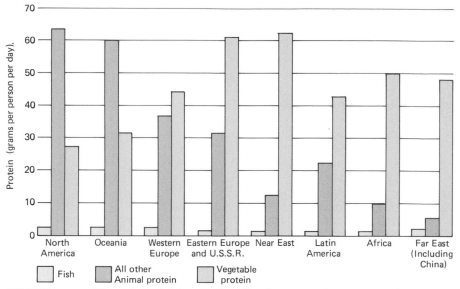

FIGURE 21.6 *The roles of fish protein, all other animal protein, and vegetable protein in the diets of various parts of the world. Clearly, fish protein is a rather small part of the total protein in all parts of the world. (From S. J. Holt, "The Food Resources of the Ocean," Scientific American, Sept. 1969.)*

resulted in greatly increased fishery harvests for these two countries in recent years.

With the cost of energy high and certain to get higher, the energy required to harvest various kinds of fish is increasingly important. Some small surface fish, which congregate in large schools that can easily be harvested with purse seines, require less energy than the cultivation and harvest of cereal grains (Figure 21.10). Unfortunately, it takes a good deal more energy to process these fish than is required to mill equivalent amounts of cereal grains, which means we can expect cereal grains always to be cheaper.

C. FISHERY MANAGEMENT

In the previous section we saw that fisheries provided a small but significant portion of the world food, being about 2% of the total food harvest by weight, and supplying about 17% of the animal protein consumed by humans. Through intelligent fishery management, these numbers could conceivably be increased by a factor of two or three times. Although this would still be only a fraction of the additional food needed, the shortages brought on by our rapidly expanding population demand that we pursue all avenues of increased food production, including this one.

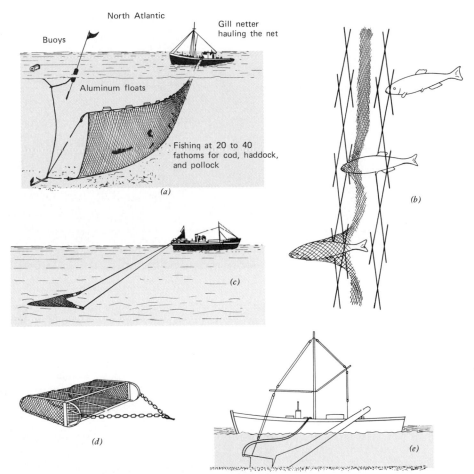

FIGURE 21.7 *Sketches of various types of fishing gear. (a) Gill net. (b) A type of entangling net that has a very fine mesh, very elastic net between two nets of much larger mesh. (c) An otter trawl. (d) One of many types of dredges. (e) Modern gear used for harvesting shellfish in shallow water, using a fancy dredge and a conveyer belt for bringing the shellfish aboard.*

The theory of proper fishery management is fairly simple. Unfortunately, in practice it has been very difficult to enforce. Needed are firm decisions, backed up by clear authority, and made on the basis of good information. All three of these requirements have been somewhat lacking in the past.

C.1 Theory

The size of any stock of fish is naturally limited by its food supply. If we harvest some of the fish, the remainder will have more food available to them, and so they will tend to grow and reproduce until they have again

(a)

(c)

(b)

(d)

FIGURE 21.8 (a) *Soviet factory ship preparing to receive catch from a small catcher boat.* (b) *Netting in Ghana.* (c) *A large purse seine has been set around a school of herring off the western coast of Norway.* (d) *About 180 tons of herring are now closely confined in the purse seine and ready for bringing aboard the mother boat.*

reached the numbers limited by their food supply. However, if we fish too heavily, we run the risk of reducing their numbers to the point where the stock has difficulty reproducing itself, and it may take years or decades to recover. That is, if we overfish in an effort to increase this year's yield, we jeopardize our harvests in future years.

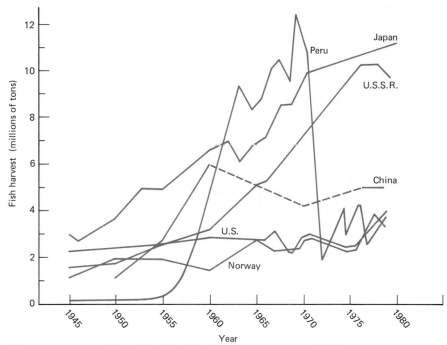

FIGURE 21.9   *Fish catch of the leading countries over the last few decades.*

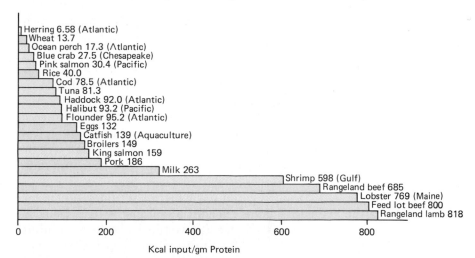

FIGURE 21.10   *Energy (in kilocalories) used in producing a gram of protein in unprocessed foods. (From Mary Rawitscher and Jean Mayer, Science, **198** (Oct. 1977), pp. 261–264. Copyright 1977 by the American Association for the Advancement of Science.)*

Since we don't anticipate that the world will end this year, the most intelligent use of our fish resources would demand that we harvest only as much this year as will not unduly jeopardize next year's harvest. The maximum amount of any stock that may be harvested year after year, is called the "maximum sustainable yield."

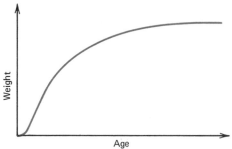

FIGURE 21.11   *Plot of weight vs. age of fish, showing that the youth gain weight much more rapidly than the adults. The calibration of the two axes depends on the fish species, but the shape of the curve is roughly the same for all.*

It is very difficult to determine the maximum sustainable yield of any species for many reasons. One is that to estimate how many of a species may be harvested, we need to know its original population, and that is difficult to determine. Also, trying to estimate the maximum sustainable yield from observing the effect of this year's harvest on next year's supply is difficult, because the size of a stock depends on many more variables than just how much was lost to human fishing. Climate, food supply, predators, and disease are just a few of many other influential variables. As a result of all these variables, it is quite difficult to determine exactly what effect last year's fishing has had on this year's crop. Understandably, estimates of maximum sustainable yields vary greatly from one expert to the next.

The problem facing fishery management is how to maximize the amount of fish that can be harvested year after year. The solution to the problem is based on the observation that youth are much more efficient at putting on weight and increasing their size than are adults, as is shown in Figure 21.11. That is, if a ton of young fish and a ton of adults are given the same amount of food, the young will put on more weight than will the adults. The more weight that the stock of fish gains per year, the more tonnage of that stock may be harvested per year.

Therefore, efficient management of our fisheries would keep the fish stock young, as the young are more efficient at gaining weight from the limited food supply. One way of doing this would be to harvest adults only, leaving the youth to fatten up for future years. In principal this is easy to do. For those fish harvested by nets, the net mesh size could be regulated to allow smaller youth to be able to escape, whereas the larger adults couldn't. The fish caught on line gear are usually larger, such as tuna and salmon, and are harvested one at a time rather than *en masse*, so they could be individually inspected as they are landed.

However, human self-interest can make even simple things complicated. International negotiations on regulation of net mesh size have gotten bogged down in such details as how the mesh size should be related to the size and type of cord from which the nets are made. Some fishermen

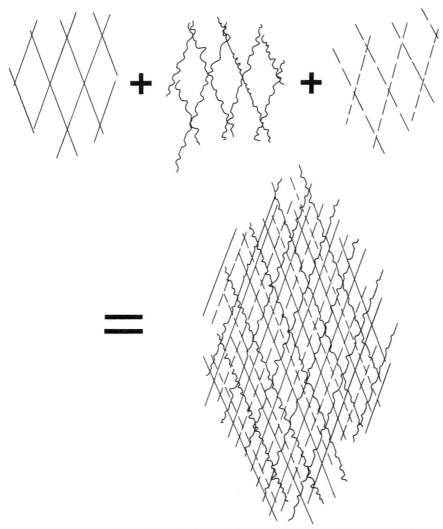

FIGURE 21.12 *Illustration of how fishermen can use three regulation nets together to get an effective mesh much smaller than regulation size.*

have circumvented the net mesh size regulation by using several nets together (Figure 21.12). Each net by itself, of course, conforms to the regulation, but all nets together have a much smaller effective mesh. Fishermen often are fiercely independent people, and balk at any policing of their behavior. It leads to the well-known problem of the "tragedy of the commons," where the long-term welfare and livelihood of the group is lost to individual greed.

However, there is another way of controlling the age of the fish stock. Heavier fishing means the average life expectancy of a fish will be shorter. That is, with heavy fishing, very perceptive fish would know their days were numbered, whereas with lighter fishing, they might expect to reach

TABLE 21.6 Illustration of How the Average Age of a Fish Stock Is a Function of the Fatality Rate. For this hypothetical stock, 64 new fish are added per year, and surviving fish die of old age just after their fifth birthday. Listed are the numbers of fish of each age in the stock for the given fatality rate. The fatality rate is measured in fraction of the fish lost per year.

| Fatality Rate | Age in Years | | | | | | Average Age |
	0	1	2	3	4	5	
0	64	64	64	64	64	64	2.5
¼	64	48	36	27	21	15	1.7
½	64	32	16	8	4	2	0.9
¾	64	16	4	1	0	0	0.3

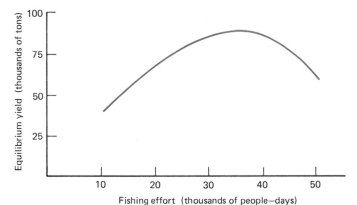

FIGURE 21.13 *Plot of annual equilibrium yield of yellowfin tuna as a function of the annual fishing effort. It is seen that if too much effort is expended, the annual yield is reduced, as there are too few mature adults left to carry on sufficient reproduction.*

a ripe old age before dying. As a result, the more heavily a particular stock is fished, the younger will be the average age of the stock (Table 21.6). So controlling the total tonnage of fish harvested is another way of controlling the average age of the stock.

Although it is advantageous to keep the stock young, some adults must be left in order to carry out reproduction. A single spawn has typically several thousand to several million fertilized eggs. The vast majority of these never reach adulthood, being lost to a variety of fates, including often being eaten by their own adults. Because of the large number of eggs in a spawn, a tiny change in their casualty rate will have a big effect on the number of youth that survive. The exact interrelationships are too numerous and complicated to be thoroughly understood, but it is known that harvesting a large fraction of adults in a stock has little effect on how well the stock regenerates. However, there is a point beyond which the removal of additional adults will cause a noticeable drop in the rate of reproduction and regeneration in the group. (See Figure 21.13.)

Therefore, the job of fishery management is to allow enough fishing

(a) *(b)*

FIGURE 21.14 (a) *Sketch of a krill, enlarged about 2 times.* (b) *Sketch of a species of copepod enlarged about 20 times.*

to ensure that the fish stocks are young, but not so much fishing that too few adults are left to carry out the necessary reproduction and regeneration of the stock. This optimum amount of fishing varies from one species to another, but usually it is such that the abundance of the species in the ocean is somewhere between ⅓ to ⅔ its original virgin abundance.

We can apply this understanding to our present fisheries to see that there are two ways to increase our present harvest from the oceans. One is to restrict the harvest from presently overfished stocks. Among these are the bottom-dwelling molluscs and crustaceans, which are presently very heavily fished because we place a high value on them and are willing to pay a high price to get them. The long-term harvest could be greatly increased if they were not overfished for short-term individual gain. Among the finfish there are over 30 varieties that are overfished now. Among the more important of these are herring, cod, and ocean perch from the North Atlantic, and anchovies from the southeastern Pacific (near Peru).

The other way to increase the production of our fisheries is to encourage the harvest from presently underfished stocks. Of the present world fish catch, about 53% comes from the Pacific Ocean, 40% from the Atlantic, and less than 5% from the Indian Ocean. This is quite disturbing since measurements of nutrients and primary productivity indicate the Indian Ocean is in no way inferior to the other two in its support of life. Clearly, we are leaving a major source untapped. With modern long-range fishing fleets, there seems to be no excuse for not using this resource more intelligently.

Another underfished stock are the Antarctic krill (Figure 21.14). Our fishing technology has developed to the point where harvesting these is becoming feasible. Intelligent harvesting of these alone could double our present total ocean harvest. It is also very tempting to try to develop our fishing technology to the point where we could efficiently harvest copepods and other small inhabitants of the deep scattering layer.[2] Their small size makes it difficult, but their large abundance makes it tempting to accept the challenge.

Finally, there are "trash fish," such as hake, that for reasons of habit or taste are currently underfished. There is a process now developed that

---

[2]The "deep scattering layer" is the layer of microscopic grazers that show daily migrations in response to sunlight. (See Figure 20.9.) It derives its name from the scattering of sonar signals.

FIGURE 21.15   *(left) The mesh size of a Russian cod fish net being checked by a U.S. officer. (right) A net in a Bremerhaven factory being checked for compliance with an agreement regarding German trawlers in Icelandic waters.*

could reduce these fish to fairly tasteless powder called "fish protein concentrate," or "FPC." It does not spoil, so there are no problems with storage or transportation, as there are with some other fishery products. It has a very high nutritive value, and so it could be added to bread flour or other foods to enrich our diets.

C.2   Practice

Most people are in agreement with the above thoughts regarding efficient management of our fisheries. However, it has been very difficult to actually carry out these ideas. Nations, as well as individual fishermen, are frequently willing to forfeit long-term mutual good for short-term individual gain. Both fish and fisherman are quite unpredictable. Fish don't respect boundaries drawn on maps, and many fishermen often don't either. International negotiations too frequently get bogged down in detail. Even when agreements have been made, there is distrust that the "others" (sometimes even when they are from the same country) are abiding by them. The fishermen are much more likely to break the rules themselves if they think others are breaking them. Heavy policing (Figure 21.15) would help remove suspicions, but it seems to be human nature that policing offends an individual's pride, and he or she balks at it. Anyway, policing would be quite difficult for fish processed aboard a factory ship, unless some sort of international police were on board when the trawlers unloaded.

When catches are restricted, there is always disagreement regarding who is entitled to the catch, and how much each can have. Limiting the "season" for catching fish is an efficient use of capital, as the equipment needed to harvest the fish cannot be used for the rest of the year in many cases. For example, one boat fishing for the entire year may catch as much as 12 boats fishing for one month, yet the latter requires 12 times the capital investment, and so the fish will be more expensive.[3] The United States has used a season for harvesting Pacific salmon, for example, and it is one of the reasons why this fish is expensive. Selling international licenses or levying international taxes on fish catches is also a means of controlling the harvest, but it also raises the price of fish, and would, therefore, be more damaging to the fishing industry in poorer countries, where people cannot afford to pay the higher prices. These are some of the problems that make regulation of the fishing industry difficult, both nationally and internationally.

If these problems could be overcome and we could have some coordinated international fishery management, then we could expect the oceans to yield a good deal more food. How much more this would be is not quite certain, but most experts feel that our harvest could be raised by somewhere between two and three times its present value, or somewhere around 200 million tons per year.

D.   THE SCOPE OF THE PROBLEM

Although in certain countries the birthrate is declining, this is not true for the world as a whole. Right now the rate is such that the population will double every 35 years. Even if some miracle should happen and the birthrate should suddenly drop to below two children per couple, there are enough children alive right now, who have not yet reached their parenting stage, to ensure that our population will keep right on growing well into the next century.

This means that in 30 years we can expect the food demand to be nearly twice what it is now. How can we meet the demand? More than half the tillable land in the world is already under cultivation, and the rising costs of fertilizers means we cannot expect increased production from presently cultivated lands. So the world's agricultural production may be increased, but it would be impossible to keep up with the population for more than another decade or two. Certainly it cannot double. Certainly the population will.

With this depressing outlook for the near future, we are tempted to look to the ocean for salvation. Unfortunately, as we have seen, seafood presently constitutes only about 2% of the gross weight of food harvested,

---

[3]*This is a simplified example, of course. In reality, the boats could be used for something else the rest of the year, and would not be idle. Only part of the fishing gear would be directed specifically at that one species of fish.*

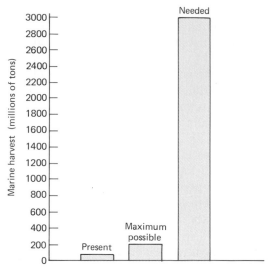

FIGURE 21.16 *Illustration of our present marine harvest, the maximum possible harvest using modern techniques and good management, and the amount that would be needed in three decades if the oceans were to solve the food problem then.*

and only about 1% of our diet. Even with excellent management of our present fisheries, the seafood production would only be able to increase by a factor of two or three from its present low value. This would hardly make a dent in the size of the food shortage the world will be facing in the next three decades (Figure 21.16). In terms of weight, we might be able to increase our annual ocean harvest by a hundred million tons or so, but this is far short of the many *billions* of tons of additional food production that will be needed.

In short, the food problem created by our expanding population is becoming so huge that within the next three decades it will far outstrip any effort we could make to cope with it, both from agricultural or conventional fishery resources. If there is any hope for salvation from the sea, then we must return to fundamentals and reexamine our use of the ocean from first principles. We will do this in the following sections.

E. ENERGY EFFICIENCY AND THE FOOD CHAIN

The basic problem is that our bodies need energy to perform our biological functions. The ultimate source for this energy is the sun, and it gets transformed into chemical energy through photosynthesis. However, there are usually many intermediate steps that this energy is channeled through before it is used by our bodies, and the losses in these intermediate steps are heavy.

Of the sunlight incident on the phytoplankton, only about 2% eventually become available to the herbivores that graze on it. From this point

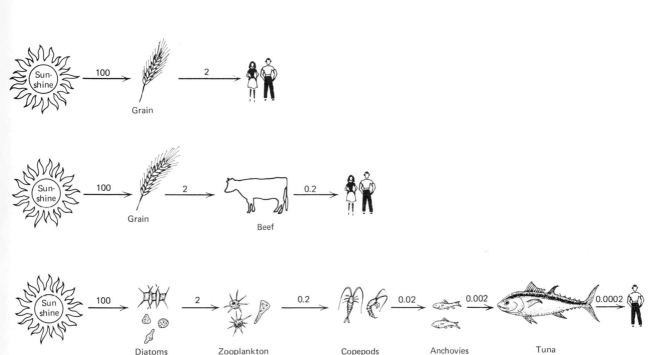

FIGURE 21.17 *Illustration of the amount of energy being passed on to successive trophic levels from the original 100 calories of solar energy incident on the primary producers. Clearly, it is most efficient to harvest the lowest trophic levels possible.*

on, the efficiencies improve slightly. At each step in the food chain, or at each "trophic level," about 10% of the available energy is passed on to the next step, and 90% is lost on the average, as is illustrated in Figure 21.17. For example, the herbivores feeding on the phytoplankton incorporate about 10% of the available energy as organic material in their bodies, and the other 90% goes into their own locomotion, reproduction, digestion, waste, and so on.

Suppose we eat a secondary carnivore, such as a tuna fish. Of the solar energy incident on the original phytoplankton, 2% of it was passed onto the herbivores, 10% of that went onto the carnivores, 10% of that to the secondary carnivore, and 10% of that is available to us. So we get 10% of 10% of 10% of 2%, or about 0.002% of the original solar energy available. This is quite inefficient. Clearly the farther down the food chain that we get our food, the more efficient it will be. In the case of agriculture, for example it is about 10 times more efficient for us to eat the grain than to eat the animals that eat the grain. Being omnivores, we have a wide choice of foods to choose from, more than any other earth creature, in fact.

Our harvest from the land is at low trophic levels, usually the zero trophic level, such as cereal grains and other vegetables. Even beef cattle

TABLE 21.7 Summary of Ryther's Results, Showing the Strong Inverse Correlation between Trophic Level Harvested and the Size of the Harvest. (From J. H. Ryther, *Science,* **166** (Oct. 1969), pp 72–76. Copyright 1969 by the American Association for the Advancement of Science.)

Province	Productivity per m² (g Carbon/ m²/yr)	Percent of Ocean Area	Total Productivity, Billion Tons of Carbon per Year	Percent of Total Ocean Productivity	Percent of Total Fish Production	Trophic Level Harvested
Oceanic	50	90.0	16.3	81.5	<1	5
Coastal	100	9.9	3.6	18.0	50	3
Upwelling	300	0.1	0.1	0.5	50	1.5

are only at the first trophic level, as they feed directly on hay and grains. Our ocean fisheries, on the other hand, are after animal protein at high trophic levels. As a result, 98% of our food production comes from land, and only 2% from our fisheries, in spite of the fact that the total primary productivity in both regions is nearly the same.

An interesting analysis demonstrating the relationship between trophic level and food production in the ocean has been carried out by J. H. Ryther.[4] It shows that less than 1% of the ocean fish catch comes from the open ocean, although more than 80% of all the primary productivity takes place there. [See Table 21.7 and Figure 21.18.] By contrast, 50% of all fish catch comes from upwelling areas, although less than 1% of all the primary productivity occurs there. The difference is that the open ocean harvest is taken at the *fifth* trophic level on the average, whereas the harvest in upwelling waters comes from the first and second trophic levels. As we saw in Chapter 20, small size and efficiency are important in oceanic waters where food is relatively scarce, or well dispersed. As illustrated in Figure 20.7, this means that the food passes through many more trophic levels before reaching the larger predators which we harvest. Consequently, the harvest from upwelling areas is a much more efficient use of the original food energy made available in photosynthesis.

From the above analysis of efficiencies, we might quickly jump to the conclusion that we should simply harvest the phytoplankton rather than the fish, as this could provide about a thousand times the food value to us, which would more than solve the world's food problem for the forseeable future. Unfortunately, the problem is more complicated than this for several reasons.

One reason has to do with the mechanics of harvesting. Most of the phytoplankton is microscopic. Furthermore, these microscopic organisms are quite dispersed, not conveniently schooling for us as well as do higher forms of marine life. They can be removed from seawater by forcing the water through filters. However, this is very slow and tedious, and a great deal more energy must be expended in making this kind of harvest than would ever be economically feasible.

[4]*J. H. Ryther, Science,* **166** *(1969), pp. 72–76.*

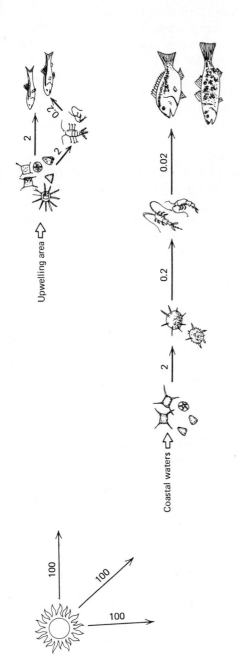

FIGURE 21.18 *Illustration of Ryther's analysis, assuming 1000 calories incident on the primary producers, 2% of which gets passed on, and 10% passed on at each higher trophic level. The longer the food chain, the greater will be the losses along the way. Although the surface waters of the deep ocean have the largest total productivity, they produce the least food for humans, because we must harvest the fifth trophic level there.*

It turns out that nature has a more efficient harvesting mechanism. The grazers feed on these phytoplankton, and then themselves get eaten by larger organisms. We have seen that this is quite inefficient, with about 90% of the food value lost at each step. However, bad as it is, it is better than we can do at the moment. We hope that with improved technology, we will be able to harvest farther down the food chain, at lower trophic levels. Krill are attracting lots of attention now, and hopefully copepods might be harvested in the future as well. However, it seems unrealistic to expect that humans will ever harvest much of the zero trophic level in the sea as is done on land.

Another reason involves the very high plytoplankton productivities. Even if we could harvest the phytoplankton, it may be wiser not to try. If we remove half of the phytoplankton today, most will have been replaced again by tomorrow because they reproduce and grow so quickly when nutrients are sufficient. So to maximize the ocean's food output, we would have to harvest a good fraction of the phytoplankton in all parts of the ocean *each day*. Imagine what a supernatural effort that would take! Even if we had the technology, the sheer volume of the water involved would ensure that we couldn't do it in so short a time.

By contrast, the tiny grazers can, and do, harvest a large fraction of all the phytoplankton each day. Even though there are large losses as this food is passed on up the food ladder to the species we harvest, the fact that it is based on *daily* phytoplankton harvests may make the total food yield greater than if we tried to harvest the phytoplankton ourselves.

Another expressed concern involves minimizing the ecological side effects of our aquaculture. The best way to make a pyramid crumble is to remove blocks from the *bottom* layer, as they support all the overlying layers. Similarly, if we harvest one of the lower layers of the biomass pyramid, we affect all the overlying layers, which depend on this layer for their food supply. Removing the top layer, however, has less effect on the lower ones.

Some people have pointed out that we can, perhaps, be too sensitive to these environmental considerations. In populating this continent, for example, we changed much of the base of the terrestrial biopyramid by removing forests and replacing them with agricultural crops. The wild animals making the upper layers of the previous biopyramid were partially replaced by domesticated animals and people. We didn't destroy the pyramid, but we did change it a bit. Now we're facing a similar situation in aquaculture. If we begin farming the lower layers, we will certainly cause a change in other layers.

F. INCREASING FISHERY HARVESTS

With the background provided by the preceding sections, we now wish to examine the possible ways of increasing our harvests from the sea, making rough estimates of the maximum possible increase from each. There are two basic ways of increasing these harvests. One is by becoming better "hunters," so we do a better job of catching what is naturally there. The other is by becoming "farmers," whereby we actually nourish and control the crops to be harvested.

F.1   Improved Hunting

We have seen that through good fishery management, we should be able to improve our ocean harvests by two or three times. That would increase them to somewhere between 4% and 6% of the total world food output. By itself, this would by no means solve the problem, but it is certainly one avenue that should be followed. It would involve restricting the harvests of overfished species and increasing those of underfished stocks, so that all are nearer their respective maximum sustainable yields.

Hunters can do better not only by hunting more intelligently, but also by using better weapons. New and emerging technologies are making it feasible to hunt previously untapped stocks. Among these, the krill in Antarctic waters are extremely attractive. They come in schools, typically several hundred meters in diameter and 40 m thick, which would be convenient for harvesting. Furthermore the schools seem to be segregated according to age, adults in some and juveniles in others. This would make it easy to harvest adults only, keeping the stock young for more efficient growth.

Estimates of possible krill harvests vary because the size of the standing crop is poorly known. Through several methods, such as estimates of krill consumed by whales, seals, porpoises, and other natural predators, it looks like the standing crop may be 1 to 5 billion tons, and an annual human harvest of a half a billion tons may be feasible. This alone is seven times greater than our present total ocean harvest, and could make our fisheries contribute roughly 15% of our total world food harvest. Again, this will not come close to solving the food problem, but it is still an avenue worth pursuing.

The krill's tiny cousins, the copepods, make up an even larger biomass in the world's oceans. We are still a long way from having the technology to make their harvest feasible, but as the need for new food sources increases, more attention is sure to be directed toward these tiny grazers.

F.2   Farming

You can see that through improved management and technology, the production of our world fisheries can be increased significantly. As welcome as these increases would be, they would still not come close to solving the food problem caused by our expanding population. If our oceans are to make a major contribution to the reduction of the world food shortage, then we must make some fundamental changes in the way we use our ocean. In particular, we must change from hunters, who harvest whatever Nature happens to provide, to farmers, who cultivate and nourish particular crops for harvest.

Aquaculture has been carried out on a relatively small scale to date. The Chinese have done most of it. Their annual fresh water aquaculture

FIGURE 21.19 *A three-dimensional oyster bed in a Rhode Island oyster pond. The oysters are artifically "seeded" with bits of shell on strings. This way they are not confined to the bottom and may make use of the food available at all depths in the shallow water. One-year weight gains of 3000% have been recorded. (Photo by Luther H. Blount, Blount Marine Corp.)*

harvest is approximately 17 million tons, or approximately ¼ as big as the total world ocean harvest. This is quite impressive when you compare the rather small volume of the Chinese fresh water ponds to the volume of the world ocean. In the West there has been some aquaculture on a small scale (Figure 21.19), but no large-scale effort yet.

As we have seen, it is technologically unfeasible to harvest the phytoplankton. We must be satisfied to harvest the higher trophic levels. The way to increase the populations of these higher trophic levels is to increase the size of the crop of phytoplankton that supports them. How can we do this?

Phytoplankton productivity is usually restricted by either nutrients or sunlight. We have no control over sunlight, but we can do something about the nutrients. In tropical waters, the permanent thermocline prevents warm sunlit surface waters from mixing with the nutrient-rich cooler waters beneath. Thus, in tropical waters, phytoplankton productivity is low due to lack of nutrients. A similar situation exists in temperate waters where the thermocline and nutrient supply are seasonal. If we could bring these nutrient-laden deeper waters to the sunlit surface (Figure 21.20), the phytoplankton productivity would increase immensely, supporting a greatly increased population at higher trophic levels for human harvest.

If done on a large scale, this pumping of cool deep water to the surface would undoubtedly have a significant impact on our environment, as

FIGURE 21.20   *The relatively small productivity throughout most of the ocean is a result of its depth, which allows the sinking organic detritus to remove the nutrients from the surface waters. If we could bring these nutrient-laden deeper waters to the sunlit surface, we could increase productivity immensely.*

has the burning of fossil fuels in recent decades. You may wish to specu- late on what these effects might be.

In summary, if there is to be any hope for the oceans to play a signif- icant role in fighting the imminent food crisis, we must change from hunter to farmer and herder in the oceans as we did on land. In agricul- ture, we learned to plow under last years organic detritus to make the nutrients available for this year's crop. Also, the recycling of nutrients via fertilizers is important when cropland becomes depleted in nutrients. In the oceans we must learn to do the same, returning nutrients from the organic detritus of previous years to the surface, where they can help life flourish again.

G. SUMMARY

As our demand for resources grows, and as we deplete our terrestrial sources, the ocean resources become increasingly attractive.

Our population is increasing at a rate that will soon outstrip our ability to feed it. The present distribution of food resources is inequitable, and the economics of food ensures that the situation will get worse as food reserves dwindle.

Seafood is an excellent source of animal protein, but the microscopic size and the dispersed distribution of most oceanic plant life make it unfeasible as a source for vegetable material in our diet. The most efficient use of our food resources would be to try to replace terrestrial animals in our diet by seafood, releasing more terrestrial vegetation for human con- sumption. Good management of world fisheries could increase seafood production two or threefold, but to make a sizeable dent in the world food problem, we will have to make fundamental changes in our farming of the oceans.

Of the solar energy originally incident on plant life, there are great losses at each step along the food chain. Therefore, it is advantageous to take our harvest from the lowest trophic level possible. Although we cannot expect to harvest the microscopic oceanic plant life on a large scale, we can increase the productivity at all trophic levels by recycling nutrients lost to deeper waters back to the surface.

QUESTIONS FOR CHAPTER 21

1. Zero population growth will eventually happen, whether we like it or not. Explain why.

2. Why is it difficult to estimate the dollar value of many resources? Why is the determination of the quantity of a resource often ambiguous? Give examples.

3. Roughly what is the world's population? At the present time, how much food is produced per person per day?

4. If we produce ample food, why is there so much hunger and malnutrition in the world?

5. What is the average number of kilocalories of food energy consumed per person per day in the world? How does average consumption in the United States compare to this?

6. What is meant by an "inelastic demand curve" for food?

7. In times of anticipated food shortage, who suffers first, and why?

8. Will the distribution of food get better or worse in future years?

9. Compare the increase in population and in food supplies between 1950 and 1980. How did poor countries fare, on the average?

10. Why is it that increased food production and exportation in wealthier countries has little benefit for the poorer countries?

11. Roughly how long will it take for the world's population to double? Can food production double to keep pace?

12. How can education help make an inadequate food supply stretch farther?

13. What do you suppose is meant by changing from "hunter" to "herder" in the oceans?

14. Why is it that not much of the ocean's plant life is harvested?

15. Why would it be a more efficient use of our land resources to replace the land animals in our diet with fish?

16. Why are fish a better source of animal protein in our diet than land animals?

17. Why is it advisable to have some animal protein in our diets. What

is the minimum amount recommended? How much would this be per day? Compare this to a typical fish dinner with ¼ to ⅓ kg of fish meat.

18. Give an example of a mollusc. Of a crustacean.

19. How much food is produced from the sea per year? How does this compare to the total world food production?

20. Why are some people concerned over the conversion of fish to fish meal?

21. Approximately what fraction of the annual fish harvest goes into fish meal? Of the part intended for human consumption, about how much is lost in processing?

22. Fish are about what fraction of the animal protein in the diet of the world as a whole? In the diet of the United States?

23. Do nets or line gear account for the bulk of the world's fishing harvest? Why?

24. What countries have had the biggest gains in fish harvests in recent years? Why?

25. Does it take more energy to harvest herring or an equivalent amount of corn? Why will fish always be more expensive than cereal grains?

26. What are three requirements needed for executing good fishery management?

27. How does overfishing jeopardize the harvests in future years?

28. What is the "maximum sustainable yield" of a fish stock? Why is it difficult to determine?

29. Why is it advantageous to keep fish stock young? How might this be done?

30. Explain how heavy fishing insures a young stock of fish.

31. Why must some adults be left in the otherwise young stock of fish?

32. Why is it that for most species, the number of adults may be reduced considerably from their original "virgin" abundance without any noticeable effect on the rate of regeneration of the stock?

33. What are some of the presently overfished stocks?

34. What ocean is obviously underfished?

35. What are some of the presently underfished stocks? How may fish with poor taste be processed for human consumption?

36. How is it that using "seasons" to limit the harvest of fish raises their price? What is wrong with using licenses or taxes on catches as a means of internationally controlling fish harvests?

37. With good fishery management, by about how much could the total ocean harvest increase? What fraction would the fisheries then contribute to the present total world food harvest (5 billion tons per year)? Would this solve our food problem in the next few decades?

38. At the present rate of growth for the world population, what is the doubling time?

39. Explain why in a steady state the average birthrate per couple should be slightly greater than two offsprings.

40. Explain why it is that our population would go right on increasing for a while, even if the birthrate were to suddenly drop to two per couple. Explain why this increase would continue through more than one generation; that is, explain second and third generation effects.

41. Consider a population of some kind of animals that were youths during their first year, produced two offspring per couple on their first birthday, and died on their third birthday. Suppose their population right now was five; four youths (in their first year) and one adult in its second year. Complete the following table to trace the population for the following two years.

Number of Animals in Each Age Group

	0–1	1–2	2–3	Total Population
Start	4	1	0	5
After 1 yr				
After 2 yr				
After 3 yr				

How many generations did it take for this initially young population to reach steady state?

42. Explain why the world's agricultural output will not be able to double.

43. What is the ultimate source of the energy used by our bodies?

44. Roughly how much of the original solar energy falling on the original plants makes it to our bodies when we eat a salmon (secondary carnivore)? Beef cattle? Explain.

45. Why is it that such a small fraction of our total available food harvest comes from the ocean, in spite of the fact that the primary productivity of the ocean is nearly as large as that of the land?

46. Why is it that less than 1% of our total ocean harvest comes from open ocean waters, although most primary productivity takes place there?

47. What are the two characteristics of most phytoplankton that make them impossible to harvest on a commercial scale?

48. Why must most of our harvest from the ocean be at higher trophic levels than most of our harvest from land? How can we increase the "crop" size of these higher trophic levels?

49. Why is it that harvesting phytoplankton would have a bigger effect on the ocean's ecology than harvesting higher trophic levels? What were some of the ecological effects of our populating the North American Continent?

50. How do we speed up the natural recycling of nutrients on agricultural lands? How could we do this in the ocean? Would this be done most advantageously at high latitudes or at low latitudes? Why?

*51. How would you go about estimating the world population of a particular species of fish?

*52. What would you think would be some of the environmental effects of pumping cool deep waters to the surface in the tropics, if done on a large scale? Explain.

SUGGESTIONS FOR FURTHER READING

1. Terry N. Barr, "The World Food Situation and Global Grain Prospects," *Science*, **214**, (Dec. 1981), p. 1087.

2. Lee Cooper, "Seaweed with Potential," *Sea Frontiers*, **27**, No. 1 (1981), p. 24.

3. D. John Faulkner, "The Search for Drugs from the Sea," *Oceanus*, **22**, No. 2 (Summer 1979), p. 44.

4. John Frye, "Menhaden—Fish on the Run," *Sea Frontiers*, **26**, No. 1 (1980), p. 44.

5. S. J. Holt, "The Food Resources of the Ocean," *Scientific American* (Sept. 1969).

6. J. Anthony Koslow, "Pacific Pollack—Already Overfished?," *Sea Frontiers*, **22**, No. 2 (1976), p. 98.

7. "Food From the Sea," *Oceanus*, **18**, No. 2 (Winter 1975).

8. "Marine Biomedicine," *Oceanus*, **19**, No. 2 (Winter 1976).

9. Ronald L. Smith, "The Valuable Alaskan Herring," *Sea Frontiers*, **26**, No. 3 (1980), p. 144.

10. Larry Wood, "Farming Giant Kelp," *Sea Frontiers*, **23**, No. 3 (1977), p. 159.

22
OCEAN ENERGY AND MINERAL RESOURCES

An oil-drilling platform being towed to a site in the North Sea.

The world of today is quite different from that of our great grandparents. Aside from the larger population, the biggest difference involves lifestyles based on the rapid use of energy and mineral resources. In order to maintain our present lifestyle, we undertake extensive searches for new supplies of these things as our terrestrial sources are exhausted. Naturally, this search turns towards the oceans.

A. ENERGY RESOURCES

At the moment, we rely heavily on fossil fuels for our energy needs, and we are exhausting the high-grade reserves at an alarming rate (Table 22.1). It is clear that within the next few decades we will have to make major changes in our energy sources (Figures 22.1 and 22.2). This may mean resorting to more expensive and lower grade fossil fuel deposits, or it may mean we will look for alternatives to fossil fuels.

One attractive alternative energy source is the sun. It is a "clean" source as it doesn't add carbon dioxide, smog, soot, or poisonous gases to our atmosphere, as do the fossil fuels. It also has the advantage that it is there, whether we use it or not. That is, when using it we don't add any extra heat to the environment that wouldn't have been there anyhow. This is clearly not the case with fossil fuels or nuclear reactors.

Our ocean is a gigantic, ready-made collector of solar energy, which it stores in several different ways. Because of the potential importance of this huge resource, this section is devoted to our energy needs and how we might meet them using the ocean.

A.1 Energy Measurement

Energy is the ability to do work, or, equivalently, the ability to move something. It comes in a variety of forms. A car speeding down the highway, or a ball flying through the air, has energy of motion, or "kinetic energy." The rapid random thermal motions of the individual atoms or molecules in a substance is called "thermal energy," which is energy of motion on an atomic scale. The energy of electrons moving through a conductor is called "electrical energy," and energy stored in a stretched spring is called "potential energy" because it gets quickly converted into energy of motion when the spring is released. "Chemical potential energy" is stored in many substances, including fossil fuels and organic matter, which is released upon oxidation (burning).

Energy can be converted from one form to another. Chemical potential energy stored in gasoline becomes thermal energy as it is burned, and the heated gases expand against a piston in the cylinder of your car. Through a series of mechanical contraptions this pushes your car down the road, giving it kinetic energy. The chemical potential energy in coal is released upon burning, transformed into thermal energy, heating up steam and making the turbine of a power plant turn.

605

TABLE 22.1 Depletion of Fossil Fuel Reserves. The years indicated are the averages of various projections of when the reserves will be 80% gone.

	Petroleum and Natural Gas	Coal
World	2030	2300
United States	2010	2400

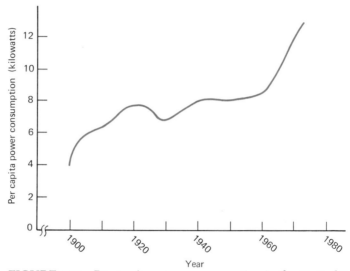

FIGURE 22.1 Per capita power consumption in the United States since 1900.

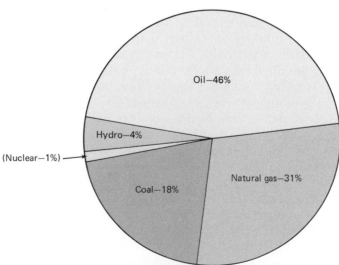

FIGURE 22.2 United States energy consumption by source. Fossil fuels account for 95% of all our energy.

TABLE 22.2 Efficiencies of Various Power
Plants, Engines, and
Lightings.

*For the power plants, the efficiencies mea-
sure useful energy output compared to
energy input, but do not include losses in
transformers and transmission lines. For the
engines, the efficiencies measure useful
energy output to energy input, but do not
include losses due to friction in transmis-
sions, wheel bearings, tires, wind resistance,
and so on when these engines are in cars.
For the lighting fixtures, the efficiencies
measure light energy output compared to
electrical energy input. In all cases, the
energy lost leaves as waste heat.*

	Efficiency
Power plants	
Fossil fuel power plants	38%
Nuclear power plants	31%
Solar power plants	28%
Solar cells	10%
Engines	
Steam turbine	45%
Diesel	35%
Internal combustion	22%
Wankel engine	18%
Lighting	
Fluorescent	20%
Incandescent	5%

This energy of motion is then converted to electrical energy in the generator, which is turned back into thermal energy in your stove burner,
which makes steam again from the water in your teapot. Each time energy
is converted to a different form, some is lost. We speak of the "efficiency"
of a machine as the ratio of useful energy coming out to energy in. Because
of losses, the efficiency is always less than one (Table 22.2).

Energy is measured in a variety of units. (See Table 22.3.) Among the
more common units are the "British Thermal Unit," or "BTU," the
"joule," and the "calorie." We have already encountered the "calorie" in
this book, which was the amount of energy required to raise the temperature of 1 g of water by 1°C. In terms of this unit, one BTU equals 252
calories, being the amount of heat needed to raise one *pound* of water by
1° *Fahrenheit*. A joule is the common unit of energy in the metric system,
and equals 0.24 calories.

The rate at which energy is used is called "power." The standard metric unit of power is the "watt," which amounts to one joule of energy used

TABLE 22.3   Units for Measuring Energy and Power

	Energy Units
1 calorie	The amount of heat energy required to raise the temperature of 1 g of water by 1°C.
1 BTU	(British Thermal Unit) The amount of heat energy required to raise the temperature of 1 lb of water by 1°F.
1 joule	The amount of kinetic energy of a 2 kg mass which is moving with a speed of 1 m/sec.
1 kilowatt hour	The amount of energy used, if used at a rate of 1000 watts for 1 hour.

Conversions

1 BTU = 252 calories

1 BTU = 1054 joules

1 joule = 0.24 calories

Power units

1 watt = 1 joule/second

1 kilowatt = 1000 watts

1 horsepower = 746 watts

per second. For example, a 60-watt lightbulb uses energy at a rate of 60 joules per second. A kilowatt is a thousand watts.

When you pay your electric bill, you pay for the number of "kilowatt hours" of electrical energy used. A "kilowatt hour" is the product of rate times time, and is a measure of total energy used. Since there are 1000 watts in a kilowatt and 3600 seconds in an hour, there are (1000 watts) × (3600 seconds) = 3,600,000 joules of energy in a kilowatt hour.

The total amount of energy used is the product of the rate at which it is used, times the time. For example, you can use 1 kilowatt hour of energy by using it at a rate of 1000 watts for one hour. But you would use the same amount altogether if you used it at half the rate for twice as long, or one tenth the rate for ten times as long. That is a rate of 1000 watts for one hour uses the same amount as 500 watts for 2 hours, 100 watts for 10 hours, 1 watt for 1000 hours, and so on (Figure 22.3).

This completes the scientific definition of these units. They can be made much more meaningful if related to our everyday experiences. When you turn a stove burner on "high," you are using energy at a rate of about 2 kilowatts, or 2000 joules per second. The burner on a *gas* stove may put out slightly more than this when turned on "high." Alternatively, one kilowatt is the rate at which energy is given off by 10 ordinary 100-watt light bulbs. (See Table 22.4).

Of course, most of our energy usage is not in the form of electricity. (See Table 22.5). When driving a small car down the road, using most of the 55 horsepower capacity of its small engine, we are using energy at the rate of about 40 kilowatts. Actually, cars are inefficient, so about five times this energy is lost to friction and other waste heat. So we might waste

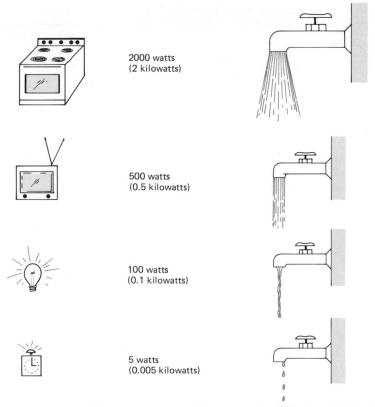

2000 watts
(2 kilowatts)

500 watts
(0.5 kilowatts)

100 watts
(0.1 kilowatts)

5 watts
(0.005 kilowatts)

FIGURE 22.3 *The amount of water in your bathtub depends both on the rate at which water comes out of the faucet, and on how long the faucet is open. Similarly, the amount of electrical energy used depends both on the rate at which it is used, and the amount of time used. Some analogics for typical appliances are shown above.*

TABLE 22.4 Approximate Power Consumption By Various Things When in Use

	Kilowatts
Electric alarm clock	0.03
Electric sewing machine	0.07
Incandescent light bulb	0.1
Average metabolism of an adult	0.1
Black and white television	0.6
Toaster	0.8
Electric iron	1.2
Electric stove burner on "high"	2.0
Automobile	240.0

TABLE 22.5   Major Uses of Energy in the United States

	Percent of Total
Transportation (fuel, but not oil and grease)	24.9
Space heating (residential and commercial)	17.9
Industrial steam	16.7
Direct heat (industrial)	11.5
Electric drive (industrial)	7.9
Feedstocks, raw materials	5.5
Water heating (residential and commercial)	4.0
Air conditioning (residential and commercial)	2.5
Refrigeration (residential and commercial)	2.2
Lighting (residential and commercial)	1.5
Cooking	1.3
Electrolytic processes (industrial)	1.2

TABLE 22.6   Per Capita Power Consumption By Various Types of Societies[a]

Society	Power Consumption (kilowatts)
Primitive hunter and gatherer	0.1
Advanced hunter and gatherer	0.2
Primitive agricultural	0.6
Advanced agricultural	1.3
Early industrial	3.5
Advanced industrial	10.0
(World average today)	(1.9)

[a]These figures include the 0.1 kilowatt needed in human metabolism.

energy at a rate of roughly 240 kilowatts when using 40 kilowatts to drive us down the road.

A.2   Our Energy Consumption

The average rate of energy consumption in the United States is about 12 kilowatts per person. (See Tables 22.6 and 22.7.) That is, we use energy at a rate roughly equivalent to each of us keeping 120 of the 100-watt light bulbs burning all the time, day and night. Equivalently, we use energy at a rate roughly equal to each of us keeping six stove burners turned on "high" all the time.

This energy usage is divided fairly evenly among domestic, transportation, and industrial sectors. That is, we use roughly 4 kilowatts per person for such things as heating and lighting our homes, cooking our meals, and running our color T.V.s. We use energy at a similar rate in our cars, trucks, trains, airplanes, and so on. Finally, about 4 kilowatts per person are used in the industrial sector for manufacturing or processing the items we buy, or in services we hire.

Although the United States has only about 6% of the world's population, it uses nearly ⅓ of the world's energy production. Although the per capita energy consumption in the United States is at a rate of about 12

TABLE 22.7 Per Capita Power Consumption in 1973 by Country[a]

Country	Kilowatts per Person	Country	Kilowatts per Person
United States	10.8	Lebanon	0.81
Kuwait	9.8	China	0.54
Canada	9.0	Brazil	0.50
Czechoslovakia	6.4	Egypt	0.28
Sweden	5.9	India	0.18
United Kingdom	5.3	Indonesia	0.12
Germany	5.0	Pakistan	0.08
Norway	5.0	Nigeria	0.06
U.S.S.R.	4.4	Ethiopia	0.04
France	3.8	Afghanistan	0.03
Japan	3.1	Nepal	0.01
Yugoslavia	1.6	World average	1.86
Mexico	1.2	World average	
Iran	1.0	without the	
Saudi Arabia	0.93	United States	1.33

[a]These figures do *not* include the approximately 0.1 kilowatts per person food energy used in metabolism.

kilowatts, the world average per capita consumption is less than 2 kilowatts. Consequently, the world as a whole uses energy at a rate of about 7 billion kilowatts, of which nearly 2½ billion kilowatts are used in the United States.

In addition to the shortage of energy resources, other problems facing us involve converting energy to the form needed, transporting it to the place needed, and storage until the time needed. These are especially critical concerns for ocean energy resources because the supply would normally be long distances from the continental population centers. It would have to be converted to forms that could be stored or transported to the continents without much loss.

A.3 Ocean Energy Resources

There are several different ways through which solar heat is collected and stored in the oceans. Because the oceans are so immense, the total amount of energy stored in each of these forms is quite large. Special interest groups sometimes try to impress us by reciting these large numbers indicating the total amount of energy available from one resource or another.

However, our long-range welfare depends not on the total amount of energy available, but rather on the maximum rate at which it can be continuously extracted. For example, if one resource is large enough to supply our energy needs for 20 years, but it does not get replaced, then after 20 years the resource is exhausted and we are in trouble. This is the predicament we have gotten ourselves into through our reliance on fossil fuels. The long-term welfare of our present type of society depends on our use of *renewable* resources. The maximum rate at which we can continuously extract energy from these resources depends on the rate at which they are

TABLE 22.8   Various Forms of Energy Stored in the Oceans. For each is given an estimate of the total energy stored in that form and an estimate of its replacement time. By dividing the two, we have an estimate of the maximum rate at which energy could be provided by this source.

Energy Source	Total Energy (joules)	Replacement Time (seconds)	Power (billions of kilowatts)
Thermal energy	3×10^{24}	3×10^7	100,000
Wave energy	10^{19}	10^6	10
Tidal energy	2×10^{17}	4×10^4	5
Surface currents	3×10^{18}	3×10^7	0.1
Deep currents	10^{14}	3×10^7	0.000003
(Present world consumption)			(7)

renewed. Clearly, we could not continuously extract energy at a rate faster than it is being replenished.

In this section on ocean energy resources, we will make estimates of the maximum rate at which energy could be continuously extracted from each of the various sources. This maximum rate is the ratio of total energy available divided by the replacement time.

$$\text{rate} = \frac{\text{energy available}}{\text{replacement time}}$$

For example, if 10 joules of energy were available from some resource, and it could be replaced every 2 seconds, then the maximum rate at which it could be continuously extracted would be 5 joules per second, or 5 watts. Of course, in the ocean, both the energy available and the replacement time will be much larger than the numbers of this example. Estimates of these numbers for various sources of ocean energy are summarized in Table 22.8.

(a) Thermal Energy. One way the ocean collects and stores energy is the direct absorption of sunlight. As we have seen, surface waters are considerably warmer than deeper waters, especially at low latitudes. From thousands of measurements made at sea, we know the thickness and extent of these warm surface waters, and we can make a quick calculation of how much energy they hold by virtue of their elevated temperatures.[1] This figure is about 3×10^{24} joules.[2]

To estimate the replacement time, we notice that in temperate latitudes the thermocline appears and disappears seasonally. Consequently, we guess that if we removed all the extra heat from the surface waters

---

[1] For each cubic centimeter of warm surface water, and for each degree Celcius of elevated temperature, there is one calorie (or 4.18 joules) of energy.

[2] The notation 10^{24} means 10 multiplied by itself 24 times, or, equivalently, a "1" with 24 "0's" following it. Because we don't like to write all these zeros, scientists often use this convenient notation, 3×10^{24} is the product of 3 times 1 with 24 "0's" after it, or 3,000,000,000,000,000,000,000,000.

today, the sun could easily replace most of it within 1 year. Since there are 3×10^7 seconds in 1 year, the maximum *rate* at which we could possibly extract this energy, according to the above equation, would be

$$\text{rate} = \frac{3 \times 10^{24} \text{ joules}}{3 \times 10^7 \text{ seconds}} = 10^{17} \text{ watts} = 10^{14} \text{ kilowatts}$$

In reality, we could not extract energy at this rate, because both the laws of thermodynamics, and the shortcomings of manmade machines demand that much energy will be lost in conversion. The efficiency of this conversion would be particularly low due to the relatively small temperature difference between surface and deep waters. Experts estimate efficiencies of only a few percent for this conversion. But even 1% of this figure would be 10^{12} (or one trillion) kilowatts, which is more than a hundred times the world consumption of 7×10^9 (or 7 billion) kilowatts. Because of this, this potential source of energy is attracting a great deal of attention at the moment.

The extraction of this energy would be accomplished through the use of a fluid that would be gaseous at the temperature of the surface waters, but a liquid at the cooler temperatures of the deeper waters. The warm surface waters would be used to heat this fluid, causing it to boil. The vapors would then shoot through a turbine, turning it and the generator coupled to it. The cooler deeper waters would then be used to cool these vapors until they condense into the liquid form, and then they would begin the cycle again. The process is called "OTEC," which is an acronym for "Ocean Thermal Energy Conversion." An artist's conception of a large OTEC plant is shown in Figure 22.4.

(b) Waves. Solar energy is stored in the waves, as the solar heating drives the winds that create the waves. Since we know the ocean's wave spectra from measurements made, we can make quick conservative estimates of all the wave energy for the whole ocean and find that about 10^{19} (ten quintillion) joules of energy are there on the average. To get the replacement time we notice that most of this energy gets dissipated onto some shoreline within a few weeks after the wave was created, so two weeks is a conservative estimate of the replacement time. In two weeks there are about 10^6 (one million) seconds, so dividing the energy available by the replacement time, we arrive at our estimate of the maximum rate at which this energy could be extracted.

$$\text{rate} = \frac{10^{19} \text{ joules}}{10^6 \text{ seconds}} = 10^{13} \text{ watts} = 10^{10} \text{ kilowatts}$$

This figure of 10^{10} (10 billion) kilowatts is only a very rough estimate. It could easily be off by a factor of two or three. Losses in conversion would reduce the amount actually available to us. Nonetheless, the figure is close to our total rate of consumption of 7 billion kilowatts, so there is sufficient energy available from this source to make it worthy of attention.

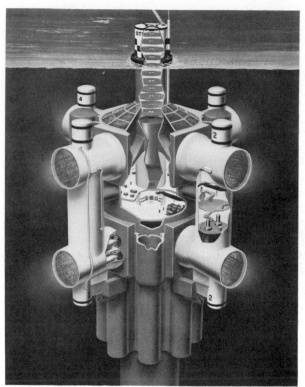

FIGURE 22.4   *Artist's rendition of an OTEC plant. The tube below brings up cold water from a depth of about 2.5 km.*

It is particularly attractive for shorelines where we wish to reduce wave activity anyhow. Instead of building breakwaters, we might as well make use of this wave energy. The design of one such device is shown in Figure 22.5.

(c) Currents. The winds also drive the ocean surface currents. From measurements of their velocity and extent we can estimate the kinetic energy available in them. The replacement time is difficult to guess, however. Clearly we can't perform an experiment where we actually stop one of these currents and see how long it takes to start up again. The best we can do is notice that there are seasonal changes in their volumes, so maybe a year is a reasonable time for regeneration. Doing this, we find the surface currents fall far short of our energy needs, allowing approximately 0.1 billion kilowatts to be removed (Figure 22.6).

The deeper currents are much more voluminous than surface currents. However, kinetic energy depends on velocity as well as mass. Since the deep currents move more slowly, they would have even less energy than the surface currents.

(d) Tides. The oceans are also collectors of "lunar" energy, or more properly, the gravitational energy of the earth–moon system. Knowing the typical height of the midocean tides, we can quickly calculate that about 2×10^{17} joules of energy is contained in them. The replacement time is

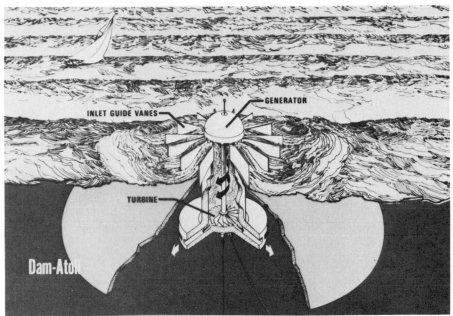

FIGURE 22.5   *One possible device for removing wave energy is the "dam-atoll."
In this device, waves refract toward artificial shoals, created by the
large underwater shell. The water is then guided through the generator's turbine blades before exiting beneath the shell.*

FIGURE 22.6   *Artist's rendition of a large turbine being towed to its mooring. Such
a turbine would be anchored to the ocean bottom and used to gen-
erate electricity from ocean currents.*

easy to find, since we know that if we remove all the energy from one tide
in a gigantic ocean power station, there will be another one coming ½ day
later as sure as the sun will rise tomorrow. In ½ day there are about 40,000
seconds, and dividing this into the 2×10^{17} joules of energy available gives
5 billion kilowatts as the maximum rate of energy extraction.

FIGURE 22.7 *The world's only tidal power station at the mouth of the Rance River Estuary, France.*

In practice, it would be very difficult to extract much of this tidal energy. In most cases, huge capital investments would be required to build the breakwaters and dams needed to extract the energy from rising and falling tides. They would also cause serious modifications of coasts and coastal processes. However, some narrow estuaries with large tidal volumes could be tapped. One of these is shown in Figure 22.7.

(5) Others. Other possible ocean energy resources include those referred to as "biomass conversion" and "salt power." Biomass conversion involves the growth of organic materials in the ocean to be harvested for organic fuels. The outlook for this particular resource doesn't seem very promising, however, because the use of the ocean's biomass for food is a much more pressing problem than its use as a fuel.

"Salt power" is derived from the strong electrostatic attraction between the electrically polarized water molecules, and the charged ions of dissolved salts. We have seen that this strong attraction causes water molecules to diffuse through semi-permeable membranes from the fresh water side to the salt water side. Consequently, when fresh water and salt water are separated by such a membrane, the fresh water will flow through to the salt side, raising the water level on that side until the pressure is sufficient to stop the fresh water flow. The pressure required to do this is actually immense, amounting to a head of several hundred feet of water.

In estuaries, the fresh water could be separated from the salt water by semi-permeable membranes and walls, as illustrated in Figure 22.8. The large difference in water level from one side to the other could be used to run hydroelectric power stations. The rate at which energy could be extracted from this source is quite large, in theory—larger than that

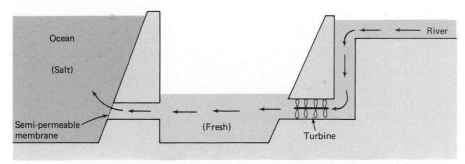

FIGURE 22.8   *One way of using salt power to generate electricity in an estuary. The tendency of fresh water to diffuse through the semi-permeable membrane toward the salt water side can be used to maintain a difference in water levels. This difference in water level can be used to generate hydroelectricity.*

from waves. Unfortunately, we do not yet know how to construct semipermeable membranes that are sufficiently large, rugged, and efficient to handle the large volumes of water involved.

Clearly, it is simply technologically not feasible to make a gigantic power plant that somehow straddles the ocean and removes a large fraction of the available energy. Any power plant we could make would be localized, inefficient, and tap only a small fraction of the available energy. If it proved profitable, then additional similar plants would be built, and in this way we would gradually build up our use of that particular resource, and the environmental consequences would gradually become apparent. These considerations are important, but far too numerous and complicated to be treated in a survey course such as this. The object of this section was just to give you an idea of the magnitude of the problem and an appraisal of the magnitude of the ocean resources available. Whether or not these resources should be tapped involves other considerations as well.

B.   MINERAL RESOURCES

B.1   Overview

As the consumption of our mineral resources continues, we exhaust the richest ore deposits, and we are forced to extract ores of poorer and poorer quality. Both the increasing demand, and the decreasing quality of the remaining ores drive the prices up, and it becomes increasingly attractive to search for alternate sources.

In addition to the economic incentive to look for alternate sources, we are gaining in our knowledge of what is available from the ocean. Searching for sources from surface ships is analogous to searching for terrestrial resources from a blimp floating 4 or 5 km high in the air and at night. Clearly our exploitation of terrestrial resources would have been considerably more difficult had we had to resort to these exploratory methods.

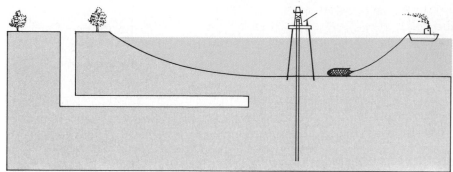

FIGURE 22.9 *Sketch of the three principal means of removing resources from the ocean floor and subfloor—through mine shafts, through bore-holes, or with dredges.*

Nonetheless, probes are being made of the ocean floor and subfloor, and we are slowly finding and mapping what is available. Although we have just barely scratched the surface, and our knowledge is pitifully sketchy, we still have learned enough to begin exploitation of some of the resources found.

In addition to the economic incentive for looking to the ocean and our increasing scientific knowledge of what is available for exploitation, we are also making progress in our technological ability to extract some of these resources. The initial capital investment is usually much higher in ocean technology. Exploitation by individual prospectors, who start out with a shovel, wheelbarrow, and a shoestring, is not possible in the ocean as it was with some terrestrial resources. The initial investment in ships, platforms, pipes, and so on, for extracting and transporting the materials, plus plants for processing them, typically runs into several hundred million dollars. Few private entrepreneurs could be expected to get into such a business.

The mineral resources of the ocean are conveniently classified according to whether they are extracted from the water itself or from the ocean bottom. Among the economically important resources removed from the water are common salt, magnesium and magnesium salts, bromine, and fresh water, which is produced in desalinization plants.

Those that are extracted from the bottom can be subdivided into those extracted from mine shafts, those removed through boreholes, and those dredged from the seafloor (Figure 22.9). Of those mined, coal and iron are the most important, although some copper–nickel ores, tin ores, and limestone are extracted from undersea mine shafts. The largest operations are in Great Britain and Japan, and are usually extensions of mine shafts originating on land.

By far, the most important ocean resource (in terms of dollar value) is oil and natural gas, which is removed through boreholes. Also removed through boreholes, although of comparatively lesser importance, are sulfur and potassium salts. Those resources dredged from the ocean bottom include sand and gravel, calcite oozes, heavy-metal-enriched sediments, and manganese nodules.

In the table below are listed the most important resources in terms of the dollar value of the annual yield. The costs are difficult to estimate in terms of U.S. dollars, so the reader is cautioned that they are approximate and certain to change with time. Nonetheless, the list does give an idea of their approximate relative importance.

Mineral	Worth in Millions of US \$/yr	Approximate Fraction of Total World Production of This Mineral
Petroleum	50,000	$\frac{1}{6}$
Sand and gravel	500	$\frac{1}{10}$
Salt	300	$\frac{1}{3}$
Magnesium and magnesium compounds	200	$\frac{1}{2}$
Bromine	100	$\frac{2}{3}$
Nickel[a]	450	$\frac{1}{20}$
Copper[a]	200	$\frac{1}{200}$
Cobalt[a]	300	$\frac{1}{6}$
Manganese[a]	100	$\frac{1}{25}$

[a]Projections for 1990

There is considerable international debate regarding who owns and who has the right to extract the ocean's mineral resources. Naturally, only those in short supply are contested. (See Tables 22.9 and 22.10.) The rich, mineral-depleted nations want these resources to belong to whomever can get them, as the rich nations have the advantage there. The poorer nations tend to claim the ocean bottom should belong to all, and extracted minerals should be taxed by the United Nations, or a similar international body, with the revenues being distributed to all. Finally, the poor mineral-exporting countries would naturally prefer that the problem not be resolved, and that no one be allowed to extract anything until it is.

Of course, many of the important resources, such as sand and gravel or the dissolved substances, are in virtually unlimited supply and no one contests them. Their extraction is determined by the demand and the economic competition from land sources. Extraction of many substances from seawater requires a great deal more energy than their extraction from land sources.

B.2 Oil, Natural Gas, and Sulfur

The world uses about 20 billion barrels of petroleum per year, or about five barrels per person per year. Of this, about 4 billion barrels come from marine sources, and the fraction is increasing. As with other resources, the easiest and most accessible fields have been exploited first. Although considerable oil remains in spent terrestrial oil fields,[3] it becomes too expensive to extract the remainder, and so ocean drilling becomes increasingly attractive.

Most marine oil wells have been drilled in less than 200 m of water

[3]Often around 50% of the original reservoir.

TABLE 22.9 Per Capita Copper Consumption for the Years 1960 and 1970

	1960	1970
World	1.5 kg	2.0 kg
United States	6.8 kg	9.5 kg

TABLE 22.10 Fraction of World's Resources Used By the United States in 1971

Petroleum	32%
Natural gas	57%
Coal	16%
Steel	19%
Aluminum	35%
Copper	27%
(Population)	(6%)

FIGURE 22.10 *The Glomar Challenger, (left) as seen from the starboard side, and (right) looking forward through the derrick from the wheel house.*

from fixed platforms. Wells have been drilled in water up to 500 m depth, but so far little production has come from these depths. The reason for this is not that there is little oil at these greater depths, but rather that the most easily exploited (and therefore cheapest) resources are extracted first. A very sophisticated exploratory ship, the *Glomar Challenger* (Figure 22.10) has drilled in greater than 6 km of water, which is a remarkable feat technologically. So far no oil has been found at these great depths. We do not yet have the technology to be able to extract it on a commercial scale, even if some should be found.

There is considerable optimism that there should be large reserves of undiscovered oil under the ocean. There is considerably less optimism,

FIGURE 22.11 *Illustration of the steps involved in creating petroleum. (a) Organic detritus sinks to the bottom, and (b) must be covered up with sediment before it is completely oxidized. (c) After millions of years, deep within the bottom sediment layers, and being subjected to high temperatures and pressures, it gradually transforms into petroleum. (d) Being lighter than the sediments, it rises into the ridges in the folds in the sediment strata until it encounters an impermeable layer, and there it is trapped.*

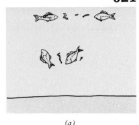

(a)

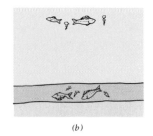

(b)

(c)

(d)

however, concerning whether these finds will be able to keep pace with the large and increasing demands of our society even in the next few years. Within a few decades from now, they certainly cannot.

Petroleum is derived from the decomposed remains of prehistoric organisms, mostly microscopic and marine. The exact chemical processes involved in their transformation are not completely understood, but the general idea is as follows. (See Figure 22.11.)

Organic detritus falls to the ocean bottom where it decomposes, being worked over by bacteria and other organisms. But it must be buried by additional sediment before undergoing complete oxidation. Otherwise, it wouldn't burn for us. That is, petroleum begins as partially decomposed, partially oxidized organic detritus. The complete transformation requires heat and pressure found deep beneath overlying layers of sediment, and it requires a long time. The exact amount of time required is not known, and probably depends on many factors. But it is known that very little petroleum is found in sediments younger than 2 million years, so this gives us some feeling for the time scales required for the transformation.

Time and temperature "crack" the long organic molecules into shorter ones. If the organic matter is cooked too long, the product is methane gas ("natural gas"). If not cooked long enough, the result is oil shale, which must be heated more for us to extract oil from it. Coal and graphite are left over as the residues of the natural cracking process.

The deeper sedimentary layers are older and hotter.[4] Therefore, deeper deposits have more methane and less oil. At temperatures above about 160°C, there is mostly methane and little oil. This means that the maximum depth for oil deposits is somewhere between 2.4 km and 7.6 km, depending on how quickly the temperature increases with depth in the sediment. Oil and gas can both rise above this, and might be found all the way up to the surface. But below this depth, only methane gas is found. Since most gas deposits are deeper than oil, the wells are more expensive. At depths greater than 3 km, drilling costs increase very rapidly.

Petroleum is considerably less dense than the sediments it is found in, so it tends to rise. When geological processes wrinkle the originally flat sediment strata, the petroleum rises until it fills the ridges beneath relatively impervious material. Consequently, the ridges in the wrinkles of deep sediment strata are likely regions to explore for oil.

The little we do know about the processes producing petroleum gives

[4]*The temperature increases by about 3°C for every 100 m depth, on the average.*

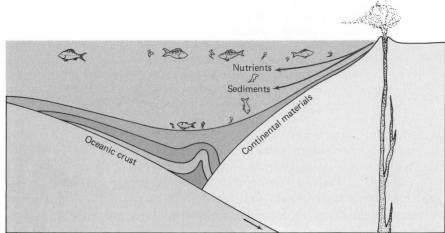

FIGURE 22.12   *Ocean trenches may be good places to look for oil for several reasons. One is that they are bordered by volcanic land masses or island arcs, which ensure large influx of nutrients and sediments. This causes both the flourishing of biological communities and the rapid burial of the organic detritus. Also, tectonic activity in some regions may create folds in the deep sediment layers, which serve as traps for the rising petroleum.*

us further insight into likely places for it to be found. We see that it is likely to be found where sediments accumulated so fast as to cover up the detritus before it was completely oxidized. Also, areas of especially high productivity in ancient times should have produced especially high amounts of organic detritus on the bottom for burial and conversion. Ancient continental margins and ancient river deltas are especially likely places to look, in view of these two considerations.

Narrow, enclosed seas are also places where oil deposits were likely to have formed. The inflow of nutrients and sediments from nearby land run-off insured both high biological productivity and rapid burial of the detritus. These places also frequently have thick salt deposits caused by heavy evaporation from waters whose communication with the open ocean are cut off or restricted. Pressures deep in the sediment layers sometimes force upward cylindrical plugs of salt from these deep salt beds, called "salt domes." Resulting distortions in the sediment layers often create traps for rising petroleum. For these reasons, petroleum deposits are often associated with salt domes.

The Red Sea is a modern example of such a semienclosed sea. Geologically, it is quite young, and is opening up as the seafloor spreads from the ridge that runs down its center. Layers of rock salt up to 5 km thick can be found beneath the Red Sea, with organic muds found beneath them. Perhaps some oil deposits along the Atlantic coasts of continents are left over from when the Atlantic was first forming, as is the Red Sea now.

There is some possibility that oil might be found in ocean trenches (Figure 22.12). There are three reasons for this speculation. First, they pro-

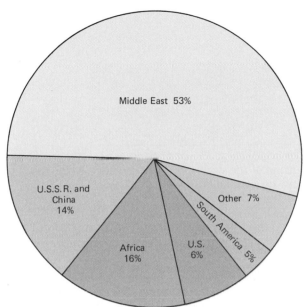

FIGURE 22.13   *Fraction of the world's crude oil reserves in the various parts of the world.*

vide good sediment traps and are frequently close to continental sediment sources. Second, the restricted current flow in these basins means that there is little dissolved oxygen in the water, and the oxidation of detritus would proceed only very slowly. Third, the tectonic movement in these regions is likely to put folds in the sediment layers, providing traps for concentrating the rising petroleum.

The continental rises hold more total sediment than the continental shelves, and in some places are up to 20 km thick. These may also cover undiscovered oil deposits. In any case, there are undoubtedly numerous untapped oil reserves in the ocean, and the above are a few examples of where we might look when the reserves of the continental shelves have been exhausted. (See Figure 22.13.)

Sulfur is also removed from the cap rock in salt domes. As the dome bulges up through overlying sediments, it eventually encounters the water, and the salt begins dissolving. However, the calcium sulfate remains, being less soluble than the other salts. It reacts with the water, petroleum, and some bacteria to release calcium and oxygen and form relatively pure sulfur deposits. It is extracted by forcing superheated water under pressure through the inner of two concentric pipes in a bore-hole. The sulfur melts and is pushed up to the surface through the outer pipe. With its salt domes, the Gulf of Mexico is a large source of the sulfur extracted in this fashion.

B.3   Sand and Gravel

More than 100 million tons of sand and gravel are dredged from the ocean bottom each year, which makes it the second most valuable marine

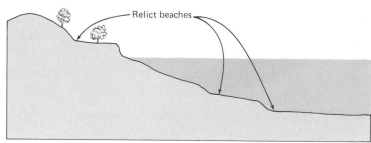

FIGURE 22.14 *Relict beaches are likely to be enriched in denser sediments due to the winnowing of the waves in times gone by. These deposits enriched in the heavy metals are called "placers."*

resource. This is about 10% of the total sand and gravel production of the world, and is enough to build about 2000 km of a modern four-lane expressway.

There are, of course, some environmental problems associated with the removal of these sediments. If done near the shore, it can interfere with the longshore flow, resulting in the erosion of beachfront property downstream. Clearly, such removal should be done either at the downstream end of a beach compartment, or safely far out from the beach. Another result is that the dredging increases the turbidity of the surrounding waters, decreasing the penetration of sunlight, and therefore decreasing productivity. Local fishermen may not like this.

B.4 Other Loose Sediments

Other loose sediments that are exploited commercially include a calcium carbonate, which is used for cement, barite, which is used as a drilling mud, and sands enriched in heavy metals. The calcareous oozes are essentially limitless in comparison to our needs, although they aren't used very heavily due to competition from land sources of lime and limestone. Nonetheless, the United States mines about 20 million tons of these oozes annually. These calcium carbonate deposits are pulverized and heated to drive off the CO_2, leaving calcium oxide, which is used in cement.

When waves work over the beach sediments, there is a tendency for the sediments enriched in heavier minerals to stay and for the lighter sediments to be removed. It is especially true when the water level changes; the heavier sediments tend to be left behind. These relict beaches, whose sands are enriched in heavy metals, are called "placers." (See Figure 22.14.) Although present beaches can be more advantageously used for recreational purposes, these placers can be mined for some of the heavier metals.

B.5 Manganese Nodules

In the chapter on sediments we have already encountered these curious deposits, greatly enriched in iron and manganese. The most common form of these deposits is spherical or acorn-shaped nodules ranging between 1 and 20 cm in diameter. The deposits are also found as coatings on rocks, or as hard pavements on the ocean bottom. They are found mostly in water deeper than 4 km, although some deposits have been found as shallow as 60 m in the Great Lakes.

Their heavy metal composition is roughly as follows:

Manganese	15%
Iron	15%
Nickel	0.4%
Cobalt	0.4%
Copper	0.3%

These are average values, and the proportions in any one deposit may vary in either direction from these values by as much as a factor of 2. They will not be mined for their manganese or iron, as continental ores can be much richer, being typically 40% to 60% manganese or iron. However, their content of other metals, especially copper, makes their harvest economically feasible. One American firm has already invested more than 200 million dollars in ships, equipment, ore carriers, and a processing plant. (See Figure 22.15.) Others are getting into the business, too.

One interesting puzzle surrounding these nodules is they are often found in regions where the rate of sediment accumulation is much greater than the rate of growth of these nodules, sometimes more than 100 times greater. How can these nodules grow? Why aren't they buried by sediment before they get started? How do they stay at the surface? Perhaps organisms groveling about the bottom roll them around, which would keep them atop the sediments and also account for their spherical shapes.

B.6 Other Possibilities from the Ocean Bottom

Other resources of the ocean bottom attracting some attention include phosphorite deposits and deep sea sediments enriched in some metals. Phosphorite deposits tend to form where upwelling causes nutrient-laden deeper water to be warmed as it surfaces. These deposits are enriched in P_2O_5, which has value in its use in fertilizers. The marine sources are not yet mined, as phosphorite from sources on land is still cheaper. Some day these marine sources will undoubtedly be tapped.

There has been some excitement in the discovery of hot brine pools at a depth of 2 km along the oceanic ridge in the Red Sea. These pools have been found to have salinities up to 250‰, and temperatures up to 56°C. Their formation undoubtedly is related to the heat sources along the active ridge, and the hydrothermal vents.

More than a curiosity, the bottoms of these pools have been found to have sediments 20 to 100 m thick, which are rich in metal sulfides. The lateral dimensions of these pools have not yet been completely mapped, but from similar deposits on Cyprus, which is a portion of an ancient ridge raised above sea level, it appears they may extend a kilometer or so horizontally. The metal sulfide content of these deposits must somehow result from the extraordinary temperature and salinity of the pools.

(a)

(b)

(c)

FIGURE 22.15   *Deep ocean mining. (a) The Research Vessel Prospector, operated by Deepsea Ventures, Inc., explores for manganese nodule mining sites. (b) A television camera, mounted on a tripod, is launched over the side of the R.V. Prospector, and towed a few feet off the ocean bottom, using high-intensity lights for illumination. (c) The Deepsea Miner II is a converted ore carrier used by Deepsea Ventures as a platform from which to conduct deep ocean mining tests.*

Analysis shows that these deposits contain 29% iron, 3.4% zinc, 1.3% copper, 0.1% lead, and also some enhancement in the concentrations of nickel, chromium, vanadium, cobalt, and manganese. Silver and gold are found in trace amounts. Of greatest economic interest among the metals is copper, followed by zinc. Adjacent to the pools are also found sediments enriched in copper, vanadium, and zinc.

There is high metal sulfide sediment along all active ridges. Most of this is too low in concentration for commercial interest, but there are some local regions of especially high concentration, perhaps associated with former brine pools like those observed now in the Red Sea. We cannot escape the speculation that if we can see these deposits now forming along

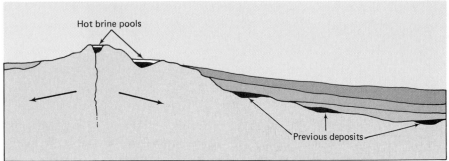

FIGURE 22.16 *Hot brine pools are found filling some depressions in presently active oceanic ridges. Beneath them are forming sediment deposits rich in metal sulfides. Probably there are many other similar deposits along the ocean floor, formed earlier when these parts were on the ridge. Of course, they would be covered by subsequent sediments.*

the active ridge beneath the "young" Red Sea, then it is likely that they have been forming throughout history along all previously active ridges. Since all oceanic crust has originated in these ridges, there may well be such deposits throughout the deep ocean bottom that have been covered up by subsequent sediments (Figure 22.16). Exposed regions of oceanic ridge, such as Cyprus and Iceland, are attracting increased interest for answers to these questions.

"Brown clays" cover much of the ocean bottom. These sediments are enriched in aluminum (up to 9%), iron (up to 6%), and also in copper, nickel, cobalt, and titanium. They are not rich enough to be yet competitive with other sources, but their volume ensures that they are virtually an unbounded source for these metals. In fact, for most of these metals, they are accumulating in the ocean bottom sediments faster than our society could use them up.

B.7 Elements in Solution

The ocean has been used as a source of salt for thousands of years. Today, we are extracting other minerals from the ocean as well, but salt still ranks third in value among them (Figure 22.17).

Magnesium, magnesium salts, and bromine are also extracted. Magnesium is a lightweight metal, like aluminum, but much stronger. It is used in construction where light weight and strength are important. In salts, magnesium has a valence of $+2$, like calcium, but magnesium is much more soluble. Consequently, magnesium salts are frequently used where solubility is important, such as in medicines. One of the principal uses of bromine has been in antiknock compounds in gasoline.

B.8 Fresh Water

Fresh water is also one of the valuable resources extracted from the ocean. However, the price paid for fresh water varies so greatly from one part of

FIGURE 22.17 *A salt recovering operation.*

the world to another that it is difficult to put a dollar value on it. Clearly, water is one of the necessities in life, and people would be willing to pay an extremely high price for it if they had to.

Fresh water may be produced extremely cheaply from seawater if gravity flow and solar energy are used in the distillation (Figure 22.18). Unfortunately, these plants require large areas to produce fresh water at relatively small rates. The use of fossil fuels to distill the water faster raises the price comparably. There are a wide variety of engineering designs for desalinization plants, with a comparable wide range in the price of the product. It has been suggested that desalinization plants may be built adjacent to nuclear reactors, so the waste heat in the reactor's cooling water may be used by the desalinization plant. In some regions of the world, this might solve two pressing problems at once.

The cost of water depends also on how pure the product must be. For drinking, continued reliance on water of salinity greater than $1\%_{00}$ is considered unhealthy, although people may drink water much saltier than this in a pinch. For agricultural irrigation, the water may be a little saltier than $1\%_{00}$, but not much. This is because evaporation from the soil will leave the salt residue behind, and the saltier soil hurts most terrestrial plants. Some research efforts are now being directed toward improving the salt tolerance of various crops. For cooling water, salt water serves the purpose as well as fresh.

There are now nearly 1000 desalinization plants in the world, more than 50 of these in the small but wealthy country of Kuwait. The first desalinization plant in the United States is in Key West, and produces about 2½ million gallons per day.

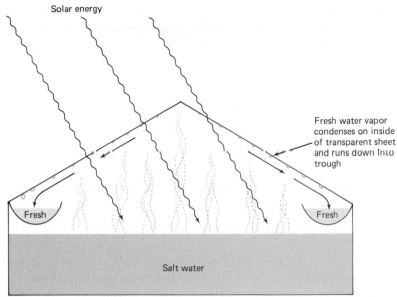

Solar energy

Fresh water vapor
condenses on inside
of transparent sheet
and runs down into
trough

Fresh

Fresh

Salt water

FIGURE 22.18   *One design of a solar still for producing fresh water from salt. The large salt water ponds are covered by tents of plastic or glass, which keep the solar heat inside. The water evaporates, condenses on the inside of the tent, and drips into the fresh water troughs along the sides. The hot fresh water exits through pipes running through the incoming salt water, helping to heat it in advance.*

The present cost of desalinized water in the United States is about 85 cents per thousand gallons. For comparison, water from city water systems costs typically 35 cents per thousand gallons, and agricultural irrigation water from wells or large irrigation systems is considerably cheaper than this. Desalinized water is much too expensive right now to be used for agricultural purposes.

C.   SUMMARY

The decreasing reserves of conventional energy resources, coupled with our increasing demand for them, has made the ocean suddenly come into prominence as a possible energy source. It stores solar energy in many ways, and it also stores gravitational energy of the earth–moon–sun system in the tides. Estimates of available energy indicate that the stored thermal energy is an especially attractive source. Energy stored in waves and tides is also receiving some interest.

Petroleum is presently the most valuable mineral resource being removed from the ocean. This source supplies about one fifth of our total petroleum consumption. Petroleum is derived from the remains of marine organisms that have been buried before complete decomposition.

Sand and gravel are economically the second most important mineral resource removed from the ocean. Other sediments exploited commercially include limestone and areas enriched in heavy metals, such as placers. Manganese nodules are being mined for their copper and zinc content, and recently discovered hot brine pools on the oceanic ridge are receiving attention for their heavy metal deposits. Salt has always been an important marine product, and more recently other dissolved substances, and the water itself, have been removed commercially.

QUESTIONS FOR CHAPTER 22

1. What advantages has the exploitation of solar energy over fossil fuels and nuclear energy?

2. Discuss the units commonly used for measuring energy. What is a watt?

3. Give examples of some electrical appliances around your house, and the number of kilowatts it takes to keep them running. (Usually this is printed somewhere on the appliance.)

4. Discuss the approximate power usage of an automobile in terms of kilowatts.

5. What is the average per capita rate of energy consumption in the United States? How is this distributed among transportation, domestic, and industrial sectors?

6. What is the average per capita rate of energy consumption in the world? What is the total rate of energy consumption by the world population? How much of this is used in the United States?

7. Discuss why an estimate of "replacement time" is important in deciding the maximum rate at which energy could be extracted from the ocean.

8. What would you guess to be a reasonable "replacement time" for the ocean's thermal energy? Why? How about its wave energy? Energy in surface currents? Tidal energy? (Explain why you guessed what you did for each.)

9. Briefly discuss some factors that drive up the prices of mineral resources extracted from conventional land sources.

10. Besides economic incentive, what other two factors are important in our increased production of mineral resources from the ocean?

11. Why is it that individual prospectors will not play a significant role in initiating the extraction of minerals from the ocean like they did on land?

12. What are a few of the most important resources extracted from seawater?

13. What are some minerals extracted through mine shafts? Boreholes? What are dredged?

14. What is the approximate dollar value of the three most valuable minerals presently extracted from the ocean?

15. Discuss the views (or view-tendencies) of the wealthy nations, poor nations, and poor but mineral-exporting nations in the international debates regarding ownership and control of the ocean's mineral resources.

16. Why do oil companies find and develop new oil fields rather than extracting the remaining oil from some old oil fields?

17. Why is it that only continental shelves have been exploited for petroleum, and not deeeper regions?

18. Where does petroleum come from originally? Discuss the processes involved in its transformation.

19. Why is it that petroleum is most likely found in regions where sediment accumulated rapidly?

20. Why is petroleum most likely found under ridges in folded sediment strata?

21. Why are ancient narrow, enclosed seas likely places for oil deposits to form?

22. Why are petroleum reserves often associated with salt domes? Is there a modern example of a semienclosed sea beneath which organic muds are found covered with layers of salt? How do you suppose this happened?

23. Why is it thought that some ocean trenches might have oil deposits beneath them?

24. Why are sulfur deposits associated with salt domes? How is the sulfur removed?

25. What are two possible environmental problems associated with the dredging of sand and gravel?

26. What is the largest commercial use for calcium carbonate (limestone) deposits?

27. How do placers become enriched in sediments with heavy metals?

28. What forms do manganese deposits take?

29. What are some of the important metals contained in manganese nodules?

30. What is phosphorite used for?

31. Why does the water in hot brine pools stay at the ocean bottom? Being hotter, shouldn't it be lighter and rise to the surface?

32. Even before Roman times, Cyprus was known and exploited for its rich mineral ores. What is it about Cyprus that makes it geologically quite different from most land masses?

33. What is exceptional about the sediments beneath hot brine pools?

34. Why is it likely that there may be localized sediment deposits, sim-

ilar to those beneath present brine pools, to be found throughout the ocean bottom? Would these probably be covered up by more recent sediment layers?

35. Brown clays are also enriched in some metals. Is it the quantity or the quality of these reserves that is impressive?

36. What is an important commercial use for magnesium? For magnesium salts? For bromine?

37. How is the price of desalinized water related to the rate of production and to the purity of the product?

38. Roughly how pure must fresh water be for drinking purposes? For agriculture?

39. Why is desalinized water not used for agriculture now?

*40. How can hydroelectric power be considered a form of solar energy?

*41. What form of energy stored in the ocean seems most likely to help us out? Why?

*42. Manganese nodules grow very slowly. Yet they are often found in regions where sedimentation rates are significant. Why is this puzzling? Can you think of any possible explanation other than the one offered in this chapter?

SUGGESTIONS FOR FURTHER READING

1. Roger H. Charlier, "Ocean-Fired Power Plants," *Sea Frontiers,* **27,** No. 1 (1981), p. 36.

2. Roger H. Charlier, "Tides and Turbines," *Sea Frontiers,* **26,** No. 6 (1980), p. 355.

3. Luc Cuyvers, "Dividing the Continental Shelf Pie," *Sea Frontiers,* **25,** No. 3 (1979), p. 165.

4. John C. Fine, "The Continuing Ocean Debate," *Sea Frontiers,* **22,** No. 2 (1976), p. 77.

5. Daniel P. Finn, "Georges Bank: The Legal Issues," *Oceanus,* **23,** No. 2 (Summer 1980), p. 28.

6. Bea Howell, "Safeguarding Our Oceans in Metal Mining," *Sea Frontiers,* **24,** No. 5 (1978), p. 265.

7. John M. Hunt, "The Origin of Petroleum," *Oceanus,* **24,** No. 2 (Summer 1981), p. 52.

8. F. L. LaQue, "Nickel from Nodules?," *Sea Frontiers,* **25,** No. 1 (1979), p. 15.

9. Michael J. Mottl, "Submarine Hydrothermal Ore Deposits," *Oceanus,* **23,** No. 3 (Summer 1980), p. 18.

10. "Ocean/Continental Boundaries," *Oceanus*, **22,** No. 3 (Fall 1979).

11. "Ocean Energy," *Oceanus*, **27,** No. 4 (Winter 1979/80).

12. "Oil in Coastal Waters," *Oceanus*, **20,** No. 4 (Fall 1977).

13. "Seaward Expansion," *Oceanus*, **19,** No. 1 (Fall 1975).

14. Pontecorvo, Wilkinson, Anderson, and Holdowsky, "Contribution of the Ocean Sector to the United States Economy," *Science*, **208** (May 1980), p. 1000.

15. Charles T. Schafer, "Petroleum Under the Sea," *Sea Frontiers*, **26,** No. 5 (1980), p. 258.

16. F. G. Walton Smith, "Oceans of Energy," *Sea Frontiers*, **26,** No. 4 (1980), p. 194.

17. F. G. Walton Smith, "Power from the Oceans," *Sea Frontiers*, **20,** No. 2 (1974), p. 87.

18. F. G. Walton Smith and Roger H. Charlier, "Saltwater Fuel," *Sea Frontiers*, **27,** No. 6 (1981), p. 313.

19. F. G. Walton Smith and Roger H. Charlier, "Turbines in the Ocean," *Sea Frontiers*, **27,** No. 5 (1981), p. 301.

20. F. G. Walton Smith and Roger H. Charlier, "Waves of Energy," *Sea Frontiers*, **27,** No. 3 (1981), p. 138.

21. Edward Wenk, Jr., "The Physical Resources of the Ocean," *Scientific American* (Sept. 1969).

Absolute humidity A measure of the water content of air, stating what percentage of the air is water vapor.

Abyssal Of or pertaining to those regions of the ocean bottom that lie at depths between about 3.5 and 6 km, which include the bottoms of all basins and most abyssal plains.

Abyssal hills All hills found on the deep ocean bottom that rise no more than 0.9 km above the bottom. Also called seaknolls.

Abyssal plains Extensive flat regions of the deep ocean floor, usually bordering on continental margins, caused primarily by the deposits of terrestrial sediments in turbidity currents.

Acid A solution in which the concentration of the positive hydrogen ion, H^+, is greater than that of the negative hydroxide ion, OH^-. In water solutions, this means an H^+ ion concentration greater than one part in 10^7.

Adiabatic Occurring without the addition or extraction of heat.

Advection The transfer of heat or matter by horizontal movement of water masses. Horizontal convection.

Algae A common name applied to all the members of the subkingdom Thallophyta, which contains the most primitive, photosynthetic plants.

Algal ridge The raised windward edge of a reef built up with the help of algae having calcareous skeletons.

Alkaline Having a hydroxide ion (OH^-) concentration greater than the positive hydrogen ion (H^+) concentration. For water solutions this means having an H^+ concentration less than one part in 10^7.

Alluvial fans The large fan-shaped deposits of sediments and rock debris found at the base of mountains, especially where ravines empty out into the lowlands.

Altitude Height above sea level.

Amino acids Complicated molecular structures that are the building blocks in the construction of even more complicated protein molecules.

Amorphous Having no recurring order to the arrangement of molecules. Not crystalline.

Amphidromic motion The large-scale circular motion of an elevation and depression of the water surface, resulting from the Coriolis deflection of the tides in large masses of water, such as in large embayments, seas, or in the open ocean.

Amphidromic point The central point in a water mass undergoing amphidromic motion, whose elevation remains unchanged.

Anion A negatively charged ion.

Annelida An animal phylum containing the segmented worms, such as angleworms and tube worms.

Anticyclonic   Clockwise in the Northern Hemisphere and counterclockwise in the Southern Hemisphere, as seen from the top.

Aphotic   Without light. Of or pertaining to those regions of the ocean below the sunlit surface waters.

Aquaculture   The farming and cultivation of life in the oceans.

Aragonite   A crystalline form of calcium carbonate having an orthorhombic lattice (i.e., the $CaCO_3$ groups are ordered in a three-dimensional rectangular array).

Arrow worm   A common name applied to any of the members of the animal phylum Chaetognatha, including small thin worms that have small fins for locomotion and small primitive jaws.

Arthropoda   A phylum containing animals with jointed legs and external skeletons (e.g., crustaceans and insects).

Asthenosphere   The rather plastic region in the upper mantle, extending from approximately 100 to 200 km below the earth's surface, over which the lithosphere slides.

Atmospheric pressure   The pressure exerted by our atmosphere on surfaces due to the weight of the air above. At sea level this amounts to 14.7 lb/in.^{2} or 1.01×10^6 dyne/cm^2.

Atoll   A ring-shaped coral reef having a central lagoon and that results from coral growth on a sinking volcanic structure.

Atomic number   The number of protons in the nucleus of an atom.

Atomic weight   The total number of nucleons (protons plus neutrons) in the nucleus of an atom. Also, the average mass of an element measured in atomic mass units.
(The two definitions are nearly the same.)

Attenuation   Loss in intensity or strength.

Authigenic   Derived from seawater and precipitating directly onto the ocean bottom sediment, as opposed to falling through a column of water first.

Autotrophic   Able to manufacture its own food.

Backshore   The portion of a beach above high tide level.

Backwash   The water returning back down the beach face toward the surf zone after a wave has washed up on the beach.

Backwash diamonds   Diamond-shaped patterns on the beach that are created by the backwash.

Bar   An offshore reservoir of beach sand in the form of a long low ridge paralleling the shore. Bars are generally located in or just beyond the surf zone.

Barrier beach   A long thin peninsula or island paralleling the mainland, being separated from the mainland by a lagoon.

Barrier reef A reef whose shallowest portions are separated from the mainland by a lagoon.

Basalt A fine-grained igneous material characteristic of oceanic crust, being enriched in ferromagnesian minerals relative to continental materials.

Base A solution whose hydroxide ion (OH^-) concentration is greater than the hydrogen ion (H^+) concentration. For water solutions, this means having an H^+ ion concentration less than one part in 10^7.

Basins The large deep portions of the oceans extending from continental margins to oceanic ridges, with depths normally in the range of 4 to 5 km.

Bathyal Of or pertaining to the portion of the ocean bottom that has intermediate depths ranging from 0.2 km to about 3.5 km. Characteristic of continental slopes and oceanic ridges.

Baymouth bar Any bar extending across a portion of the mouth of a bay. It usually begins on a headland and extends in the direction of a longshore current.

Beach cusps Side-by-side "U"-shaped indentations along the beach, typically ranging in width from about 2 to 200 m each.

Beach face The surface of the beach. Sometimes used to mean only the beach surface extending from the surf zone to the berm crest.

Beach rock Rock formed when beach sediments become cemented together by the salt residue left behind as sea spray falls on them and evaporates.

Beaufort numbers A scale of numbers, ranging from 1 through 17, used to describe the state of the sea, with 1 representing the calmest conditions and 17 representing the most violently stormy conditions.

Benthic Of or pertaining to the ocean floor.

Benthos All organisms that are attached to, live on, or live in the ocean bottom.

Berm The large reservoir of loose dry sediments on the beach above the high tide line.

Bight A large open bay or indentation in the coastline.

Biogenic Derived from biological organisms.

Bioluminescent Giving off light through biological processes.

Biomass All living organic matter.

Boiling point The temperature at which further addition of heat to a liquid will not raise its temperature but rather will transform some if it into the gaseous state.

Boulder A rock of dimensions greater than 26 cm (about 10 in.).

Brackish Of salinity less than that of sea water, resulting from the mixture of fresh and salt waters.

Breakers Waves that are breaking, as the crests get ahead of their support.

Breakwater A man-made wall or other barrier built in coastal waters and used to protect some area from incoming waves.

Brine Water having salinity greater than that of sea water.

Buffer Anything that reduces an impact.

Buoyancy The ability of an object to rise or float due to the forces of the fluid it is in or on.

Caballing The process of surface water sinking in regions where surface waters converge.

Calcareous Made of calcium carbonate.

Calcite A crystalline form of calcium carbonate having a hexagonal lattice structure (i.e., the arrangement of calcium carbonate groups resembles the structure of a tiny hexagonal honeycomb).

Calorie A unit for measuring energy defined as the amount of energy required to raise the temperature of one gram of water by one degree Celcius.

Capillary waves Those small waves on the water surface having wavelengths less than 1.73 cm, often arising as wind-generated ripples.

Carbohydrates A class of molecules formed of carbon, hydrogen, and oxygen, including sugars and starches.

Carbon-14 A radioactive isotope of carbon, having 14 nucleons in the nucleus (6 protons and 8 neutrons) as opposed to the normal 12. It decays into nitrogen-14 with a half-life of 5700 yr.

Carbon assimilation The incorporation of carbon into new organic material during photosynthesis.

Carnivore Any animal that feeds on other animals.

Cation A positively charged ion.

Cay A small island.

Celcius A temperature scale calibrated by defining the freezing point of fresh water to be 0°C and its boiling point to be 100°C.

Centrifugal force The outward force exerted by an object traveling a curved path upon the object that is making its path curve (e.g., the force of the car against the pavement on a curve). According to common but incorrect usage, it is an apparent outward force on an object in a rotating reference frame.

Chaetognatha A phylum of small thin marine worms having small fins for locomotion and small primitive jaws. Arrow worms.

Change of state Any transformation between solid and liquid, liquid and gaseous, or gaseous and solid states of matter.

Chlorinity A measure of the halogen content of seawater, measured in parts per thousand. The salinity is derived from it by multiplying by 1.80655.

Chordata   A phylum of animals that have had gill slits and a cartilaginous skeletal rod at some stage in their development.

Clastic   Derived from rock fragments.

Clay   Sediments having dimensions in the range of 2.5×10^{-4} mm to 4×10^{-3} mm.

Coast   The strip of land bordering the ocean including the beach and extending landward as far as the environment is noticeably affected by the ocean's proximity.

Cobbles   Sediments having dimensions in the range of 6 to 26 cm.

Coccolithophores   A class of microscopic single-celled marine planktonic plants having exoskeletons made of tiny calcareous plates.

Coelenterata   A phylum of primitive marine carnivores, each having a gut lined with protoplasm and tentacles for capturing and maneuvering prey into the gut, such as a jellyfish or anemone.

Colloid   An extremely fine sediment having dimensions less than 2.5×10^{-4} mm and capable of being held in suspension in seawater indefinitely.

Compensation depth ($CaCO_3$)   The depth beneath which calcium carbonate deposits can no longer form, due to the increased acidity of deeper waters.

Compensation point (biological)   The depth beneath which photosynthesis and production of oxygen by plants cannot support their respirational needs.

Compound   A substance whose elementary constituents are stable groups of two or more atoms.

Condensation   The transformation of a vapor into a liquid state.

Conduction   The transfer of heat through contact during which thermal energy is transferred from one molecule to the next through their mutual interactions.

Continental crust   The upper layer of the earth's surface beneath continents that has a thickness of typically 30 to 50 km and is composed of materials more deficient in ferromagnesian minerals than those characteristic of the crust beneath the oceans.

Continental drift   The relative motion of the various continents.

Continental margin   The edge of a continent bordering an ocean basin, usually considered to include the coast, continental shelf, continental slope, and continental rise.

Continental rise   The wedge of sediments deposited at the base of the continental slope.

Continental slope   The region of the ocean bottom beginning at the outer edge of the continental shelf and plummeting downward at a relatively steep angle to a depth of typically 3 or 4 km.

Contour line   A line drawn on a chart or cross-sectional profile connecting continuously all points for which a certain variable has the same value. For example, it could connect all points on the ocean having the same depth, or it could connect all points on the ocean surface having the same salinity.

Convection   The transport of heat from one place to another via the motion of the heated medium. Also, any net movements in a medium.

Convergence   A region where two or more surface currents converge.

Copepods   Small arthropods resembling very tiny shrimp that feed heavily on microscopic plants.

Core (earth)   The central regions of the earth extending out to about 2900 km beneath the surface. It consists of a liquid outer portion and a solid inner portion.

Core (sediment)   A vertical column of sediment retrieved by certain types of sediment samples. *See* corer.

Corer   A sediment sampling device that operates by plunging a hollow tube, usually of circular or square cross section, into the bottom sediments.

Coriolis effect   The apparent deflection of moving bodies from their expected straight-line motion caused by the rotation of the observer's reference frame.

Cosmogenous   Derived from outer space.

Crest   The highest part of the wave.

Crust   The outer layer of the earth's surface, extending to depths of roughly 10 km beneath the ocean and 30 to 50 km beneath the continents, and which is composed of materials that are lighter, cooler, and more brittle than those of the mantle beneath.

Crustal plates   The twelve or so pieces of the earth's crust, each one moving as a unit and outlined by the large amount of seismic activity around its perimeter where it engages neighboring plates.

Crystalline   Having some periodic recurring order in the arrangement of its constituent atomic groups.

Ctenophore   A phylum of small luminous jellyfishlike animals.

Current   Any large-scale sustained movement in a fluid mass.

Cyclonic   Counterclockwise in the Northern Hemisphere and clockwise in the Southern Hemisphere, as seen from above.

Dark Ages   The period in the history of Western Civilizations extending from about A.D. 476 to the Renaissance (roughly A.D. 1200).

Decibar   One-tenth of standard atmospheric pressure. Depth in the ocean can be measured in decibars as the pressure due to the weight of overlying layers increases by one decibar for every meter of depth.

Deep   A deep portion of the ocean bottom.

Deep ocean channels   Channels found on the deep ocean bottom resembling large river beds, being typically 2 km wide and 100 or 200 m deep.

Deep scattering layer   The layer of microscopic grazers that show daily vertical migrations in response to sunlight. It derives its name from the scattering of sonar signals.

Deep waters   Those water masses originally formed in polar regions that now are found in the deepest regions of the ocean, typically below 2 or 3 km depth.

Deep water waves   Waves in water of depth greater than one-half their wavelength.

Delta-front trough   A large canyonlike feature carved in the shelf and slope, similar to a submarine canyon, but "U"-shaped in cross section.

Demand curve   A plot of demand versus price for any product, illustrating what the demand will be at any particular price.

Demersal   Residing on the ocean bottom.

Density   The ratio of mass to volume for any material, often measured in units of grams per cubic centimeter.

Density current   Any oceanic current that is driven by gravity, which makes denser water masses tend to sink or flow downhill.

Deposition   The settling out onto the bottom of sediments previously held in suspension.

Desalinization   Removing the salt.

Detrital   Resulting from the breakup of rock.

Dew point   The temperature below which water will begin condensing from air on any available condensation nuclei (e.g., blades of grass).

Diatomaceous earth   Sediments composed of the microscopic siliceous skeletons of diatoms.

Diatoms   Single-celled microscopic marine planktonic plants having siliceous exoskeletons and responsible for the bulk of the ocean's primary productivity.

Diffraction   The bending or spreading out of waves as they pass through openings into protected waters, or as they pass by an obstacle, which is not attributable to changes in wave speed.

Diffusion   The net transport of materials from regions of higher concentration to regions of lower concentration, due to the thermal motion of the individual molecules.

Dinoflagellates   Microscopic single-celled organisms that can propel themselves with the use of tiny whiplike flagella. Some can photosynthesize.

Discontinuity   An abrupt change.

Dispersion   The sorting out of waves in a group as the longer faster waves take the lead, and the shorter slower waves fall behind.

Dissociation   The splitting of molecules into fragments.

Dissolution   Entering into solution; dissolving. Also, breaking into parts or elements.

Distillation   The purification of a substance by first heating and vaporizing it, and then cooling and recondensing it. As used in this book, the separation of materials with lighter molecules and lower melting or boiling points, by heating rock at some depth beneath the surface to the point where these materials become fluid and rise, and the subsequent recondensation of these released materials in the cooler surface or near surface environment.

Diurnal   Daily.

Divergence   A region from which surface waters are flowing away.

DNA   A very long thin protein molecule, in the shape of a long twisted ribbon, that resides in a cell nucleus and contains all the information necessary to govern the activities of the cell.

Doldrums   The region of the earth near the equator characterized by relatively light breezes, rising air masses, and frequent rainstorms.

Dolomite   A sedimentary rock composed of calcium and magnesium carbonates and formed under conditions when enclosed seas are subjected to heavy evaporation.

Downwelling   The flow of surface waters downward.

Dredge   A basketlike apparatus that is dragged along the ocean bottom in order to retrieve biological or geological specimens.

Earthquake epicenter   The point on the earth's surface directly above the origin of a seismic disturbance.

Earthquake focus   The position within the earth where the seismic disturbance originated.

Ebb tide   The part of the tidal cycle when the water level is lowering, and water is flowing out of estuaries.

Echinodermata   An animal phylum containing spiny-skinned radially symmetric bottom dwellers with internal skeletons, such as starfish and sea urchins.

Eckman spiral   The pattern of wind driven water flow where the Coriolis deflection causes the water at any depth to flow in a direction slightly to the right (Northern Hemisphere) of the water just above it.

Eddy   A small localized current flowing in a circular or whirling motion.

Edge waves   Waves moving sideways to the shore, rather than coming directly in, usually caused by the superposition of incoming and reflected waves.

Electron A negatively charged elementary particle having about ½₀₀₀ the mass of a neutron or proton and having a charge equal but opposite to that of the proton. In an atom it orbits the positively charged nucleus in dimensions of roughly 10^{-8} cm, or about 10^5 times larger than the nuclear dimensions.

Electron cloud A term applied to the atomic regions where electrons are found.

Element Any of the approximately 106 chemically different fundamental constituents of matter. The atoms of any one element are characterized by having a certain fixed number of protons in each nucleus.

Ellipse A planar geometrical figure formed by the locus of all points for which the sum of the distances to two fixed points (the foci) is a constant. Equivalently, it is the figure seen when a circle is tipped so that the plane of the circle is not perpendicular to your line of sight.

Energy The ability to do work, such as to move something by applying a force to it. Energy comes in many forms such as thermal, electrical, kinetic, and so on.

Energy efficiency The ratio of useful energy output to energy input.

Enzyme An organic substance produced in living cells that aids in certain chemical reactions.

Eolian Airborn.

Erosion The removal of debris, especially rock debris, from its original location through the action of water, wind, gravity, animals, and so on.

Estuary Any embayment or partially enclosed body of water that opens to the ocean somewhere and (normally) also has some fresh water inflow.

Eustatic World-wide (as opposed to local changes due to local isostatic readjustments, for example).

Evaporation The process of changing from liquid to gas.

Evaporites Sedimentary deposits formed by the evaporation of seawater, which leaves a more salty solution and causes the precipitation of those salts that reach saturation.

Excess volatiles Those light elements abundant in our atmosphere and oceans whose presence cannot be accounted for by the weathering of the crust.

Exoskeleton A skeleton covering, or partially covering, the exterior of the organism.

Fahrenheit A temperature scale for which the freezing point of water is 32 degrees Fahrenheit (°F) and the boiling point of water is 212 degrees Fahrenheit (°F).

Fan valley A current-cut channel in an alluvial fan or on the continental rise.

Fathom A unit of length equal to 6 feet, or 1.83 meters.

Fault A crack in the earth's crust along which there has been some displacement.

Ferromagnesian Relatively rich in iron and magnesium silicates.

Fetch The horizontal distance over which a wind blows.

Fission (biological) The reproductive process during which one biological cell splits into two.

Fission (nuclear) The splitting of one atomic nucleus into two or more, which is accompanied by a release of energy for heavy nuclei.

Fjord A glacially cut valley that empties into the sea.

Flood tide That part of the tidal cycle when water level is rising, and water is flowing into estuaries.

Flushing time A measure of the turnover time for fresh water in an estuary that is calculated by comparing the total fresh-water content of the estuary to the rate at which fresh water enters.

Fold A twisting, bending, or doubling over of sedimentary layers.

Food chain The sequence of organisms through which organic matter passes after being produced by a plant.

Foraminifers Single-celled animals belonging to the order Foraminifera of the phylum Protozoa, having calcareous tests that form some calcareous sedimentary deposits.

Foreshore The intertidal portion of the beach extending from the low tide line up to the high tide line.

Fossil fuel Any fuel, such as coal, petroleum, or natural gas, derived from the remains of organisms.

Fracture zones Faults oriented roughly perpendicular to the oceanic ridge along which there has been lateral displacement of the ridge crest.

Freezing point The temperature at which a liquid solidifies.

Fringing reef A reef immediately adjacent to a mainland and not separated from it by a lagoon.

Fully developed sea A state of the sea reached when the wind-generated waves will not grow any bigger even if the wave-generating wind should continue to blow.

Fusion (nuclear) The process during which two atomic nuclei stick together to form one large nucleus, which is accompanied by a release of energy if the nuclei are small ones.

Fusion (solid) Melting.

Geopoetry A term coined by H. H. Hess to describe his thoughts on the causes and mechanics of continental drift.

Geostrophic flow A large-scale flow of water in the ocean resulting from elevation or depression of the ocean surface. This elevation (depression) is maintained by the opposing effects of Coriolis deflection and gravity.

Graded bedding   The sorting of sediment within sedimentary layers having the largest sediments at the bottom of each layer, successively finer sediments above, and the finest sediments at the top of each layer.

Gradient   The rate of change of a variable with distance and oriented in the direction of greatest change. For example, a salinity gradient would be greatest at a halocline and oriented in the direction of greatest change.

Granite   A coarse-grained igneous rock somewhat deficient in ferromagnesian minerals and characteristic of continental crust.

Granule   A sediment whose dimensions lie in the range of 2 to 4 mm.

Gravity   A force with which two bodies pull on each other and which is in proportion to their masses.

Gravity waves   All waves of wavelengths greater than 1.73 cm that are driven by gravity and their own inertia, regardless of how they are generated.

Grazers   Animals feeding on plants; herbivores.

Great circle   A "largest possible circle" that can be drawn on the surface of a sphere and divides the sphere surface exactly into two equal hemispheres.

Greenhouse effect   The warming effect caused by our atmosphere's transparency to sunlight coming in and its opacity to the earth's infrared radiation going out, therefore making it easy for heat to come in and difficult to leave.

Gyre   A large mass of water having a circular or rotary motion.

Hadal   Of or pertaining to the very deepest portions of the ocean bottom below 6 km in depth (i.e., trenches).

Half-life   The length of time during which half the original nuclei will have decayed and half will still remain.

Halocline   A region over which there is very rapid change in salinity.

Halogens   The family of chemical elements that are one electron short of the preferred inert gas electronic arrangement, including fluorine, chlorine, bromine, and iodine.

Headland   Any point or piece of land sticking out into the ocean from the mainland.

Heat capacity   A measure of the ability of a body to hold heat, described according to how much heat energy must be added to the body to increase its temperature by one degree.

Herbivore   Any animal that feeds on plants.

Heterotrophic   Unable to manufacture its own food.

Holoplankton Organisms that are plankton their entire lives.

Horse latitudes Those regions around 30° to 35° North and South latitudes characterized by light surface winds, falling air masses, and arid conditions.

Hurricane A violent storm having wind velocities in excess of 110 km/hr.

Huygen's Principle A principle of physics stating that every point on a wave is a point source for generating new waves.

Hydration The addition of water.

Hydrocarbon Any molecule composed primarily of carbon and hydrogen.

Hydrogenous Derived from the materials in solution in seawater.

Hydrologic cycle The cycle of water in our hydrosphere during which water in the ocean evaporates, precipitates, and returns back to the ocean through any of a variety of routes.

Hydrosphere The outer portions of the earth where water is plentiful, including ocean, atmosphere, ice, lakes and streams, and groundwater.

Hypertonic Having body fluids more salty than the water of the environment.

Hypotonic Having body fluids less salty than the water of the environment.

Ice age Any period of earth history lasting thousands of years during which temperatures were below normal and ice sheets covered large portions of continents, even in temperate latitudes.

Igneous rock Rock derived from molten magma.

Indicator species Any species of organism that is very sensitive to the conditions of the environment and whose remains can be found among older sediments to indicate environmental conditions in times past.

Inelastic demand The situation described when the demand for a product does not change much with a change in price.

Inertia The tendency for a resting body to remain at rest and for a moving body to maintain its motion in a straight line. A quantitative measure of this property is called the mass.

Inertial currents Currents following looped trajectories characteristic of a body trying to pursue inertial straight-line motion along the surface of the earth, but being influenced by the earth's spin and its equatorial bulge.

Inertial force A force resulting in the apparent deflection of a moving body arising because the observer's reference frame is rotating.

Infrared Pertaining to that portion of the electromagnetic wave spectrum having wavelengths longer than the longest visible waves (red light).

Infrared window   Any of several small portions in the infrared region of the electromagnetic spectrum to which our atmosphere is transparent.

In situ   In place.

Interference   The process by which two or more sets of waves combine to create a wave pattern different from any of the original sets.

Intermediate waters   Water masses formed in high latitudes that flow beneath the surface waters but above the deep waters.

Intertidal   Of or pertaining to the region of the shore that is between low and high tide levels.

Ion   Any atom or group of atoms that carries a net charge, either positive or negative.

Island arc system   An arcuate chain of volcanic islands found alongside an oceanic trench that is created by the volcanic distillation of oceanic crust plunging into the subduction zone.

Isostasy   The tendency of the lithosphere to float atop the more plastic asthenosphere, with surface elevations reflecting the buoyancy of the crustal material below.

Isostatic readjustment   The readjustment of surface elevations until the weight of that portion of the lithosphere is equal and opposite to the upward buoyant force from the asthenosphere below.

Isotonic   Having body fluids of the same salinity as the water of the organism's environment.

Isotopes   Atoms of any one element having different numbers of neutrons in the nucleus. For example, $_6C^{12}$ and $_6C^{14}$ are two isotopes of carbon.

Jetty   Any man-made wall or barrier projecting into the sea so as to provide protection from waves, a guide to water flow, and so on.

Joule   A unit of energy amounting to 0.239 calories and is equal to the kinetic energy of a 2-kg mass moving at 1 m/s.

Kelvin   A temperature scale having degrees of the same size as the Celsius degree, and for which absolute zero is defined as zero degrees Kelvin (K). On this scale, water freezes at 273.15K and boils at 373.15K.

Key   A long low island or reef.

Kilometer   A unit of distance equal to 1000 m, or 0.62 mi.

Knot   A unit of speed equal to one nautical mile per hour, or 1.15 mi/hr, or 1.85 km/hr.

Krill   An arthropod, resembling small shrimp (typically a few centimeters in length), that is very abundant in cooler waters of higher latitudes and is a favorite food for many larger carnivores, including several species of whales.

Lagoon   An area of shallow water separated from the open ocean by barrier beaches or shallow banks.

Latent heat   That heat energy which goes into changing the state of a substance (e.g., solid to liquid or liquid to gas) without changing its temperature.

Latitude   A measure of north-south position on the earth relative to the equator, measured in degrees, with the equator being 0° latitude and the poles at 90° North and South, respectively.

Lattice   The imaginary periodic framework of the atomic groups in a crystal.

Lava   The molten rock that flows out of volcanic areas and onto the earth's surface.

Leading edge   The margin of a continent on the side toward which it is moving, relative to neighboring crustal plates.

Leeward   Pertaining to the side facing away from the wind.

Limestone   A sedimentary rock made of calcium carbonate.

Lithification   Solidifying or cementing together to form rock.

Littoral   Along the shore.

Load   The sediment carried in suspension.

Longitude   A measure of east-west position on the earth measured in degrees east or west of Greenwich, England.

Longshore bars   Bars located in or just beyond the surf zone, oriented approximately parallel to the shore.

Low tide terrace   The portion of the beach exposed and damp at low tide, extending from low tide level to high tide level. The foreshore.

Lunar tide   That component of the tide caused by the moon's gravitation.

Macroplankton   Plankton large enough to be easily visible to the unaided eye, generally considered to be those having dimensions greater than 1 mm.

Magma   Molten or partially molten rock capable of fluid flow.

Magnetic anomaly   Any local deviation from the normal pattern in the earth's magnetic field.

Magnetic declination   The angular difference between true north and magnetic north, as seen, for instance, by comparing the direction indicated by a magnetic compass needle to the direction of the North Star.

Magnetic dip   The angle that the earth's magnetic field makes with the horizontal, ranging from $+90°$ at the south magnetic pole, to 0° at the equator, and to $-90°$ at the north magnetic pole.

Magnetic polarity   A term used to indicate whether the earth's magnetic field pointed toward the north or toward the south in any certain period of earth history.

Magnetic reversals   Reversals in the direction of the earth's magnetic field that occur rather suddenly on geologic time scales.

Magnetometer.   Any sensitive instrument used to measure the earth's magnetic field.

Major constituents   The most abundant salts in seawater whose concentrations can appropriately be expressed in parts per thousand.

Manganese nodules   Hard nodular deposits of typically a few centimeters diameter that are found in some places on the ocean bottom and are enriched in heavy metals such as manganese, iron, nickel, cobalt, and copper.

Mantle   The interior region of the earth between the crust and the core, extending from roughly 30 to 2900 km beneath the surface and encompassing about 84% of the earth's total volume.

Maria   Large grey-colored regions of the moon's surface.

Marine   Of or pertaining to the oceans and seas.

Maximum sustainable yield   The maximum weight of any kind of organism that may be harvested year after year. Harvesting more than this in any one year will diminish the amount that can be harvested in succeeding years.

Mean sea level   The level of the water surface averaged over the last 19 years, so that transient effects like tidal patterns and the effect of seasonal barometric pressures are averaged out.

Mediterranean   Surrounded by land.

Medusa   Free-floating coelenterates, such as jellyfish.

Megaplankton   Any plankton large enough to be easily seen with the unaided eye, commonly defined as having dimensions greater than one millimeter.

Meroplankton   Plankton that spend only a portion of their lives as plankton, such as many juveniles of nekton.

Metamorphic rock   Rock that began as sedimentary or igneous rock but which has subsequently been subjected to sufficient heat and stress to have changed character noticeably.

Meteorite   Any solid object of extraterrestrial origin that is found on the earth's surface or in the sediment. Before encountering the earth they are called meteoroids, while falling through our atmosphere they are called meteors, and after landing on the earth's surface they are called meteorites.

Microplankton   Plankton whose size ranges from about 0.06 to 1.0 mm.

Mid-ocean ridge   A very long, high range of submarine mountains, typically 1500 or 2000 km wide and 65,000 km long, running through all the world's oceans and occasionally even reaching the surface.

Millimole One-thousandth of a mole, or the weight of a material in milligrams equal to its molecular weight. For example, a millimole of N_2 (molecular weight = 28) is 28 milligrams.

Mineral A substance occurring in nature having uniform chemical composition and crystal structure.

Mixing time A measure of the time it takes for a body of water to become thoroughly mixed.

Mohorovicic discontinuity (Moho) The discontinuity in materials and properties that occurs at the interface between the crust and mantle.

Molecular weight The sum of the number of nucleons (protons and neutrons) in the nuclei of all the atoms in a molecule. Also, the mass of a molecule measured in atomic mass units. (The two definitions are nearly the same.)

Mollusca A phylum of animals each having a soft foot used for locomotion and most having calcareous protective shells, including clams, snails, slugs, mussels, and squid.

Monsoon A seasonal wind pattern.

Nannoplankton The smallest plankton having dimensions less than 0.06 mm.

Nansen bottle A type of water-sampling device that is lowered on a cable and inverted when it reaches the depth desired for taking the sample.

Neap tide The two times during the month when tidal variations are smallest, occuring when the moon is near first quarter or third quarter.

Nearshore Close to the shore.

Nekton Marine organisms capable of fast sustained motion, as opposed to plankton, which are not.

Neritic Of or pertaining to those regions of the ocean over the continental shelves.

Neutron A strongly interacting uncharged particle found along with equally massive protons in atomic nuclei. It is roughly 2000 times more massive than the electrons found in the electron cloud.

Newton's second law The net force on an object equals the product of its mass times its acceleration ($F = ma$).

Nitrogen fixation The process of producing nitrates from free molecular nitrogen, N_2.

Nonferromagnesian Deficient in iron and magnesium (applied to minerals).

Nuclear   Of or pertaining to either an atomic nucleus or a cell nucleus.

Nucleus (atomic)   The tiny dense central portion of an atom containing the neutrons and protons, having positive charge, and having dimensions of about 10^{-13} cm, which is about 10^5 times smaller than the dimensions of the atomic electron cloud.

Nucleus (biological)   A differentiated mass of protoplasm in a biological cell that is separated from the rest of the cell by a membrane and plays an important role in regulating the cell's behaviors.

Nutrient   Anything other than the elements carbon, hydrogen, and oxygen that is needed in the synthesis of organic matter, and whose scarcity may limit biological productivity. Common nutrients are nitrates and phosphates.

Oceanic   Of or pertaining to the deeper regions of the oceans beyond the continental shelves.

Oceanic crust   The type of crust underlying the oceans that is typically 8 km thick and somewhat richer in iron and magnesium minerals than the continental crust.

Omnivore   An organism that is able to feed on either plants or animals.

Ooze   Any sediment whose composition is more than 30% biogenic.

Orbital (atomic)   A path followed by an electron in the electron cloud as it orbits the atomic nucleus.

Orbital (wave)   A path followed by a water molecule as a wave passes overhead.

Osmosis   The diffusion of a material through a membrane from a region of higher concentration toward a region of lower concentration.

Osmotic pressure   A quantitative measure of the tendency of water molecules and dissolved salts to diffuse in opposite directions through a membrane separating solutions of differing salinity. It is measured in terms of how much pressure must be applied on one side to oppose the diffusion from the other.

Oxidation   The chemical process of combining with oxygen.

Oxide   A compound formed by combining anything with the 0^{--} ion.

Oxygen minimum layer   The region of the ocean immediately beneath the productive photic zone that is depleted in dissolved oxygen because of the respiration of the many animals that reside there.

Pack ice   Ice cover formed of chunks of sea ice jammed together and obstructing navigation.

Paleomagnetism   The study of the magnetization of ancient rocks.

Pancake ice   Small thin plates of sea ice that have been rounded through collisions with neighbors.

Partial melting   The heating of rock to the point where the more weakly bound minerals melt, thus lubricating the flow of the crystals of more strongly bound, unmelted minerals.

Pebble   A sediment having dimensions in the range of 0.4 to 6.0 cm.

Pelagic   Of or pertaining to the waters of the oceans, as opposed to benthic. For example, pelagic organisms inhabit the waters, whereas benthic organisms live on or in the ocean bottom.

Permeability   The ability to allow water (or any fluid) to flow through.

pH   A quantitative measure of the hydrogen ion (H^+) concentration, being the negative of the exponent of the hydrogen ion concentration. For example, if the pH of a solution is 5, then the hydrogen ion concentration is 10^{-5}, or one part per 10^5.

Phase diagram   A plot of temperature versus pressure for any certain material showing all the temperatures and pressures for which it is a solid, liquid, and gas.

Phosphate   Any compound formed by combining anything with the phosphate ion (PO_4^{---}).

Photosynthesis   The process of producing organic matter by using carbon dioxide, water, nutrients, and sunlight.

Phytoplankton   Plankton that can photosynthesize.

Pinhole   A small beach feature formed when air, trapped beneath sand overrun by a wave, escapes upward through small holes.

Piston corer   A device used for retrieving sediment samples by plunging a tube into the bottom sediments and using a piston as a syringe to help pull the sediment sample up into the descending tube.

Placer   A beach or relict beach enriched in heavy metal deposits by the winnowing action of the waves.

Plankton   Marine organisms having little or no means of self-propulsion, as opposed to nekton, which are capable of fast extended motion.

Plathyhelminthes   An animal phylum containing flatworms.

Plunging breaker   A breaking wave with the crest plunging in front of the wave, as opposed to spilling down its face.

Plutonic   Formed by the intrusion of magma into cracks and other subsurface openings.

Polar   Of or pertaining to the region near the North Pole or South Pole.

Polarization   The separation of electrical charges to form one region of net positive charge and another region of net negative charge.

Polymorphism   The ability of a single species to appear in two or more very different forms (aside from sex-linked differences).

Polyp   An attached coelenterate, such as an anemone.

Porifera   A phylum of multicellular bottom-dwelling animal filter feeders having very little organization among their cells, including all sponges.

Porosity A measure of what fraction of the volume in a sedimentary deposit is empty space.

Potential temperature The temperature of a water sample when the pressure is adiabatically reduced to atmospheric pressure.

Power The rate of energy use.

Precipitation (atmosphere) Rainfall, snow, or hail.

Precipitation (ocean) The process by which dissolved materials are removed from solution and join the bottom sediments.

Pressure Force per unit area.

Pressure points Points on the sediment grains where they are in contact with their neighbors.

Primary conductivity The rate at which organic matter is produced in photosynthesis by plants.

Primordial Existing in the very beginning, or the earliest days.

Productivity The rate at which organic matter is produced in photosynthesis by plants. (Usage in this book is the same as primary productivity.)

Propagate To travel across or through a medium.

Protein A class of complicated molecular structures formed from chemical alteration of carbohydrates, including the addition of nitrogen, phosphorus, and trace elements.

Proton A strongly interacting, positively charged particle found in atomic nuclei, equally massive as neutrons, and about 2000 times more massive than electrons.

Protozoa A phylum of primitive one-celled animals, including foraminifers and radiolarians.

P-waves Seismic waves in which the direction of the vibrations parallels the direction of propagation of the wave.

Pycnocline A region in the ocean where there is a rapid change in density.

Radiation The transport of energy from one region to another without help from the intermediary medium, such as with electromagnetic waves, alpha particles, beta particles, or other energetic particles.

Radiolarians Members of an order of protozoa that are single-celled animals and have siliceous tests.

Red clay A broad class of red to brown inorganic clays found over wide regions of the deep ocean bottom below the calcium carbonate compensation depth in temperate and tropical latitudes.

Reef A localized region of shallow water that creates a navigational hazard.

Reef front The edge of a reef facing the open ocean.

Reflection   The process by which a wave will bounce off of sufficiently large objects it encounters.

Refraction   The process by which the direction of propagation of a wave or a group of waves is changed due to changes in the wave speed. For water waves this is caused by variations in the water depth.

Relative humidity   A quantitative means of describing atmospheric water content by comparing how much the air does hold to how much it could possibly hold under ambient conditions.

Relict   Left over from previous times.

Residence time   A measure of the average time a particular component of seawater spends in solution, between the time it first enters and the time it is permanently removed from the ocean.

Resonance   A condition of extra large wave amplitude obtained when the frequency of an external wave-generating force matches a natural frequency for waves bouncing back and forth along or across an enclosed space, such as a protected harbor or estuary.

Respiration   The oxidation of organic matter in organisms to produce useful energy.

Rhyzome   A special rootlike stem produced by some spermatophytes that can be extended along the ground and produces new roots and leafy stems at some distance from the parent.

Rift mountains   The rugged, heavily faulted mountains found bordering the rift valley near the crest of the oceanic ridge.

Rills   Small stream-cut channels in the beach sediments.

Ring of fire   A term applied to the periphery of the Pacific Ocean, being characterized by its volcanic activity, which results from the subduction of oceanic crust on all its margins.

Rip current   A swift narrow seaward-flowing current in the surf zone allowing water to return to sea when it is being stacked up on the beach during periods of heavy wave activity.

RNA   Any of a variety of long thin protein molecules found in biological cells that form the scaffolding on which complicated molecules are constructed.

Salinity   The relative mass of all dissolved substances with all carbonate converted to oxide, all bromine and iodine replaced by chlorine, and all organic matter oxidized to 480°C. It is measured in grams per kilogram of seawater, or parts per thousand.

Salt water wedge   The wedge of salt water that intrudes into an estuary from the ocean along the bottom.

Sand   Sediment with dimensions in the rage of 0.06 to 2.0 mm.

Sea   An enclosed or semienclosed body of salt water that is large compared to most estuaries and embayments but small compared to the oceans. This term is also used to describe the usual chaotic pattern of small- and medium-sized gravity waves encountered at sea.

Sea ice Ice that is formed on the sea. It differs from land ice in that it is thin (seldom more than two meters thick and usually only a few centimeters thick), and it is about one percent salt.

Seaknoll Any hill in the deep ocean bottom extending upward less than 0.9 km.

Sea spray The spray that is created by breaking waves or strong winds.

Sediment Any extended granular deposit that has been transported by, precipitated from, or fallen from suspension in a fluid such as water or air.

Sedimentary rock Rock resulting from the cementing together of sediments.

Sedimentation The process during which sediment is deposited on the bottom.

Sediment maturity A qualitative measure of how much a sediment has been moved and sorted since breaking away from its original position in a rock. The more rounded and uniform the grains in a deposit, the more mature it is.

Seiche A resonant sloshing of water in a confinement, such as a harbor or estuary.

Seine A fishing net that floats vertically in the water with floats along one edge and weights along the other.

Seismic Generated by movement within the earth.

Seismic sea wave A wave generated by movement of the ocean bottom, such as an earthquake, slumping of the ocean bottom, and so on. Also called a Tsunami.

Seismic waves Vibrations that travel through the earth.

Sensible heat Heat that goes into raising the temperature of a substance, as opposed to latent heat, which goes into changing the physical state without raising the temperature.

Sessile Attached.

Shallow water wave Any wave traveling on water of depth less than $\frac{1}{20}$ of its wavelength.

Shelfbreak The very outer edge of the continental shelf having an abrupt change of slope, indicating the beginning of the continental slope.

Shoal Shallow water.

Sialic Of or pertaining to rocks that are relatively deficient in ferromagnesian minerals, as is characteristic of the materials of the continental crust.

Sigma-t A quantitative measure of the density of a seawater sample measured at atmospheric pressure, defined as (density $- 1) \times 1000$, where the density is measured in gm/cm³.

Significant wave height   The average height of the highest one-third of the waves.

Silicate   Any compound formed with the silicate radical (SiO_4^{----}).

Siliceous   Composed of silica (SiO_2).

Sill   Any shallow area or ridge that separates one basin from another.

Silt   Sediment whose grains have characteristic dimensions ranging from 0.004 to 0.06 mm.

Simatic   Of or pertaining to rocks that are enriched in ferromagnesian minerals, as is characteristic of the materials of oceanic crust.

Slack water   That time during the tidal cycle when the water level is neither rising nor sinking, and when water is neither entering nor leaving estuaries.

Slicks   Long thin concentrations of floating debris on the water surface resulting in reduced wave activity and a shinier appearance.

Slope gullies   Gullies on the continental slope often caused by the slumping of poorly packed sediments.

Slough   A protected embayment that has filled in with sediments to the point where they are alternately exposed and inundated during the tidal cycle.

Slump   To sink or slide down.

Solar tide   That component of the tide caused by the sun's gravitation.

Solubility   A quantitative measure of the ability of one substance to be dissolved in another, usually measured in terms of how many grams, moles, or cubic centimeters of the substance can be dissolved in one liter of the solvent.

Solvent   The medium in which something is or could be dissolved. For example, water is the solvent in seawater.

Sounding   A measurement of water depth.

Specific heat   Heat capacity per gram measured in terms of how many calories of energy are required to raise one gram of the substance one degree Celsius.

Specific volume   A measure of the volume occupied by one gram of the substance in units of cm³/gm.

Spectrum   The entire range of wavelengths present.

Spermatophytes   Members of the subkingdom Spermatophyta, including all flowering and seed-producing plants having true roots, stems, and leaves.

Sphere depth   The depth to which that amount of water would extend if the solid earth surface were smooth and covered evenly.

Spilling breaker   A breaking wave having water from the crest spilling down the face of the wave.

Spit   A narrow peninsular bar, or shoal, extending into the ocean from the shore.

Spring tide   The two times during the month when tidal variations are greatest, occurring when the moon is either new or full.

Standing wave   The wave pattern created by the interference of two groups of waves having equal amplitude and wavelength but traveling in opposite directions. The resulting standing wave has regions where the water surface is oscillating up and down, separated by regions where there is little or no water motion at all. It is called a standing wave because although there appears to be much wavelike motion, these waves do not appear to be moving in either direction.

Stand of tide   Any period during the tidal cycle when the water level is neither rising nor lowering.

Station (oceanographic)   Any location at sea where a measurement is made.

Still-water level   The position the water surface would have if the waves were flattened.

Stipe   A stemlike structure with which some plants are attached to the bottom.

Storm surge   A coastal rise in water level due to water being blown shoreward during a storm.

Stratification   The formation of horizontal layers.

Subduction zone   Any region of the earth where an oceanic plate is slowly plunging beneath another crustal plate and into the asthenosphere and upper mantle.

Subkingdom   A subdivision of organisms according to criteria that are more general than those of phyla but less general than those of a kingdom.

Sublittoral   Of or pertaining to those regions of the continental shelf extending from low tide level all the way to the outer edge of the shelf.

Submarine canyons   Very large steep-walled canyons carved in the continental shelves and slopes.

Sulfide   Any compound formed with the sulfide ion (S^{--}).

Superposition   The placement of one atop another.

Supply curve   A plot of supply versus price showing how large the supply of any good would be at any particular price.

Surface tension   The cohesive force with which the molecules on the surface of a fluid pull on each other, forming a very thin elastic film.

Surf beat   The periodic change in amplitude of waves coming ashore, with the repeating pattern of a series of larger waves followed by a series of smaller waves, followed by a series of larger waves, and so on.

Surf zone The nearshore region where there are breaking waves.

Surging breakers Breaking waves that break on the shore.

Swash The moving mass of water that washes up on shore after an incoming wave has broken.

Swash marks Beach features resembling large fish-scale patterns created by each swash depositing a line of sediments at the highest point it reaches on the beach face.

S-waves A type of seismic wave in which the vibrations are in a direction perpendicular to the direction of propagation of the wave.

Swell Large long parallel waves arising after waves have left the generating storm center and have begun to sort themselves out, with the longest, fastest waves taking the lead.

Symbiosis The living together of two different species, which is not harmful to either. In many cases it is beneficial to both.

Tectonics The study of the history, structure, and motion of the earth's crust and upper mantle. Also, the motion itself.

Temperate Of or pertaining to intermediate latitudes, usually considered to be those lying between the tropics (23°N or S) and the Arctic or Antarctic circle (67°N or S).

Terrace A level, flat region on the side of a hill or other slope.

Terrestrial Of or pertaining to the earth.

Terrigenous Of, pertaining to, or derived from the continents.

Tests Tiny skeletons or skeletal debris of microorganisms.

Thallophytes Members of the subkingdom Thallophyta, commonly called algae, including the most primitive plants.

Thermocline Any region of rapidly changing temperature.

Thermohaline circulation Any circulation of ocean waters caused by differences in water density. The name is derived from the fact that the density of a water mass is determined by its temperature and salinity.

Tidal bore A slowly moving, steep-walled front of water created as an incoming tide stacks up in shallow water, especially in rivers and estuaries where there is a slight seaward current.

Tidal prism The volume of water contained between high tide level and low tide level in a estuary, which is the amount of water flowing in and out of the estuary during a tidal cycle.

Tide pool A pool of water left above sea level as the tide goes out.

Topography Surface features or relief.

Trace elements Those elements present in seawater in only very small amounts, and whose concentrations are best measured in parts per billion or less.

Trade winds   The prevailing winds that blow toward the west in the tropics, having a slight southward component north of the equator, and a slight northward component south of the equator.

Trailing edge   The margin of a continent that faces away from the direction of motion of the continent relative to neighboring crustal plates.

Trajectory   The path followed by a moving object.

Trawl   A large bag or pouch-shaped net towed behind a boat to catch fish.

Trophic level   A step in the food chain measured in terms of how many different organisms the organic material has been in since its original synthesis in plants.

Tropical   Of or pertaining to the region of the earth lying near or between about 23°N and 23°S latitudes.

Troposphere   The lower portion of the atmosphere extending to an altitude of about 13 km.

Trough   The portion of a wave where the water surface is depressed.

T-S diagram   A plot of temperature versus salinity showing the temperature and salinity of all points in a vertical column of water.

Tsunami   A wave generated by movement of the ocean bottom, such as an earthquake, slumping of the ocean bottom, and so on. Also called a seismic sea wave.

Turbidite   A sedimentary deposit formed of the deposits of turbidity currents.

Turbidity current   A muddy slurry of water and sediment that flows downhill along the ocean bottom due to its greater density. Usually, it originates in the head of a submarine canyon and flows down through the canyon and out onto the ocean bottom at the base of the slope.

Turbulence   Irregular, unsystematic, or small eddylike motions in a water mass causing local mixing.

Ultraviolet   Of or pertaining to those wavelengths of light bordering on the short wavelength side (shorter than violet) of the visible spectrum.

Upwelling   The rising of deeper waters to the surface.

Vagrant   Having the ability to move about.

Valence   An indication of the number of electrons that an atom or group of atoms tends to give to or take from others in forming compounds.

Vaporization   Conversion from liquid to gaseous form.

Viscosity   The ability of a liquid to resist shear stress, or a measure of the amount of drag exerted on one layer of the fluid by motion in the layer next to it. For example, syrup has a high viscosity and air has a low viscosity.

Volatile Easily vaporized. Having a relatively low boiling point. Also, as a noun it means any volatile substance.

Vorticity A measure of the amount of rotary or circular motion in a fluid.

Wake The group of surface waves generated by something moving near or on the surface.

Watt A measure of power or rate of energy usage that is equal to one joule per second.

Wave amplitude The maximum vertical displacement of the water surface from its average height during the passage of waves, being equal to one-half the wave height.

Wave group A group of waves having roughly the same wavelength and wave speed and moving together in the same direction.

Wave height The vertical distance between the lowest point in a trough and the highest point on the neighboring crest.

Wavelength The horizontal distance between any point on one wave and the corresponding point on the next wave, such as from one crest to the next.

Wave period The time interval between the passage of two successive waves past a stationary point.

Wave steepness The ratio of wave height to wavelength.

Wave of translation The moving mass of water created when a wave breaks, differing from normal gravity waves in that the individual molecules no longer have oscillatory orbital motions but rather move along with the wave.

Weathering The processes by which solid rock is broken down into fragments.

Westerlies Winds blowing from the west.

Western intensification The tendency of currents along the western edges of all oceans in both hemispheres to be particularly strong, swift, and narrow.

Windward Of or pertaining to the side facing into the wind.

Zooplankton Those members of the plankton that are animals.

Chapter 1

Opener Challenger Office Report, Great Britain, 1895. **Figure 1.1** NASA. **Figure 1.2** Hale Observatories. **Figure 1.4a** Courtesy E. S. Barghoorn. **Figure 1.4b** Smithsonian Institution. **Figure 1.5** Smithsonian Institution. **Figure 1.7** From David Eugene Smith, *History of Mathematics*, Vol. 1, 1951, Ginn & Co. **Figure 1.8** Challenger Office Report, Great Britain, 1895. **Figure 1.9** Viking Ship Museum, Oslo. **Figure 1.10** Sigurd Stefansson, 1570. **Figure 1.11** National Archives. **Figure 1.12a** and **1.13** Library of Congress. **Figure 1.14** Official U. S. Navy photo. **Figure 1.15** Library of Congress. **Figure 1.16** Fram Museum, Oslo. **Figure 1.17** Scripps Institution of Oceanography, University of California, San Diego.

Chapter 2

Opener NOAA. **Figure 2.1a** and **2.1b** Official U. S. Navy photos. **Figure 2.3** NASA. **Figure 2.5** Photo by D. J. Miller, U.S. Geological Survey. **Figure 2.6** Geological Survey of Canada, Ottawa. **Figure 2.7** Official U. S. Navy photo. **Figure 2.22** Smithsonian Institution.

Chapter 3

Opener Photo by R. E. Wallace, U. S. Geological Survey. **Figure 3.11** Courtesy Woods Hole Oceanographic Institution. **Figure 3.17** Photo by Dudley Foster, Woods Hole Oceanographic Institution. **Figure 3.25** NASA.

Chapter 4

Figure 4.2 Photo by J. R. Balsey, U. S. Geological Survey. **Figure 4.14** (all) Official U. S. Navy photos; F. P. Shephard and R. F. Dill, Submarine Canyons and Other Sea Valleys, John Wiley & Sons, 1965, **Figure 4.24** Photo by Sigurdur Thorarinsson. **Figure 4.27** NASA. **Figure 4.29** (all) Photos by K. O. Emery, U. S. Geological Survey.

Chapter 5

Figure 5.2 Photo by M. V. Adams, U. S. Geological Survey. **Figure 5.4** NOAA.

Chapter 6

Opener Courtesy Woods Hole Oceanographic Institution. **Figure 6.2** Courtesy Geological Survey of Canada, Ottawa. **Figure**

6.11 Courtesy NASA. **Figure 6.12** Courtesy NASA. **Figure 6.13** Scripps Institution of Oceanography, University of California, San Diego. **Figure 6.15** Robert C. Hermes/National Audubon Society/Photo Researchers. **Figure 6.16** (top) Russ Kinne/Photo Researchers. **Figure 6.16** (bottom) Russ Kinne/Photo Researchers. **Figure 6.16** Boisnard-Guiter/Rapho-Photo Researchers. **Figure 6.18** NOAA. **Figure 6.19** Photo by M. R. Mudge, U. S. Geological Survey. **Figure 6.20** Photo by J. C. Anderson, U. S. Geological Survey. **Figure 6.21** (all) Smithsonian Institution.

Chapter 7

Opener Robert Perron/Photo Researchers. **Figure 7.10** Official U. S. Navy photo. **Figure 7.15** U. S. Army Corps of Engineers, Los Angeles District. **Figure 7.18** Official U. S. Navy photo. **Figure 7.22** United Press International. **Figure 7.23** (left) Hydraulic Engineering Laboratory, University of California, Berkeley. Courtesy of Robert L. Wiegel. **Figure 7.23** (right) Photo by J. I. Tracy, U. S. Geological Survey. **Figure 7.24** (both) D. A. Patrick and R. L. Weigel, "Amphibian Tractors in the Surf," *Proceedings of the First Conference on Ships and Waves*, The Engineering Foundation Council on Wave Research and the American Society of Naval Architects and Marine Engineers, pp. 397–422, 1955.

Chapter 8

Opener Alexander Lowry/Photo Researchers. **Figure 8.4** (all) NOAA. **Figure 8.6** Tourism New Brunswick. **Fiugre 8.9** Tourism New Brunswick. **Figure 8.11** (top) NASA. **Figure 8.17** U. S. Coast Guard Service.

Chapter 9

Figure 9.3 (both) National Film Board Phototéque/Photos by G. Bloudin. **Figure 9.4** NOAA. **Figure 9.5** NOAA. **Figure 9.7** Ray Atkeson. **Figure 9.12** (both) From *Geology Illustrated* by John S. Shelton, W. H. Freeman and Company, 1966. **Figure 9.28** U. S. Army Corps of Engineers, Los Angeles District.

Chapter 10

Opener Paolo Koch/Photo Researchers. **Figure 10.1** NASA. **Figure 10.2** NASA. **Figure 10.3** NASA.

Chapter 11

Opener Photo by Susan Kadar/Woods Hole Oceanographic Institution. **Figure 11.1** NASA. **Figure 11.14** Courtesy Woods Hole Oceanographic Institution. **Figure 11.15** (left) Oregon State University, Department of Oceanography. **Figure 11.15** (right) Woods Hole Oceanographic Institution. **Figure 11.16** Challenger Office Report, Great Britain, 1895. **Figure 11.18** Bill Cella/Photo Researchers.

Chapter 12

Opener NOAA. **Figure 12.13** Official U. S. Navy photo. **Figure 12.14** U. S. Coast Guard.

Chapter 13

Opener NASA.

Chapter 14

Oepner U. S. Coast Guard Photo. **Figure 14.9** NOAA.

Chapter 15

Opener U. S. Naval Photographic Center. **Figure 15.5** U. S. Coast Guard photo. **Figure 15.12** U. S. Naval Photographic Center. **Figure 15.18** (both) Courtesy Woods Hole Oceanographic Institution.

Chapter 16

Opener Skyviews. **Figure 16.4** Dr. John S. Shelton. **Figure 16.5** NASA. **Figure 16.8** Dr. John S. Shelton. **Figure 16.9** Joe Nunroe/Photo Researchers. **Figure 16.10** Courtesy Norwegian Tourist Bureau. **Figure 16.18** Maurice E. Landro/National Audubon Society/Photo Researchers. **Figure 16.19** Grant Heilman.

Chapter 17

Opener Russ Kinne/Photo Researchers. **Figure 17.1** Russ Kinne/Photo Researchers. **Figure 17.3** (both) NOAA. **Figure 17.5** Paul E. Hargraves, University of Rhode Island. **Figure 17.14a** and **17.14b** Runk/Schoenberger/Grant Heilman. **Figure 17.14c** Eric V. Gravé. **Figure 17.15** United Press International. **Figure 17.19** (left) Courtesy Woods Hole Oceanographic Institution. **Figure 17.19** (right) Environmental Science Services Administration. **Figure 17.21c** Food and Agriculture Organization of the United Nations.

Chapter 18

Opener Runk/Schoenberger/Grant Heilman. **Figure 18.7** Mary M. Thatcher/Photo Researchers. **Figure 18.9** David Donoho/Photo Researchers. **Figure 18.10b** Hugh Spencer/National Audubon Society/Photo Researchers. **Figure 18.13** (left) Walter Dawn. **Figure 18.13** (right) F. J. Taylor, Institute of Oceanography, Department of Botany, University of British Columbia. **Figure 18.14** C. C. Maxwell/National Audubon Society/Photo Researchers. **Figure 18.16** American Museum of Natural History. **Figure 18.18** American Museum of Natural History. **Figure 18.19a** Robert C. Hermes/National Audubon Society/Photo Researchers. **Figure 18.19b** Russ Kinne/Photo Researchers. **Figure 18.20** Jen and Des Bartlett/Photo Researchers. **Figure 18.21a** Grant Heilman. **Figure 18.21b** Walter Dawn. **Figure 18.21c** Gordon S. Smith/National Audubon Society/Photo Researchers. **Figure 18.21d** A. W. Ambler/National Audubon Society/Photo Researchers. **Figure 18.22a** Robert C. Hermes/National Audubon Society/Photo Researchers. **Figure 18.22b** Photo by Hulger Knudsen, Marine Biologisk Laboratory, Helsingor, Denmark. **Figure 18.22c** American Museum of Natural History. **Figure 18.23** Walter Dawn/National Audubon Society/Photo Researchers. **Figure 18.24** G. Clifford Carl/National Audubon Society/Photo Researchers. **Figure 18.25a** Karl H. Maslowski/National Audubon Society/Photo Researchers. **Figure 18.25b** Robert C. Hermes/National Audubon Society/Photo Researchers. **Figure 18.25c** Jen and Des Bartlett/Photo Researchers.

Chapter 19

Opener Grant Heilman. **Figure 19.1** Jeanne White/Photo Researchers. **Figure 19.2** Courtesy Dr. Robert L. Wood. **Figure 19.6** From J. Wolfson, D. Dressler and M. Magazin, Proc. National Acad. Sci., U.S. **69**:499, 1972. Original micrographs courtesy of D. Dressler. **Figure 19.9** (left) Courtesy The United Nations. **Figure 19.9** (right) United Press International. **Figure 19.11** F. E. Round, University of Bristol, Department of Botany, United Kingdom. **Figure 19.13** Runk/Schoenberger/Grant Heilman. **Figure 19.14** Jen and Des Bartlett/Photo Researchers. **Figure 19.15** (both) Courtesy of Pacific Gas and Electric Company. **Figure 19.17** Douglas Faulkner. **Figure 19.18** Hugh Spencer/Photo

Researchers. **Figure 19.21** Gordon S. Smith/ Photo Researchers.

Chapter 20

Opener Wometco Miami Seaquarium. **Figure 20.1** Runk/Schoenberger/Grant Heilman. **Figure 20.2** T. D. Lovering/Stock, Boston. **Figure 20.4** Dr. John D. Dodge, Department of Botany, Birkbeck College, University of London. **Figure 20.5c** W. D. Russell-Hunter. **Figure 20.6** (top) Runk/ Schoenberger/Grant Heilman. **Figure 20.6** (bottom) Courtesy Scripps Institution of Oceanography. **Figure 20.11** Ron Church/ Photo Researchers. **Figure 20.13** Tom McHugh/Photo Researchers. **Figure 20.15** Jeanne White/Photo Researchers. **Figure 20.16** Runk/Schoenberger/Photo Researchers. **Figure 20.18** Runk/Schoenberger/Grant Heilman. **Figure 20.19** American Petroleum Institute, Cities Service Photo. **Figure 20.20** Georg Gerster/Rapho-Photo Researchers.

Figure 20.21 Courtesy Woods Hole Oceanographic Institution.

Chapter 21

Opener George Bellrose/Stock, Boston. **Figure 21.8** (all) Food and Agriculture Organization of the United Nations. **Figure 21.15** (left) Photo by Brooks Townes, National Fisherman Magazine. **Figure 21.15** (right) United Press International. **Figure 21.19** Photo by Luthter H. Blount, Blount Marine Corporation.

Chapter 22

Opener Exxon Photo. **Figure 22.4** Courtesy Lockheed Missiles & Space Company. **Figure 22.5** Courtesy Lockheed Missiles & Space Company. **Figure 22.6** Courtesy AeroVironment. **Figure 22.7** Phototeque EDF, courtesy of the French Embassy. **Figure 22.10** (both) Global Marine, Inc. **Figure 22.15** (all) Deep Sea Ventures. **Figure 22.17** Fred Lyons/Rapho-Photo Researchers.

THE FLOOR OF THE OCEANS